D0677619

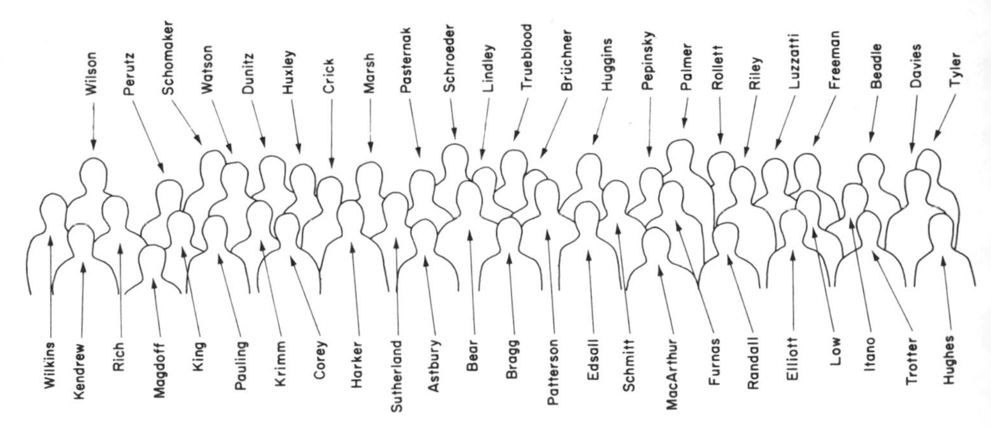

1 *Frontispiece. Pasadena Conference on the Structure of Proteins 21 to 25 September 1953.*

THE PATH TO THE DOUBLE HELIX

The Discovery of DNA

ROBERT OLBY

Professor of the History and Philosophy of Science
The University of Pittsburgh

FOREWORD BY

FRANCIS CRICK

DOVER PUBLICATIONS, INC.
NEW YORK

Copyright

Published in Canada by General Publishing Company, Ltd., 30 Lesmill Road, Don Mills, Toronto, Ontario.
Published in the United Kingdom by Constable and Company, Ltd., 3 The Lanchesters, 162–164 Fulham Palace Road, London W6 9ER.

Bibliographical Note

This Dover edition, first published in 1994, is an unabridged, corrected and enlarged republication of the work published by the University of Washington Press, Seattle, 1974 (published in the U.K. by The Macmillan Press, London, 1974). For the Dover edition, to which the subtitle *The Discovery of DNA* has been added, the author has corrected a number of typographical errors, corrected and expanded the bibliography, added death dates to entries in the Index where applicable and feasible, and provided a new "Note to the Introduction" and "Postscript."

Library of Congress Cataloging-in-Publication Data

Olby, Robert C. (Robert Cecil)
The path to the double helix: the disovery of DNA/Robert Olby.
p. cm.
Originally published: Seattle: University of Washington Press, 1974.
Includes bibliographical references and index.
ISBN 0-486-68117-3
1. Molecular biology—History. 2. Dna. I. Title.
QH506.045 1994
574.87'32'09—dc20 94-3665
CIP

Manufactured in the United States of America
Dover Publications, Inc., 31 East 2nd Street, Mineola, N.Y. 11501

Foreword

It is a pleasure to write a foreword to *The Path to the Double Helix.* As the reader will easily discover for himself it contains a fascinating series of related scientific histories told in a lively, perceptive and scholarly way. In spite of my overfamiliarity with parts of the story I have found it absorbing reading.

It is inevitable that comparisons will be made between this book and Jim Watson's earlier volume, *The Double Helix,* in spite of their obvious differences. Watson's book was really a fragment of his autobiography. Not only did he attempt to describe the discovery of the DNA through the eyes of the young man he was at that time but he included many lively personal details not strictly essential to his main theme. His Chapter 15, for example, relating how he lay on the floor at Carradale (the book was dedicated to Naomi Mitchison) and attempted to grow a beard tells us little of purely scientific interest. But then Watson's principal aim was to show that scientists were human, a fact only too well known to scientists themselves but apparently not, at that time, to the general public. Hence the book's enormous success. Even if the science could only be glimpsed it *sounded* exciting and the gossip was irresistible.

To achieve his aim Watson could not avoid simplifying the scientific side of the story. Indeed it is surprising how many technical arguments he managed to slip in without making his text too heavy. It is also clear, as Medawar has pointed out, that Watson is not really interested in history as such. His interest lies in the future—in what will soon be discovered—but time, in this case, has turned his future into his past and like the Ancient Mariner he could not rest till he had relived his story by communicating it.

Robert Olby has set himself quite different aims. As a historian of science he is interested in both the science and the history. This is not to say that he is in any way indifferent to personality. Who can resist the story of the ultra-shy Fred Griffith being tricked into a taxi by his friends in order to get him to attend a meeting, or the picture of Darlington, during a rough night in mid-channel, teaching Bernal "all the genetics and cytology anyone needs to know". The character of Maurice Wilkins, in particular, seems to me to come over more vividly in Olby's book than in Watson's, partly due to Wilkins' letters which Olby has often quoted *in toto.* This is no accident. It springs from Olby's care and industry in collecting original documents and taping interviews, together with a good eye for a striking incident or a telling phrase.

The major difference between the two books, which is clearly reflected in the greater length of the present volume, is that Olby treats the science more thoroughly and at a higher intellectual level; in addition, he traces both the ideas and the methods back to earlier periods. I am not always clear why he has followed some historical lines further back than others. His choice is often rather personal but in each case the story he presents is full of interest. Together, these earlier studies set the stage for the attempt on the structure itself.

As a professional historian, Olby has been at pains to check and cross-check meticulously the details of the historical record. This by itself is no more than careful scholarship—it corresponds more to getting coordinates than to solving the structure—but Olby's professionalism has more depth than this. He develops the story from many points of view and he asks questions. Watson is perhaps inclined to regard an incorrect conclusion as a pure act of folly. Olby is more concerned to track down the roots of the mistake.

Sometimes the results of his investigations will come as a complete surprise even to those who were actively working in the field. Who would have suspected that Beighton, in Astbury's laboratory, had obtained very good X-ray photos of the B form of DNA in 1951 and had not thought it worthwhile to publish them. At other times the facts were well known, at least in outline, but Olby has shed new light on them. In every case his reconstruction has been both imaginative and thoughtful. One follows with fascination the many aspects of the story of how the idea that genes were made mainly of protein was gradually replaced by our present ideas.

This, then, is the first full scholarly account of how the structure of DNA was discovered, set against its proper historical background. This does not mean it will be the last. The story—and what a good story it is—has too many ramifications for that. Some topics which Olby has touched on could well receive a longer treatment. If Watson had never come to Cambridge, who would have discovered the structure? More important, how long would it have taken? After all, the structure was there waiting to be discovered—Watson and I did not invent it. It seems to me unlikely that either of us would have done it separately, but Rosalind Franklin was getting pretty close. She was in fact only two steps away. She needed to realize that the two chains were anti-parallel and to discover the base-pairing. Wilkins, after Franklin left, could well have got there in his own good time. Whether Pauling would have made a second attempt (as we then feared) I am less certain. Olby demonstrates what was not clear to us at the time, that in the absence of our structure Pauling was not inclined to accept the more obvious criticisms of his own. Or, as Olby speculates, would the biochemists have hit on it in the end? And, if so, what difference would it have made?

Then there is the question of the motivation of the principal scientists involved, which often seems to contrast somewhat with the ostensible motives of the people providing them with money. Although Olby gives us much of the raw data on this topic it might deserve a more extended discussion. But that is the nature of history. What is I think certain is that no future historian of science in this area will be able to ignore the present volume, both for the thoroughness of Olby's investigation and for the good judgment he brings to the task.

FRANCIS CRICK

Preface

I cannot recall the word DNA ever being mentioned when I was a student at London University in the early fifties. How times have changed! Now this abbreviation for desoxyribonucleic acid is known far beyond the confines of university walls. J. D. Watson has provided us with an absorbing account of the discovery of its structure and others have added to the growing literature on the history of nucleic acid researches; but we still lack a broadly based study which will offer an overview of the whole intellectual and institutional movement in experimental biology which has yielded a physical and chemical account of the gene.

Here I attempt to contribute to filling this gap and, in so doing, to offer the scientist, sociologist and historian a reference work divided into five sections, any one of which may prove useful on its own. The reader who finds the sections dealing with X-ray crystallography difficult may profit more from those given to the chemistry and biology of the gene, and *vice versa.* The sociologist of science may wish to confine his attention to those chapters dealing with the development of research schools. The events leading to the discovery of the Watson–Crick model have been described in chapters 19 to 21 in some detail, supported by many quotations from original source material. This presentation was deliberate in view of the considerable publicity which this phase of the work has been accorded. The final chapter surveys the state of molecular biology from the vantage point of 1973. This may reveal the limitations of one who is caught in the paradigm of the molecular biology of the 'sixties. I hope it does not. It should convey my recognition of the limited extent of present-day knowledge, and of the possibility of further unexpected, even revolutionary, developments. At the same time I have not felt it necessary to comment on such discoveries as peptide synthesis without nucleic acids (Laland and Zimmer, 1973).

It has not been my aim to trace all historical antecedents to discoveries. Thus I have not referred to Araki's work of 1903 on what later became recognized as DNase, or to the extraction of DNA in small quantities from yeast by Dellaporte in 1939, or the extraction of nucleic acid from the tubercle bacillus by Ruppel in 1898–9. Nor have I dealt adequately with several important aspects and contributors to the subject matter of the book. My assessment of H. J. Muller's contribution to molecular biology is brief and critical. For a more appreciative and detailed study the reader is referred to the forthcoming biography of Muller by Carlson which the University of Indiana Press will publish. My account of the cytochemistry of nucleic acids does not include the work of the Belgian scientist, Jean Brachet. This omission resulted from my decision to exclude studies in the field of chemical embryology. For a fuller account of the history of nucleic acid chemistry the reader is referred to the forthcoming book by Cohen and Portugal to be published by M.I.T. Press. The wide-ranging studies of Fruton and of Florkin are also available for those who wish to learn about the history of biochemistry.

This book would not have been possible without the very substantial help which so many scientists have given me. I am greatly indebted to all of them, and particularly to Francis Crick for his unfailing enthusiasm, constructive criticism, and willingness to explain what I failed to understand. Many insights in the book have originated either directly from him or from discussions with him by letter, telephone or in person. I owe a similar debt to Linus Pauling and Maurice Wilkins. In the writing of some of the earlier chapters I have received much encouragement and help from Cyril Darlington and Sir Hans Krebs. For advice on recent developments in molecular biology I am indebted to Simon Baumberg, Tony North and Peter Speakman, all of Leeds University.

The later chapters contain many quotations from recorded interviews. Some of these are somewhat colloquial in style. The reader is asked to make due allowance both for this and for the direct, informal style of the extracts from personal correspondence. It is a pleasure to express my thanks to those concerned for the time they gave me and for permission to publish these extracts, also to those who gave me background material not referred to in the text. They are:

United Kingdom:

G. S. Adair, V. W. Allison, E. S. Anderson, W. F. Astbury, G. R. Barker, E. Beighton, Mrs. E. Bernal, Lady W. L. Bragg, G. L. Brown, A. J. Caraffi, C. H. Carlisle, W. Cochran, C. Coulson, F. H. C. Crick, Mrs. O. Crick, C. D. Darlington, M. Davies, Mrs. A. Dickens, S. D. Elliott, R. G. Gosling, J. S. Griffith, F. Happey, Sir H. Himsworth, D. Hodgkin, Sir R. D. Keynes, A. Klug, I. MacArthur, R. Markham, J. Needham, P. Pauling, M. Perutz, M. Polanyi, M. R. Pollock, E. Posner, C. Preston, R. D. Preston, Sir J. T. Randall, K. M. Rudall, Mrs. A. Sanderson, W. E. Seeds, P. T. Speakman, A. R. Stokes, Sir A. Todd, C. H. Waddington, M. H. F. Wilkins, H. J. Woods.

United States of America

G. Brawerman, E. A. Carlson, S. S. Cohen, J. F. Danielli, M. Delbrück,

T. Dobzhansky, J. Donohue, R. Dubos, H. Fraenkel-Conrat, Mrs. D. Fankuchen, F. Haurowitz, A. D. Hershey, R. D. Hotchkiss, M. L. Huggins, J. Lederberg, Mrs. G. Lewis, S. Luria, A. Lwoff, M. McCarty, C. M. MacLeod family, H. Mark, E. Mayr, A. E. Mirsky, Mrs. H. J. Muller, G. Oster, L. Pauling, F. Payne, B. Post, A. W. Ravin, Mrs. F. O. Sawyer, V. Schomaker, J. Schultz family, R. Sinsheimer, T. M. Sonneborn, Mrs. N. Sonneborn, G. S. Stent, D. D. Van Slyke family, P. A. Weiss, S. Wright, G. R. Wyatt, S. Zamenhof.

European Countries [other than the U.K.]
T. Caspersson, B. Ephrussi, A. F. Frey-Wyssling, S. Furberg, A. Lwoff, O. Maaløe, G. Melchers, R. Signer, M. Staudinger, R. Vendrely, E. Vischer, D. von Wettstein.
I also thank F. H. C. Crick, S. Furberg and R. G. Gosling for loaning me their Ph.D. theses.

This work has benefited from the use of archive collections. It is a pleasure to thank the staff concerned, especially W. Bell and J. Goodstein. The fourteen archives used are in the following institutions: American Institute of Physics, American Philosophical Society, Bancroft Library of the University of California at Berkeley, Biophysics Department of Leeds University, California Institute of Technology, Columbia Oral History Office, Darlington Correspondence in the Botany School Oxford, Leeds University Archive, Lilly Library of Indiana University at Bloomington, Medical Research Council, Needham Correspondence at Gonville and Caius College Cambridge, the National Foundation, Rockefeller University and Rockefeller Foundation.

I acknowledge grants for travel and research from the universities of Leeds and Oxford, the American Philosophical Society, Royal Society and Wellcome Trust. I thank the American Academy of Arts and Sciences for inviting me to conferences at Bellagio and Boston, and the Committee of the International Congress of Genetics for their invitation to Berkeley.

Small sections of the book have already appeared in an abbreviated form in *Daedalus*, *Journal of Chemical Education*, *Journal of the History of Biology* and *Nature*. I thank the editors for permission to expand these papers here.

Several of my friends have read large sections of this book. I am grateful to all of them, and especially to Peter Speakman for the great care and attention he has given to this task.

May 1974 **Robert Olby**

Acknowledgements for the Dover Edition

The work of preparing the 1974 text for reprinting was carried out whilst in receipt of a Mellon Research Fellowship in the programme organized by Dr. Lederberg at the Rockefeller University and Dr. Zuckerman at the Rockefeller Foundation. I thank them and my hosts, Dr. Wiesel, President of the University, and Dr. Gilbert, with whose laboratories I was associated. I also thank the staff of Dover Publications, especially Alan Weissman, for their excellent work editing the 1974 text for the Dover edition.

August 30, 1994 **R. O.**
Pittsburgh, Pa.

Note on References to Source Material

Published papers, books, reviews and newspaper articles are referred to in the text in the form: author, date and page, enclosed in brackets. Where an address was not published in the year in which it has been described as publicly read there is an obvious discrepancy of dates which does not require explanation each time it occurs. Unpublished documents have been treated in the same way as the published sources, and are therefore not separated from them in the list of sources at the back of the book. Copies of the American interviews will be presented to the American Philosophical Society. Information obtained by telephone or through informal conversation is simply referred to as "personal communication". There are only two exceptions to the above procedure, that of Watson's book *The Double Helix*, which is referred to simply as *Helix*, and Schrödinger's *What is Life?*, referred to as *Life*.

The sources of passages quoted and diagrams redrawn from published material is given in all cases. The author is grateful to the authors and publishers concerned for permission to use this material. The publishers are: Acta Chemica Scandanavica; Academic Press Inc.; Akademische Verlagsgesellschaft; The American Academy of Arts & Sciences; American Physiological Society; Annual Reviews Inc.; American Chemical Society; Blackwell Scientific Publications Ltd.; Cambridge University Press; Chemical Society; Churchill Livingstone; Cold Spring Harbor Laboratory; Elsevier Publishing Co.; Institute of Physics; International Union of Crystallography; Munksgaard International Publishing Co.; National Academy of Sciences; Pergamon Press Ltd.; Royal Society; Science; Springer Verlag; University of Chicago Press; Weidenfeld & Nicolson Ltd.; Wykeham Publications (London) Ltd.

All minuted and recorded interviews were made possible by financial support from the Royal Society.

Contents

List of Plates

hitherto unpublished, and therefore unknown to Watson, Crick and Pauling.
18 "Sheet" pattern from calf thymus DNA made at King's College London. [Material prepared by M. J. Fraser and Photographed by Gosling.] It shows the "Cross-Ways", meridional spot at 3.4 Å and spurious intensity along the meridian inside this arc.
19 X-ray diagram of "crystalline" DNA taken by Gosling in June 1950. This was the first example of a single-phase diagram, and shows the wealth of data which the A form yields.
20 X-ray diagram of *Sepia* sperm, showing features of the B pattern. Photograph taken by Wilkins in the Winter of 1951/2.
21 X-ray diagram of *E. coli* DNA, taken by Wilkins about the same time as (20)—a good B pattern.
22 Franklin's and Gosling's first B pattern (R.H. = 92%, 6 hour exposure of several fine fibres).
23 Franklin's remarkable B pattern of May 1952 (R.H. = 75%. 62 hours exposure of a single fibre, diameter = 50μ).
24 The pattern which showed "double-orientation' (R.H. = 75%. 100 Hrs. exposure of a single fibre, diameter = 40μ).
25 The A pattern, showing the meridional arc on the eleventh layer line. [Photograph taken with the tilting camera.]
26 Watson & Crick at the Cavendish Laboratory beside their model, May 1953.

For permission to publish these plates the author is grateful to the California Institute of Technology (plate 1), Springer Verlag (plate 4), Leeds University Textile Industries (plates 7 to 11), the Royal Society (plate 13), Professor Sonneborn (plate 14), Dr. Beighton (plates 16 and 17), Dr. Gosling, Professor Wilkins and the International Union of Crystallography (plates 18 to 25).

Introduction

> The century of biology upon which we are now well embarked is no matter of trivialities. It is a movement of really heroic dimensions, one of the great episodes in man's intellectual history. The scientists who are carrying the movement forward talk in terms of nucleoproteins, of ultra-centrifuges, of biochemical genetics, of electrophoresis, of the electron microscope, of molecular morphology, of radioactive isotopes. But do not be fooled into thinking this is more gadgetry. This is the dependable way to seek a solution of the cancer and polio problems, the problem of rheumatism and of the heart. This is the knowledge on which we must base our solution of the population and food problems. This is the understanding of life.
>
> (Weaver, 1949)

With these words the director of the Rockefeller's natural sciences programme underlined the importance of the subject which in 1937 he had called "Molecular Biology". Only four years later we find Max Delbrück, the biophysicist, writing to Niels Bohr, the physicist, "Very remarkable things are happening in biology. I think that Jim Watson has made a discovery which may rival Rutherford in 1911" (Delbrück, 1953e). He was referring to the model for DNA which James D. Watson and Francis H. C. Crick devised in Cambridge that year. Now, twenty years later, the state of our knowledge before that great event is a thing of the past. Yet it was in those now distant times that Weaver's "movements of heroic dimensions" gathered strength. When did this movement start and when did there come into existence a clear knowledge of the nature of DNA?

I went back to the year 1869 when Fritz Miescher extracted a substance, corresponding to what we now know as DNA, and called it "Nuclein". I pored over the writings of this Swiss physiologist who founded the histochemistry of the nucleus: was his conception of nuclein anything like his successor Levene's conception of nucleic acid, or like that of the later nucleic acid chemists, Gulland, Chargaff and Todd? No! Was Miescher really the man who deliberately went in search of the chemical substances which determine our heredity (Glass, 1965). No. On the contrary, all that Phoebus Levene, the great nucleic acid chemist, had said in the historical sections of his book *The Nucleic Acids* (1931) was confirmed. I therefore concluded that the difference between the nineteenth century conception of these substances and the precise picture of the, albeit ill-famed, tetranucleotide which emerged from Levene's brilliant work on the nucleosides and nucleotides of

RNA and DNA between 1909 and 1929 was so great that it was advisable to exclude the pre-1900 period of work on DNA.

Levene was himself initially a victim of the old approach to the nucleic acids. This was largely motivated by the hope that a metabolic link existed between cell proteins and the nucleic acids. Failure to remove all the protein bound to nucleic acid in the nucleus gave credence to this link. The continued use of the umbrella term *nuclein,* both for nucleoproteins and nucleic acids, perpetuated earlier ambiguities. Phospho-proteins continued to be referred to as nucleins after Kossel had shown that true nucleic acids can be distinguished from these "pseudo" or "para-nucleins" of the white yolk platelets. Nor was there general agreement at the turn of the century on the extent of the contribution of nucleic acid and protamine to the composition of the sperm head. Whereas Miescher computed it as 80 per cent, the editor of his posthumous paper on the subject, Ostwald Schmiedeberg, presented a quantitative interpretation of Miescher's data which gave the figure as 96 per cent. The latter was accepted by Kossel's student A. P. Mathews, and by Mathews' colleague E. B. Wilson, for a time. It was also favoured by Richard Burian and Theodor Boveri, but on the significance of the Schmiedeberg value these two were cautious. Thus Burian wrote: "As we see, so far sperm chemistry has achieved nothing important towards the understanding of hereditary phenomena. Indeed, it is even unlikely that the results of *chemical* research will help the attainment of real progress in this sphere. To bring the form-determining qualities of the chromosomes into a closer causal relationship with their chemical constitution still seems an unattainable goal today. Nevertheless, it would be an important step in this direction if one succeeded in putting the experimentally established inequality of chromosomes and chromosome segments alongside a demonstration of their *chemical* diversity" (Burian, 1906, 846). He rightly saw this as a task for the distant future and he ended by quoting the following words of Theodor Boveri: "For it is not cell nuclei, not even individual chromosomes, but certain parts of certain chromosomes from certain cells that must be isolated and collected in enormous quantities for analysis; that would be the precondition for placing the chemist in such a position as would allow him to analyse [the hereditary material] more minutely than [can] the morphologists. . . . For the morphology of the nucleus has reference at the very least to the gearing of the clock, but at best the chemistry of the nucleus refers only to the metal from which the gears are formed" (Boveri, 1904, 123). Nor was Jacques Loeb happy with the evidence which indicated that "the nucleic acid is of more importance for heredity than protamine . . . It is impossible," he concluded, "to draw any far-reaching inference concerning the nature of the substances which transmit hereditary qualities from these meagre data" (Loeb, 1906, 180).

The time span covered by this book has therefore been confined to the period 1900 to 1953, when the Double Helix was discovered. It is true that the axioms of molecular biology, as we now know them, were not clearly enunciated until several years later, but in the "Conclusion", which follows the last chapter, later developments are described briefly. A fuller account of Crick's involvement in them has already been published (Olby, 1970; 1972). The aim here has been to keep within the specified period, but to broaden the scope of the account beyond the confines of the structure, chemistry and biological functions of DNA: to include the more general conceptual and technical changes which lay between the physiological chemistry of the period 1910–1930 and the macromolecular chemistry which succeeded it. In this way the reader will find out not only why nucleic acids were originally thought less important than proteins, but how scientists in the 1930s conceived of the relationship of nucleic acids and proteins to the functions of the gene. Of necessity, this broad treatment has resulted in certain areas of research being excluded, but other scientists and historians are working on these subjects. In their forthcoming study of the chemistry of nucleic acids, Cohen and Portugal, at the National Institutes of Health, will do justice to the contribution of Lord Todd's group in Cambridge. Others are at work on the history of chromatography. The challenging but fascinating story of Fourier theory will call for an independent study, but one aspect has been treated in some detail in this account because it was of such value in the interpretation of the fibre diagrams produced by proteins and nucleic acids—the Fourier transform of a helix.

The material in this book has been arranged in the form of five sections. In the first of these the work leading to the concept of long-chain macromolecules is described; it concludes with an account of Astbury's work in the 1930s on the structure of the proteins and nucleic acids. The second section is devoted to the nucleoproteins as they were studied between 1900 and 1939 by organic and physiological chemists and by cytologists, geneticists and virologists. Their work established what we may call the "Nucleoprotein Theory of the Gene" and "The Protein Version of the Central Dogma". We turn in the third section to the subject of bacterial transformation and we see how research in this curious corner of science provided the impetus to Erwin Chargaff's discovery of base ratios in DNA and to Vendrely's discovery that the DNA content of nuclei was constant, whereas it halved in the formation of the germ cells. The fourth section is given to those intellectual migrations which brought physicists and structural chemists into biology, starting with Max Delbrück in 1933 and ending with Francis Crick in 1947. Finally, there is a fifth section in which we examine the events which brought these several approaches together in the discovery of the molecular structure of DNA.

Although the areas of knowledge which contributed to the discovery of the Double Helix were many, as can be seen by the division of this book into five sections, I see behind all this work a single path into which flowed one after another of the subjects here considered. This path began with the concept of long-chain macromolecules, which was applied first to carbohydrates and proteins, then to DNA. The X-ray crystallographers determined the configurations of these chains, first in an approximate manner using Polanyi's layer line concept, then more accurately, with co-ordinates, and using Fourier methods. We then see how such structures were used to account for the hereditary codescript and the duplication of the gene in the 1930s, a period of intense and international debate over the nature of the gene, which we may call the "first phase" of molecular biology. Its central axioms were that the gene can be described in molecular terms and that all information flows from protein. The "second phase" of molecular biology began in 1950 and is distinguished from the first phase by the increasing conviction that most if not all information flows from nucleic acids, by the availability of more highly polymerized DNA, and better methods for interpretating the X-ray data.

I do not claim a straight path of continuity from the first phase to the second in terms of the development of knowledge of DNA and the gene: only in the sense that the aim was to account for the properties of the gene in molecular terms. Between the two phases lay not only the curtailment and isolation of research activities due to the Second World War, but also a conceptual contrast in the way the molecular theory of the gene was envisaged, a contrast so striking as to constitute a transformation of paradigms from what I call "The Protein Version of the Central Dogma" to the "DNA Version of the Central Dogma". *The central theme of this book is therefore this continuity in aspiration and transformation in conception by which we have arrived at the Double Helix.*

In the collection of data I have turned to both written and oral sources. The former consist of published papers, private correspondence and grant applications. Information on the oral sources will be found in the Note on references to Source Material on p. xiii. By balancing the two kinds of data against one another I hope I have avoided some of the pitfalls which all too easily trap the unwary who use only one kind of source. As a first attempt on a difficult and complex story, however, this account is bound to need correction and amplification in many places. I present it not as a definitive account but as a first attempt which will stimulate others to do better. My main hope is that it will afford the historian, sociologist and scientist a guide to the source material relating to the origins of molecular biology, with special reference to DNA.

The story told here opens with the debate over the existence of macromolecules, which developed into a major conceptual conflict during the first three decades of this century. During this time the basic structural features of DNA were being worked out, the polypeptide chain theory for proteins was announced and developed, Mendel's theory of heredity was rediscovered and extended, and the chromosome theory of inheritance was formulated. Our first conceptual problem can be stated thus: Should we rely on organized [hierarchical] structures above the level of the chemist's molecules to account for such vital processes as biological specificity, the duplication of the gene, hereditary transmission, the properties of cytoplasm and chromosomes, or can we refer these to the characteristics of huge and highly complex molecules or polymers? If we need to introduce supra-molecular entities, what is their nature? Are they, as was so widely held, identical with the particles of the colloid scientist? Should we, then, leave the organic and structural chemists to their own devices, and seek to introduce the principles of colloid science. Should we champion a "colloidal biophysics" of the cell?

Note to the Introduction (1994)

The notion of a "path to the double helix" has been criticized by those historians who judge it as "presentist" or "Whiggish", because it suggests a view of the historical developments from the vantage point of later knowledge. The result is often an unjustified emphasis and selection of data from the historical record that appear to be leading teleologically towards our current knowledge. The attentive reader will note, on the contrary, that I have placed emphasis on so-called "failures"—the colloid, the protein theory of the gene and virus, the repeating-sequence model of protein and nucleic-acid structures, the Lamarckian theory of bacterial transformation, the reasons for adherence to integral rather than non-integral helices for proteins, etc. Moreover, I am prepared to stand by the view I took in 1974 that a research programme was established in the 1930s and 'forties which attributed the many aspects of biological specificity to the molecular structures of macromolecules. This programme contrasted with the rival programme of attributing such phenomena to dynamic processes—the liquid crystalline state, the rates of reactions, the embryological concepts of fields, polarity, evocation, etc. The focus was on the proteins, but it also embraced the nucleoproteins and its focus moved in time towards DNA. In this sense there was a path.

In speaking of a transformation of paradigms I wished to stress the discontinuity in the conceptual development. At the same time, by introducing the term "Protein Version of the Central Dogma," I was signalling the continuity in the reductionist programme of explaining biological phenomena in terms of the concepts of chemistry and physics. Naturally I was aware that the central dogma of the 1950s possesses a precision and strictness which the older protein view lacked. Nevertheless, the expectation that all forms of biological specificity in organisms were to be traced ultimately to the proteins was as strong as the current view today that they should be traced to the nucleic acids. It might have been wiser to use the phrase "Protein Dogma", but I wished to emphasise the fact that the conception favoured in the 1930s and 'forties involved the notion of sequence as *a* determinant of biological specificity, even if not uniquely so. Whilst it is clear that the "Central Dogma" of the 1950s embodied a much more explicit conception of biological specificity than did the protein dogma, the latter raised the very questions concerning specificity that 1950s developments answered, and, moreover, introduced the appropriate techniques for doing so.

R. O.
Visiting Professor
Rockefeller University
New York

SECTION I

FROM COLLOIDAL PARTICLES TO LONG-CHAIN MOLECULES

Bergmann, Staudinger, Svedberg, Polanyi, Mark, Astbury

CHAPTER ONE

The Macromolecule

The 'growth' of organic matter is seen forcibly in the almost endless carbon chains with their most varied arrangements, as they are formed in the bodies of plants. These chains have originated from quite separate carbon atoms, which earlier were present in carbonic acid. Thus carbon has a great tendency in living molecules to cause growth by chain formation. Cyanogen also has this tendency in a high degree, especially for [other molecules of] cyanogen. The important elements of living protein [cyanogen radicals] thus have the most marked tendency to attract radicals of the same kind and in this way to produce ever larger molecules, i.e., to grow.

(Pflüger, 1875, 342)

. . . since the molecules of organized substances contain on an average about fifty of the more elementary atoms, we may assume that the smallest particle visible under the microscope contains about two million molecules of organic matter. At least half of every living organism consists of water, so that the smallest living being visible under the microscope does not contain more than about a million organic molecules. Some exceedingly simple organisms may be supposed built up of not more than a million similar molecules. It is impossible, however, to conceive so small a number sufficient to form a being furnished with a whole system of specialized organs.

Thus molecular science sets us face to face with physiological theories. It forbids the physiologist from imagining that structural details of infinitely small dimensions can furnish an explanation of the infinite variety which exists in the properties and functions of the most minute organisms.

(Maxwell, 1875, 42)

[Walter Morley] Fletcher had a bias towards chemistry, which he had acquired from his father. He asked Michael Foster whether he thought that there was any future in the application of chemistry to physiology. At this enquiry, Foster rolled his large beard over his mouth with both hands, to smother his hilarity.

(Crowther, 1968, 313)

A very distinguished organic chemist, long since dead, said to me in the late eighties: 'The chemistry of the living? That is the chemistry of protoplasm; that is superchemistry; seek, my young friend, for other ambitions'.

(Hopkins, 1933, 245)

One reason which has led the organic chemist to avert his mind from the problems of biochemistry is the obsession that the really significant happenings in the animal body are concerned in the main with substances of such high molecular weight and consequent vagueness of structure as to make their reactions impossible of study by his available and accurate methods. There remains, I find, pretty widely spread, the feeling—due to earlier biological teaching—that, apart from substances which are obviously excreta, all the simplest products which can be found in cells or tissues are as a class mere *dejecta*, already too remote from the fundamental bio-chemical events to have much significance.

So far from this being the case, recent progress points in the clearest way to the fact that the molecules with which a most important and significant part of the chemical dynamics of living tissues is concerned are of a comparatively simple character.

(Hopkins, 1913, 144)

No upper limit is usually assigned to molecular magnitude. E. Fischer has synthesized a polypeptide with the molecular weight 1212, and in the case of colloids, molecular weights of the order 10^4, and even 10^5, are commonly spoken of. A difficulty arises, however, in admitting that molecular weights can exceed a certain value, unless the density increases as the molecular weight increases.

For suppose that a compound can exist, such as a protein, with a density at 0° not much greater than that of water, and with a molecular weight of rather more than 30 000, the grammolecule of such a compound at 0° would occupy about 30 000 cc. The gram-molecule of a perfect gas under the standard conditions occupies only 22 400 cc, and we should therefore have a solid compound, at 0° and under a pressure that cannot be less than one atmosphere, occupying a greater molecular volume than that of any gas.

That the molecules of liquids and solids should occupy greater volumes than those of gases under similar conditions, seems at first contrary to the usual conceptions of the gaseous, liquid, and solid states. It is true that at sufficiently low temperatures this condition must arise for all substances, but a simple calculation shows that for the majority of chemical compounds it would only occur at temperatures not far removed from the absolute zero.

Two suggestions appear to be indicated. The first is that under the ordinary conditions there is an upper limit to molecular magnitude, and that for most substances, more especially colloids, the molecular weight cannot exceed a value of about 20 000. The second is that our ordinary kineto-molecular conceptions no longer apply when for a given temperature the molecular magnitude exceeds a certain critical value. The latter view seems most in keeping with our present knowledge, and perhaps serves to throw some light on the behaviour of colloids.

(Crompton, 1912, 193–194)

. . . Of course for oxyhaemoglobin, which crystallizes in the well-known beautiful manner, a molecular weight of 16 000 has been deduced from the iron content, but against such calculations the objection can be raised that the existence of crystals in no way guarantees by itself chemical individuality; on the contrary one can have to do with an isomorphous mixture such as the mineral kingdom offers us in such variety in the silicates. Such doubts cease in the case of synthetic products whose production by analogous reactions can be controlled.

(Fischer, 1913, 3288)

. . . In the application of the condensation principle of albuminous compounds, to molecules of the order of magnitude of 4000 and far higher, chains of truly fantastic proportions would be yielded, the existence of which can be assumed very improbable according to our ideas.

(Hess, 1920, 232)

The concept of very large molecules is not a twentieth century idea. It arose in the previous century from the consideration of the chemistry of carbon compounds and from the attempt to describe physiological processes in chemical terms. In 1857 the quadrivalency of carbon had been evident to Kekulé in the case of methane, and in 1858 he made it a law for all carbon compounds (Kekulé, 1858, 152). At the same time he arrived at the conception of the carbon–carbon link. Carbon atoms, he said, "are themselves joined together, so that naturally a part of the affinity of one for the other will bind an equally great part of the affinity of the other" (Kekulé, 1858, 154). This idea was, he declared, distinct from the earlier picture of atoms associated in a centric or mutually connected pattern. As he later recalled: "The separate atoms of a molecule are not connected all with all, or all with one, but, on the contrary, each one is connected only with one or a few neighbouring atoms, just as in a chain link is connected with link" (Kekulé, 1878, 212). Such linkages when made between polyvalent atoms could lead to the production of "net-like" and "sponge-like" "molecular masses which resist diffusion, and which, according to Graham's proposition, are called colloidal ones" (*Ibid*, 212). The same hypothesis, he argued, could be used as Pflüger had done, to produce the elements of the form of living organisms.

The Application of Kekulé's Polymer Concept

This concept was used by Kekulé's colleague, Eduard Pflüger, in his theory of intracellular respiration. Pflüger believed that the energy released during respiration was liberated by the decomposition of the peculiar, highly unstable, polymeric, *living* protein molecules of protoplasm (Pflüger, 1875). W. Pfitzner applied the polymer concept to the chromosome structures which he called "chromomeres" in 1882. Edmund Montgomery in 1885 and George Hörmann in 1899 applied it to a number of morphological structures. All these authors saw the polymeric state as distinct from that of the compounds analysed by the organic chemists. In Pflüger's opinion, living polymeric compounds were to the molecules of the chemists as the sun was to the smallest meteor. In studying albumin extracted from the organism "one has to do mostly with torn-off fragments of those vast molecules, which may well be as large as an entire creature" (Pflüger, 1875, p. 343). At the close of his life he commented on Emil Fischer's synthesis of polypeptides: "in spite of the great exploits of Emil Fischer the synthesis of protein will take up another century and the synthesis of living protein is hardly likely ever to succeed" (Cited in Cyon, 1910, 1).

Emil Fischer's attitude to the polymer concept was, no doubt, influenced by the vitalism it had acquired. But it was also natural that he should try to construct his version of the chemistry of life only on the solid foundation of organic chemistry. He was able to characterize some small polypeptides by the demanding criteria of this science; therefore such chains of up to thirty amino acids were acceptable to him. Of such, he thought, were proteins constituted. Native proteins were mixtures of these. To support this view he worked out the potential isomerism of a polypeptide consisting of thirty different residues [today only 20 are recognized in the amino acid alphabet].

It came to 2.635×10^{32} (Fischer, 1916). What irony that Fischer's recognition of the sequence hypothesis for proteins served to consolidate the case against their macromolecular nature, for, he implied, if the isomerism of small polypeptides was so great, what need had nature for giant polypeptides!

The Colloid–Aggregate Theory

The fact that the polymer theory enjoyed some popularity at the close of the nineteenth century was due rather to the lack of a clear alternative than to the cogency of the arguments in its support. When the aggregate theory of colloidal particles was developed in the early part of the twentieth century the polymer theory was quickly dropped. Most so-called polymeric molecules (with molecular weights above 5000) were then regarded as aggregates of much smaller molecules. Haemoglobin, for instance, was probably an aggregate of four globin molecules of molecular weight $\sim$ 4000, which, when combined with the prosthetic group gave a particle with the well known empirical weight of 16 700.

There were three developments which made this suggestion plausible. Alfred Werner, founder of the co-ordination theory, had introduced the concept of two kinds of combining forces in chemical compounds—*Hauptvalenzen* or primary valency forces, and *Nebenvalenzen* or secondary valency forces (Werner, 1902, 268). He claimed that atoms united by primary valency forces [covalent bonds] still possessed varying degrees of "residual affinity" whereby several molecules became united into "compound molecules" or aggregate molecules. This idea was applied by Karrer and Hess to starch and cellulose, by Pummerer and Harries to rubber, by Bergmann to proteins, and by Hammarsten to thymonucleic acid or DNA.

A second support came from the estimates of the unit cells of these substances made by the X-ray crystallographers. Many such scientists claimed that the molecule could not be larger than the unit cell, and since their unit cells were small, so were the molecules (see Chapter 2).

The third and most influential support was provided by the promising young subject of colloid science. This bridge subject between chemistry, physics and biology achieved such recognition that many reputable textbooks on physiology and biochemistry took the principles of colloid science for their foundations. This tendency has been labelled "Biocolloidology", a "dark age", the "age of micellar biology" (Florkin, 1972, 279). This was no crank science, but a subject which fascinated the most powerful minds. Was it not from colloidal particles that Jean Perrin gained his evidence for the reality of the molecule? It was, surely, through the paths of colloid science that Svedberg was led to design his ultracentrifuge and demonstrate the unique character of macromolecular species of proteins.

The chief axiom of colloid science was that there is a state of matter, the colloidal state, to which the ordinary laws of chemistry—the laws of constant and multiple proportions (Hardy, 1903, p. xxix) and the law of mass action—are not applicable. This is because the particles are large aggregates of molecules, many of which are not therefore accessible to the other reactants. The surfaces of such large particles show special properties: they adsorb ions, double electrical layers form around them, they often function in catalysis. Their "autoregulative properties" were attributed to "the surface field around the particle" (Svedberg, 1928, 17). Colloidal solutions showed abnormally high viscosities and gave abnormally low osmotic pressures. Some of these properties were shown by the thymonucleic acid or DNA extracted by Einar Hammarsten in 1924.

Not surprisingly, experimentally-minded cytologists fastened upon colloid science in their biophysical attack on the living substance of the cell. Protoplasm was a colloidal system. What organic chemistry could not explain about the behaviour of the proteins of the living substance colloidal science would clarify. In the absence of the electron microscope and the technique of differential centrifugation and with the neglect of the ultraviolet microscope there was a "neglected dimension" which colloid science, it seemed, could fill. There thus arose in opposition to the old morphological cytology a new experimental cytology founded upon colloidal biophysics.

The Debate over Macromolecules

This debate came to a head in the period 1926 to 1930. The man who coined the term *Macromolecule* was the German chemist, subsequently Nobel Laureate, Hermann Staudinger. He began his opposition to the aggregate theory by devising a crucial experiment to test the theory as it had been applied to rubber (Harries, 1904), and by the time of his farewell lecture to the Zürich Chemical Society in 1926 he was convinced that Kekulé's polymer concept was right and the aggregate theory wrong. Chain molecules held together by Kekulé [co-valent] bonds could be of fantastic lengths without the need to invoke secondary valency forces. This was the import of his lecture, but it was not well received, as the following report of a witness shows:

> I remember Staudinger's lecture to the Zürich Chemical Society in 1925 on his high polymer thread molecules with a long series of Kekulé valency bonds. It was impossible to accommodate his view in the unit cell as established by X-ray analysis. All the great men present: the organic chemist, Karrer, the mineralogist, Niggli, the colloidal chemist, Wiegner, the physicist, Scherrer and the X-ray crystallographer (subsequently cellulose chemist), Ott, tried in vain to convince Staudinger of the impossibility of his idea because it conflicted with exact scientific data.
>
> The stormy meeting ended with Staudinger shouting '*Hier stehe ich, ich kann nicht anders*' in defiance of his critics.
>
> (Frey-Wyssling, 1964, 5)

Another witness of this meeting was Rudolph Signer, now professor of organic chemistry at Bern. At that time (1925) he was working under Staudinger for his doctoral thesis. Throughout his stay in Zürich, Signer

> . . . was very impressed by Staudinger. He was completely sure that his idea of the existence of macromolecules was right and he had practically all his colleagues against him and his opinions. And so it was a very interesting situation to see this man already having a great experience in this field, having a special conviction, and having against him all his colleagues. . . . The crystallographer Niggli said that each substance in the pure state should crystallize and if these materials of polystyrenes and other polymers which Staudinger had synthesized were pure they should form crystals. After the meeting Wieland told Staudinger that in his opinion molecules with more than forty carbon atoms should not exist.
>
> (Signer, 1968)

According to Staudinger's own recollections, Niggli's contribution to the discussion was simply to oppose Staudinger's suggestion of the macromolecule with the words: "Such a thing does not exist!" (Staudinger, 1961, 85). And his friend Wieland had advised him "at the end of the 1920s", presumably by letter to Freiburg:

> Dear Colleague, leave the concept of large molecules well alone: organic molecules with a molecular weight above 5000 do not exist. Purify your products, such as rubber, then they will crystallize and prove to be lower molecular substances.
>
> (Staudinger, 1961, 79)

The result of this encounter was that the debate between the polymer and aggregate conceptions became widely known in Germany and the *deutsche Naturforscher und Aerzte* devoted a session of their Düsseldorf meeting a year later to the subject. Staudinger, Max Bergmann, Hans Pringsheim, and Mark were the principal speakers. Mark's neutral stand and wisdom in not directly attacking the macromolecule we will discuss later. Pringsheim took the view that we have to distinguish between *Hauptvalenz* and *Molekularvalenz*, the former being Kekulé bonding and the latter representing forces of aggregation whereby colloidal particles are formed. Bergmann, in his impressive and lucid paper, began by declaring that the classical structure theory and Avogadro molecule were inappropriate to the complex carbohydrates and proteins. This body of theory had reference only to simple molecules which could be evaporated and studied in the gaseous state. In the liquid and solid states he turned elesewhere for a structural theory and found it in the "higher-order compounds" of Alfred Werner. As his model he took potassium platinichloride. The $PtCl_4$ represented what he called the "individual group", the forces holding together the K_2, Cl_2 and the $PtCl_4$ the "aggregating bonds", and the crystal of K_2PtCl_6 the "pseudo-high molecular substance". The association of individual groups to generate the aggregate was reversible, as Pringsheim had shown for inulin and Hess for cellulose; it was the individual groups which gave rise to osmotic pressure and retained their integrity after the pseudo-high molecular substance had been broken down. He closed his brief but telling speech with the words:

> If a carbohydrate like inulin has only the modest individual group weight of 324, then the cause of its non-volatility and high-colloidal behaviour lies less in the sphere of its individual groups and much more in the relative firmness of the aggregating forces, which bind together the individual groups. *Pseudo-high molecular substances* are distinguished from readily dispersing substances by the higher share taken by the aggregating forces in the binding energy as a whole. . . . Therefore, what is at the present time especially necessary for the chemistry of pseudo-high molecular substances is the development of a structural and spacial chemistry the object of which lies outside the molecule, outside the individual group—a structural chemistry, a spacial chemistry of aggregating forces and of aggregates.
>
> (Bergmann, 1926, 2981)

The organizers of the meeting knew their job when they allowed Bergmann to open the case for the aggregate theory. His thesis was supported by Pringsheim from Berlin, whose evidence for the reversibility of the breakdown of inulin into its individual group molecules (molecular weight 324) appeared convincing. This noted chemist was clearly as confident as Bergmann that the Wernerian distinction between main and residual valency forces was applicable to colloidal polymers. What concerned him was the question of how to distinguish between these two forces. Perhaps, he suggested, optical activity was determined by the main valencies, the residual or molecular valencies having no influence upon this property.

Hermann Mark, who knew Max Bergmann from the time they were colleagues at the Kaiser Wilhelm-Institut für Faserstoffchemie in 1920–21, proceeded cautiously. In a long speech he sought to clarify the intricacies of X-ray diffraction analysis. He was careful to distinguish between strong and weak lattice forces, using as his examples diamond and hexamethylenetetramine. In such examples the size of the molecule in the crystal was established, but in crystals of high polymers it was not. Here he called for more accurate measurements on really good crystals. Wisely, Mark was "sitting on the fence", but at least he did not promote the crystallographers' mistaken idea that the molecule cannot be larger than the unit cell of the crystal.

Finally, Staudinger spoke in defence of the macromolecule. He was able to confront his audience with an impressive array of data on polymerization, hydrogenation, comparisons of viscosity, melting point and solubility of fractionated polymers which had been carried out by his doctoral students in Zürich between 1923 and 1926. He was able to draw upon his studies of polyoxymethylene begun in 1920, of polystyrene homologous series begun three years later. In the conversion of polystyrene to hexahydropolystyrene, and polyindene into hexahydropolyindene, reactant and product had the same average molecular weight and the reaction was a normal chemical one, uninfluenced by the so-called non-stoichiometric behaviour of colloids. This, he declared, was clear evidence that in these polymers "the monomers are united by main valencies" (Staudinger, 1926, 3034). Proudly he referred to the term *Makromoleküle* which he had first used in connexion with rubber in 1922 and defined two years later in terms of *Hauptvalenzen*:

> For such colloidal particles in which the molecule is identical with the primary particle, where the individual atoms of the colloidal molecule are those bound by normal valency activity, we suggest the term *Makromoleküle*.
>
> (Staudinger, 1924, 1206)

He admitted to his audience in Düsseldorf that molecular weights for high polymers can only be *average* values for a mixture of chains of different lengths. He very rightly saw that the molecular concept might lose its significance in the crystal state (e.g. NaCl), but Mark's distinction between lattice forces of the order of strength of intra-molecular bonds and those much weaker was not lost on him. There was a world of difference between a crystal of monomeric formaldehyde and one of a polymer formaldehyde.

Yet after all this, only Richard Willstätter who presided over the meeting came down clearly on Staudinger's side, having been convinced by the hydrogenation experiments on polystyrene. Bergmann's lecture, on the other hand, received a warm welcome. Mark has stated that the macromolecule concept would have been established more easily if Freudenberg had expressed himself in a more forthright manner in his paper on the long chain structure of cellulose in 1921, and if Freudenberg, H. K. Meyer and Staudinger could have agreed with each other instead of disagreeing on points of detail. "On certain occasions", he wrote, "they argued with each other more vigorously than with the defenders of the association theory" (Mark, 1965, 16). There was undoubtedly considerable bitterness between Meyer and Staudinger, the latter feeling that the former had no sooner given up opposing him than he adopted his ideas as if they were his own. In the meantime a most important event took place in Sweden which profoundly altered the climate of opinion—the first molecular weight estimations with the ultracentrifuge were made.

CHAPTER TWO

The Ultracentrifuge

Attempts to measure the weight of colloidal particles by sedimentation and diffusion date back to the first decade of this century (Perrin, 1908; Millikan, 1910; Svedberg and Estrup, 1911). The use of sedimentation equilibrium is a little later (Westgren, 1913). These estimates were all made with the large dense particles of gold sols, where the rate of fall in the earth's gravitational field was perceptible after a few days. The smaller particles of proteins and nucleic acids called for much stronger gravitational fields. Although ordinary bench centrifuges were used during the second decade of this century to arrive at the dimensions of smaller particles, these efforts were vitiated by the occurrence of convection currents. The fields were, in any case, too weak for the average protein, and it was not until 1924 that Svedberg completed the construction of the first "ultracentrifuge" with which he succeeded in determining the weights of proteins, and later of nucleic acids (see Table 2.1). He turned first to ovalbumin which was supposed to have a molecular weight of 34 000 based on its sulphur content. The results were difficult to interpret. Haemoglobin, which he had ruled out at first because its molecular weight was supposed to be only 16 000, now seemed the obvious choice, and on Robin Fåhraeus' advice Svedberg agreed to try it. He told Tiselius and Claesson that he "was awakened in the middle of the night by a telephone call from Fåhraeus who was watching the run and shouted "The., I see a dawn". Svedberg rushed to the institute, and there was indeed a marked lightening of the colour at the top of the cell; the haemoglobin was sedimenting" (Tiselius and Claesson, 1965, 4).

Because it yielded a sharp boundary (with the absorption optical system), unlike the heterogeneous systems of man-made gold colloids, Svedberg and Fåhraeus reckoned they were dealing with a homogeneous molecular species —a monodisperse system. This was strongly supported when they carried out an equilibrium run. In this case the tendency to sedimentation was balanced by the tendency to diffuse, with the result that no boundary was formed. Instead, the concentration of haemoglobin in the cell was estimated optically at varying distances from the axis of rotation, and the molecular weight of solute at these distances calculated. Svedberg and Fåhraeus argued that if they were dealing with a polydisperse system they should have encountered

TABLE 2.1

Molecular Weights of Proteins

Date	Author	Substance	Method	Mol. Wt.
1886	Zinoffsky	haemoglobin	Q.A.	16 700
1890	Jacquet	haemoglobin	Q.A.	16 669
1891	Sabanjeff & Alexandrov	egg albumin	D.P.	14 000
1908	Herzog & Kasernovsky	egg albumin	D	17 000
		pepsin	D	13 000
		emulsin	D	45 000
1910	Herzog	egg albumin	D	73 000
1917	Sørensen	egg albumin	O.P.	34 000
1920	Dakin	gelatin	Q.A.	10 000
1922/3	Cohn & Hendry	casein	Q.A.	12 800
1925	Adair	haemoglobin	O.P.	66 700
1925	Sørensen	egg albumin	O.P.	43 000
1926	Svedberg & Fåhraeus	haemoglobin	E.S.	68 000
1926	Svedberg & Nichols	egg albumin	E.S.	45 000
1928	Svedberg & Sjögren	serum albumin	E.S.	67 000
1928	Svedberg & Chirnogar	haemocyanin	E.S.	5 000 000

D = diffusion; D.P. = depression of freezing point; E.S. = equilibrium sedimentation; O.P. = osmotic pressure; Q.A. = quantitative analysis.

TABLE 2.2

Molecular Weight Determination of carbon-monoxide–haemoglobin by Sedimentation Equilibrium

Distance along cell	Mol. Wt./16 700
4.61	4.27
4.56	4.05
4.51	3.49
4.46	4.02
4.41	4.37
4.36	3.65
4.31	4.59
4.26	4.16
4.21	3.98
Average	4.06

(from Svedberg and Fåhraeus, 1926, Table 1)

a *systematic* variation in molecular weight, but they did not (see Table 2.2). There seemed little doubt that haemoglobin did really have a precise molecular weight about four times the minimum molecular weight of 16 700. Further proteins were studied, amongst them the haemocyanin of *Helix pomatia* which gave the value of 5 000 000 although it was thought on the basis of its copper content to have a molecular weight of between 15 000 and 17 000.

Svedberg's Multiple Law

Over the next decade the conviction that proteins were genuine molecular species grew strong, so that in 1938 the Royal Society saw fit to hold a one-day meeting on "The Protein Molecule". In the opening speech Svedberg declared that the proposal to hold such a meeting "would have looked preposterous" a few years ago. But now, he continued:

> We have reason to believe that the particles in protein solutions and protein crystals are built up according to a plan which makes every atom indispensable for the completion of the structure. The removal of even a single atom means loss of individuality . . . Protein reactions are therefore elementary acts which must of necessity obey the laws of quantum mechanics.
>
> (Svedberg, 1939, 40–41)

Unfortunately Svedberg became a little too confident about the meaning of his ultra-centrifugal data, and proceeded to overinterpret them. As early as 1929 he had become greatly excited to find that he could arrange the molecular weights of the proteins for which he had sedimentation data in a sequence of multiples of the value for egg albumin (34 500). From a Swedish interview two years earlier it is clear that he was even then thinking along the lines of a general plan for protein construction (see Pederson, 1940, 406). The British crystallographer, W. T. Astbury, quickly took this idea up and tried to give the Svedberg classes of n, $2n$, $3n$ and $6n$ (where $n = 35\,000$) a crystallographic significance (Astbury and Woods, 1931). The South American mathematician, based in England, Dorothy Wrinch, devised her "cage" or "cyclol" structure corresponding to Svedberg's basic unit. But there was worse to come. Bergmann and Niemann tried to fit Svedberg's unit, consisting of 288 amino acids, and the numbers of each individual amino acid therein, to a mathematical law expressed by the product $2^m \times 3^n$, where m and n could be zero or positive integers (see Chapter 7). So strong was the desire to discover order that speculative short-cuts were taken and difficulties swept aside.

At the protein meeting of the Royal Society in 1938 objections were expressed by Neuberger and Bernal. The latter repeated his criticism of the cyclol theory at a Royal Institution lecture in 1939. Then came the masterly attack in *Science* by Pauling and Niemann (who had by this time come to

Caltech and rejected Bergmann's frequency hypothesis). At the close of this paper Svedberg's unit of 288 residues was mentioned. The authors did not think it likely that "this number will be adhered to rigorously. Some variation in structure at the ends of a peptide chain might be anticipated". It seemed to them:

> very unlikely that the existence of favoured molecular weights (or residue numbers) of proteins is the result of greater thermodynamic stability of these molecules than of similar molecules that are somewhat smaller or larger, since there are no interatomic forces known which could affect this additional stabilization of molecules of certain sizes. It seems probable that the phenomenon is to be given a biological rather than a chemical explanation—we believe that the existence of molecular weight classes of proteins is due to the retention of this protein property through the long process of the evolution of species.
>
> (Pauling and Niemann, 1939, 1867)

The Shape of Macromolecules in Solution

When Staudinger used viscosity as an indication of molecular size he was going against the tradition of the colloid scientists according to which there is no linear relationship between viscosity and concentration for such solutions. The Einstein–Smoluchowski equation which related molecular weight to viscosity applied only to spherical molecules. Colloidal particles deviated in shape and in other respects from the conditions for Newtonian flow and therefore did not conform to this equation. Viscosity, under these circumstances, could not be related to molecular weight.

Staudinger attacked the opposition on its own ground. If ordinary low-molecular weight compounds are to be distinguished from high-molecular compounds as regards the cause of their viscosity in solution, do they all obey the Einstein law? As he expected, low-molecular linear chain compounds did not. Instead, there was a *proportionality between viscosity and chain length.* In 1929 and 1930 his students, R. Nodzu and E. Ochiai, reported their experiments on paraffin and soap solutions in support of this relationship. What Staudinger now required in order to establish this relationship was an independent method of molecular weight estimation. The obvious choice was Svedberg's ultracentrifuge. Signer had just returned from visits to Stockholm and Manchester. In Stockholm he had become familiar with the use of the ultracentrifuge. But the Nothgemeinschaft der deutschen Wissenschaft refused Staudinger's request for a grant to purchase this instrument. Magda Staudinger can recall a beautiful Sunday walk she took with her husband in the autumn of 1929 on the nearby Schönberg. Staudinger had just learnt the decision of the authorities. Evidently they considered work on the size of polymers, using this new-fangled and expensive instrument, a waste of time. He was very angry. On the walk their conversation turned to viscosity, end-group analysis and osmometry. When they returned home around 6 p.m. Staudinger "sat at his table and started to write and think. It was at two o'clock in the morning that the viscosity formula was on the table in front of him" (M. Staudinger, 1969). This was the well-known Staudinger Law expressing the relationship between molecular weight and specific viscosity η_{SP} which represented the relative increase in the viscosity of a liquid due to the addition of the solute molecules per unit of concentration.

Flow Birefringence

A knowledge of the shape of macromolecules was of crucial importance both for viscosimetry and for ultracentrifugation. In the latter technique, sedimentation rates could only be used to calculate molecular weights if the length/breadth ratio was known (this is not so for equilibrium sedimentation). Staudinger's colleague, Rudolf Signer, therefore devised a very simple technique based on flow birefringence. Freundlich had shown that Vanadium pentoxide particles in a V_2O_5 sol are needle-shaped when observed with the ultramicroscope, and that they showed positive flow birefringence, that is when the solution was illuminated the light transmitted through it was polarized in two planes perpendicular to one another, and the refractive index of the extraordinary ray was greater than that of the ordinary ray (Freundlich, 1916). This is what Wiener had predicted for a system of rods, their axes all parallel, immersed in a medium of different refractive index (Wiener, 1912). These two arrangements are shown in Figure 2.1. For disc-like structures packed one above another he had predicted negative birefringence. Because his apparatus was too crude Freundlich failed to observe birefringence in cellulose derivatives, which, being true molecules, were far smaller than the colloidal particles of vanadium pentoxide. This was overcome in the apparatus designed by Signer. Here very high flow gradients could be generated between two closely fitting coaxial cylinders.

It was probably in 1936 that the Swedish cytochemist, Caspersson, made the journey from Stockholm to Bern to use Signer's apparatus on the highly polymerized thymonucleic acid (DNA) extracted at the Karolinska Institute. He had already seen a crude demonstration of the flow birefringence of thymonucleic acid by Runnström in 1927. Now he was able with Signer to measure the length: breadth ratio. It was 300:1, and from a knowledge of its vicosity they reckoned the molecular weight must be in the region one half to one million. They also noted that the birefringence was negative, thus suggesting Wiener's pile of discs. The molecules, they concluded, "must contain strongly double-refracting components arranged in a definite pattern. Apparently the purine and pyrimidine rings lie in planes perpendicular to the longitudinal axis of the molecule" (Signer, Caspersson & Hammarsten, 1938). This experience was not lost on Signer. After reading the proceedings of the symposium on nucleic acids held in Cambridge in 1946 by the Society for Experimental Biology he set about improving the extractive procedure

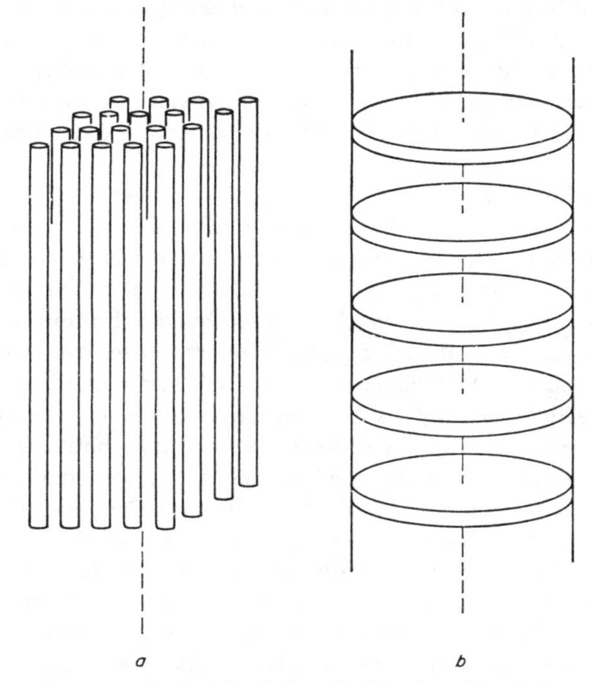

Fig. 2.1. Composite system of rods (a) and discs (b). The dotted line indicates the direction of the fibre axis (from Oster, 1955, 446).

for DNA which led to the remarkable material Schwander and he were able to distribute in London in 1950 (see Chapter 19).

Again it was from flow birefringence that the first evidence came that not all virus particles were spherical. Using a capillary flow technique Takahashi and Rawlins arrived at a rod-shape for tobacco mosaic virus from the positive birefringence of particles in infected sap (1932; 1933) and from particles of the extracted nucleoprotein (1937). Subsequent work using viscosity gave the length: breadth ratio, on the basis of which the known sedimentation rate could be used to give a reliable molecular weight (Lauffer and Stanley, 1938). Finally we may note that it was from birefringence studies that Wilkins gathered his first ideas on the intramolecular structure of DNA (see Chapter 19).

The Union of Classical and Macromolecular Chemistry

It is revealing to see how little macromolecular chemistry was esteemed in the eyes of organic chemists in the 1920s. Staudinger has put on record the serious attempts which his friends made to dissuade him from devoting his career to this subject. "My colleagues were very sceptical about this change [from low molecular to high molecular compounds], and those who knew my publications in the field of low molecular chemistry asked me why I was neglecting this interesting field and instead was working on very unpleasant and poorly defined compounds, like rubber and synthetic polymers." (1970, 77.)

Throughout these debates on the macromolecule there had been one persistent question: how can such huge molecules possess a single molecular weight? Surely they must be mixtures of varying sizes. It could not be argued that constant osmotic pressures, viscosities, or even sedimentation constants gave *the* molecular weight, only that they gave a *number average* value. Thus Meyer and Mark regarded as acceptable the term *average chain length*, but not the term *molecular weight* (Meyer and Mark, 1930, 83). Maybe this view was justified for substances like collagen, but not for haemoglobin. Adair thought "it was hope rather than anything else" which kept so alive the belief in the ultimate resolution of giant molecules into manageable subunits which would turn out to be the true homogeneous molecules (Adair, 1968).

The growing knowledge of macromolecules was welcomed by many biologists. Thus Frey-Wyssling gladly embraced long chain molecules in his study of cell ultra-structure. Like him, Nicoli Koltzoff saw the X-ray studies of such molecules as a vindication of Nägeli's belief in the crystalline nature of biological substances. The old micellar theory was now a theory about macromolecules which was fundamentally opposed to the false conception of "pure protoplasm" as advocated by Chambers (Koltzoff, 1928, 362). In a remarkable address delivered in 1927, Koltzoff explored the genetical implications of macromolecular chemistry. Such high molecular weights as Adair, Cohn and Conant discovered brought molecular and microscopic dimensions close together. Maybe chromosomes were single molecules, or bundles of just a few. And if the chromosomal proteins in man represented a polypeptide chain ten microns in length, then the potential isomerism was 10^{600}! If 50 000 volumes, each containing 100 pages, were written every year this variety of amino acid combinations would, he declared, take from archaic times to the present to print (Koltzoff, 1928, 362).

At this time Alexander and Bridges in America were still picturing the protein molecules of the gene as relatively small, maximum length 5 nm the length of genes being from 20 to 70 mμ. This left a large gap which they filled with "molecular groups" (i.e., aggregates of molecules) at $\frac{1}{2}$ to 10 nm, and "primary colloidal particles" at 2 to 20 mμ (Alexander and Bridges, 1929, 508). The need for such units intermediate between the molecule and the gene had already been rejected by the Viennese zoologist, Hans Przibram, in 1927, and in 1929 he told the Royal Society of Edinburgh that recent molecular weight estimates for proteins did allow the possibility that a single protein could extend to the full length (10 mμ) of a chromosome (Przibram,

1929, 227) and that racial characteristics could be referred to "substitutions of atoms in the molecules of the chromosome" (Przibram, 1929, 226). Meyer and Mark, who as yet were not convinced of the case for such long protein molecules, threw doubt on Przibram's suggestion (Meyer and Mark, 1930, 238).

Not until 1936 did Muller describe the gene as equivalent in size to a protein molecule with molecular weight "of the order of several millions." (Muller, 1936, 213.). Ten years earlier the diameter of the gene was "in the range of size for 'colloidal' particles, but", added Muller, "it might still contain hundreds of typical protein molecules." (Muller, 1926, 189.)

Richard Goldschmidt was only too glad to apply the macromolecule to the concept of the gene. For him it was an ally in his stand *against* the particulate conception of this unit.

> Authors who assume that the gene is a molecule or a small group of molecules are naturally interested in proving that the size of the gene is of the order of magnitude of protein molecules. The calculations of this type may not seem so significant now that it is known that chain molecules exist having a magnitude of 800 to 1000 Å length and more . . .
>
> (Goldschmidt, 1938, 283)

If the macromolecule was considered as a compound with fixed chain length, then qualitative changes were invoked to account for genetic diversity. On the other hand, variable chain length macromolecules were used by those who wanted to account for diversity in terms of different quantities of the genetic substance. Thus, a notable geneticist described mutations as polymerizations (Baur, 1924), and Goldschmidt, who championed the quantitative theory of gene action from 1917 wrote:

> The latest developments of organic chemistry suggest a different relation with respect to multiple allelism which is also of a quantitative type and nevertheless capable of being produced by the addition of a single quantum: changes in the length of a chain molecule in both directions.
>
> (Goldschmidt, 1932; cited in Goldschmidt, 1938, 301)

This interest in the macromolecule amongst biologists was in contrast to the opposition of the organic chemist. Not even the achievements of W. H. Carothers, in his study of polymerization at the Dupont laboratories, were able to disturb them in their well-defined world. Mark recalled that:

> the developments from 1920 to 1940 did not create very much impression on the classical organic and physical chemists. Of course some enlightened leaders—Haber, Haworth, Rideal, Svedberg, Tiselius, Willstätter—realized the importance of this new branch of chemistry but many organic chemists of great stature and overwhelming influence remained unimpressed and were not ready to accept polymer chemistry as a truly scientific discipline which deserved to be promoted and incorporated into the main stream of modern science.
>
> (Mark, 1965, 18)

The Biological Significance of Macromolecules

I have sought to show that the aggregate theory constituted a barrier in the way of reduction of physiological phenomena to chemistry. The colloidal theory of protoplasm offered a way out of this impasse which proved a snare and a delusion. From the start of the macromolecular debate in 1926 Staudinger must have been aware of the biological implications of his work, for at the meeting in Düsseldorf that year he said:

> Hitherto the organic chemist has worked largely on substances stable at relatively high (100–200°C) temperatures. This is connected with the procedures employed up till now for isolation and identification. Organic structures which can only exist at lower temperatures are far more numerous and complex. Their study is made difficult by their increasing sensitivity. Despite the large number of organic substances which we already know today we are only standing on the threshold of the chemistry of true organic compounds and have got nowhere near a conclusion.
>
> (Staudinger, 1926, 3042)

It was no accident, he argued, that biological molecules were so resistant to the techniques of the classical chemists, for these large "eucolloidal" substances could exist only at normal living temperatures.

> Such eucolloidal molecules are probably present in protein compounds, in enzymes, thus in the substances important for life processes. It seems likely to me that life processes are bound up with such molecules. These structures are destroyed even by small temperature increases, and with them the possibility of life—itself dependent on the permutations of very labile structures—is lost.
>
> (Staudinger, 1926, 3042–3043)

Later, Staudinger was to appreciate the importance of the macromolecule in relation to the vague ideas of plant physiologists like Haberlandt, whose teaching his future wife Magda Woit described to him when they met in 1927. With this growing interest in biological problems, Staudinger hoped to take up the structure of proteins at his institute in Freiburg, but the destruction of the latter during the Allied bombing of the town in 1944 meant beginning from scratch which by the time it became possible was too arduous an undertaking for the ageing Staudinger. He had realized that such a venture required special equipment which he lacked. But he had anticipated that the structure of proteins *would be easier* to unravel than that of homopolymers. It ought to be possible he thought to localize the several amino acids within the molecule more easily because of their chemical distinctness. But he did not know how to do it. "*Es ist möglich, aber wir haben keine Methoden*", he is reported to have said (M. Staudinger, 1969). In his book on macromolecular chemistry and biology of 1943 we find the author realized how slight differences in the construction of proteins would yield different biological properties. Accordingly he described the first attempts at amino acid sequence determination by Linderstrøm-Lang in Copenhagen. Moreover, Staudinger had read and appreciated Schrödinger's famous book *What is Life?* But, most significant, is the fact that he drew attention to the potential diversity in the nucleic acids, although as yet possessing but four types of residue: "with the potential multiplicity of structural possibilities the likelihood that each living type constructs its own nucleic acid is not excluded" (Staudinger, 1943, 48–49).

Svedberg and Staudinger

Whereas Staudinger came to the study of polymers by way of research into the synthesis and properties of rubber, Svedberg was drawn in the same direction by his interest in colloidal particles. He came to the university of Uppsala in 1904 to read chemistry and in his spare time he read Nernst's famous book *Theoretical Chemistry from the Standpoint of Avogadro's Rule and Thermodynamics* (1894). The colloidal chemistry of those days fascinated Svedberg and the transition between crystalloid and colloid appeared to him highly significant in connexion with the debate on the existence or non-existence of molecules. Perhaps it was the separate origins of their work which accounts for the fact that they first met when Staudinger came to Sweden to receive the Nobel Prize in chemistry in 1953 at the age of 75. Thirty years earlier his opponent Karrer had received the Marcel Benoit prize for his work on the aggregate structure for polysaccharides, followed by the Nobel Prize in chemistry for his work on vitamins in 1946. Svedberg became a laureate twenty-seven years before Staudinger. This we may attribute to the fact that Svedberg's work was in physics, and the theory of sedimentation upon which his technique was based, was well founded on physical theory. Staudinger's viscosimetry was not so well founded and his work came under the Nobel Foundation's committee for chemistry, not physics.

When it came to practical applications Staudinger's viscosimetry was quick, practical and cheap whereas the ultracentrifuge was expensive, intricate and time-consuming. Industry, particularly the cellulose industry, took up viscosimetry rapidly but not the ultracentrifuge. In 1940 Svedberg and Pederson could cite the Lister Institute, Oxford's biochemical department, and the chemical department at Wisconsin, as places where Svedberg's design of the ultracentrifuge had been installed. By this time the air-driven ultracentrifuge had been developed in the United States and was in use in the Rockefeller Institute for protein, virus and cell research (Bauer and Pickles, 1936).

The belated recognition of Staudinger was no doubt in the mind of Fredga when he addressed Staudinger in Stockholm:

> It is no secret that for a long time many colleagues rejected your views which some of them regarded as laughable. Perhaps this was understandable. In the world of high polymers, almost everything was new and untested. Long-standing established concepts had to be revised or new ones created. The development of macromolecular science does not present a picture of peaceful idylls.
>
> (Fredga, 1953, 395)

The victory of the macromolecule and the demise of the aggregate theory provides a good example of the conflict of paradigms described by Thomas Kuhn. But it would be untrue to say that either the aggregate theory or colloid science has become redundant. It would be even more absurd to claim that the earlier emphasis on membranes and electrolytes by the biophysical cytologists has gone right out of fashion. From experimental cytology came tissue culture. At the Rockefeller Institute, Albert Claude used this technique in his important work on cell organelles. At last the neglected dimension between the molecular and the microscopic could be filled not by vague colloidal systems but by macromolecules and by organelles, some of which had, after all, been seen in stained sections under the microscope long before. But it was only after the development of the methods described in this chapter for the determination of molecular weight, chain length, and length: breadth ratio, and *after their successful application to the proteins* that the macromolecular character of nucleic acids was recognized, the extractive procedures improved and material won which was sufficiently undegraded to give good X-ray patterns. It was no accident that the finest X-ray pictures of DNA available in 1953 had been taken using material supplied by a former student and colleague of Staudinger, Rudolf Signer.

By the 1940s thymonucleic acid had indeed come to mean something very different from the nuclein which figures in Zsigmondy's textbook, *Kolloidchemie*. There the reader only learned how well it demonstrated the swelling property of colloids in contact with water (Zsigmondy, 1920, 115–116).

CHAPTER THREE

The Fibre Diagram and the Long-chain Molecule

> So far nothing is known with certainty about the inner structure of colloid particles. . . . Typical organic colloids: albumen, gelatin, casein, cellulose, starch, etc., all showed an amorphous structure, a fact which makes it likely that these colloid particles are either individual molecules or they consist of molecules placed randomly beside one another.
>
> (Scherrer, 1918, 98, 100)

> The number of amorphous solid substances is no doubt much smaller than has been supposed up to now. Thus many powders regarded as amorphous (e.g., amorphous boron) have proved to be finely dispersed crystalline powders. . . . Thus we arrive at the remarkable result that the individual crystallites in the fibre [of ramie] are deposited in an ordered fashion. As a result we confirm the beautiful optical investigations of H. Ambronn (*Kolloid-Zeitschrift*, 1916) according to which the double refraction of ramie fibres consists of rod birefringence and intrinsic birefringence. The rod birefringence derives from the oriented deposition of the individual particles, the intrinsic birefringence from the fact that we have before us crystalline birefringent particles.
>
> (Scherrer, 1920, 395, 409)

> It is impossible in practice to determine the position of the atoms in cellulose from the symmetry of the unit cell and the intensities of the spots, without taking into consideration other data. The atoms in the cell are too numerous and all have approximately the same scattering power. Furthermore, the intensity of the reflections cannot be determined with sufficient accuracy owing to the coincidence of reflections. We must therefore take advantage of existing data on interatomic distances and valence angles in carbon compounds, in order to prepare a model of the crystalline structure.
>
> (Meyer, 1942, 241–242)

The discovery that crystals can act as 3-dimensional gratings which diffract X-rays was made in Arnold Sommerfeld's Institute of Theoretical Physics in Munich in 1912. The suggestion to test such a possibility had been made by Max von Laue and was carried out by two young physicists, Friedrich and Knipping, against Sommerfield's wishes (Friedrich, 1922, 365; Laue, 1944, 294). In the course of this work Laue had pinned his hopes on the possibility of diffracting the secondary fluorescent X-rays produced by the crystal specimen when irradiated from an X-ray tube. Friedrich later admitted that their preoccupation with fluorescent rays motivated their experiments "in an unfavourable direction" (*Ibid*, 366). As a result, Laue did not think of the crystal lattice as "selecting" certain wavelengths from the incident primary radiation, but of the oscillating atoms in the crystal specimen generating secondary X-rays of certain discrete wave lengths. A year after the discovery he still denied the evidence in the diffraction patterns for this selective action of the crystal lattice and reasserted his original view that the observed periodicities were due to the characteristic atomic oscillations (Laue, 1913, 1002).

As a supporter of the wave theory of X-rays, Laue, of course, correctly attributed the patterns to the diffraction of waves, but his erroneous explanation of the periodicities in them allowed the correct interpretation to be made by others—the Braggs, father and son. Whilst father Bragg (Sir William) wanted to give the discovery an interpretation in terms of corpuscular X-rays, his son (Sir Lawrence) accepted Laue's conclusion that the effect

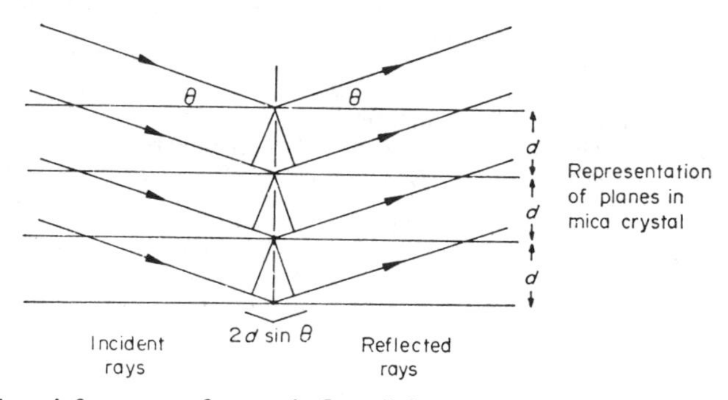

Fig. 3.1. The reinforcement of waves 'reflected' from successive planes of a crystal lattice when the path difference (2dSinθ) is equal to an integral number of wave lengths.

was due to the diffraction of waves. Both men, however, rejected Laue's fluorescent theory. They pondered the question of the periodicities in the patterns and the younger Bragg arrived at the very simple and visual model of crystal diffraction in which the planes of atoms were treated as "reflecting" planes and the periodicities in the patterns were attributed to the selective action of the lattice of such planes on the incident X-rays (see Fig. 3.1).

It was C. T. R. Wilson's treatment of optical diffraction which had set the younger Bragg thinking along the right lines, and it was the same Cavendish physicist who suggested to him that he should try to reflect X-rays from a cleavage face. It worked. Father and son then set to work to explore the rich field which opened up to them. The X-ray spectrometer was invented to measure the intensity of X-rays at varied angles. The result was the determination of the wave length of X-rays, of the spacings between lattice planes in crystals, the first correct crystal structure determinations, and the enunciation of the Bragg Law: $n\lambda = 2d \sin \theta$ (where n = an integer, the order of reflection, λ = the wave length of the incident radiation, d = the spacing between reflecting planes, θ = the glancing angle; see Fig. 3.1). The Bragg analysis of crystal structure based on this remarkly simple conception proved immensly fruitful. In addition, the beautiful but complex Laue diagrams obtained with a heterogeneous X-ray beam were replaced by the simpler Bragg diagrams with a monochromatic beam.

Amorphous Substances

Neither the Braggs nor the Munich physicists had expected in 1912 that powders, fibres and liquids were sufficiently crystallinc to act as diffraction gratings for X-rays. Had not Friedrich and Knipping shown the absence of a diffraction pattern when, instead of the entire crystal, they used the powder produced from grinding it up (Friedrich, Knipping and Laue, 1912). To be sure, the nineteenth century botanist Carl Nägeli, followed by his student, Ambronn, had inferred an ordered packing of crystalline micelles in starch and cellulose from optical and other data, but their opinion was treated with widespread scepticism. The new technique of X-ray diffraction did not therefore appear to be a valuable aid in the scientist's attack on substances of biological importance. A whole new world, therefore, opened up when Debye and Scherrer discovered the powder diagram in 1915, and Scherrer, Herzog and Jancke followed with the fibre diagram in 1920. Before the latter event it was still possible for so informed a crystallographer as Paul von Groth in Munich to reassert the long-established distinction between the crystalline and amorphous states. In the former state the periodicities of the constituent particles in three dimensions were different—anisotropic (excepting those in the cubic class); in the amorphous state they were the same—isotropic. Birefringence, which we described in Chapter 2, was to be expected of crystals. Where it was observed in glass, resins, and solutions it was said to be due to the preferred orientation of molecules under strain, and could therefore be discounted as evidence of true crystallinity. Hence the second Scherrer quotation at the head of this chapter. Groth admitted that those lattice forces known as *Richtkräfte* (directional forces) were present in haemoglobin crystals and in Emil Fischer's crystals of a sugar derivative (phenylmaltosazone), and in other proteins, but they were so weak in these giant molecules that external forces easily overcame them, causing the strange swelling phenomena which Nägeli had studied and which had led him correctly, as Groth believed, to set protein crystals apart from true crystals, calling them *Kristalloide.* Substances like silk and cotton had weak *Richtkräfte* and under tension their constituent molecules readily assumed orientations which gave rise to birefringence. Accordingly the anisotropy of these substances was to be distinguished from that due to true crystals (Groth,

1919, 651). Only a year passed before Groth's assertions were overthrown by the first fibre diagrams of natural cellulose fibres.

The Discovery of Powder and Fibre Diagrams

We shall pass over the early powder diagrams (Friedrich 1913) and fibre diagrams (Nishikawa and Ono, 1913). They were taken with the old gas-filled Röntgen tubes. Only with the more powerful evacuated tubes introduced by Scherrer in 1915, and Coolidge in 1917, was it possible to get really striking and convincing evidence of crystallinity in what were held to be amorphous substances. With the fibre and powder diagrams obtained with these tubes there came into existence a series of X-ray diagrams: the single crystal diagram showing the existence of a 3-dimensional lattice throughout the crystal; the fibre diagram, also giving evidence of such a lattice, but restricted to smaller units, the crystallites, these being oriented in a regular manner with respect to only one dimension, the fibre axis; and powder diagrams showing similar crystallites with random orientation in all dimensions. When the first successful pictures were obtained from the sodium salt of DNA in 1938 they were found to be fibre diagrams. Single crystal pictures of DNA have never been obtained (but see p. 438).

The systematic use of the Scherrer-type tube on compounds of biological importance, including DNA, was first undertaken by Herzog and Jancke at a newly founded institute—the Kaiser Wilhelm-Institut für Faserstoffchemie [chemistry of fibres]. As a result of this programme, which owed its support to the vigorous attempt of the German nation to promote applied scientific research, the fibre diagrams of silk and cotton were discovered. (DNA gave them no pattern, probably because they used Steudel's highly depolymerized material.) A simple method for interpreting these diagrams was introduced. Moreover, the characteristics of the fibre diagram suggested to the Institute's staff the invention of the cylindrical camera and X-ray goniometer, with which to take rotation and oscillation pictures. This was only one of the many research institutes of the Kaiser Wilhelm-Society, which contributed so much to the subjects described in this book. A brief account of this society, now renamed the Max-Planck-Society, is therefore appropriate here.

Kaiser Wilhelm-Institut für Faserstoffchemie

On October 11, 1910, the centenary of the foundation of Berlin University was celebrated amidst much pomp and ceremony. All foreign representatives, including J. J. Thomson, were given a rehearsal the day before "filing past and bowing before the empty throne, and practising the gesture of handing over their offerings to the Rector's empty chair" (Willstätter, 1965, 208).

The next day the Rector presided in his crimson robes and Kaiser Wilhelm II, dressed in the uniform of a Hussar, addressed the assembled audience.

Reminding the great concourse of people in the newly built auditorium that the university of Berlin had been founded "to make good by intellectual forces what the state had lost in material strength", he announced his intention to form a society to be known as the Kaiser Wilhelm-Gesellschaft für Förderung der Wissenschaft" (*Times*, 1910, 6). He had mentioned the idea to about two hundred people in the managerial classes of German society and he could now state that he had received offers totalling nearly 10 000 marks. In the discussions which had been going on between the Minister of Education, Schmidt-Ott, Berlin's influential academic, Adolf von Harnack, Emil Fischer and others, it had been pointed out that Sweden had its Nobel Foundation, the United States had its Carnegie, Rockefeller and Henry Phipps Institutes and its Thompson and Yale Laboratories, England its Royal Institution, British Museum of National History, and Lister Institute, and France its two Pasteur Institutes. But in Prussia

> . . . the foundation of such institutes had not kept pace . . . with the development of universities, and this lacuna, especially with regard to the natural sciences, was felt more and more with the growth of knowledge. They needed establishments for pure research in close touch with the Academy and the University, but unhampered by the giving of instruction.
>
> (*Times*, 1910, 6)

The Kaiser's wish to do something grand for science had been first expressed on a most unlikely occasion—at the Frankfurt singing contest of 1909—when he was thinking of the approaching Berlin University centenary. Once the suggestion had been made the project was handed over to Berlin's eminent theologian, Adolf von Harnack, who had access to the Kaiser, knew wealthy bankers and industrialists, and as official historian of the Prussian Academy of Sciences was well informed on the many provinces of *Wissenschaft*, including the natural sciences. According to his son Axel, it was Harnack who wrote the Kaiser's speech for October 11, the monarch having been forced by earlier *faux pas* to have all his speeches written for him (A. Harnack, 1963, 437), and according to the detailed account of the foundation of the society by Harnack's daughter, Agnes, its character was greatly influenced by her father. He disliked the idea of a very large society, for he held that "in itself science is an aristocratic structure—and this [feature] may not be obliterated without damaging the thing itself" (Zahn-Harnack, 1936, 424).

Thus it came about that a rather exclusive society which, by 1912, had only 186 members, was formed under the presidency of Harnack himself, Krupp von Bohlen and Halbach, and Ludbrück acting as vice-presidents. Those who had the good fortune to work in the Society's institutes enjoyed very considerable privileges. Those invited to become directors were able to dictate their terms. There was a wealth of space, equipment and technical assistance, not to mention freedom from teaching duties. When Emil Fischer visited

Willstätter to persuade him to come as section chief to the Society's chemical institute he said: "You will be completely independent. Nobody will bother you or interfere with you. You can go walking in the Grünewald for a few years or, if you wish, think up some beautiful new thing" (Willstätter, 1965, 212). Alas, he was soon to find things very different!

Before the outbreak of war in 1914 three institutes were established in what was then open country outside the city of Berlin: the Chemical Institute, where Beckmann worked on molecular weight determinations by depression of freezing point, and Willstätter continued his study of plant pigments; the Institute for Physical Chemistry and Electrochemistry under Fritz Haber, where Freundlich worked on colloids, and the Institute for Biology under Carl Correns, the plant geneticist. These institutes were all intended as centres for pure research in the spirit of Wilhelm von Humboldt's *Hilfsinstitute*; places where no teaching duties would stand in the way of research (Glum, 1921, 293).

This pure research ideal was shattered by the 1914–18 war during which time work at Haber's institute was taken up entirely with the development of offensive and defensive aspects of gas warfare. After the war Germany was subjected to reparations which she could not pay. This prompted Haber to try to extract gold from sea water. To cover the costs of the war Germany had increased the supply of Deutschmarks to five times the pre-war quantity. So the mark slipped and slipped, until in 1923 it collapsed completely.

Fortunately, before this happened the Kaiser Wilhelm-Gesellschaft recognized the need to stimulate industry and to supplement their institutes for pure research with similar institutes for applied research and in 1920 the Institut für Faserstoffchemie was established. R. O. Herzog, a friend of Haber's, whose immense *Chemische Technologie der organischen Verbindungen* of 1912 gives evidence of its author's interest in applied science and in textiles, became its first director.

Very likely it was Herzog who had written the memorandum on textile research which Adolf von Harnack studied (Schmidt-Ott, 1921, 292) before this institute was founded. Soon Herzog had gathered together a brilliant team of scientists: Michael Polanyi from Budapest, his assistants Hermann Mark and Karl Weissenberg, both from Vienna, and his student Rudolf Brill from Eschwege. Herzog, also Viennese, who had been at the German Technical University in Prague since 1912, directed this team from 1919 until the institute was taken over by the Institute for Silicate Research in the 1930s. In their first few years at Dahlem these men layed the basis of X-ray diffraction analysis of fibres, thus initiating a new era in textile research from which emerged the important conception of long-chain molecules made up of regularly repeating residues.

Polanyi's Layer Line Concept

In the summer of 1920 Herzog and his assistant Willie Jancke were encouraged by their success in obtaining powder patterns from crushed natural fibres to irradiate entire fibres in the hope that evidence of a regular structure along the length of the fibre might be revealed. Using an X-ray tube of the type Jancke had been taught to build by Scherrer in Göttingen, they obtained neither clear spots, nor powder rings but something between the two —smeared points placed symmetrically in groups of four. This Four-point-Diagram reminded Herzog and Jancke of the diffraction pattern obtained from thread-shaped tungsten crystals (Herzog and Jancke, 1920, 2163). They described the unit cell as rhombic with the axial ratio of 0.6935:1:0.4467. One can imagine the excitement evoked by a picture of ramie fibre showing twenty-six spots! The old and unappreciated tradition of crystalline structure for fibres, originating in Nägeli's micellar theory, had some truth in it after all. Competition with Scherrer, at the E.T.H., Zürich, who had obtained similar if poorer pictures of cellulose (Scherrer, 1920), heightened the urgency of the task of interpretation.

It was at this stage that Polanyi joined Herzog's staff and was given the "Four-point-Diagram" to solve. In a week, or maybe a fortnight, he came up with the answer. When he began he knew no X-ray crystallography, not even Bragg's law. The moment he had solved the diagram, the concept of the layer-line was there in principle. Simple though the problem may appear in retrospect, its solution marked an important step into the world of the structure of fibres. Herzog was delighted with Polanyi's success and showered him

> ... with every facility for experimental work, most precious of which were funds for employing assistants and financing research students. In this I was incredibly lucky. I was joined by Hermann Mark, Erich Schmid, Karl Weissenberg, all three from Vienna, by Erwin von Gomperz and some others; the place was soon humming. It was the time of runaway inflation and poor Herzog found it difficult to pay all these people. Protest meetings were held, resolutions passed, Weissenberg in the lead; the Institute earned the name of an "Assistenten-Republik". We had a glorious time.
>
> (Polanyi, 1962, 630)

Polanyi's success was achieved using a rather clumsy geometrical argument. When he had thus indexed the cellulose diagram he noted that all spots or arcs with the same index in the fibre direction lay on hyperbolae (see Fig. 3.2). This made

> ... the calculation of the identity period parallel to the fibre direction much easier. As soon as the hyperbolae have been arrived at from the diagram the identity period *c* can be read off immediately. Having fixed one of the identity periods in this way the calculation of the other two is facilitated, and the problem loses much of its uncertainty." (Polanyi, 1921, 340.)

The term *hyperbolae* was altered to *Schichtlinie* (layer line) a year later. (Polanyi and Weissenberg, 1922, 125).

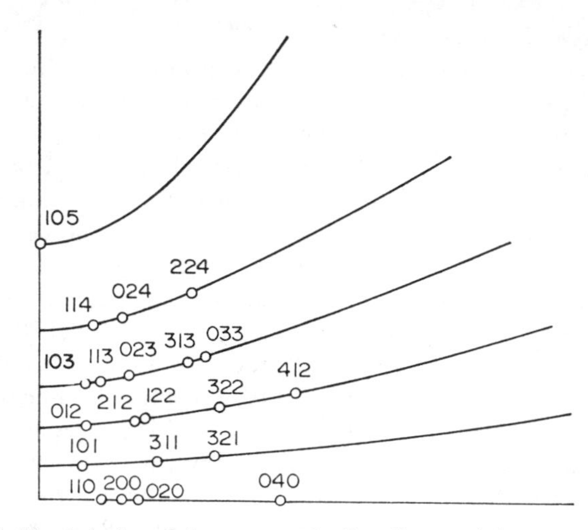

Fig. 3.2. The indexing of the spots on the flax diagram (after Polanyi, 1921).

By 1922 the features of a fibre photograph had been thoroughly analysed and such questions as why the spots should become less discrete towards the poles, why some spots should be lacking, how varying the angle of inclination can reveal fresh spots, and so on, were all dealt with. In the case of the cellulose fibre they found that an inclined picture gave many more spots, and they were able to plot nine instead of five layer lines. What they did not realize was that the diffraction data on all layer lines except the equator *represented the transform of the molecules*, and therefore could be indexed reliably only in the fibre direction.

The Number of Molecules in the Unit Cell

For us it is obvious that the molecule can be far larger than the unit cell of the crystal. To the crystallographers of the 1920's—especially those who were mineralogists—it was far from clear, indeed to some it was obvious that the molecule *could not* be larger than the unit cell. When Polanyi presented his conclusions in one of Fritz Haber's colloquia in the crowded little lecture hall in his institute in 1920 a storm broke loose.

> The assertion that the elementary cell of cellulose contained only four hexoses appeared scandalous, the more so, since I said that it was compatible both with an infinitely large molecular weight or an absurdly small one. I was gleefully witnessing the chemists at cross-purposes with a conceptual reform when I should have been better occupied in definitely establishing the chain structure as the only one compatible with the known chemical and physical properties of cellulose. I failed to see the importance of the problem.
>
> (Polanyi, 1962, 631)

Some of the reasons for opposition to the concept of long-chain polymers have already been described in Chapter 1. Here I will only discuss the crystallographers' objections. To begin with, the assumption that the chemical molecule is never larger than the crystal molecule was a long established tradition. Thus, Lord Kelvin in his paper "On the Elasticity of a Crystal according to Boscovitch" wrote:

> The crystalline molecule [unit cell] . . . may be the group of atoms kept together by chemical affinity, which constitute what for brevity I shall call the chemical molecule; or it may be a group of two, three, or more of these chemical molecules kept together by cohesive force.
>
> (Thomson, 1893, 59)

The fact that tartaric acid showed the same optical activity whether in the form of a crystal or in solution assured Kelvin that in this case the chemical and crystal molecules were identical. The contrary result for sodium chlorate, he remarked, had convinced Sir George Stokes that there the crystal molecule consisted of two or more chemical molecules. Similarly, Muthmann thought it unlikely that in sodium chloride the two were identical and he added: "The opinion, fairly widespread among crystallographers, that crystal molecules are polymers of the chemical molecules is much more probable . . ." (Muthmann, 1893, 499).

Jumping to a much later occasion, Sir William Bragg gave what he regarded as quite obvious reasons why the chemical molecule had to be smaller than the crystal molecule when he addressed the Chemical Society in 1922:

> . . . X-ray analysis shows that the unit [cell] nearly always contains the substance of more than one molecule; generally of two, three, or four.
>
> The crystal unit must contain the substance of an integral number of molecules; this is a simple consequence of the fact that the atoms of the different elements are present in the same proportion in both solid and liquid.
>
> (W. H. Bragg, 1922, 2767)

Is it surprising, therefore, that several authors, including Herzog, equated the mere facts of crystallinity and small unit cell in rubber and cellulose with small molecular volumes?

Whilst Mark, Polanyi, Weissenberg and Brill kept open alternative models of fibre structure—small molecules or bundles of long chains—their director, Herzog, became more and more convinced of the case for small molecules. Speaking of their number n in the unit cell he declared that "Since n cannot be less than one, the molecular weight [M] can in no case be smaller than the product nM . . ." (Herzog, 1924, 958). With a unit cell for cellulose of 680 $Å^3$ and for silk of 675 $Å^3$ this put severe upper limits on M. Within a year he had moved more definitely into the camp of Max Bergmann and his followers and could add his data on small particle size of cellulose nitrate based on diffusion estimates (Herzog, 1925). Svedberg's estimates of molecular weights of proteins based on the ultracentrifuge in 1926 stimulated Herzog to determine the particle size in solution of silk. To get this fibre into solution he had to immerse it in resorcinol at 100°C for thirty hours! Not surprisingly, his estimate of M from diffusion was small—320. Therefore for n he favoured 2, 4, or a larger number (Herzog, 1928, 531).

Long-chain Molecules in Crystals

Opposition to Herzog came from his own staff. At first the alternative model of long chains contributing so many residues to each unit cell was merely stated on an equal footing with small molecules. Brill, for instance, concluded that there were eight amino acid residues in the unit cell of silk fibroin, either in the form of four parallel chain polymers, or in the form of four

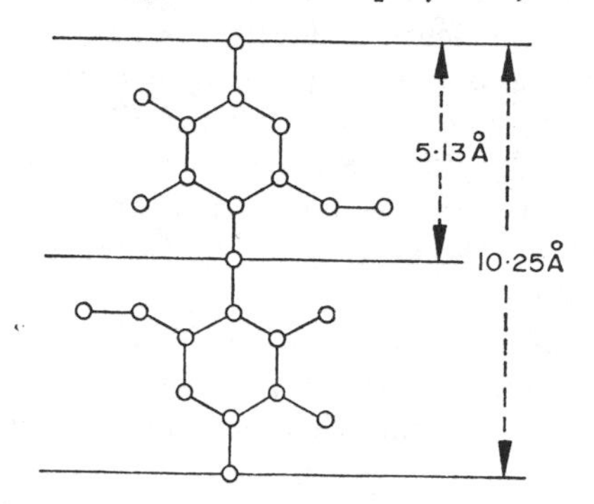

Fig. 3.3. The model of the chains in cellulose in which the wrong linkage was used (from Sponsler and Dore, 1926).

dipeptide molecules (Brill, 1923, 216–21). The equation by which Brill arrived at the number of molecules or residues n in the unit cell of volume V was:

$$V = \frac{N\,n\,M}{\rho} \text{ where } N = \text{Loschmidt number, and } \rho = \text{crystal density.}$$

In one form or another this equation crops up again and again in the analysis of fibre structure, including DNA.

In an important paper on the principles of stereochemistry, Weissenberg distinguished between micro and chain building blocks (*Mikro-baustein* or *Ketten-baustein*). All crystals, he pointed out, must contain the former atoms or molecules, but not all crystal symmetries are compatible with the latter (Weissenberg, 1926, 1534). The *Ketten-baustein* was championed first by Sponsler and Dore in America (see Fig. 3.3) and then by Weissenberg's colleague Mark, and by Mark's Swiss colleague in Ludwigshafen, Kurt H. Meyer. They rejected Herzog's diffusion estimate for silk on the grounds of the rough treatment. They noted a low angle interference spot on the X-ray diagram of silk, indicating a long spacing of 150 Å. This suggested four chains running through the unit cell, each made up of 40 amino acids (Meyer and Mark, 1928 b, 1933). When they moved to the main laboratory of Germany's greatest industrial giant, I. G. Farben Industrie, at Ludwigshafen on the Rhine, Mark and Hengstenberg, assisted by Brill, observed low angle spacings which were suggestive of crystallites containing several hundred unit cells, and long enough in the direction of the fibre to accommodate extended chain molecules. In cellulose these crystallites were reckoned to be > 600 Å by 55 Å. So chains made up of 120 glucose residues *could* occur in them. Neither Hengstenberg and Mark in 1928, nor Meyer and Mark in 1930, however, made this assertion. For rubber, the length of the crystallites was also put at 600 Å; in the case of artificial silk it was 305 Å (Hengstenberg and Mark, 1928, 280–282).

In the early thirties opinion moved in favour of molecular chains as long as the crystallites. But the suggestion that these chains might go beyond one crystallite into another was rejected because it was naturally assumed that the maximum length of the crystallite was determined by the maximum length of its constituent molecular chains. Moreover, end group analysis by the reliable and eminent Walter Haworth put these chains at 100–200 residues in length (Haworth and Machemer, 1932). In contrast to this, Staudinger's molecular weight estimates for cellulose demanded far longer chains and as a result, disagreement between the X-ray crystallographers at Ludwigshafen and Staudinger in Freiburg continued, Herzog himself disagreeing with both Staudinger and with his friends in Ludwigshafen. Their positions are best represented by reference to cellulose:

Herzog: The molecule is not larger than the unit cell (680 $Å^3$); the crystallites or micelles are 117 Å long and 66 Å wide and contain hundreds of molecules. Such micelles are also present in cellulose nitrate in solution.

Staudinger: The molecules must exceed the length of the crystallite. Both in solution and in the fibre their weight must be in the region of 40 000, chain length 4000 to 5000 Å. The crystallite does not persist as such when dissolved.

Meyer and Mark: The molecule is less than 600 Å in length. Molecular weight estimates of cellulose in solution refer in fact to the crystallites which persist in solution as micellar aggregates, each containing about 50 molecular chains.

These were the differences of opinion which prevented the X-ray crystallographers in Ludwigshafen from fully embracing Staudinger's conception of the macromolecule until after 1935. It did not matter that 600 Å was only the *minimum* length of the crystallites. Indeed Hengstenberg and Mark recognized this by the notation > 600 Å.

Evidently it required courage to believe that these long molecular chains can pass through amorphous regions from one crystallite to another. Accordingly, the claims of macromolecules were met first by the device of "micellar fringes", according to which the near-random "tails" of molecular chains at the ends of crystallites became entangled as they grew, some chains passing via such amorphous "fringes" from one crystallite into another (Herrmann, Gerngross and Abitz, 1930, 389–390). The idea was applied to cellulose by Frey-Wyssling in 1936. Much later Meyer introduced his more simple picture of "micellar ropes" in which the sharp distinction between crystalline and amorphous regions was rejected and replaced by bundles of 100 to 150 very long chains packed with varying degrees of order, like the chains in an asbestos fibre subject to deformation (Meyer, 1942, 257).

Achievement of the German School

Many were the assumptions which had to be uprooted and the techniques which needed improving before macromolecular chains several thousand Å in length could be accepted in the model of fibre structure. Like the concept of geological time, the dimensions of the chain molecule were extended cautiously, step by step, first to neighbouring unit cells, then to micelles and finally beyond. In this process the young team under Herzog played the decisive part, stimulated by him from the beginning, though he himself rejected the younger men's conclusions.

In the late 1920s the lively atmosphere in Dahlem was transported to Ludwigshafen where collaboration with Meyer began. What emerged was a fundamentally similar model for all fibres investigated: that of long chains packed parallel to the fibre axis in crystallites (micelles), these in turn being held together by "secondary valencies". Herzog brought crucial evidence for the restoration of the nineteenth century micellar theory of Nägeli but Meyer and Mark transformed it into what we know today as the X-ray crystallographers' conception of the micelle in fibres, a bundle not of short but of very long molecules. The tensile strength of the fibres, they attributed to main-valency (covalent) bonds and the lateral cohesion to micellar, or Van der Waals, forces. They emphasized the important point that these "secondary forces" can "achieve the order of magnitude of primary valencies, and so give strength, insolubility and tenacity to the whole structure, only when the main-valency chains themselves are of sufficient length" (Meyer and Mark, 1928a, 614).

As for the orientation of the chains within and without the unit cell, Meyer and Mark, like Sponsler and Dore in America before them, built molecular models to test out alternative possibilities. In rubber and in cellulose the symmetry indicated a screw axis, and this Meyer and Mark expressed in the 180° rotation per glucose residue in cellulose, and per isoprene residue in rubber. So convinced were they of nature's unified scheme that they expected the structure of cellulose to throw light on the structure of the proteins, collagen and silk. For silk they favoured a zig-zag chain in which the tetrahedral angle of 109°–28° was preserved as in the similar models for aliphatic long-chain hydrocarbons (Shearer, 1925). They admitted that silk lacked a screw axis (see Fig. 3.4) but pointed to the special form of the glycylanalyl anhydride residue which had, they said, a translation very similar. Therefore the overall character of the structure of silk came very close to those of cellulose and rubber (Meyer and Mark, 1928b, 1934). This expectation of the wide application of the cellulose structure had an undoubted influence on Astbury's work on keratin.

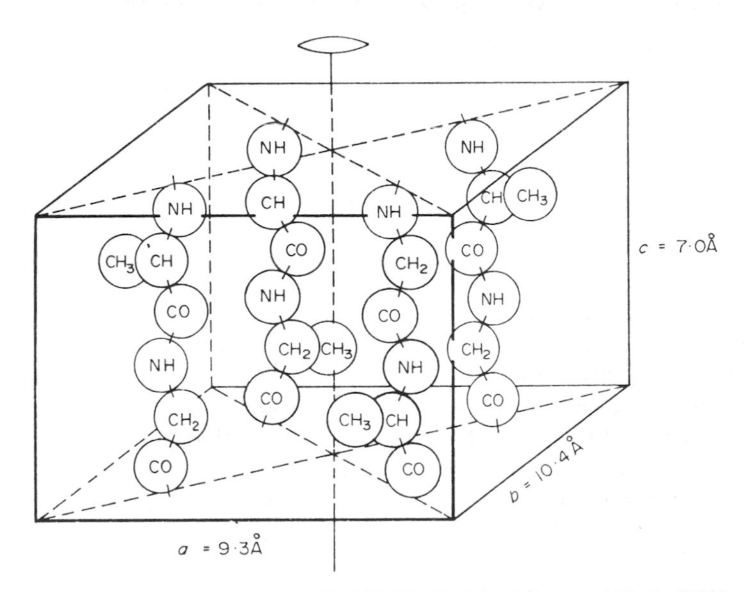

Fig. 3.4. Schematic picture of the unit cell of silk fibroin (after Meyer and Mark, 1928b, 1934).

The Cellulose Model

Meyer's and Mark's model for cellulose was published in a paper in 1928 which very rapidly became a classic. Their figure reproduced on page 601 of the *Berichte* has been reproduced many times. A side view of it is given in Figure 3.5a. Their improved diagram of 1930 is shown in Figure 3.5b.

The models for cellulose and rubber in 1928 agreed in the possession of a screw axis of symmetry. They differed in that the rotation in rubber was split up between eight C–C bonds, whereas in cellulose the rotation involved only the oxygen bridge between successive rings. Model building,

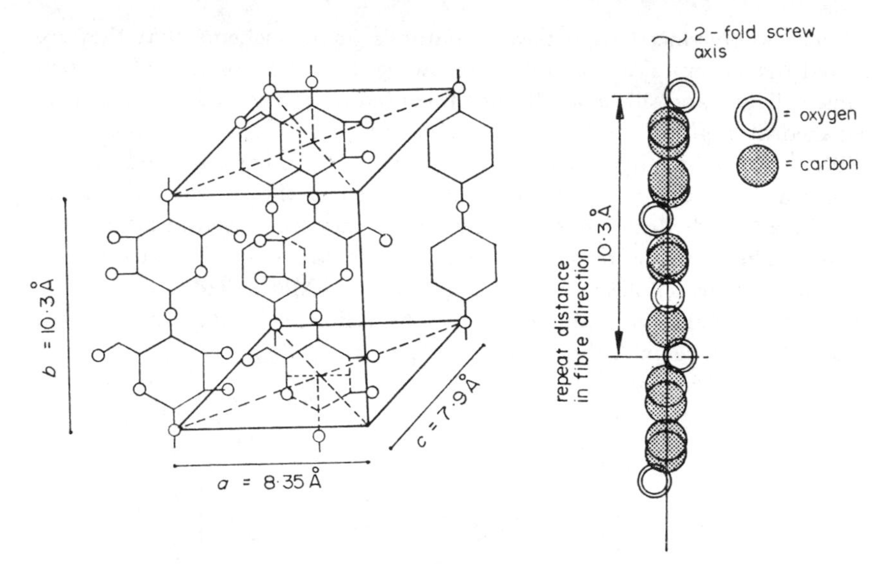

Fig. 3.5. The unit cell of native cellulose (from Meyer and Mark, 1930, 111) and three residues of the cellulose chain viewed from the side (from Meyer and Mark, 1928a, 601).

using the analogy of rubber, more easily led in the direction of coiled conformations than did cellulose. We shall see that it was in this direction that Meyer went when he studied plastic sulphur and polyphosphonitric chloride. The cellulose model was associated instead with *ribbon chains*, all folds being in one and the same plane. If it be objected that cellulose is helical since it has a screw axis of symmetry, then let us remember that this is so only in the formal sense that the operation of going from one residue to the next involves a translation along the axis of the molecule and a rotation about its axis. It is not what one means by a staircase-like molecule such as the Watson–Crick model for DNA, because in cellulose each step is separated by a rotation of 180°—not only has it a two-fold axis of symmetry, but also the centre of each step lies on the axis of the helix. (The latter condition is characteristic of cellulose, but is not in fact essential to two-fold symmetry itself.) *This distinction between a ribbon chain model with two-fold symmetry as in cellulose and a helical chain model also possessing a screw axis of 2, 3, 4 or 6-fold symmetry* is of the greatest importance in the early phase of structural investigations which followed the German work. Meyer's and Mark's papers in the *Berichte* set up cellulose as the paradigm for natural fibres. We may call it the "ribbon chain paradigm". Their famous book *Der Aufbau der hochpolymeren organischen Naturstoffe* (1930), ensured this paradigm international recognition. Astbury conformed to it in his work on α-keratin, although we should mention that there were other reasons for so doing. As late as 1950, the Cavendish group of crystallographers were forced to admit that Astbury's modified ribbon model fitted the known data on proteins best. Indeed, it is noteworthy how widespread was the tendency to build ribbon chains where now helical chains are accepted.

In their book, Meyer and Mark also discussed the arrangement of the micelles in the cotton thread. As long ago as 1921 Weissenberg, acting on a suggestion of Polanyi's, had attributed the spread of certain spots in the cellulose diagram to *helical packing* of the micelles, the gradient of which he

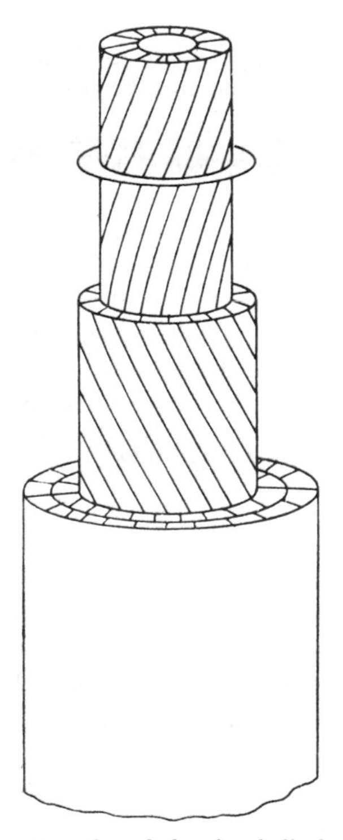

Fig. 3.6. Spiral model of the cotton thread showing helical packing of the micelles (from Meyer and Mark, 1930, 117).

reckoned could be calculated from the diagram (see Fig. 3.6). Further examination confirmed this interpretation for which, incidentally, the botanists had long possessed quite independent evidence. The historian might seek in this imposition of a secondary helical conformation upon these linear chains a further reason why workers like Astbury kept to the model of the planar or ribbon configuration of the cellulose molecule and reserved helical arrangements for secondary structure. Weissenberg believed such secondary coiling to be present not only in cotton, but in hair, and in the fibres of muscle, tendon and nerve (Weissenberg, 1922, 31).

The State of Structural Analysis

This, then, was the state of work on the structure of fibres when W. T. Astbury came to Leeds in 1928 and began his adventure into molecular biology. What were its weaknesses? Clearly no structures were *solved* in the sense that accurate determinations of the positions of individual atoms were made. Nevertheless, work along the same lines was later to get very close to this. In 1945, C. W. Bunn could say of the analysis of rubber and gutta-percha that although

> the coordinates of atoms in chain-polymer crystals cannot be determined with the precision attained in single crystal work, on account of the smaller number of reflections available and the overlapping of some of them [yet] the amount of evidence available is sufficient (for polymers of the degree of complexity of those considered here) to leave little doubt of the general arrangement, as well as the approximate coordinates of the atoms.
>
> (Bunn, 1945, 328)

In the organic field as a whole, structures were rarely solved, and in the few successful cases there were always special conditions—the existence of a very simple structure, as in the long-chain ketones studied by Müller at the Royal Institution and in the long-chain hydrocarbons which Shearer studied in the same laboratory (Shearer, 1925; Müller, 1928). Outside the class of long-chain molecules, success rested on high space group symmetry, as in hexamethylbenzene (Lonsdale, 1928) where all the atoms lie in one plane, and in hexamethylene tetramine with only two molecules in the cubic unit cell. In such structures few parameters were needed to define the positions of the atoms. The same holds for long-chain polymers. These early workers could therefore make up for the scanty information of fibre diagrams by other means. The chief tool here was model building, and after having indexed the diagram and derived the dimensions and shape of the unit cell, the first task was to find out in how many different ways the chains could be fitted into it, and to attempt to eliminate as many as possible. In this way the *cis* configuration for rubber was eliminated. An aid in the determination of the structure of rubber was the availability of several forms of it. For the same reason cellulose was particularly attractive to Mark and his colleagues. The slightly different diagrams could be correlated with the changes in chemical constitution. The same sort of additional clues were later afforded by the two forms of DNA.

These achievements in fibre analysis were only made after laborious and long drawn out studies. An improvement in this situation was not to be had until the introduction of Fourier methods. For dates in the development of this subject see Table 3.1. They opened the way to the solution of structures from the diffraction pattern alone; *but only in those cases where the phases as well as the intensities of the spots on the diagram were known.* Normally, only the *intensities can be determined directly from the diagram.* In some cases the special con-

TABLE 3.1

Important dates in the Application of Fourier Methods

Date	Author/s	Innovation
1879	E. Abbé	Introduced the diffraction theory of image formation
1906	A. B. Porter	Represented this theory by a Fourier Series
1915	W. H. Bragg	Used Porter's equations to represent the density of scattering matter in a crystal by Fourier Series
1925	W. Duane	The same idea based on equation of Epstein and Ehrenfest
1925	G. Shearer	1-dimensional Fourier series analysis of long-chain ketones
1925	R. T. Havinghurst	1-dimensional Fourier series analysis in rock salt [along certain lines only]
1926	A. H. Compton	1-dimensional analysis [in sheets parallel to a given plane]
1927	J. M. Cork	Isomorphous replacement in the alums [full significance not realized]
1929	W. H. Bragg	Introduces the 2-dimensional Fourier series
1934	Beevers & Lipson	Rapid method for the summation of 2-dimensional Fourier series
1934	A. L. Patterson	Patterson vector analysis
1935	Beevers & Lipson	Isomorphous replacement in the alums
1936	D. Harker	Harker synthesis
1936	J. M. Robertson	The first "direct" solution of a structure—phthalocyanine, using isomorphous replacement and 2-dimensional Fourier series
1938	R. B. Corey	The solution of the structure of diketopiperazine using 2-dimensional Fourier series
1939	C. W. Bunn	The structure of the CH_2 group in long-chain hydrocarbons using 3-dimensional Fourier series

ditions of symmetry and orientation enabled the phases to be guessed. In others, the introduction of a heavy atom (heavy atom method) and the replacement of one such atom by a different one (isomorphous replacement method) furnished this information (J. M. Robertson, 1935; 1936; Robertson and Woodward, 1937).

These methods could not be applied to long-chain polymers. Here, as we shall see, it was by an *indirect* method that success came. This involved *a more rigorous application of model building* to the task of solving the structure than that of fitting a chain into the axial repeat and packing the chains into the unit cell. The chosen model had to be tested against the observed diffraction pattern by calculating the pattern such a model ought to give, using the method of Fourier series. This model could, as a result, "be refined by

Fourier series methods until a highly accurate picture of the molecule was obtained, from which bond lengths to within a few hundredths of an Angström unit might be estimated" (Robertson, 1962, 152).

The work described in this chapter is also to be distinguished from later work by the absence of the Patterson vector method. This innovation in Fourier theory (Patterson, 1934; Harker, 1936) yielded a map showing a series of peaks in the electron density distribution. These peaks did not represent the actual positions of atoms. The spatial separation of the peaks however, did represent the distances and directions of the lines which could be drawn between atoms in the structure (interatomic vectors). This information was obtained without any knowledge of the phases of the reflections. We shall see that some crystallographers became too enthusiastic for the use of Patterson anlysis in cases where the data—from fibre diagrams—was too crude. Others used it wisely as the best guide to the formulation of an initial model.

It was while Patterson was in Berlin at Herzog's Institute in Dahlem, trying to determine the particle size of cellulose, that he came to learn of the existence of the Fourier Transform. As a result of this stay in Berlin the notion that something was to be learned about structural analysis from Fourier theory became an obsession with him (Patterson, 1962, 617). The right ingredients for the development of molecular biology were, it seems, in Dahlem—chiefly a powerful school of theoretical and practical X-ray crystallography. But by 1933 Hitler had come to power. The Kaiser Wilhelm Institutes, which had already come increasingly under State control due to the loss of the Society's capital in the depression, were subjected to State directives and their employees became civil servants. No time was wasted by the Nazi government in drafting legislation to implement its racist aims. April 1933 saw the passing of the Government Employees' law under which employment was not to be given to persons of non-Aryan descent. Within three weeks of this event Haber resigned and with him Freundlich and Polanyi. Herzog left for Istanbul in 1934 and died a year later. Mark, who had been at the I.G. Farbenindustrie since 1925, had already been advised to leave the country and in 1932 had become professor of chemistry in Vienna. Six years later he left Austria and went to work for the Canadian International Paper Company, before settling at the Brooklyn Polytechnic Institute, New York. For the sequel to this first phase of work on polymer structure in natural fibres we must turn to England and to Leeds.

CHAPTER FOUR

The Leeds School under Astbury

The arrival of Astbury in Leeds in 1928 was just one of the developments which resulted from the general shake-up administered by the University's dynamic Vice-Chancellor, Dr James Baillie, later Sir James. Naturally, when he came to Leeds he took an interest in the University's largest department—textile industries—under Professor Aldred F. Barker, M.Sc. As an academic Dr Baillie was taken aback by what he found. He had not been in a university like Leeds before, where the pattern of teaching was largely determined by the ties with local industry—especially textiles—to which the Yorkshire College (predecessor of Leeds University) owed its existence. The only graduate on the teaching staff of the textile industries department apart from the professor was J. B. Speakman, M.Sc., later D.Sc., and professor. The teaching was devoted almost entirely to the diploma in textile industries and to preparation for the City and Guilds examination in woollen and worsted spinning and weaving. Of the 175 students in the department in 1924, 96 were evening students. It was not therefore to be expected that the staff would have the time and motivation to undertake fundamental scientific research. They were fully occupied teaching applied science courses in what was after all an applied science department.

Astbury and Speakman were later to give the impression that they had had to fight to establish fundamental research in the department and that no one else was interested. Certainly Speakman's researches were among the first fundamental studies of the wool fibre at Leeds, and in those early days there was little money available for such research. True, F. W. Dry, as Ackroyd Fellow, was carrying out his fine study of the genetics of the Wensleydale breed of sheep, but support for other research had been meagre. It was a well-known fact that before the First World War the textile industry had been remarkably conservative and complacent, and as far as Leeds University was concerned Aldred Barker said: " . . . The industry had felt that it had far more to give to the Yorkshire College than the College could possibly be expected to give in return." The First World War gave the industry a jolt and:

> the idea gradually gained ground that science might have important bearings on textile processes and should be given a chance to make good.
>
> This fairly represents the position towards the end of the War period and also the position to-day, excepting that the Scientist has been given a chance during the past ten years and, upon the whole, has not fulfilled the perhaps too great expectations raised at the beginning of the research movement.
>
> (Barker, 1927, 3)

At the same time, the Council for Scientific and Industrial Research came to the conclusion that the textile industries "have had greater difficulties to contend with, in the absence of trained men and of fundamental knowledge, than most other industries" (C.S.I.R., 1927, 22).

From 1917 until 1919 the British Research Association for the Woollen and Worsted Industries supported research at Leeds University. Then it altered its policy, perhaps for the reason hinted at by Barker in his 1927 memorandum. Instead, it founded a research institute in a house by the name of "Torridon" in Headingley, Leeds, with the result that applications to undertake fundamental research within the department of textile industries tended to be referred to Torridon. This decision of Central government should be born in mind when Speakman's difficulties are described.

The first research scientist to work in the department was the able young physical chemist from Rothamsted, E. A. Fisher, who had made comparative studies of tensile strength and moisture relations of wools of various kinds and of the nature of shrinkage during drying, he had discussed his results by comparing them with the behaviour of the colloidal soil particles which he had studied at Rothamsted. The wool fibre was therefore a colloidal particle and his paper to the Royal Society's *Proceedings* was entitled: "Further Observation of the Ratio of Evaporation of Water from Wool, Sand and Clay" (Fisher, 1923).

Fisher's appointment in 1920 as research assistant in textiles, subsequently lecturer in textile chemistry, had been the deliberate act of the University to maintain the tradition of research established within its walls between 1917 and 1919. Fisher left Leeds in 1923. About a year later Speakman was appointed. He was, he wrote "compelled to study the industry *de novo* but commenced with the determination to form a research school for the application of science to textiles" (Speakman, 1927, 2). His greatest need was for well-trained graduates to carry out research and for funds to support them. If he applied to the D.S.I.R. for grants his applications were passed on to the British Research Association, but this organization was already supporting a programme at Torridon. "In the end only one source of money remained, namely, what is known as the accumulated fund of the Clothworkers' Scholarships", and this was exhausted by 1928. Speakman's appeal for help in 1925 had led to two scholarships being established by Messrs. Wolsey and a Mr Wilson but to no awards for research unrelated to specific problems in industry (Speakman, 1927, 5).

What a heaven sent opportunity it was when Dr Baillie, impressed by Speakman's researches into the physico-chemical properties of the wool fibre, asked him to prepare a memorandum on a scientific research unit for the textile department. In the resulting document Speakman spoke of his profound disappointment at the capabilities of textile students, of the inability of the department to attract first class students, and of the need to build up a nucleus of able men to staff the department of the future, which should contain graduates in physics, mathematics, physical and organic chemistry (*Ibid*, 14). Until such staff could be trained within the department, a graduate physicist should be appointed and postgraduate scholarships and funds for apparatus should be established. A week later Barker submitted a memorandum in which the posts of Lecturer in Mathematics and Mechanics, and Assistant for Mechanics were requested, and a recurrent budget of £2900. No doubt Barker was able to find out what sort of budget was likely to be granted. There was but one source to look to—the Worshipful Company of Clothworkers—but although their funds (from interest on city investments) were considerable, they had already supported the University in a very big way.

That summer the Advisory Committee to the Clothworker's departments (textiles and colour chemistry) met and approved an application for the following budget:

	£
8 Research Scholars @ £150	1200
New Lecturer (Physics) £500	500
2 Assistants @ £300	600
Superannuation	100
Apparatus etc.	500
	2900

(Leeds University Advisory Committee . . ., 1927, 6)

By May, 1928, the clothworkers had agreed to make an annual grant of £3000. Barker wrote to Bragg at the Royal Institution for suggestions on candidates for the post of physicist, and received the following reply:

My dear Barker,

I have a man here who might possibly make you the research physicist that you want—W. T. Astbury. He comes from Stoke-on-Trent district, made his way to Cambridge where he became a scholar of Jesus, took 1st class Honours in Physics and then was recommended to me on the score of his originality when I wanted research assistants. He is really a brilliant man, has done some first class work which is quoted everywhere and has had a number of papers in the Royal Society Proceedings and the Philosophical

> Transactions. He has considerable mathematical ability and a very good knowledge of physics and chemistry, including organic chemistry and he has been immersed in the problems of crystalline structure. He published with Miss Yardley, now Mrs. Lonsdale, who is working with Whiddington, a very valuable summary of X-ray tests of crystalline structures which we all make use of. He is very energetic and persevering, has imagination, and, in fact, he has the research spirit. Although not trained in the workshop he is sufficiently skilful with his hands.
>
> He is a most loyal colleague to me and I do not want to lose him at all but it is good for these people to make a move from time to time. He can lecture, though I do not call him a very good lecturer, he can write a great deal better than he can speak. Mrs. Lonsdale would tell you all about him if there is more that you would like to know.
>
> (Bragg, W.H., 1928)

Astbury replied by return. Only recently he had failed to get Hutchinson's new lectureship at Cambridge; Bernal had been appointed. This time Astbury was successful. In his correspondence with Astbury the Vice-Chancellor had written:

"I ought to add that the Lecturer would have the services of a trained assistant, and that £500 a year has been allocated for equipment for research in this subject" (Baillie, 1928). If Astbury assumed that all £500 would go to textile physics, he surely had good grounds.

The new research unit with H. J. Woods, the mathematician, as Astbury's assistant, seemed set for a happy course. It would have started without a cloud in the sky had not the Vice Chancellor failed to inform Professor Barker in advance of his dealings with the lecturer in textile chemistry—Speakman—and had he not assured the chosen candidate for the lectureship in textile physics—Astbury—that nobody in the University was working on the X-ray diffraction of textile fibres at that time. Indeed, apart from Astbury's friend, Kathleen Yardley, who had moved to Leeds when she married Thomas Lonsdale in 1927, there was probably no public knowledge of any work in the X-ray crystallography of wool. Since Thomas Lonsdale was working in the Silk Research Association's Laboratory in the university his wife had taken this opportunity to get a Bedford College scholarship to work in the Leeds Physics department and collected together bits and pieces surviving from W. H. Bragg's era in Leeds (1908–1915). With the blessing of Bragg's successor, Richard Whiddington, then in the Cavendish chair at Leeds, and the gift of large crystals of hexamethylbenzene from Christopher Ingold (Leeds' professor of organic chemistry) Kathleen Lonsdale had demonstrated the planar configuration of the benzene ring shortly before Astbury's arrival in October, 1928. But this had called for a Royal Society equipment grant and visits to the Royal Institution to use equipment there.

Astbury, who had taken a fibre picture of human hair for W. H. Bragg's Royal Institution lecture "The Imperfect Crystallization of Common Things" a few years before, was looking forward to developing this field with the skills he had acquired at University College and the Royal Institution. He had given up the stimulating environment of the R.I. for the unknown world of textiles at Leeds only after Sir William, whom he so much admired had persuaded him. Having made this sacrifice he expected to have the new field to himself, and his cross-examination of the Leeds' Vice Chancellor had yielded the assurances he demanded. One can imagine his annoyance when he arrived and found that a lecturer in Whiddington's department, J. Ewles had been taking X-ray pictures of wool fibres at Speakman's request over the previous eighteen months! They had first shown such pictures at the British Association's meeting in Leeds in 1927 when there were special sessions for textile subjects.

Astbury was highly disconcerted, Speakman recalled, "... he appealed to me on the matter and we agreed to hand over this [work] to him. It was quite amicably settled" (Speakman, 1968). And in the first Report on the Research Scheme on textile physics we find the following statement:

> Although the preceding investigation is of so promising a character, its continuation and extension were abandoned by Mr. Ewles and the writer, out of courtesy to Mr. Astbury, soon after his appointment.
>
> (Speakman, 1928, 8)

Thus it came about that Ewles and Speakman's paper, communicated to the Royal Society in 1929 by Whiddington was the last they wrote on the fine structure of wool (Ewles and Speakman, 1930). This story reminds one of the difficulties encountered at King's College, London, when Rosalind Franklin arrived in 1951 to be, as she thought, the sole worker on DNA. Fortunately, on this occasion the situation was quickly resolved and there was soon no doubt that the new recruit from the Royal Institution was a power to be reckoned with. Nothing was ever the same again in Barker's department after Astbury penetrated its walls. He had the extensibility of the wool fibre handed to him on a plate. With Speakman he shared the Worshipful Company of Clothworkers' generous provision of six research fellowships. Behind him he had the goodwill of the Advisory Committee on the departments of textile industries and colour chemistry and dyeing and of the University in which "for some time past, there has been a strong desire ... to develop the research side of the work of the Textile Department" (Leeds University: Report of the Advisory Committee ..., 1928, 1).

It is true that Astbury, his research assistant, Woods, his research students R. Lomax, and A. Street, and a year later his research fellow, Miss Thora Marwick, all worked together in the "Conservatory". It is also true that Astbury spent his first three month in Leeds converting this room from four bare walls into a research laboratory. But this was a large room, and in 1929 when, in a successful bid to keep the Silk Research Laboratory in its fold, the textile department moved this research group from "the Hut" to better accommodation in the department's buildings, Astbury had a "hut" into which he could spread. Later he was able to occupy the rooms vacated by the Silk Research Association's group when it was finally moved to the Shirley Institute in Manchester. No, on these counts Astbury could not complain. Nor should he really have complained about the "conservatory" and the paltry equipment grant which he stated was only £150 p.a. In fact, as the University archives show, within four months Astbury had spent nearly £550. But soon the colour chemists claimed a share of the £500—they had two research students on the scheme—and Speakman had a claim. After an initial imbalance, therefore, Astbury's share probably fell to £150 p.a. The Worshipful Company of Clothworkers can hardly be blamed for this state of affairs. Nor can Barker be held up as a scapegoat, for he went out of his way to put credit from other accounts into the research fund for Astbury and Speakman, thus increasing it from £500 to £700 (Leeds University, Textiles Sub-Committee on Clothworkers' Grants, 1929, 4). If anyone was to blame it was the Vice Chancellor who clearly stated that the £500 equipment allowance was for textile physics, and who made no mention of the claims of textile chemistry, let alone colour chemistry. To Astbury, used as he was to the riches of the Royal Institution, a share in £700 seemed no doubt paltry.

It is my aim to show how radically Astbury changed this state of affairs once he received an international reputation. In so doing he stamped his own style of work very deeply on the fine structure analysis of fibres carried out in this country. Both research on the fibrous proteins and on the nucleo-proteins came under his spell. To understand the state of knowledge in these areas before the period covered in Watson's book *The Double Helix* it will be necessary to examine Astbury's personality and his contribution to molecular biology in some detail.

Bill Astbury

"... I am alpha and omega, the beginning and the end of the whole thing."

Such a statement could only have come from that irrepressible enthusiast and egoist, W. T. Astbury. He made it on the spur of the moment when asked what he was doing at a symposium on coal, petroleum and their derivatives (Astbury, 1948a, 99). This remark was typical of the mixture of candid statement and facetious delivery which left many of his contemporaries nonplussed. His extrovert enthusiasm and naive adoration of nature's wonders led many scientists, especially those of the cold, cautious, plodding type, to mistrust his scientific judgement. As for Astbury himself, he knew very well that his work was speculative. Thus, he likened the study of the structure of keratin to the sort of problem that Peter Cheyney's character Lemmy Caution talks about in the following way:

> You gotta realise that in a job like this you gotta start something. It is no good just hangin' around an' lookin' for clues. The great thing to do is to throw a spanner into the works an' get everybody sorta annoyed with each other.
>
> (Astbury, 1955, 234)

Some of his successors even go so far as to call him an amateur and dismiss his work with scorn. Such an attitude shows a lack of historical perspective.

William Thomas Astbury was born and bred in the pottery town of Longton where his father was a potter's turner. In the "smooth" atmosphere of Cambridge after the First World War he felt ill at ease and it was not until he came to work under Sir William Bragg at University College London that the exuberant cheerful Astbury blossomed forth. Two years later when Sir William took over the direction of the Royal Institution and established X-ray crystallography in the Davy–Faraday Laboratory, Astbury and Miss Yardley went with him. After five years spent in the stimulating environment of the Royal Institution Astbury returned to the north of England and plunged cheerfully into the world of textile fibres. Soon he was haranguing those who dismissed natural fibres as amorphous mixtures or colloidal aggregates. A phrase like "wool is an 'amphoteric colloid' called keratin—a biochemically lifeless and uninteresting protein which is some kind of polypeptide" was enough to set him off declaring how very interesting wool really was. Many years later he looked back on these early days and spoke gratefully of Speakman:

> . . . he it was who showed us right from the very beginning that wool was *different.* All those other miserable textile fibres made from cellulose, and even natural silk when it came to elastic properties, were simply nowhere compared with wool, which not only stretched very much further in any case, but always came back again, all the way, if wetted. That was wool's greatest technological asset, and it was from there that our structural enquiry set out.
>
> (Astbury, 1955, 221)

These words, spoken at a time when his health was breaking down, are witness to his unquenchable enthusiasm, an enthusiasm which R. D. Preston felt when he worked in Astbury's department in the early 1930s. "Every photograph was exciting and Astbury was full of enthusiasm all day long" (Preston, R. D., 1968).

As a physicist, Astbury accepted the rational order of the universe and was convinced that this concept could be carried over to the realm of living organisms. Although not in any sense a religious man:

> . . . he almost worshipped order in nature and was always exceedingly excited to think that he was working in the realm of living things in which the order was a bit obscure, but he was doing his little bit to discover it. He was convinced of the intense simplicity of this order and he had an almost mystical belief that it rested on a very simple structure of a protein which is universal . . .
>
> (Preston, R. D., 1968)

The discovery of this order was for him the adventure of his life. "To the day he died Astbury was a boy who lived in terms of high adventure", said Preston. Then, too, he was for Preston an artist rather than a scientist: "He always wanted a beautiful answer". When he had a two hour conversation with Yehudi Menuhin he came away with a conviction of the common nature of their very different endeavours—Menuhin discovering the truth playing on his violin, Astbury by "playing on" the proteins. In the fine lecture he gave to the Science Masters' Association in 1948 he gave full expression to his idealistic and creative conception of science.

> There is no need for everyone to be Nobel prizemen or even experts. The deep joy that science has to offer is a simple, elementary joy. It is sufficient to sense what it is all about, to feel the urge of divine curiosity and to be moved always by the spirit of impartiality and reasonableness.
>
> . . . science is truly one of the highest expressions of human culture—dignified and intellectually honest, and withal a never-ending adventure. Personally, I feel much the same with regard to the more ecstatic moments in science as I do with regard to music. I see little difference between the thrill of scientific discovery and what one experiences when listening to the opening bars of the Ninth Symphony.
>
> (Astbury, 1948b, 278, 279)

Astbury, it appears, fully accepted the statement attributed to Sir William Bragg that: "Science is not a religion, but scientific research is a principal act of religion" (Quoted in Astbury, 1948b, 280).

This brief character sketch of Astbury has been given not with a view to discrediting him as a scientist, or to discussing him as an amateur, but in order to show how admirably he was equipped to tackle a fresh field, to overcome inertia and scepticism and to generate enthusiasm. Not for him was the dull plod of research in a well-worn path once he had had his first taste of adventure. His first glimpse into the molecular structure of the wool fibre gave him the ambition to solve the structure of proteins as a whole not by detailed analytical studies of their building blocks, but by *comparative* studies of spacings in natural polymers. "The structure of a substance", he wrote, "is revealed in its true significance only in relation to the structure of innumerable other substances. From this point of view, though working in a textile laboratory, we hope to take our place . . . in the fullest scientific activities of the University" (Astbury, 1930, 14–15).

Structural crystallography at Leeds, he hoped, would become a powerful influence linking up the work of the chemists and biologists. He soon shed the purely technological studies, which had been undertaken in 1928, in order to concentrate on the molecular structure of fibres, a central problem in biology, but of long-term significance to technology. "Our aim", he wrote in 1931, "has been above all things to build up a laboratory for the investigation of fundamental structural principles, so that the foundations of textile technology may be well and truly laid. The ultimate advantage of such a plan to industry is far greater than the temporary gains of specific and limited investigations. The method may sometimes appear slow, but it is the only sure one if an industry is to live and thrive in an age of science" (Astbury, 1931, 23).

The Bragg Approach

In Sir William's work in the 1920s, according to Dame Kathleen Lonsdale "there was little certain about the structures then, in fact in those days we did not think much about agreement between experiment and theory. All you had to do was to get something plausible" (Lonsdale, 1968). Maybe this was an overstatement of the situation, but as she pointed out:

> Bragg's structure for naphthalene and anthracene was wrong. It was because he published it that somebody else saw that it was interesting and corrected it. He got one feature—that the extension of the molecule was along the *c* axis, which was a very important thing—but he got the orientation wrong (see Fig. 4.1c). He supposed it was a puckered molecule when in fact it was planar.
>
> (Lonsdale, 1968)

Nevertheless, these early attempts by the Bragg school in London were of profound significance. In diamond, the bond angles were tetrahedral—109°—28′ (see Fig. 4.1a), in graphite, planar—120°. What were they in anthracene, benzene, the paraffins, and fatty acids? From the fine work of Bragg's colleagues, Müller and Shearer, the tetrahedral angle was obtained in the long-chain paraffins and fatty acids (see Fig. 4.1b). It was as if the diamond structure could be opened out into a long chain in which each carbon atom preserved its tetrahedral character. The early X-ray pictures of silk from Germany seemed to support a very similar conformation for the structure of this fibrous polypeptide. The silk fibre consisted of chains like those of the fatty acids. When Astbury went to Leeds he extended this zig-zag chain paradigm from silk to stretched animal hair, and at the same time referred the chains in the unstretched fibre to a folded configuration similar in its architecture to the cellulose model of Meyer and Mark. The great importance of this work was not lost on Sir William who described it with much admiration in his R.I. lecture "Crystals of the Living Body" (1933). Nor did Astbury himself fail to appreciate its significance. Indeed he became so excited that he was never able to sit down and apply himself to the task of establishing these structures with the care he had displayed in his work at the Royal Institution.

Inevitably, as his very able assistant, H. J. Woods admitted, Astbury became a "cream-skimmer" who searched around for interesting problems, in the hope of finding "something fairly spectacular."

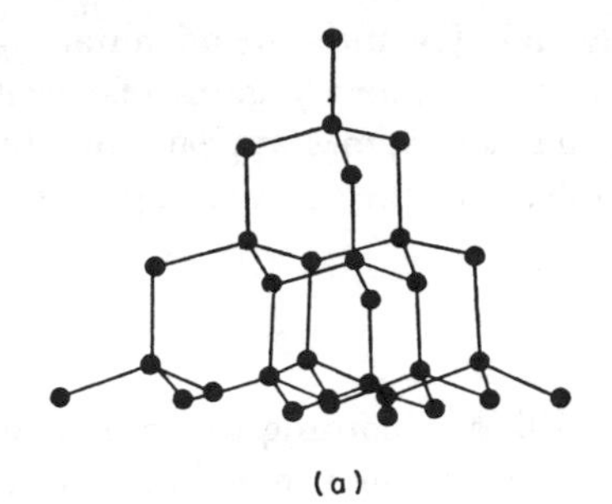

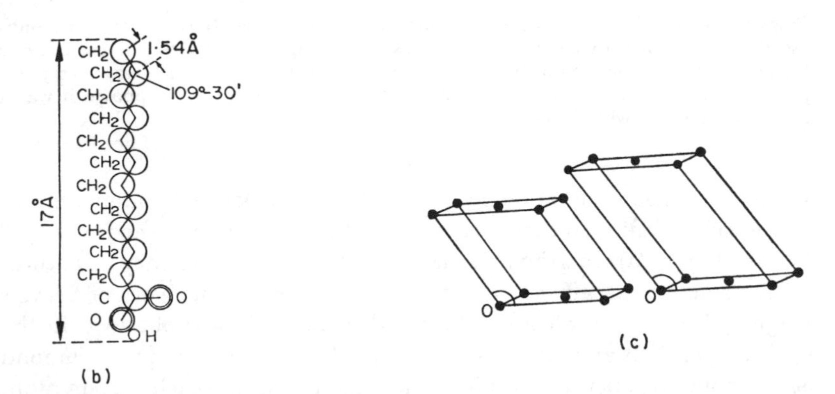

Fig. 4.1.(a) The diamond structure composed of carbon tetrahedra (b) the zig-zag chain for laurinic acid (from Meyer and Mark, 1930, 9) (c) unit cells of napthalene and anthracene (from Bragg, 1922).

> It was not his way to go plugging on at an unrewarding subject. His importance lies in the way he made things appear important. He was able to blast a way through the jungle for other people to follow. It does not matter that he was on the wrong compass bearing.
>
> (Woods, 1968)

Astbury's Models for Keratin

On Astbury's arrival in Leeds, Speakman could show him his detailed studies of the load extension curves for the wool fibre (i.e. keratin) and, Ewles' X-ray pattern for the unstretched fibre showing a 5.15 Å meridional spacing and two equatorial spacings, and the loss of the former spacing on stretching the fibre. This loss Speakman could account for provisionally in terms of the rotation into line of fibrillae which in the unstretched fibre lay at an angle to the fibre axis. Speakman postulated a longitudinal periodicity of 5.15 Å, these periodic units only "reflecting" X-rays when the fibrils were oriented at an angle to the X-ray beam, as in the unstretched fibre. The idea of taking X-ray pictures of the fibre under tension had no doubt been suggested by Katz' famous discovery that rubber gives a crystalline picture when stretched (Katz, 1925). Although Elwes had wanted to account for the elasticity of wool by analogy with rubber "there was", said Speakman,

> some difficulty at the time with fitting the rubber picture to the behaviour of wool; not enough was known about the constitution and molecular configuration of wool to make it easy to adapt the rubber picture to the wool case, and Ewles finally moved over with me to the picture of segments rotating into line. We had not got a picture of an [intramolecular] fold at all.
>
> (Speakman, 1968)

There was a very good reason why the analogy with rubber seemed shaky. Rubber only became crystalline on stretching by at least 80 per cent of its original length, and was most crystalline at 500 per cent extension. Meyer's and Mark's helical conformation for rubber applied to the stretched form. A study of the energetics of its extension supported the conclusion that in the unstretched state the molecular chains were completely disordered. A transition on stretching from random to helical conformation was conceivable in natural rubber (as opposed to vulcanized rubber) where no cross-linkages were known between molecular chains. In wool, on the other hand, we shall see that such linkages played a decisive part in dictating the sort of structure Astbury could admit.

Astbury wasted no time on Speakman's earlier explanations of the elasticity of wool but started afresh and took over a hundred X-ray pictures, while Speakman continued his study of the chemistry of wool and revealed the presence of disulphide and "salt" linkages. Woods complemented these studies with his researches into the mechanics and energetics of fibre extension. By the time Astbury had been in Leeds for little more than two years he had definite evidence of two distinct X-ray pictures of keratin. In this early work with his research student, A. Street, Astbury did for keratin what Franklin and Gosling were later to do for DNA. In 1929 Astbury claimed:

> ... already from the X-ray photographs obtained it has been possible to draw striking conclusions with regard to the molecular structure of the protein matter of wool, the arrangement of the small crystals in the fibres, and their alignment on stretching. We may also mention what appears at the moment to be the discovery of a reversible protein change, perhaps of the nature of the protein changes that physiologists have concluded must form an important part of the phenomena of muscular action.
>
> (Astbury, 1929, 11)

From the records of equipment bought, it seems that Astbury had not been able to start taking X-ray pictures until July, 1929, and that it was not until June, 1930 that he purchased a Hyvac pump. About this time, too, he got a continuous current X-ray generator with a Royal Society grant of £200. This equipment, together with a £24 X-ray camera, yielded in the hands of

Astbury and Street the magnificent X-ray patterns of keratin which appeared in the *Philosophical Transactions of the Royal Society* in 1931.

The principal features of the two forms distinguished by Astbury in 1930 as the unstretched α and the stretched β forms are listed in Table 4.1. Like Ewles before him, Astbury was impressed by the analogy with rubber, and

TABLE 4.1

Principal X-ray data for wool

	α-Keratin			β-Keratin	
Spacing	Intensity	Plane	Spacing	Intensity	Plane
27 Å	s	100			
9.8 Å	vs	001 and others	9.8 Å	vs	001
5.15 Å	vs	020	4.65 Å	vs	200
5.05 Å	vs	120			
			3.75 Å	s	210
			3.32 Å	s	020
Unit cell $a = 27$ $c = 9.8$ Å		$b = 10.3$	Unit cell $a = 9.3$ $c = 9.8$ Å		$b = 10.3$

like Ewles he did not pursue it. Instead he seized on the analogy between the pattern for β-keratin and that for silk, first obtained by Brill in Berlin, Dahlem, but subsequently studied by Meyer and Mark, who postulated extended polypeptide chains with an axial periodicity of $3\frac{1}{2}$ Å (Meyer and Mark, 1928b). "It soon became apparent", wrote Astbury,

> that natural silk finds its counterpart, not in normal wool, but in *stretched* wool. There is a close similarity between the X-ray photograph of silk and of β-wool, from which we may make the deduction that silk does not show the long-range elastic properties of wool *because it is already in the extended state* . . .
>
> (Astbury, 1930, 13)

Astbury also seized on two other numerological analogies:

1. The glucose residue repeat in cellulose = 5.15 Å
   Principal meridional spacing in α-keratin = 5.15 Å
2. Powder photographs of cystine suggest the long axis of the molecule is approximately = 9.4 Å
   Equatorial spacing which persists through α–β transformation = 9.8 Å

The work on cellulose had been done in Germany, that on cystine by Astbury in Leeds (see W. H. Bragg, 1930, 319). On the basis of these three identity, or near-identity, spacings Astbury built up a picture of how an extended polypeptide chain, as in silk, folds up into a series of hexagonal residues as in cellulose (see Fig. 4.2), the whole structure being held together by permanent cystine disulphide bridges between neighbouring chains. Since the folding was in one plane only, Astbury could have these bridges in another plane so that they would be unmolested by the folding up of the α chains, a situation which was incompatible with a spirally folded molecule. When Astbury and Woods discovered that with steam the wool fibre could be stretched not just by 30 per cent as in water, but by nearly 100 per cent, they felt convinced that the hexagonal fold was right, for the 5.15 Å hexagons when opened out gave the required extension and approximately the right sub-unit repeat corresponding to the meridional spot on the X-ray diagram at 3.4 Å. Whilst reminding his readers of the meagre data and the vague information yet available on the chemistry of protein fibres he took the positive view that "... it

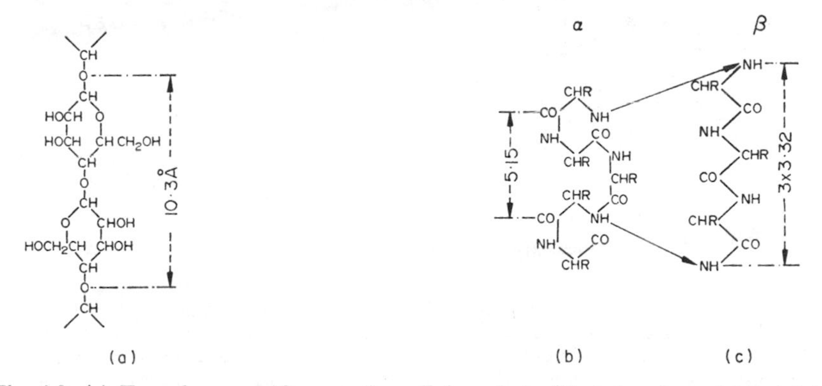

Fig. 4.2. (a) Two glucose residues on the cellulose chain (b) Astbury's model for alpha keratin (c) Astbury's model for beta keratin.

would be unreasonable to neglect even the faintest hint as to what is the basis of any particular structure type" (Astbury and Street, 1932, 89). If he had been more cautious he would have achieved nothing. If he had been studying some other substance—polyoxymethylene, or a synthetic polypeptide, no doubt he would have arrived at a different solution. But he was, as H. J. Woods recalled, "completely interested in keratin ..." All the ideas were designed to try to explain keratin, not polyoxymethylene, rubber, or the then unknown synthetic polypeptides. The α-keratin picture did not look helical and the requirement that the disulphide linkages between polypeptide chains in keratin must remain unbroken during the transformation seemed to rule out helice (see Fig. 4.3). On this point Woods said:

> The difficulty arose here because as this idea [of the α-keratin fold] developed it did so by way of two separate lines of research, the one purely-crystallographic, the other based upon what limited knowledge we could get from the mechanical behaviour or the chemistry of wool. In the '30s the chemistry seemed to lead very very strongly to the idea

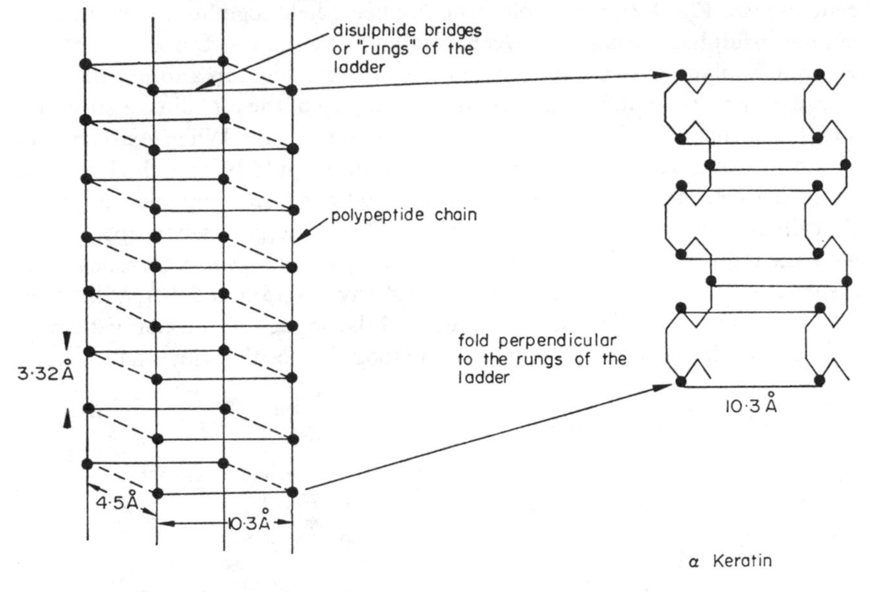

Fig. 4.3. The alpha-beta transformation with preservation of disulphide bridges (author's figure).

that there was a permanent cross-linking between polypeptide chains which could not be significantly disrupted without destroying altogether the characteristic mechanical behaviour of the fibres. The fact that you could get the α–β transformation without, as the chemists insisted, breaking these disulphide links holding the polypeptide chains together seemed to imply a rather simple kind of chain-folding. The planar chain fold was the obvious form, for it was quite clear in β-keratin that the system had a sheet-like form. But at that time the sheet-like form was associated not so much with hydrogen bonding as it is now, but with disulphide linkages of a permanent nature between the protein chains. In the β structure these linkages were believed to hold the chains together in sheets *via* the side chains. So you get what we used to call the "keratin grid" with the ladder-like structure.

Now given this structure for the β form the problem was how to make this ladder collapse into something which could be accepted for the α form, and this was always the basic problem. Well, it is fairly easy to get a ladder structure and to fold it up by folds in the plane perpendicular to the rungs, but even now it is difficult to see how you can get this ladder to coil up into a helical structure without disruption of these disulphide linkages.

(Woods, 1968)

But at that time what a remarkable insight into the properties of wool was afforded by studies of the breakage and reformation of these disulphide linkages, how suggestive for improving the dyeing characteristics of wool. How suggestive, too, was Woods' discovery of the irreversible extension of steam-treated keratin and Astbury's explanation of it, for giving wool permanent set. What irony that Speakman's technique for achieving the latter by chemical means was only belatedly used in the wool industry long after it had been exploited by hairdressers! What a triumph Astbury had achieved; what brilliant intuition! With just a few spots on the X-ray plates as his clues he had built up a comprehensive picture of the molecular basis to the elasticity of wool and the distinction between the extensibility of wool and silk. Can one blame him for being supremely confident and for stating to the Worshipful Company of Cloth Workers:

> ... we are now in a position to believe that the discoveries described below constitute the first true steps towards the final solution of a problem of first-rate importance, whether for industry or for biology. The new X-ray evidence permits us to claim that we have reached the foundation of the structure of wool and hair, and perhaps of proteins in general.
>
> (Astbury, 1930, 11)

Vistas of research opened out to Astbury's gaze which tempted him to expand his programme of research in the direction of biology. In 1934 the Rockefeller Foundation, in response to his request, offered a grant of £600. A year later its emissaries, Warren Weaver and W. E. Tisdale, visited the textile physics laboratory and were well impressed. In 1938 they gave a further grant of £10 000.

Thus supported, Astbury was able to apply his keratin model to myosin (with Sylvia Dickinson, Rockefeller research assistant) to epidermin (with K. M. Rudall, Ackroyd Fellow), and to fibrinogen (with Kenneth Bailey and Rudall) thus generating the *k-m-e-f* group of proteins and affording a plausible molecular mechanism for such diverse phenomena as the motion of cilia and the contraction of muscle. Seeking fresh fields to conquer, Astbury studied fibres of the collagen group, the denaturation of proteins (with Dickinson and Bailey), he encouraged Preston (in the botany department) to attack the structure of algal cell walls by X-ray diffraction and he put Mrs. Dickinson's successor, Florence Bell, on to the structure of nucleic acids. Just as his comparison of wool and silk had led him to the correct conclusion that the keratin molecule is folded in the natural state and unfolds to give the silk conformation under tension, so his studies with Dickinson and Bailey led him to an explanation of protein denaturation which was on the right lines, and in 1935 he wrote to Dorothy Hodgkin:

> These molecules [in crystalline globular proteins] must be built up each from some very specific configuration of chains which breaks as soon as the rather narrow conditions of stability are departed from. On heating the chains they simply coagulate to form parallel bundles analogous in their structure to that of the stable natural fibres ...
>
> (Hodgkin & Riley, 1968, 18)

This was even more wonderful. The key to the proteins as well as to the carbohydrates was the long-chain molecule. In α-keratin it was folded, in silk and β-keratin it was extended, and in those magical globular proteins which were at the centre of life it was folded into a regular 3-dimensional structure. From this conception resulted Perutz' view of the haemoglobin molecule as a box in which the chains were arranged on top of one another like a pile of logs. This picture of parallel extended and planar-folded chains was so convincing because it accounted for so much. Astbury's hypothesis for the α–β transformation just had to be at the very centre of molecular biology.

Frank's Suggestion

Sooner or later the question was to be asked: what is the nature of the dotted link by which Astbury closed up the hexagon rings in his model for the folded chain of α-keratin? One such occasion was Astbury's lecture to the Oxford Junior Scientific Society in 1933.

On this occasion it must have been suggested that hexagons of the dimensions required by Astbury could not be built unless this dotted link was a covalent bond, in which case its length had to be about 1.5 Å, but that was too short.

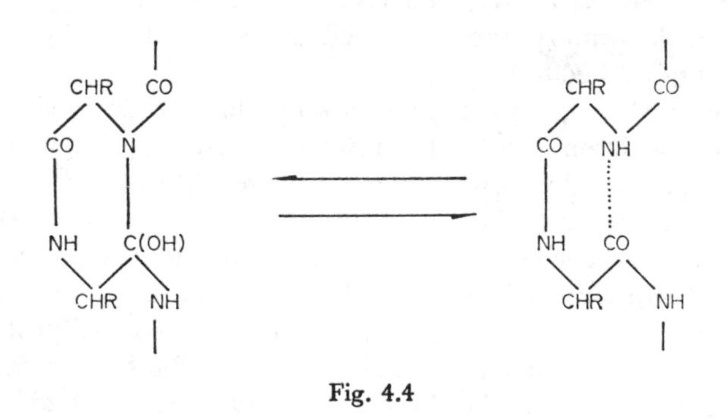

Fig. 4.4

F. C. Frank then suggested a way out of the difficulty by postulating a sort of lactam–lactim interchange between the carbonyl and amino group as in Figure 4.4. Frank went on to suggest how the super contracted form of keratin (one third the length of β-keratin) could be accounted for. Dorothy Wrinch then took up Frank's suggestion and developed "a general scheme for the structure of the 'globular proteins' which incorporates just this hexagonal fold first postulated in α-keratin and the lactam–lactim improvement proposed by Frank" (Astbury, 1936, 285). This was the notorious cyclol theory. Wrinch was convinced that some such regular cage structure must be present in these proteins, but neither Bernal nor Pauling, nor Carl Niemann, nor Huggins was in favour of Wrinch's theory. Astbury was later to side with them, but in 1936 he supported her. The wheel, he said, had turned full circle; first had come the discovery that silk and wool were "spun" from fully extended polypeptide chains, then came the suggestion that the globular proteins might represent elaborations of the sort of folded chains found in α-keratin, and finally that denaturation liberated the chains from globular proteins allowing their conversion into visible fibres which on stretching gave the fully extended polypeptide chains as found in silk.

> The wheel is even now commencing its second revolution, for . . . Wrinch has attempted to build up a consistent molecular system for the globular proteins by folding back, theoretically, these liberated polypeptide chains into series of folds that are none other than those first postulated to explain the X-ray diagrams and properties of keratin.
>
> (Astbury, 1936, 290)

Great was Astbury's pride when he could sport a pullover made of a yarn spun from a globular protein. "Only the other evening", he recalled, "I was watching my daughter knitting yarn spun from monkey-nut protein—a protein, I repeat, with round molecules that once seemed to bear no relation whatsoever to fibres—and as I touched the knitting, again the wonder of it all flooded over me . . ." (Astbury, 1948b, 278).

CHAPTER FIVE

Astbury under attack

Astbury was the scholarship boy who had worked his way up from a working-class background; he was the pride of the family, the clever one, and how he proved his promise! The smooth climate of Cambridge had not suited him but in the homely family of Sir William Bragg, first at University College London and later at the Royal Institution he had blossomed forth. His quick success with keratin at Leeds bolstered his ego. Bragg had not rated his powers as an orator highly, but how wrong Bragg proved to be! No doubt the fine example of Bragg's Royal Institution lectures served as Astbury's model. One has only to read the published version of Bragg's lecture series "Old Trades and New Knowledge" to see what interest Bragg could take in the work of the artisan (Bragg, 1926). The lecture in this series "On the Weaving Trade", it is relevant to note, discussed the elasticity of the wool fibre. Judging by the newspaper reports at that time Bragg must have been a most effective publicist both for science and more especially for X-ray crystallography. Astbury, likewise, became a most effective propagandist for X-ray crystallography both in textile research and in biology and was among those who early popularized the term "molecular biology". His skill as a public lecturer became well known. He was much in demand. As early as 1933 awards began to come, beginning with the Worshipful Company of Dyers gold medal and the University of Lille's medal. In 1937 under the auspices of the Rockefeller Foundation he visited "some fifty of the most important universities and research institutions in the United States, at nearly all of which he gave lectures or took part in colloquia on the molecular structure of fibres and biological tissues" (Leeds University: Department of Textile Industries, 1937, 4).

The early years of the war inhibited Astbury's research programme, and at the same time he became increasingly possessive, inflexible and over-confident regarding any approach to the proteins which differed basically from his own. Ian MacArthur, who joined Astbury's unit as research assistant in the '30s said of him that "He brought his findings to market in the green ear, but would not clear the weeds nor suffer the system and technique necessary for the harvest. Inevitably, the protein–structural breakthrough was lost, and it hurt" (MacArthur, 1961, 332).

So much of a perfectionist was MacArthur that Astbury's rapid and profuse publications of "green" research dismayed and alarmed him and in his obituary he went so far as to describe Astbury as an "artistic amateur not a scientific professional" (*Ibid.*). Astbury had become over-confident at just the time when the science of X-ray crystallography had moved to a fresh plateau. Fourier and Patterson analyses were becoming manageable techniques for arriving at molecular structures and for determining bond distances and angles to an accuracy of about a fiftieth of an Ångström unit. But instead of going back to consolidate the ground he had already covered, Astbury moved on to fresh polymers which he interpreted in the light of his keratin model. He never supplemented his study of powder patterns of cystine (Astbury and Street, 1932, 91) by taking single-crystal pictures. Fortunately the unit cell and space group he had then inferred were also arrived at by Bernal in his pioneer single-crystal analysis of amino acids, diketopiperazine and peptides (Bernal, 1931). Astbury made no detailed study of atomic packing in the light of fresh data on atomic radii which was yielded by the electron diffraction studies of Pauling's school in Caltech during the 1930s, to be supplemented later by Corey when he carried Bernal's work on the amino acids and peptides a stage further and demonstrated the partial double-bond character of the C–N bond in diketopiperazine (Corey, 1937).

The Rubber Model

While Astbury built on the unsound foundations of the keratin planar fold, Kurt Meyer in Geneva extended the Meyer–Mark helical structure for rubber to plastic sulphur and polyphosphonitric chloride with axial repeats of 9.26 Å (8 sulphur atoms) and 5.17 Å (two residues) respectively (Meyer, 1936) as in Figure 5.1. Two years later, Brill and F. Halle published a beautiful X-ray diffraction picture of oppanol, in *Die Naturwissenschaften* and drew attention to the growing number of polymers for which helical conformations had been advanced. To these authors it was clear from the axial periodicity of 18.50 Å in oppanol

> . . . that the molecule cannot be formed of a planar zig-zag chain but—under the influence of the side groups—its most likely form is spiral by analogy with the structure of polymethylene and polyethylene oxide, which show an identical fibre period.
>
> (Brill and Halle, 1938, 12)

This quotation takes us back to the fine work of Erwin Sauter at Freiburg on the structure of polyoxymethylene, which marks a clear break with the planar zig-zag chain established by Alex Müller at the R.I., and by Hengstenberg in Freiburg for the long chain paraffins. Hengstenberg had become convinced that no models of this type could be devised for the oxypolymers without violating the tetrahedral angle for carbon of 109° 30′. Sauter devised instead what he called the "tub form" (*Wannen Form*) as in Fig. 5.2b and c,

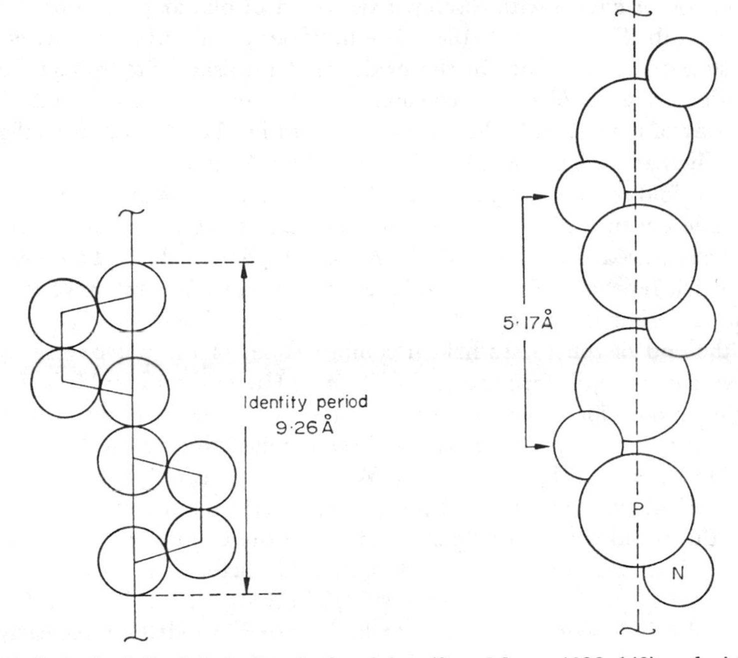

Fig. 5.1. Left, the helical chain for plastic sulphur (from Meyer, 1936, 149) and, right, the helical chain for polyphosphonitrile chloride (from Meyer, 1936, 151).

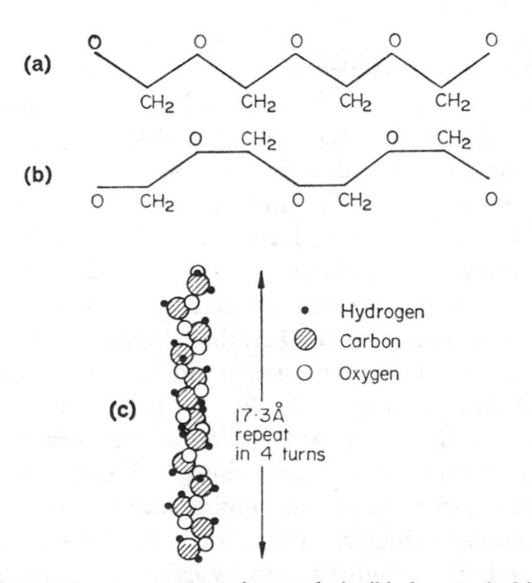

Fig. 5.2. (a) The 'old' polyoxymethylene zig-zag chain (b) the new 'tub' form (from Sauter, 1933, 188) (c) Sauter's helical model for polyoxymethylene of 1933.

a chain conformation with a screw axis which in planar projection has the form of a tub (Sauter, 1933, 188). The importance of polyoxymethylene for the historian lies not only in the evidence it furnished Staudinger for the existence of macromolecules (see Chapter 1) but in the obstruction it placed in the way of the model builder who tried to fit his data to planar configurations. Whereas rubber, though helical, could be given a planar conformation (with two-fold screw axis) polyoxymethylene could not. A ribbon model was impossible because of the tetrahedral angle, and a zig-zag chain would not have the required fibre repeat of 17.3 Å. Instead, Sauter devised a *non-integral* helix which repeated after nine residues in four turns, a rotation of 160° per residue, as originally suggested by Hengstenberg in 1927.

By the end of the 1930's helical conformations of the polyoxymethylene type were being built for proteins by Maurice Huggins and by H. S. Taylor. Huggins' model for α-keratin had three (but not necessarily exactly three) residues per 5.1 Å repeat, which could be uncoiled to give the 100 per cent extension demanded by Astbury and Woods. In the early 1940's in Oxford, Dorothy Hodgkin built a spiral polypeptide chain for the insulin molecule with a three-fold screw axis "partly led on by the trigonal symmetry of the crystal, which we soon found had been built before us, both by M. L. Huggins (1943) and by H. S. Taylor (1941)" (Hodgkin and Riley, 1968, 26). These quite independent attempts to build helical models in the early 40s indicate a growing recognition of the inadequacy of zig-zag and planar configurations. How did Astbury's fold for α-keratin come under fire and how did he respond?

Criticism of the α-keratin Model

When Preston visited Huggins in Rochester, New York State, in 1936 he was introduced for the first time to the suggestion that the polypeptide chains in α-keratin might be helical and was made aware of the shortcomings of Astbury's α and β models. In the previous year Huggins had entered the protein field on his appointment to Eastman Kodak Ltd. This Company was anxious to know the structure of the gelatin used in their photographic films and Huggins' immediate superior, Dr Sheppard, wanted him to do for gelatin what Astbury had done for keratin. Huggins, who claimed he had arrived at the concept of the hydrogen bond in 1919,* was well suited for this task. Trained, like Pauling, under Roscoe Dickinson at Caltech, he had collaborated with Pauling on the application of the Lewis theory to covalent radii and the existence of resonance structures, and shortly after his move to Rochester had introduced the planar peptide unit as a more stable structure than a non-planar unit (Huggins, 1936). Presented with the task of finding the structure of gelatin Huggins began by examining Astbury's models for α- and β-keratin and collagen. "At first", he wrote,

*This claim has been queried by Linus Pauling.

I tried to modify Astbury's models to make them accord with my principles: introducing hydrogen bonds of the NHO type, avoiding too close approach of nonbonded atoms, etc. (*J. Org. Chem.* **1**; 407 (1936), pp. 447–450). For β-keratin this seemed to lead to a satisfactory result but I soon became convinced that the model for α-keratin so obtained was not satisfactory, because the unbalanced forces on opposite sides of the chain axis would cause it to bend continuously. Only a spiral structure could avoid this. I investigated various spiral structures, hoping to find the correct ones for α-keratin, collagen, etc. When Astbury visited me in May 1937 and I showed him my models, he at first argued against hydrogen bonding. (I think that he was not previously acquainted with the concept.) Before he left, however, I think that I had convinced him.

(Huggins, 1969a)

I told him that I could agree with some of the concepts and models that he had proposed but not with certain others and that I felt sure that where he had dotted lines or vague attractions between NH and CO groups there were real hydrogen bonds, and where he had those dotted lines between other things they did not represent anything.

(Huggins, 1969b)

This meeting was during Astbury's American tour under the auspices of the Rockefeller Foundation. Two years later Astbury received the following letter from the American protein chemist, Hans Neurath:

For a long time I meant to send you the manuscript of a paper which I presented last spring at a meeting and which has not been published yet. I was rather hoping that the editors would not accept it and I forgot all about it until recently I got a note telling me that the result of a sudden "brain storm" had been accepted for publication. I'll dig up the manuscript and send it to you right away. I have an uneasy mind because I seem to have found that your alpha-keratin structure and your supercontracted keratin don't meet the requirements of atomic dimensions. I can't see how there can be enough room for accommodating the side chains and the hydrogen atoms, as judged from scaled atomic models. May I suggest that you read the paper and let me have your reaction by Trans-atlantic Air Mail so that I still can withdraw the manuscript if necessary.

(Neurath, 1939)

In his reply Astbury admitted

. . . I may say at once that I am not completely surprised at the contents of your paper, for I have realized for some time that there may be perhaps insuperable difficulty in fitting the side chains on to our α-keratin model. This model was first put forward a long time ago, and I no longer insist on the details, or even the essential accuracy, of the suggestion. In my more recent papers you will probably have noticed that I insist rather on the essential fibrous nature of all proteins, and in fact my paper on denaturation in the *Biochemical Journal* includes a proposed grid-iron structure for the globular proteins. This system is of course no other than a folded β-configuration, and several times I have pointed out that the folding of polypeptide chains must in general take place in planes transverse to the side chains, just as I have argued for the α–β transformation of keratin and myosin.

However, in spite of all these feelings I have had that some such trouble must ultimately turn up as you have crystallized out in your own paper, still I rather think that you have made the worst of the difficulty. I think that perhaps you have over-estimated some of the inter-atomic and inter-molecular distances. Perhaps these alterations will not make a great deal of difference to your general calculations, but the C–N distance in silk works out to be 1.37, and the C=O distance in urea crystals is 1.15. Again the radius you have taken for oxygen is that found in water and mineral salts, and it is possible that it is not the same in organic acids. I feel that you want this manuscript back quickly, so, as I have said, I am not taking time to make up models of my own or to search the literature on atomic data, but I suggest that you go again into the question of the latter.

(Astbury, 1940a)

Neurath's letter goaded Astbury into action and with his research assistant, Florence Bell, he arrived at a fresh model for keratin, but the folding of the chain was still planar (Astbury and Bell, 1941) (see Fig. 5.3).

To go back to an old problem and rework his solution was most uncharacteristic of Astbury and perhaps for that very reason the result did not represent a radical departure from his old model. The hexagonal cellulose analogy had been rejected but not the planarity of the fold, which he still

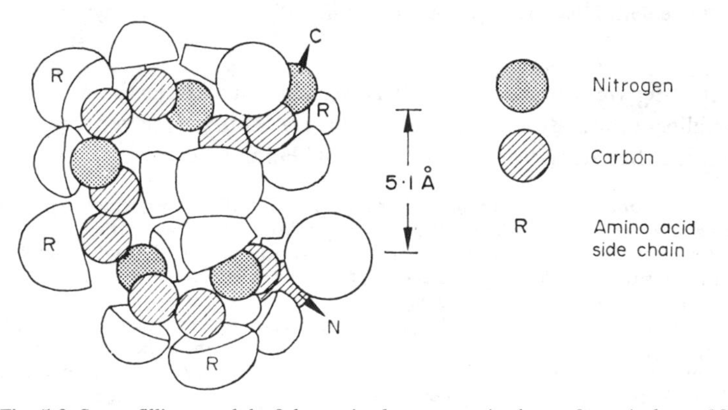

Fig. 5.3. Space-filling model of the revised structure (redrawn from Astbury, 1941).

held to be essential for the preservation of cross-linkages in keratin. In any case a protracted re-examination of the subject was not possible at a time when he had so many irons in the fire and the war looked like terminating his research programme.

Among the many uncompleted studies upon which he had embarked were the nucleic acids. I admit that his X-ray examination of these compounds occupied but a small part of his time, but they were the only studies of their kind in the late 1930s and as such they claimed the attention of all those who were concerned with the mechanism of chromosome duplication and the nature of the gene. In addition they represented for Astbury the culmination of his researches in molecular biology, which had started with forms and led up to origins.

The First X-ray Pictures of Nucleic Acids

Astbury being a scientist of international standing and of unbounded enthusiasm for fibres of biological significance, it was natural for those who succeeded in preparing highly viscous solutions from which they obtained birefringent fibres to send them to him for X-ray investigation. He had certainly not solicited nucleic acid in 1935 or 1936 when W. J. Schmidt sent him material from Giessen. From this material Astbury obtained a very imperfect fibre diagram with a meridional arc at about 3½ Å. This was an improvement on the two previous attempts, one by Herzog and Jancke in 1920 using material supplied by Herman Steudel and the other by Rinne ten years later. By the time Astbury received fresh material he had a new research assistant.

In 1937 Astbury's Rockefeller research assistant Sylvia Dickinson, resigned. Her successor was a very able and attractive young Cambridge graduate, Florence Bell (now Mrs. Sawyer), who had been an undergraduate in Cambridge before going to Manchester. This move had not proved a success, but when she came to Leeds as a timid young research student she soon found her feet and Astbury found in her a critical mind and a cautious and reliable research scientist whose criticism of his speculative leaps caused him to dub her his *vox diabolica* (Sawyer, 1967). Astbury was able to supervise her work well, as their rooms were close by on the same floor. This team of two was probably the most effective example of collaboration in the late 1930's in Astbury's department. One of its fruits was the first molecular structure for DNA.

The Pile of Pennies Model

When the cytogeneticist Jack Schultz visited Astbury on his way to the Karolinska Institute in Stockholm as a Rockefeller fellow, he formed a clear picture of the type of work going on in Leeds. This prompted him later to ask Tobjorn Caspersson to send Astbury some of the highly polymerized DNA prepared by the Hammarsten technique. Although Einer Hammarsten was not keen for material to be sent to Leeds, Caspersson complied with Schultz' request and Astbury passed the material he received on to Florence Bell. She made films of the sodium salt by quickly drying a pool of the solution on a glass plate and detaching 2 mm. wide strips. These were stretched to two and a half times their original length and placed in a frame in the X-ray beam. Some of the resulting pictures were remarkably good, but appear to have been mixtures of the A and B forms subsequently described by Franklin and Gosling.

Two features stood out—a meridional arc at 3.3_4 Å and an equatorial spot at 16.2 Å. From the position of vague diffraction spots near the origin, Astbury and Bell concluded that the true period along the axis of the DNA molecule must be at least 17 times the thickness of a nucleotide (Astbury and Bell, 1938, 113). What gave rise to the 3.3_4 Å spot? Obviously there must be strongly scattering groups in a regular array separated by this distance.

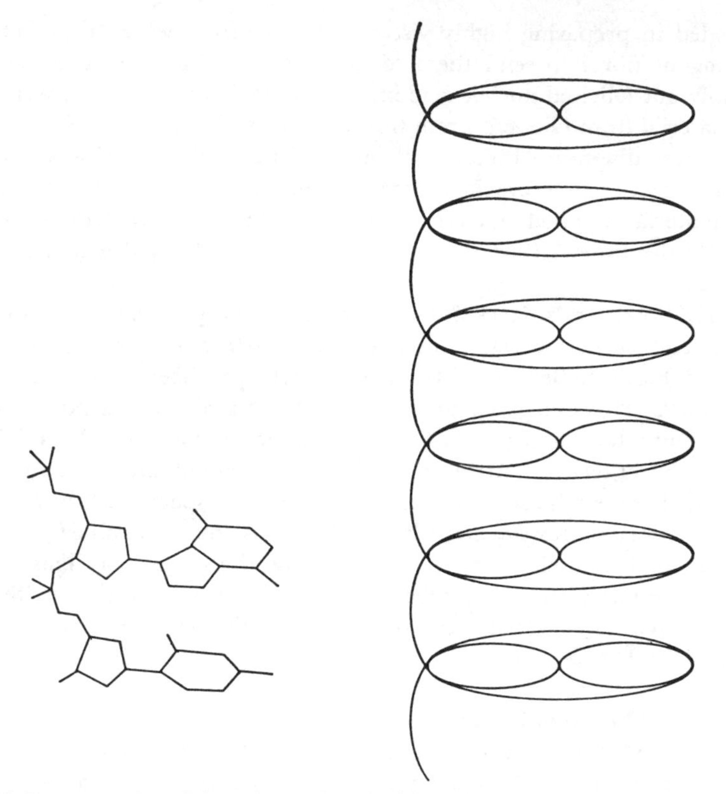

Fig. 5.4. Representation of the column of nucleotides constituting the unit of thymonucleic acid (from Astbury and Bell, 1938, 112).

W. J. Schmidt had already noted the negative birefringence of the DNA fibre in 1932. Astbury and Bell attributed this to the purine and pyrimidine bases which must be in a plane at right angles to the fibre axis. The negative birefringence of both chromosomes and of DNA fibres could now be interpreted in terms of a molecular model in which the bases were piled one on top of another like the plates in a plate rack or like a pile of pennies as in Fig. 5.4. Since the tetranucleotide hypothesis demanded, they thought, a repeat at $3.3_4 \times 4 = 13$ Å, the absence of a spot on the meridian at this distance caused them to doubt this notorious hypothesis. Clearly they attributed any higher order periodicities than 3.3_4 Å to a chemical repeat rather than a conformational one.

After their first paper on the structure of DNA appeared in *Nature* in April 1938, Astbury attended the Cold Spring Harbor Symposium on the chemistry of proteins and read a paper on his work with Miss Bell. Amongst those who commented on it was Stuart Mudd (one of those who suggested that antibody and antigen have complementary 3-dimensional structures); his comment was:

> I find your diagram of the structure of nucleic acid exceedingly reassuring, because certain chemists who have discussed this data from the point of view of analytical chemistry have questioned whether there is enough variation in composition of nucleic acids to give the possibility for serological specificity, but from your diagram it is obvious that there is more than enough variation possible. We know that changes in configuration and spatial relations of the same components do give an adequate basis for serological specificity. The rotation of the H and OH groups about an asymmetric carbon atom may give differences detectable serologically. So it seems apparent that these piles of nucleotides by slight changes in the order in which nucleotides occur, or possibly by other changes in configuration, give us an adequate basis for specificity.
>
> (Mudd, 1938, 118–119)

Astbury's reply was to repeat his belief that the "period along the nucleotide column is at least 17 times the thickness of a nucleotide. So the nucleotides do not follow each other always in the same order, and this gives you another chance of great variation" (Astbury, 1938, 119). Their structure was a single chain molecule with the sugar and base rings coplanar and at right angles to the axis of the chain.

Astbury and Bell were excited to get so much evidence of crystallinity in the Karolinska material, but it was the spacing 3.3_4 which gave special joy to them, and they commented on the parameters as follows:

> The significance of these findings for chromosome structure and behaviour will be obvious, for we cannot fail to be struck by the fact that the spacing of successive nucleotides is almost exactly equal to the *spacing of successive side-chains in a fully-extended polypeptide chain.* It is difficult to believe that the agreement is not more than a coincidence: rather it is a stimulating thought that probably the interplay of proteins and nucleic acids in the chromosomes is largely based on this very circumstance. In the mitotic cycle there is a rhythm also in the way nucleic acid makes its appearance in the chromomeres, and we can well imagine that some critical stage in mitosis, involving elongation of the protein chains, is realized under the influence of the linear period of the interacting nucleotides.
>
> The idea is equivalent to saying that the molecule of thymonucleic acid fits so perfectly on the side-chain pattern of a fully extended polypeptide chain that interaction should take place almost without any steric hindrance whatsoever; most easily between the basic side-chains and the phosphoric acid groups, but presumably, too, between the acid side-chains and the basic groups of the nucleotides. Furthermore, the products of combination should also be fibrous, like the two original constituents.
>
> (Astbury and Bell, 1938, 113)

When asked whether they had expected this equality Mrs. Sawyer (née Florence Bell) wrote:

> We were "ecstatic" to find a spacing identity—but not really surprised, because we had been hoping to find some relationship. Astbury considered that the nucleic acids were templates for protein duplication and organization, and held the polypeptide chains stretched and parallel for the division process. This was one time when his "*vox diabolica*" (as Astbury labelled me) gave little argument.
>
> (Sawyer, 1967)

They went on to test this hypothesis by forming a nucleic-acid–clupein complex in the form of a membrane between aqueous solutions of the two compounds. When rolled into a thin ribbon this complex gave an X-ray picture very similar to that of the acid alone, from which they inferred the attachment of the clupein down one side of the nucleic acid chain. The amino acid residues 3.3_4 Å apart would then support the 3.3_4 Å meridional reflection.

This was wonderful! Astbury could fit the nucleic acids into the prevailing theoretical framework which had at its centre the chromosomal proteins. In jubilation he wrote:

> Knowing what we know now from X-ray and related studies of the fibrous proteins, . . . how they can combine so readily with nucleic acid molecules and still maintain the fibrous configuration, it is but natural to assume, as a first working hypothesis at least, that they form the long scroll on which is written the pattern of life. No other molecules satisfy so many requirements.
>
> (Astbury and Bell, 1938, 114)

At this Cold Spring Harbor meeting Astbury met Wendell M. Stanley, who had crystallized TMV three years earlier. Later, Astbury and Bell studied the X-ray patterns produced by Stanley's TMV crystals. "I hope", wrote Stanley, "you were able to get some definite results before war was declared, and that you will still be able to do at least some work of a useful scientific nature" (Stanley, 1939). But by this time Astbury's attention had been diverted to collagen upon which compound he had just given the first Proctor Memorial Lecture (August, 1939). Besides, as he said to Stanley, all of them in the Leeds laboratory had volunteered for scientific service so "we may be called away at any time. The only thing to do in these circumstances is to try to get as many things finished as possible, and that is just what we are doing" (Astbury 1939a). Nor did the nucleic acid chemist, Masson Gulland, get any results out of Astbury to whom he had given samples of BDH thymonucleic acid and yeast nucleic acid (Gulland, 1939).

Before war broke out Astbury made only one further contribution to the literature on the structure of nucleic acids—his paper to the International Congress of Genetics in Edinburgh in 1939. This contains no fresh data but plenty of speculations regarding the interaction between nucleic acids and proteins which will be considered in Chapter 7. Both in this and in his previous papers, however, Astbury's context for the discussion of nucleic acids was that of their association with the proteins. Here approximately identical fibre periods played *the* decisive role in Astbury's argumentation, just as it had done in α- and β-keratin. What he clearly considered most significant was his demonstration that an artificial nucleoprotein could be produced which gave a fibre pattern almost identical with that of sodium thymonucleate.

As with keratin, so with nucleic acid, Astbury made no attempt at this time to build molecular models. Had he done so he would surely have discovered that sugar and base cannot lie in the same plane and that the backbone must form a helix. He would also have discovered that one of the side spacings, that at 18.1 Å, which Bell and he attributed to the greatest length of a *single* nucleotide, was more than twice the maximum dimension of these residues! This erroneous assumption of theirs (due to their belief in the co-planarity of sugar and base) was one cause for the following calculation coming out in support of a single-chain model. This goes as follows: Assume a single-chain, with an average diameter 18.1 Å. The total length of the molecule must be 18.1 × 300 Å since the length: breadth ratio from streaming birefringence is 300: 1. Since the average molecular weight of a nucleotide is 330, and the internucleotide separation is 3.34 Å, the molecular weight of the entire molecule should be

$$\frac{18.1 \times 300 \times 330}{3.34} = 536\ 000$$

Caspersson and Signer had arrived at a value for the same material of between 5000 000 and 1 000 000. If for 18.1 we substitute a figure of 9 Å, the molecular weight of DNA comes out at half the minimum value and quarter the maximum. In other words there had to be between two and four chains in the molecule! Astbury and Bell had happily settled for one. There was a further way Astbury and Bell could check their conclusion. Bell had measured the density of DNA when dried at 1.63 g cc. This value multiplied by the volume of a single nucleotide should have given the theoretical average molecular weight of 330. Using the following dimensions for the nucleotide: 15 × 7.5 × 3.4 Å they obtained a molecular weight of 364.5. Had they known of the true "average" dimensions of the nucleotides this calculation would have come out nearly one half the figure of 330. It is extraordinary that Astbury did not see fit to publish more of Bell's data. Thus she found the density of DNA at "ordinary humidity" to be 1.515 g cc., the water content being 21 per cent (whether by volume or by weight is not stated) (Bell, 1939, 56). She indexed the diagram of clupein and of sodium thymonucleate, giving for the latter the following dimensions: $a = 36.2$ Å, $b = 53.5$ Å, $c = 22.2$ Å (Bell, 1939, 67). The fact that this cell was so large shows that Astbury and Bell, like Bernal and Funkuchen, experienced difficulty in indexing fibre diagrams (see Chapter 14). The exceptionally long dimension in the b direction was caused by assigning an arc at 5.35 Å to a tenth order reflection. But Astbury very sensibly attributed pronounced equatorial and meridional spots to features of the *molecule* of DNA. Where he failed was in not getting Bell to work out the number of chains in the molecule from *all* her data.

The reason why more attention was not given to this point may well have been Astbury's preoccupation with what at that time was thought the most significant aspect of nucleic acids—their interaction with proteins. "Possibly the most pregnant recent development in molecular biology", wrote Bell, "is the realization that the beginnings of life are closely associated with the interactions of proteins and nucleic acids" (Bell, 1939, 63). Back in 1931, Astbury had been speculating about the mechanism of gene replication (Astbury, 1931, 21). The near identity between the residue separation in the proteins and nucleic acids was all that Astbury needed to assure him that the nucleic acid molecule could operate as a stretcher holding the polypeptide gene molecule in the extended (β-keratin) configuration, ready for the new gene to be laid down upon it.

As late as 1953 Pauling supported Astbury in regarding this spacial identity as significant (Pauling and Corey, 1953, 96). Unfortunately, Astbury's view has not been substantiated. The near identity between the separation of nucleotides in DNA and RNA and in the separation of amino acids in a fully extended polypeptide has been a false trail which Crick knocked on the head with his adaptor molecules in 1956. The near identity discovered by Astbury and Bell is significant neither for protein synthesis nor for the duplication of the chromosomes as such, though it may lie at the root of the "wrapping-up" of a DNA helix in histone, under which conditions neither duplication of DNA nor synthesis of messenger RNA can occur.

Back in 1938 Astbury and Bell, elated by the discovery of this near identity, were sure they had chanced upon the secret of secrets—the key to the master plan for the self-duplication of the chromosomes. It was this plan which Astbury believed was simple and rational and which he admired with the devotion of a youthful mystic. The spacing 3.3_4 Å between neighbouring polynucleotides was the great clue to the duplication of the proteins which he believed to constitute the gene. It had been a long way from the elasticity of wool to the molecular architecture of the gene, but Astbury believed he had got there! Oh, it was good to be alive to partake in the act of scientific discovery, to be penetrating to that master plan upon which life's programme was based! When Astbury gave the Mather lecture he entitled it "The Great Adventure of Fibre Structure", mindful of his own good fortune "in the path along which I have been led since my early training as a 'straight' physicist—through the wonder world of crystallography and X-ray structure analysis, thence to the study of textile fibres, and finally to the tremendous sweep of the science of chain-molecules of all kinds, but especially biological" (1952, 81–82). This method of approach, he claimed, in his Harvey lecture, could hardly be improved upon.

> All I have had to do has been to trace out and systematize the development and inevitable broadening of my own interests as the panorama unrolled itself to my innocent gaze. The line I propose to adopt thus starts from structure, goes on to properties, then seeks out the underlying plan, and lastly tries to delve down into the origin of things.
>
> (Astbury, 1950, 5)

SECTION II

NUCLEIC ACIDS AND THE NATURE OF THE HEREDITARY MATERIAL

LEVENE, CASPERSSON, GARROD, MULLER, DARLINGTON, STANLEY

CHAPTER SIX

Kossel, Levene and the Tetranucleotide Hypothesis

> Originally, the term "tetranucleotide" indicated the occurrence of the four appropriate nucleotides or the corresponding nitrogenous derivatives in equimolecular proportions in the decomposition products of a nucleic acid, and was thus used to describe certain of the then known nucleic acids at a period when these were regarded as having a molecule so simply composed. Now that the complex nature of nucleic acids as polynucleotides is established, the term is still applicable in this sense to such polynucleotides as conform to the tetranucleotide ratio in their nucleotide contents.
>
> Later, as a development arising from the recognition of this complex character, the name has been used to denote a unit, consisting of one molecule of each of the four nucleotides, which by recurrent combination with itself forms the polynucleotide; it seems to be implied that the mode of this union is uniform throughout the polynucleotide. Finally, it has been postulated that in each of these units the four nucleotides are always combined in a fixed manner in an unchangeable sequence. As a logical outcome of these later hypotheses there has arisen the conception of certain polynucleotides as "poly-tetranucleotides".
>
> The term "tetranucleotide" thus has a graded series of implications ranging from a statistical expression of analytical results to a definition of an exact chemical structure, and in order to avoid confusion, the sense in which the name is being used must be defined; the terms "statistical" and "structural" tetranucleotide are suggested for this purpose. It is opportune to review and assess the evidence on which these interpretations are based, all the more so because the importance of the polynucleotides in the biological fields makes it essential that the possibilities of resemblance or divergence between polynucleotides from different sources should be clearly recognized so that conjectures involving nucleic acids may be based only on accepted chemical knowledge.
>
> (Gulland, Barker and Jordan, 1945, 184–185)

Despite all the nineteenth century work on the nucleic acids under the leadership of Albrecht Kossel, their chemistry was still far from clear when Phoebus Levene entered the field in 1900. During the ensuing three decades this great chemist lifted the subject out of its confusion and extricated the nucleic acids from their conceptual entanglement with the proteins. The book in which he surveyed the chemistry of nucleic acids appeared in 1931. It is a monument to his brilliant use of the techniques of organic chemistry: in particular in the case of the pentose sugars. But the price for this achievement was the acceptance of a simple structural formula in which the bases had a fixed location which the organic chemist thought he could identify.

In this chapter we shall see how this concept arose out of the discovery of only four bases in nucleic acid molecules. When combined with the molecular weight derived from the empirical formula, the result was the tetranucleotide. It will also be seen how Levene's successful application of the first structure for a nucleoside and nucleotide—that for inosine and inosinic acid—to RNA in 1909 lent powerful support to the hypothesis, which gained in explanatory power when Levene succeeded in extending the inosinic acid structure to DNA twenty years later. So firm did the grip of this hypothesis become that Levene's and Feulgen's later recognition of the macromolecular nature of DNA and RNA failed to shake it. The tetranucleotide theory offers a remarkable example of those paradigms which start off as working hypotheses in an ill-defined field: tools with which to open up the subject. They are successful and become entrenched so firmly that a revolution is required to dislodge them. In the case of the tetranucleotide there was a two-fold revolution. First came the improved techniques of extraction introduced by Hammarsten, followed by Mirsky and Signer, allied with reliable methods for estimating high molecular weights. Then came the application of chromatography to the nucleic acids by Chargaff. Before the work of Chargaff, nucleic acid chemists were not generally concerned that their base analyses often failed to approach the requirement for the tetranucleotide. They could always say that their procedures were too crude. Here, then, as in the case of the macromolecular concept, we see the powerful influence of innovations in technique.

In the meantime, the remarkable complexity of the proteins was increasingly being recognized, with the result that what we may call "The Protein Version of the Central Dogma" came into existence. According to this theory biological specificity flowed from the proteins, the subtle variety of which was due to their amino acid sequences.

It was not that nucleic acids were reckoned unimportant. Far from it. Kossel associated them with the formation of fresh protoplasm, with growth—synthesis. But for the repository of biochemical individuality he clearly judged the histones more significant, and around them he built a biochemical systematics and a theory of protein structure and synthesis. What, then, was Kossel's conception of biochemical individuality? How was the tetranucleotide hypothesis constructed? Why was it retained so long?

Kossel's Concept of Biochemical Individuality

The schoolboy enthusiast who had directed the botanical excursions when the *Gesellschaft für Naturforscher und Aerzte* visited his home town of Rostock, who had gone to the German university of Strassburg to sit at the feet of the famous botanist, de Bary, though he had been persuaded to read medicine rather than botany by his father, was not the sort of person to lose sight of the biological significance of his work. When Hoppe-Seyler won Kossel for physiological chemistry and put him on to the isolation of the products of the hydrolysis of nuclein, can we believe that Kossel ignored the biological implications of his chemical studies? Far from it. His papers show that such considerations were constantly lurking in the background, ready to be expressed the moment some reliable chemical data was obtained. His co-worker, S. Edlbacher, has said that Kossel was eager to move from the "statistical to the dynamic" (Edlbacher, 1928, 13) that is to say, from chemical analysis to metabolic pathways. Indeed this very concern for biological implications may well have been the chief cause of his continuing involvement in the search for a metabolic pathway from the paranucleins (phosphoproteins) to the nucleic acids!

Kossel concluded correctly from his physiological studies that nucleic acids act neither as storage compounds nor as energy sources for muscular contraction but are closely involved in the synthesis of fresh tissues. As early as 1885 he accepted Zacharius' evidence for the chemical characterization of chromatin as nuclein, " . . . a substance which", remarked Kossel, "besides albumin and phosphoric acid, yields hypoxanthine, guanine and xanthine" (1885, 1929). Eight years later he included the possibility that chromatin might be the free acid: "What histologists term chromatin is essentially a compound of nucleic acid with more or less albumin, to some extent it is also quite 'free' nucleic acid [*zum Theil ist es auch wohl 'frei' Nucleinsäure*]" (1893, 158).

In the essay he contributed to *Die Kultur der Gegenwart* on the relation of chemistry to physiology he wrote:

> The occurrence of nucleic acid is confined [*verknupft*] to the cell nucleus, indeed to a part of the nucleus which has been long known to the histologists by the name "Chromatin" on account of its tendency to take up basic dyes in contrast to the remaining morphological constituents of the nucleus. This fact is of major significance for the relationship of chemistry to the cell theory, for it gives us a chemical characterization for an elementary organ of the cell in addition to the morphological characterization. The knowledge of a peculiar building block, which forms the chromatin of the cell nucleus, must be considered the foundation for investigating the chemical activity of this organ. Essentially, however, this activity is to be sought in relation to growth and to the replacement of protoplasm.
>
> (Kossel, 1913, 383)

What an anti-climax for the twentieth-century reader, but what a clue to Kossel's inner thoughts! We shall remark on this way of thinking later on; it bears the hallmark of biochemistry before the 1950s—a limitation to metabolic pathways, which was a natural development from "statistical" to "dynamic" chemistry, and was a logical extension of organic synthesis from organic chemistry into physiological chemistry. This task of erecting a synthetic and degradative chemistry mirroring the events within the cell was the programme of the day, and the genetic information was a given fact of nature into which enquiry would have been deemed premature.

Whilst Kossel was happy to have his student Heine studying the chemistry of mitosis in the hope of establishing a parallel between the stainability of the chromosomes and the state of union of nucleic acid with histone—deeply staining with basic dyes when "dissociated" or loosely linked to histone, and faintly staining when "undissociated"—this can hardly be said to indicate a serious interest in mitosis as a mechanism for duplicating the chemical information of the chromosomes. No, it was, surely, an interest in the chemical interactions which stimulate division of the cell and the formation of fresh protoplasm.

On the other hand Kossel was the author of the *Bausteine* (building block) concept of cell chemistry according to which the metabolism of the cell can be resolved into the linking together and the separation of a number of primary molecules, the *Bausteine*, by which secondary complex molecules are synthesized and degraded. In the embryo more and more varied *Bausteine* are linked together to yield the complex proteins, in the sperm these have been removed, leaving the protamines—proteins in their *simplest* form. Kossel was particularly impressed by the large number of C=N bonds present both in the nucleic acids (in the purine and pyrimidine rings) and in the protamines and histones (in arginine and histidine). The fact that such compounds were produced in large quantities wherever growth took place indicated to him the special synthetic activity of these C=N bonds. Here the influence of Pflüger's polymer theory is noticeable.

But what did Kossel make of the arrangement of the *Bausteine* in the polymer molecule? Did he see chemical individuality in terms of chain sequences as did Emil Fischer? There is no doubt he did. Structural chemistry, he stated, is involved in the understanding of proteins: "But it is not a question of the relative position of the atoms in the molecule, but of the grouping of the molecules or *Bausteine* in the larger unit—the protein" (1913, 383). Understandably, he did not get far in his attempts to extract short chain polypeptides, but this line of attack, which we have seen advocated by Fischer, culminated at the end of the 1940s in the first sequence analyses. Kossel's faith in the existence of such arrangements of the *Bausteine* was based on the finding of constant quantitative relations between the cleavage products of a given protein. He had introduced the silver salt technique, and when silver became scarce in the 1914–18 war he evolved his naphthol-yellow-S method, and he used Van Slyke's new technique for estimating nitrogen. The resulting data, he claimed,

> . . . force us to the conclusion that the chemical structure of proteins with varying properties possess entirely different structures. We must also conclude that the atomic groupings

of the various cleavage products actually exist as such in the protein molecule; for on hydrolysis of the same protein they are always obtained in constant quantity.

(Kossel, 1912, 67)

This quotation is from Kossel's fine Herter Foundation lecture which he gave to the Johns Hopkins University in 1911. Even more striking is the following passage from the Harvey Lecture he gave in 1911. Speaking of the reconstruction of proteins in the body from degraded proteins in the food he said:

I should like to compare this rearrangement which the proteins undergo in the animal or vegetable organism to the making up of a railroad train. In their passage through the body parts of the whole may be left behind, and here and there new parts added on. In order to understand fully the change we must remember that the proteins are composed of *Bausteine* united in very different ways. . . . The number of *Bausteine* which can take part in the formation of the proteins is about as large as the number of letters in the alphabet. When we consider that through the combination of letters an infinitely large number of thoughts may be expressed, we can understand how vast a number of the properties of the organism may be recorded in the small space which is occupied by the protein molecules. It enables us to understand how it is possible for the proteins of the sex-cells to contain, to a certain extent, a complete description of the species and even of the individual. We may also comprehend how great and important the task is to determine the structure of the proteins and why the biochemist has devoted himself with so much industry to their analysis.

(Kossel, 1911, 45)

But what had become of the nucleic acids? Kossel and his students had isolated the purines and pyrimidines of yeast and thymus nucleic acid, they had shown that the oxypurines (xanthine and hypoxanthine) are produced by the deamination of guanine and adenine, they had found a remarkable degree of similarity between the nucleic acids of animal origin—invertebrate and vertebrate sperm, blood corpuscles, and thymus gland—and clearly marked differences between these and the nucleic acid from yeast. In 1902 Osborne and Harris prepared the first nucleic acid from a higher plant and showed it to contain, like yeast, the pyrimidine uracil, instead of thymine as in animal nucleic acid. Like yeast nucleic acid, it appeared to contain a pentose rather than a hexose (as the carbohydrate was thought to be in animal nucleic acids). Clearly there were at least two kinds of nucleic acids. Perhaps there were more; perhaps there was a guanylic, thymic and adenylic acid; if so there might be compound nucleic acids between such homopolynucleotides and yeast or thymus nucleic acid. At the close of the nineteenth century these possibilities were invoked to account for the curious nucleic acid obtained from the pancreas. At this juncture an aggressive young Swedish chemist, Ivar Christian Bang entered the field and, besides raising many questions about the identity of pancreas nucleoprotein, sought to discredit some of Kossel's work. Kossel found himself in the situation he had always so studiously avoided of having to enter into public controversy.

This quiet, introverted, Prussian from the Baltic port of Rostock, his personality hedged in by a wall of reserve, who spoke in public only with difficulty (Edlbacher, 1928, 4) may have been encouraged by the dispute with Bang to pay less attention to the nucleic acids and more to the proteins. On the other hand, Miescher had died in 1895 leaving the protamine of salmon an ill-defined substance. These substances and the histones first described by Kossel in 1884 offered him what the nucleic acids had failed to yield—a rich and varied source of *Bausteine*.

It was not that Kossel reckoned the nucleic acids to be very small molecules. To be sure he believed they contained one base and one sugar *Baustein* for each phosphorus atom, and that there were four different bases. Therefore, he concluded, the nucleic acid molecule contains twelve or a multiple of twelve *Bausteine* (Kossel, 1913, 383), and in his Nobel lecture he described nucleic acid as "a complex of at least twelve building blocks, but in the living cell the structure is probably larger, because some observations suggest that in the organs several of these complexes are combined with each other" (Kossel, 1910, 399). But how much richer were the protamines and histones in breakdown products, and how much more reliable appeared his silver–baryta method for estimating basic amino acids than the harsh techniques required to estimate purines and pyrimidines. It was in this direction that Kossel saw hope of achieving his treasured aim to represent growth and differentiation in terms of chemical transformations, and morphological distinctions in terms of chemical distinctions (see his monograph *The Protamines and Histones*, 1928, p. vii).

Kossel's most fruitful work was done in Berlin, and in his ten happy years at the university of Marburg. There Kossel lived in the professorial residence within the magnificent physiological institute which the Prussian minister of culture had built for Kossel's predecessor, Külz. Among his graduate students were the British chemist, T. H. Milroy, working on paranucleins, Heine studying the chemistry of mitosis, the American, A. P. Mathews, analysing fish and invertebrate nucleic acids and Kutscher working on protamines. It was also to Marburg that P. A. Levene and F. P. H. Steudel came. At this time Kossel still took a very active part in his students' work and entertained informally in his appartment in the Institute (Jones, M.E., 1953, 85). When asked why he left Marburg in 1905 for Heidelberg, he is said to have replied that he disliked the domineering ways of his countrymen, the Prussians (Mathews, In: Jones, 1953, 86). In 1913 he lost his wife; in 1914 the First World War isolated him from the international circle of science and from his relatives in the United States. Not for him was there any necessity to keep "Germany's place in the sun" (Dakin, In: Jones, 1953, 93) or to pursue an aggravating submarine policy. His introversion increased, his shame at the turn of events in Germany was deep. More and more did Herr

Petri, his chief technical assistant from the Berlin days, come to dominate the laboratory in Heidelberg, a man of "Hindenburgian facies, physique and methods" (Kennaway, 1952, 394). Kossel's approach and manner

> . . . were abrupt; he would click his heels together and say "Nun, wie geht es mit ihren Untersuchungen?" One would then ask his advice; any proposed course of action was pretty sure to elicit one of his favourite phrases—"Es hat Vorteile und es hat Nachteile". One got the impression that he was not interested in any compound, or for that matter in any research worker, which, or who, did not produce "sehr interessante Salze" or at any rate something crystalline.
>
> (Kennaway, 1952, 394)

Very differently did Albert Mathews remember his days with Kossel in Marburg. For him Kossel was a venerated teacher and beloved friend (Mathews, 1936, p. viii).

Kossel's Influence

The twin techniques of amino acid analysis and of purine and pyrimidine analysis were securely established by Kossel. He is said to have pictured protamines as chains of basic amino acids by analogy with the polysaccharide chains of starch, and when asked by Mathews why he had not attempted the synthesis of such chains from amino acids he replied that "he had considered the idea but had turned it over to Emil Fischer, who was better equipped to do that work and knew far more than he about organic synthesis" (Jones, 1953, 87). Just how much Fischer owed to Kossel for the stimulus to his great work cannot be judged from this statement. When we come to the question of the identity of the hereditary material we find no explicit discussion in Kossel's writings. *The quest for a chemical systematics was as far as he went.* It was this hope which motivated his suggestion to Mathews that he study the nucleoproteins of the invertebrate *Arbacia.* The sperm of so lowly an animal should surely have a simpler chemical constitution than that of salmon and herring. But no, Mathews found the nucleic acid of *Arbacia* to be just like that from fish sperm, and its protamine (arbacin) was not less but more complex in terms of the variety of its amino acid constituents! (Mathews, 1897, 411). As for herring sperm, the constitution of the head nucleoprotamine which he regarded as the chromatin of the sperm could be represented by the formula $C_{30}H_{57}N_{17}O_6 . C_{40}H_{54}N_{54}P_4O_{27}$. But according to August Weismann chromatin "ought to be the most complicated compound in the body, since it contains the stem molecules of all the other molecules [in the body]" (*Ibid*, 411). Accordingly the chromatin in the sperm head should be the most complex form of this substance, but in fact it is the simplest yet known! From enthusiasm for nucleic acid as the hereditary material Mathews moved to a position of extreme doubt. Consider the following passages which come from his contribution to Cowdry's much used *General Cytology.*

The structure which the cytologist calls a chromosome and in which most see the bearer of all hereditary traits and some even go so far as to imagine that each trait or character is represented by a distinct unit or gene, this structure as shown in the fixed dyed section is nothing else than a salt of nucleic acid with the basic dye which has been used to stain it. Whether in addition to the dye there is also present some protein matter cannot be definitely stated. What then is nucleic acid? Are there many nucleic acids or one?

The frequent discussion of the composition of chromatin, while very incomplete owing to the poverty of our knowledge, lends no support to the hypothesis that the chomosomes are made of genes. The best and most convincing proof of this theory which can be had from a chemical point of view is given by the examination of the chromosomes of the fish sperm. Here, where there should be the most complex chromatin in the body we find the simplest; namely a salt of protamine and nucleinic acid for which a definite formula may be given . . . Now, it is very improbable that were the chromosomes constituted of widely different genes they would show so simple and definite a composition. The nucleic acid of widely different cells appears to be the same. Of course it may be different, but the fact that it shows the same physical properties, analysis numbers, rotatory power, and so on indicates that there are probably not several different nucleic acids. Certainly not a vast number. We are, therefore, forced to seek the differences in the protein moiety of the chromatin if such differences exist. Now it is remarkable that nothing of the kind which one would expect on this theory is found in those sperms which we have examined. Always we find but one sort of protamin, containing only three or four different amino acids. While a considerable difference of kind of arrangement is possible even in this one sort of protamin, there is no chemical indication that it exists . . .

At the best, or worst, they [the chemical facts] give little if any support to this view. In the author's opinion, which is here given for what it is worth, the chromatin of spermatozoa is nothing else than the chromatin of spermatozoa; and that of an egg cell is the chromatin of an egg cell. They are not nerve, muscle, epithelial chromatin, in masquerade.

While when united they may lead to the formation of all the other chromatins of the body, or at least play an important part in their formation, neither should they be regarded as a museum containing samples of all the different products which it is capable of making. For since the number of these products is in fact infinite for each chromatin, as is shown by the differences in cells produced by any change in the conditions of development, by the accidents of existence, such as galls on plants etc., there is not room in the chromosomes for all these samples. But if the chromatin can make some other chromatins without having a sample to guide it, why not make them all? Why have any samples at all? Considerations of this nature will have different weight with different minds. And it must be remembered that the onus of proof is on those who assert that the chromosomes are such museums containing samples of all the chromatin of all the cells of the body, not only all the chromatins which develop during life, but all that infinite collection of old masters inherited from the past, and all the infinite number of descendants yet to appear in the eons before us, and presenting qualities usually said to be dormant. They are concealed no doubt in the chromosomal attic, ready to be produced when the occasion arises.

(Mathews, 1924, 75, 89, 90)

By 1936 Mathews had again changed his tune and now saw in the protein moiety of nucleoprotamines and histones the expected chemical differences between cells belonging to different tissues and between spermatozoa from different species. "Each species of fish has its own kind of proteins in the chromatin; and it is perhaps this difference in protein composition which makes them different species" (Mathews, 1936, 108). This expectation to account for genetic diversity in terms of the protamines was maintained by Kossel's student Kurt Felix, director of the Proteinforschungs Institut in

Heidelberg and a great authority on the protamines. But the geneticists seem to have been influenced chiefly by A. P. Mathews, whose views were publicized by E. B. Wilson, and by Strasburger. If we ask what lay behind Mathews' statements, the obvious answer would be the tetranucleotide hypothesis and the work of P. A. Levene.

The Tetranucleotide Hypothesis

With or without this hypothesis it should be evident from what has been said that the nucleic acids just did not appear sufficiently varied to account for biochemical individuality. These compounds showed no parallel with the contrast say between silk fibroin composed chiefly of the two amino acids, glycine and alamine, and haemoglobin containing all twenty amino acids. There is, however, a clear distinction to be drawn between the conception of nucleic acids before and after the consolidation of the tetranucleotide hypothesis. The latter represented more than a structure, it embodied a particular sort of approach to structural problems. Before its formulation there was present in discussions of nucleic acid structure the distinction between the gel-forming and the non-gel-forming sodium salts of nucleic acids which Kossel's student Albert Neumann had termed the *a* and *b* forms. Using his modification of Richard Altmann's method of extraction Neumann had achieved good yields of nucleic acid rapidly. The improvements introduced by Altmann, Kossel and Neumann were aimed at yielding really *protein-free* nucleic acid with which to settle the debate over the relationship between the nucleins, paranucleins and proteins. At the same time their improvements made nucleic acids easier to extract, so that by 1900 their chemistry had become a very popular subject.

Now to us Neumann's discovery of a non-gel-forming *b* nucleic acid which he recognized as a depolymerization product of the gel-forming *a* nucleic acid is highly significant. Our enthusiasm for his observations is increased when we read that he could control the degree of breakdown by the duration of exposure to heat. And we delight to find him relating this process to the similar observations in the proteins and polysaccharides thus:

"The gelatinizability of compound *a* can be explained easily by polymerization. Here a relationship appears to hold similar to that in starch, gum and in certain respects to that in the albumin compounds" (Neumann, 1899, 375).

Seven years later Richard Burian incorporated Neumann's conclusion in the following scheme for the progressive breakdown of nucleic acids:

1. Depolymerization (Burian's term was *Depolymerisierung*) nucleic acid *a* to nucleic acid *b*.
2. Removal of guanine to yield what Schmiedeberg called heminucleic acid and Neumann called nucleothymic acid.

3. Complete removal of purines yielding what Schmiedeberg called nucleotin phosphoric acid and Kossel and Neumann called thymic acid.
4. Total resolution of the depurinized molecule into its constituent elements.

(Burian, 1906, 784)

Earlier, Burian had discussed what the molecular weight of nucleic acid might be and concluded that although the empirical formula gave a value of 1272, Miescher's demonstration of the extremely slow diffusion of nucleic acid suggested that "possibly its formula must be multiplied" (Burian, 1904, 80). Sad to relate these discussions of depolymerization fell by the wayside, not to be revived again until Feulgen's work three decades later. Meanwhile, a very much simpler picture emerged with the work of Osborne and Harris and of Phoebus Levene who used harsh extractive procedures. The latter, in describing Kossel's and Neumann's improvements on Altmann's technique, made the following revealing remark: "Thus the old traditional fear of using energetic methods for the separation of the nucleic acid from the protein was abandoned" (Levene and Bass, 1931, 251). And in the same place he spoke of their earlier work being "under the influence of the traditional belief in the unusual lability of the nucleic acids, and they accordingly avoided the use of heat . . ." To be sure vigorous methods were and still are needed to separate both pyrimidines and purines from the parent molecule, whose structure, it seems, could only be established at the expense of its polymeric nature. Without the tetranucleotide hypothesis the awful confusion of the subject could not have been resolved. If only molecular weight estimations, basicity and the number of primary and secondary ionizable phosphoric acid groups had not been made on similarly treated material! This one-sided progress begins with the work of Osborne and Harris.

Triticonucleic Acid

All nineteenth century formulae for nucleic acids show four phosphorus atoms in the empirical formula, but the first to suggest that there are but four bases in the molecule associated with the ^{4}P was the famous American protein chemist, Thomas B. Osborne, who with Isaac Harris worked at the laboratory of the Connecticut Agricultural Research Station in New Haven and introduced the term Triticonucleic acid in 1902 for the substance they isolated in large quantities from wheat embryos. Since they noted its resemblance to yeast nucleic acid we may conclude that it was RNA, but it is puzzling to read their description of the embryonal tissue from which this RNA was extracted; the nuclei, they said, were so large as to leave no room for anything else in the cells (Osborne and Harris, 1902, 86).

These authors proudly announced that they had subjected triticonucleic acid to a thorough study which allowed them "to establish the probable configuration of the molecule". Unfortunately they had been forced to break off this work before bringing it to a final conclusion. Because they published their findings thus far in *Hoppe-Seyler's Zeitschrift*, edited by Kossel, all students of nucleic acids were thus appraised of their results which inevitably had a profound influence on subsequent developments, for as Jones remarked, there was a time when triticonucleic acid "stood as the most completely and accurately examined of all nucleic acids" (Jones, 1920, 49).

Of their ten elementary analyses they reckoned that the purest gave the empirical formula $C_{41} H_{61} N_{16} P_4 O_{31}$. On the basis of the silver salt they put its basicity at six. On the proportions of adenine and guanine they wrote:

> ... doubtless the bases occur in equimolecular proportions. The small differences [too much guanine] are due to the difficulties in separating the bases completely, since a certain amount of adenine is mixed in the guanine precipitant.
>
> (Osborne and Harris, 1902, 104)

Since the yield of these two bases per gramme of nucleic acid varied by as much as 9 per cent their results were surely inadequate as a basis for proof or disproof of equimolecular proportions.

Following Kossel's technique Osborne and Harris went on to subject their triticonucleic acid to 20 per cent sulphuric acid in an autoclave at 150–160°C for two hours. They obtained a pyrimidine which corresponded with the base uracil discovered in yeast nucleic acid by Kossel's student Ascoli in 1901, the melting point of which had been determined by Emil Fischer and Hagenbach that year. On the yield Osborne and Harris wrote:

> One molecule of uracil in the nucleic acid molecule corresponds to 8 per cent of the latter. In the experiments described above 11 per cent uracil was found, from which we must conclude that in the nucleic acid molecule at least two molecules of this substance are present. That the quantity found is much less than the 16 per cent required for two molecules, cannot carry much weight when one considers the long path of the preparation and the bulky precipitates of phosphotungstic acid and of barium hydrate.
>
> (Osborne and Harris, 1902, 109)

Their conclusion of 1 guanine: 1 adenine: 2 uracil molecules per nucleic acid molecule followed. Whilst admitting the presence of a substance in insignificant quantities, which might be the base cytosine discovered by Kossel and Neumann in 1894, they settled for only one pyrimidine in their material and computed its molecular weight as follows:

	Molecular weight
1 molecule of guanine	151
1 molecule of adenine	135
2 molecules of uracil	225
3 molecules of pentose sugar	450
6 hydroxyl groups	102*
4 atoms of phosphorus	124
3 atoms of oxygen	48
	1234
Subtract 7 atoms of hydrogen for phosphate linkages	
	1227

But 1397 was the true molecular weight: "So there remains an unresolved residue of 12.2 per cent, which probably belongs to the unidentified breakdown product, but we cannot as yet demonstrate this satisfactorily" (Osborne and Harris, 1902, 119). In other words they favoured a pentanucleotide containing: one molecule of adenine, guanine and cytosine (?) and two of uracil. In the formula (see Fig. 6.1) they put forward for triticonucleic acid, cytosine (?) was represented by an X.

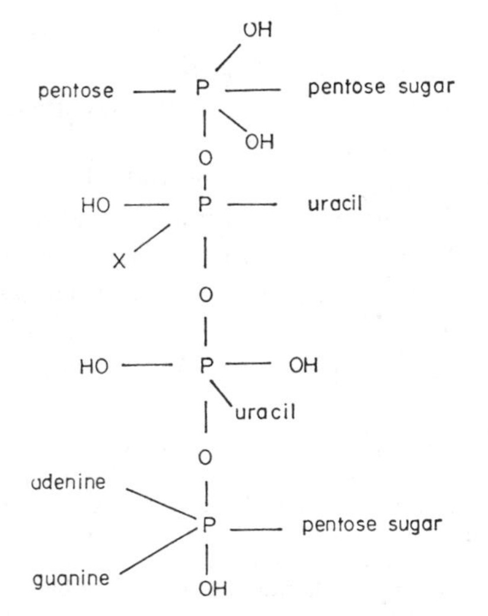

Fig. 6.1. Triticonucleic acid (according to Osborne and Harris).

This formulation—to our eyes extraordinary—allowed Osborne and Harris to account for the relative ease with which the purines can be split from the parent molecule, and for their incorrect observation that one phosphorus atom is removed in the process. To us their structure shows just how much work there was to be done before our modern conception of the linkages between the various groups could be achieved. Anyone who cares to read the closing section of their paper will not fail to see how strongly they were under the influence of the belief in a direct metabolic pathway between the true nucleins and the paranucleins. Their interest in the structure of nucleic acid appears to have been motivated by their desire to demonstrate such a relationship in the embryonic tissues of wheat—to do for the plant kingdom what Levene was attempting to do for the animal kingdom.

Cytosine in Triticonucleic Acid

Nearby, at the Sheffield Laboratory of Yale University, Henry Wheeler and Treat Johnson were working on the synthesis of uracil. They wanted to compare their synthetic product with Osborne's, so they evaporated the mother liquors from Osborne's crystallization—he no longer had any uracil crystals to send them—but the needle-like crystals they obtained were, they declared, "a mixture of apparently equal parts of uracil and cytosine" (Wheeler and Johnson, 1903, 505). This exposure of the impurity of Osborne's and Harris' uracil, their own recognition of the presence of an unidentified constituent X which *might* be cytosine, and their belief in the presence of two molecules of uracil to one of the purines did not prevent the establishment of the tetranucleotide, for other facts soon claimed attention and all that was remembered of triticonucleic acid was the clear impression given by Osborne and Harris that there were two purine and two pyrimidine molecules per four atoms of phosphorus—adenine and guanine being present in equimolar proportions. In short, the recognition that there are but four kinds of nitrogenous bases in a nucleic acid: for thymus nucleic acid: adenine, guanine, cytosine, thymine, for yeast nucleic acid: adenine, guanine, cytosine, uracil, invited the tetranucleotide formulation. To be sure, there was other evidence—the liberation of one secondary phosphoryl group for every four primary groups (Levene and Simms, 1925; 1926) was interpreted in terms of either an open four-residue chain (Levene and Sims, *Ibid.*) or a cyclic molecule which opened out into a chain (Takahashi, 1932) as in Figure 6.2. Early molecular weight estimates from diffusion gave 1360 to 1700 (Myrbäck and Jorpes, 1935), roughly in agreement with the tetranucleotide—1357. When Levene had taken the important step of advocating an ester phosphate–sugar linkage for the backbone of DNA (Levene, 1921) and for RNA (Levene and Simms, 1925) there were thus three independent lines of evidence in favour of the tetranucleotide.

One might ask what would the outcome have been if the polymeric nature of the nucleic acids had been recognized in the early days by a wider circle than that composed of Burian, Neumann and Miescher. I am sceptical, for is it likely that these workers had our conception of macromolecular size? It is far more likely that they were thinking in terms of what we today would call oligonucleotides. This was the sense in which Levene and Bass discussed the

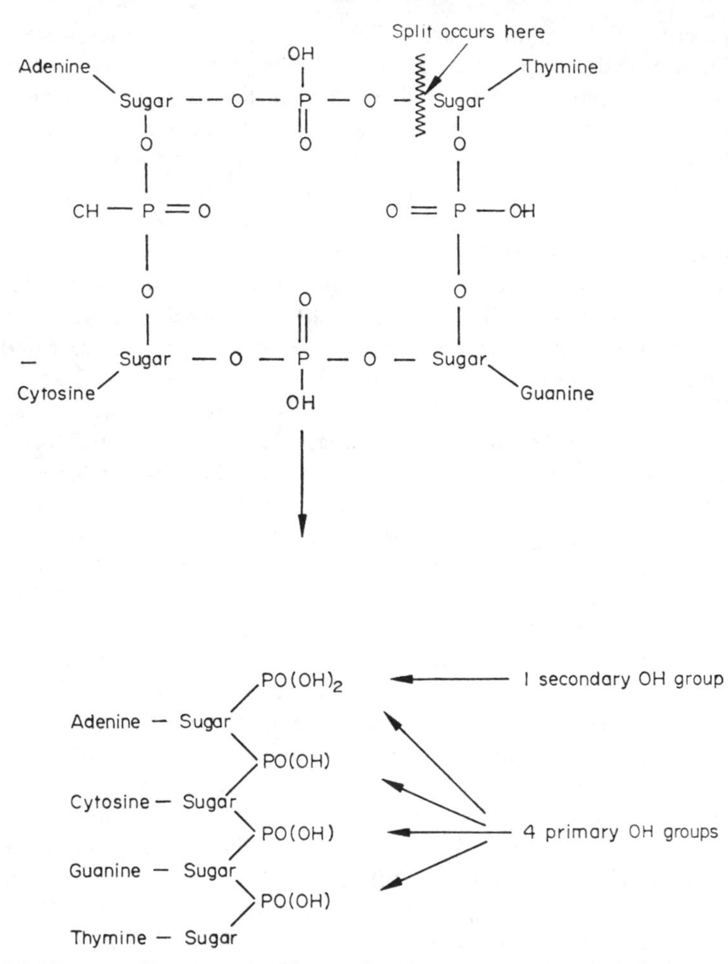

Fig. 6.2. How a cyclic tetranucleotide was thought to open up into a chain form with four primary and one secondary phosphoryl groups.

reports of Hammarsten and Feulgen on "higher order" nucleic acids in their monograph (1931, 302–303). Thus we have every reason for having paid so much attention to the macromolecular concept in Chapter 2.

Long ago, G. R. Wyatt pointed out how surprisingly good were some of the early analytical data on DNA composition based as they were on the differential solubility of the salts of the four bases. He noted they deviated to the same extent from the values predicted by Chargaff's base ratios as from those predicted by the tetranucleotide structure (Wyatt, 1950, 143). In Table 6.1 some of the figures obtained between 1905 and 1950 are given.

TABLE 6.1

Base analyses of nucleic acids from 1902 to 1950

Molar Ratios

Date	A	G	C	T/U	Authors	Nucleic acid
1902	1.0(?)	0.98	—	1.17	Osborne and Harris	RNA
1905	0.47	0.23	1.35	1.0	Levene	DNA
1906	1.30	0.98	0.63	1.09	Steudel	DNA
1908	0.98	0.98	1.02	1.02	Levene and Mandel	DNA
1909	1.0	1.3	1.7	?	Levene	RNA
1947	0.8–0.9	1.0	1.1–1.2	1.0	Gulland	DNA
1948	0.8	2.0	1.0	0.25	Chargaff	RNA
1949	1.6	1.3	1.0	1.5	Chargaff	DNA
1950	0.29	0.18	0.18	0.31	Chargaff	DNA

It was not Levene, but Steudel who first declared himself in favour of the equimolar proportions of purines and pyrimidines and used the tetranucleotide as a working hypothesis, but his data were ill-suited for it, his cytosine values being only 54 per cent of those required by the structure (Markham and Smith, 1954, 6). Unfortunately Levene, after suggesting two pyrimidines to three purines in 1908, came over to Steudel's view in 1909, and in support thereof cited Steudel's theoretical values (the expectation for a tetranucleotide) as Steudel's experimental results (Markham and Smith, 1954, 7), repeating his error in 1931 when in his monograph he declared that "the tetranucleotide structure has been substantiated by Ellinghaus on the basis of an entirely different and ingenious physical method, namely, by comparing the heats of combustion of the separate components of nucleic acid with that of the nucleic acid itself" (Levene and Bass, 1931, 277). One wonders whether in fact Ellinghaus' results were consistent with either a tetranucleotide or a huge macromolecule.

Finally, we must not omit a reference to Levene's model for the arrangement of phosphate, sugar and base in nucleic acids—inosinic acid and inosine. In 1909 he had argued that the most likely structure of inosinic acid was that of a purine attached by a glycosidic link to the sugar, this being in turn linked through a hydroxyl group to phosphoric acid by an ester bond (Levene and Jacobs, 1909a, 336). Two years later he put forward the structure illustrated in Fig. 6.3, in which the link between sugar and base is wrongly put at position N^7 instead of N^9.

It was also in 1909 that Levene and Jacobs isolated guanosine and adenosine from yeast nucleic acid (1909b; 1909c) and stated their belief that the parent molecule consisted of the four nucleosides linked by associated phosphate

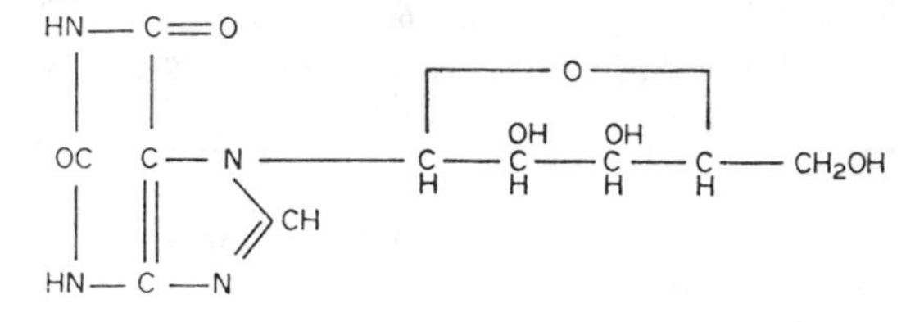

Fig. 6.3. The structure of inosine (after Levene and Jacobs, 1911).

groups. Such a tetranucleotide he termed a "polynucleotide", and he favoured a similar structure for thymonucleic acid.

We may call the Levene and Jacob's formulation of the nucleic acid molecule at this time their "four-nucleoside" structure, in which the wrong backbone linkage was assumed, as in the dinucleotide in Figure 6.4.

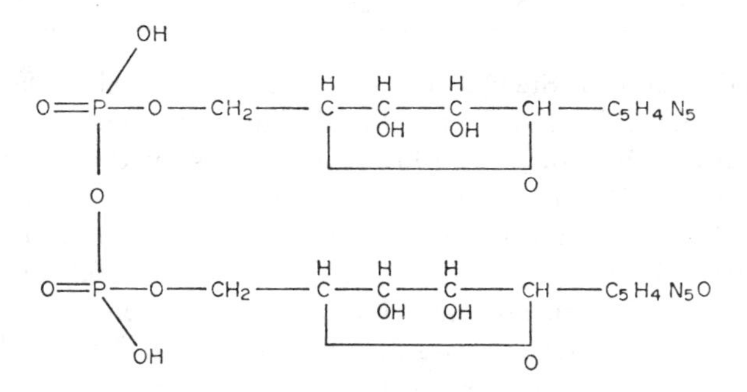

Fig. 6.4. A dinucleotide from yeast nucleic acid (from Levene and Jacobs, 1909c, 2703).

Not until Levene had obtained all four *nucleotides* from yeast nucleic acid did he introduce the present-day sugar-phosphate backbone for RNA (Levene, 1919). If he could have demonstrated the production of the four nucleotides in DNA at this time he would have succeeded in extending this structure to both acids. But the instability of DNA compared with RNA, which Levene and Jacobs attributed rightly to the sugar (1912), foiled Levene until 1929.

Plant and Animal Nucleic Acids

Recent commentators have found a source of amusement in the statements of Jones that the nucleic acids were "possibly the best understood field of Physiological Chemistry," and, "... there are but two nucleic acids in nature, one obtainable from the nuclei of animal cells, and the other from the nuclei of plant cells" (Jones, 1920, pp. vi, 9). The first remark refers to the knowledge of the metabolism of nucleic acids, in particular to their catabolism. Jones' second remark must be set against the confused background of the early literature on nucleic acid structure and variety; there were paranucleic and thymic acids (Kossel and Neumann, 1893), guanylic acid of the pancreas (Bang, 1898), the nucleotin radicle (Schmiedeberg, 1900). Then there were the debates over the primary or secondary nature of the pyrimidines—Burian long maintaining that they originated from purines—and more especially of uracil, was it produced from the deamination of cytosine? By the time that Jones wrote his remark, only the guanylic acid of the pancreas remained unresolved.

In the end it was work on this acid by E. Jorpes which broke the plant animal classification. Levene, with whom Jorpes worked in 1929 as a Rockefeller fellow, noted this point in his 1931 monograph and yet in his classification of nucleic acids he placed at the head of his texts the two columns:

Yeast nucleic acid	Thymonucleic acid
(Phytonucleic acid)	(Zoonucleic acid)

(Levene and Bass, 1931, 260)

Although we might account for this lapse by suggesting that the book had been written over a period of years, having been started in 1925 (Levene, 1925), the very fact that Levene did not remove it at a later date gives us a hint of the ingrained state of this classification. In 1920 Einar Hammarsten had stated the pentose character of the sugar in pancreas nucleoprotein, his case having been progressively strengthened by his student Jorpes (Hammarsten and Jorpes, 1922; Jorpes, 1924; 1928), and demonstrated conclusively at the Rockefeller Institute (Levene and Jorpes, 1930) before Levene's monograph was published. Do we see here the alluring character of simple assumptions or the fulfilment of the biochemists' desire to establish the separation between the plant and animal kingdoms on a chemical basis? Or was it a result of the intense relief at having cleared up the muddle over the various kinds of nucleic acid in nature? Should the man who put the weight of all his authority behind this classification as late as 1931, and behind the tetranucleotide structure until 1938, when he went over to the poly-tetranucleotide structure, be labelled with these two notorious assumptions?

Phoebus Aaron Levene

It seems hardly possible that a man like Levene could go down in history as the chief supporter of an unsophisticated structure and classification of nucleic acids. This highly intelligent young Russian Jew from Alexander Borodin's Chemical Institute in St. Petersburg (Leningrad) had come to America with his family in 1891 at the age of twenty-two, had been appointed associate in physiological chemistry in 1896, and chemist at the Saranak Laboratory for the study of tuberculosis, in New York State, where he had earlier been a patient. Finally, at the invitation of Dr Simon Flexner, director of the new Rockefeller Institute for Medical Research, he found a haven, first in the Institute's house in 127E 50th Street, which had been in use just over a year, and in 1906 on the first floor of the new Founder's Hall on the present East River site. In this position of unrivalled privilege, Levene's researches expanded year by year, as he took on fresh research programmes, obtained larger and larger grants and responded to the growing demand from young scientists to come and work under him at the Institute.

Although Levene and Flexner remained close friends, the Russian scientist's autocratic appropriation of funds irritated the American director and caused friction between them. This is seen in the correspondence preserved at the American Philosophical Society Library, in which passages like the following occur in Flexner's letters to Levene:

> At the meeting of the Executive Committee last night it was concluded not to increase your staff this year.
>
> (letter dated 3. vi. 1916)

> I would like to see you concerning the supplies and expenses of your department.
>
> (27. iv. 1918)

> I can't say . . . that I am greatly affected by the distinction between yourself and your facilities. It is poor judgement that attributes real advances to mere alcohol, ether, etc. . . .
>
> (7. ii. 1919)

> If the excessive inroads on it [your regular budget] are due to over purchase of yeast and perhaps other things in view of a possible failure in production [due to the war] you must have foreseen that the unexpected outlay would embarrass your budget. . . . You are embarrassed and you are embarrassing the Executive Committee which asks that you explain how you propose to complete the year's work. The storeroom will not honor any requisitions of any department when the budget allowance is exhausted. . . .
>
> (17. ii. 1919)

> I want to clear my mind of all doubts hereafter about your placing orders for items not contained in your budget and for which no appropriation has been made. . . . After nearly twenty years of experience of the methods of purchase you should I think know that I have as little power as you have to spend money not in hand.
>
> If I told you to place the order [for a drying oven costing $600], I exceeded my authority. I am determined this shall not happen again.
>
> (19. vi. 1923)

> There is not one rule for certain persons and another rule for other persons. . . . It is to no purpose to consider what happened in Germany.
>
> (22. x. 1924)

> It is a pity to take on more workers than you can support.
>
> (30. ii. 1931)

By 1929 Flexner had become deeply concerned at the multitude of programmes which Levene was running. Whereas in the past, Flexner wrote, "you took up first one [problem] and then another, pursuing them while they were profitable and then dropping them in favour of something else", now he found Levene carrying on with an "orientating" piece of work on vitamins for five years. "Within this period you would naturally have discovered whether it is worth your while to make a larger attack on the problem" (Flexner, 1929). Another piece of expensive work which had been worrying Flexner in 1929 and 1930 was Levene's enzymatic study of pancreas nucleic acid (Bang's so-called guanylic acid) and Levene's attempt to identify the sugar in thymonucleic acid. Flexner pointed out that between 1926 and 1930 Levene's department had cost the Foundation nearly $343 000! Levene's own salary in 1929 alone was $16 000, and between 1929 and 1930 his total departmental expenses had increased by $10 000. It was with some relief that Levene could tell Flexner of his success in at long last identifying the sugar in thymonucleic acid as 2-desoxyribose and in June 1929 he wrote:

> ... there is hope that in another year all the remaining details will be worked out. The financial end was shocking even to myself—the bill for sweetbreads alone was $1500 and the fitting up of Dr London's operating room consumed $250. The total laboratory expenses not including Dr London's salary were surely above $3000.

But what a year was 1929 for Levene. With Jorpes, who had come over from Hammarsten's laboratory in Stockholm, he collaborated to obtain base analyses and nucleosides from the much-disputed nucleic acid of the pancreas. This substance which Jorpes had claimed in 1928 as a genuine ribonucleic acid—the first to be identified in animal tissues—gave Levene the ribosides of guanine and adenine and made him sure that pancreas nucleic acid is a polyribonucleic acid (or in modern terminology an oligoribonucleic acid).

With Professor E. S. London, from Leningrad, Levene carried out enzymatic studies which led to the identification of the sugar in thymus nucleic acid as desoxyribose. This grand achievement had eluded him in 1912 in New York and in 1924 in Leningrad. On both occasions he had carried out physiological digestion of nucleic acid in a dog's intestinal fistula, hoping thereby to achieve nucleic acid hydrolysis under mild and controlled conditions, the agents being the nucleotidases and nucleosidases which he and Medigreceanu had discovered in 1911. In Pavlov's laboratory in 1924 Levene had failed to achieve thymonucleic acid breakdown, but when London came to the Rockefeller Institute four years later he prepared several dogs with gastric and intestinal fistulas such that Levene was able to pass thymonucleic acid through the stomach and collect it from the intestine. By using only a short segment of the gut in this way Levene and London obtained not only the nucleosides of the two purines and the two pyrimidines but also their constituent sugar. This turned out to be neither a hexose as had been widely held hitherto, nor glucal as Feulgen had suggested, but identical with the sugar first isolated by Kiliani in 1895 and for which he had devised the "pine-stick" test, and a test with ferrous ions and sulphuric acid (Kiliani, 1896).

Levene and London failed to form the osazone of this sugar—as Kiliani had failed before them—this being due to the absence of the hydroxyl group in position C_2. Their analyses were therefore carried out on the benzylphenylhydrazone. This gave the empirical formula for the desoxyribose derivative.

Also in 1929, with T. Mori, fellow of the International Education Board, Levene compared the desoxyribose sugar from thymonucleic acid, which he called provisionally "thyminose", with three synthetic 2-desoxyribose sugars. All four sugars gave a positive Schiff reaction and were converted by sulphuric acid into levulinic acid, thus accounting for the discovery of this acid in thymonucleic acid breakdown products.

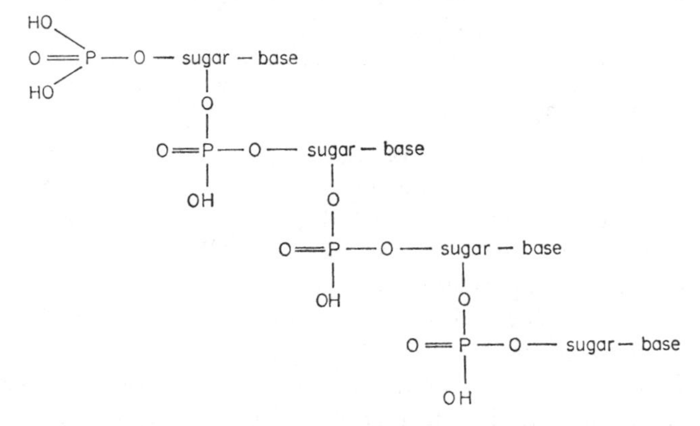

Fig. 6.5. Levene's 'Ester' structure for the tetranucleotide of RNA and DNA (from Levene, 1919 and 1921).

The importance of these achievements can be appreciated when it is recalled that a great nucleic acid chemist like Feulgen, discouraged by failures of chemists to extract nucleosides and nucleotides from thymonucleic acid using the techniques employed for yeast nucleic acid, "expressed the firm conviction that this compound [DNA] was built on a plan entirely different from that of plant nucleic acid" (Levene, 1931, 264; Feulgen, 1923, 296).

Thus did Levene crown his earlier achievements in the isolation of nucleosides and nucleotides and provide the unshakable evidence for the conception of thymo- and yeast nucleic acids as linear chains of purines and pyrimidines linked to each other through the sugar–phosphate–ester backbone as in Figure 6.5.

Gone was the phosphate–ester backbone of Levene and Jacobs (1909c, 2703) and of Steudel (1912) and the polysaccharide backbone of Jones and Read

(1917), for such structures would not yield the four nucleotides and nucleosides. The ingredients for this achievement were Levene's combination of enzymatic tools—here relying on the skills of Pavlov's laboratory—with his own sophisticated organic analysis of the breakdown products. Here his mastery of the techniques of stereochemistry was invaluable. Now surely so astute a chemist as Levene must have come across evidence of depolymerizing activity in his enzyme preparations, and so must have realized that native thymonucleic acid is a high polymer? This is so, although it should be stated that the final stimulus to pursue the early indication he had of such activity was probably given by a report from Stockholm in *Nature* of molecular weights in the region of 500 000 to 1 000 000 (Signer, Caspersson and Hammarsten, 1938).

A Depolymerase for Thymonucleic Acid

It will be recalled that Levene's failure in 1909 to isolate nucleosides and nucleotides from thymonucleic acid, although he succeeded from yeast nucleic acid, stimulated his continuing interest in nucleotidases. In 1930 R. T. Dillon and he were surprised to find that if they used precipitated intestinal preparations of nucleases instead of the raw juice the liberation of phosphate from thymonucleic acid varied, one sample working at only one tenth the rate of the others, despite its vigorous activity on glycerol phosphate. This led Levene to surmise the absence in this sample of a "nuclease" [a depolymerase] present in the others and to conclude tentatively that the degradation of nucleic acid to nucleotides must precede dephosphorylation (Levene and Dillon, 1930, 756). Eight years later Levene and Schmidt showed that the purer the nucleotidase (which liberates phosphate from nucleotides) and the more carefully prepared the thymonucleic acid, the less was the dephosphorylase activity. Using the ultracentrifuge with the help of his colleagues, J. H. Bauer and E. G. Pickles, Levene showed that nucleic acid prepared by Neumann's technique (*a*-thymonucleic acid) had a molecular weight between 50 000 and one to two million, whereas Feulgen's enzymatic preparation did not sediment at all. Levene concluded that Neumann-type nucleic acid was a mixture of different chain-length molecules whereas Feulgen-type material was highly degraded. Levene rightly emphasized the importance of this discovery and in *Science* he wrote:

> . . . This find, then, is of significance not only because it brings out an additional step in the process of biological catabolism of nucleic acids, but also because it furnishes a means of testing the purity of "native" nucleic acid, on the one hand, and of testing the purity of a nucleophosphatase by means of the native nucleic acid, on the other.
>
> It will be of significance also in connection with other questions bearing on the structure of nucleo-proteins . . .
>
> (Schmidt, G., and Levene, 1938, 173)

At last Levene was in harmony with the old conception of the polymeric *a* form of thymonucleic acid* and its degraded product, the *b* form, described by Neumann in 1899 and confirmed by Feulgen in Giessen from comparative studies of optical activity and gel formation (Feulgen, 1914). Feulgen further demonstrated the conversion of form *a* to form *b* by the action of a commercial preparation of the pancreatic juice. Because it destroyed the gel-forming activity of the *a* form he had called the active principle in this extract "nucleogelase" (Feulgen, 1923, 276 and 1935, 237).

Nor was this the only evidence of depolymerization in the nucleic acids. From far nearer home than Stockholm and Giessen—in the Rockefeller Institute itself—Levene was made aware that ribonucleic acid is easily depolymerized by an extract from the pancreas. This find was the work of René Dubos and R. H. S. Thompson, fellow of the Society of Apothecaries of London. The clue came from Dubos' and Colin MacLeod's discovery that heat killed pneumococci lose the ability to stimulate the production of antibodies and that this is associated with the conversion of the bacteria from gram-positive to gram-negative. From a number of tissues they isolated an extract which brought about this change of staining property, and of all the pure compounds tested for susceptibility to its degradative action only the yeast nucleic acid supplied by Levene was attacked (Dubos and MacLeod, 1937, 697). Dubos called the enzyme in question polyribonucleotidase (Dubos, 1937, 550) and concluded that the staining change was associated with the destruction of RNA. The detailed paper describing this work appeared at the same time as Levene's and Schmidt's full paper on desoxyribodepolymerase (Dubos and Thompson, 1938). Dubos can recall being invited to have a sandwich lunch with Levene in his office so that the great nucleic acid chemist could learn about Dubos' enzyme (Dubos, 1969). As so often happens in such cases there existed an earlier report of such a find. Thus Walter Jones had reported that an extract of pancreatic juice acted on yeast nucleic acid (Jones, 1920) but subsequent attempts to repeat his work had failed.

Now, if Dubos, Feulgen, Caspersson, Signer and Levene all recognized thymonucleic acid in its native state as a high polymer, in the modern sense of that term, what became of the tetranucleotide? It survived intact as the basic unit which by polymerization formed the polymers in native nucleic acids. In 1938 Schmidt and Levene wrote that "Native ribonucleic acid is a polymer of the tetranucleotide" (Levene and Schmidt, 1938, 425). And of thymonucleic acid they commented that: "Complete depolymerization of the [native] acid to a single tetranucleotide has not yet been accomplished by a chemical means" (Schmidt and Levene, 1938, 173). Clearly, they thought

* There is an almost reluctant admission of the possible polymeric state of DNA in Levene & Bass, 1931, p. 289.

it feasible. Feulgen was confident that: "All the evidence cited speaks for the fact that *b* thymonucleic acid, as obtained using nucleogelase, still shows the tetranucleotide structure, and that the action of the ferment is a depolymerization . . ." (Feulgen, 1935, 266). Finally, lest the reader imagine that Feulgen had obtained base ratios in support of this conclusion, it is worth noting that the yields of bases he recorded from *a* and *b* thymonucleic acid in 1935 were 0.5 G:0.9 A:0.8 C:1.9 T! Needless to say, Feulgen had not worked out these molar ratios. His concern was to show that the base constitution of *a* and *b* forms was the same.

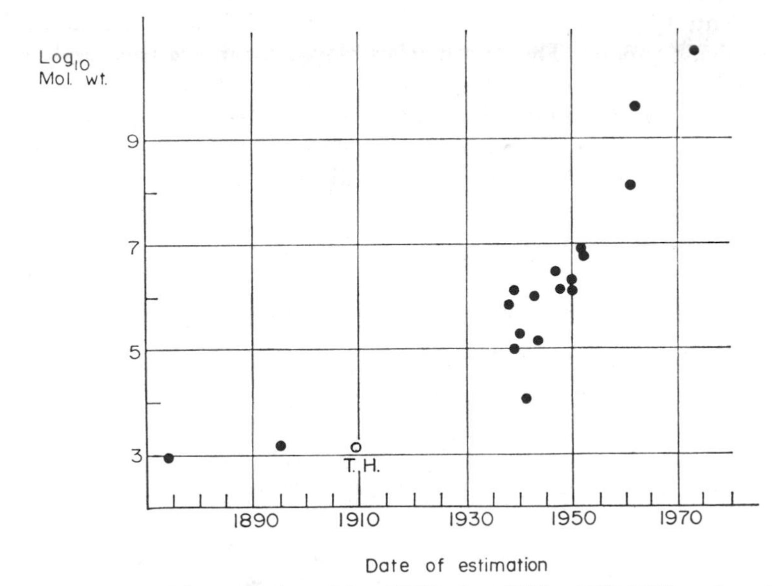

Fig. 6.6. Estimates of the molecular weight of DNA from 1874 to 1973 (T.H. = the value for the tetranucleotide).

The graph of molecular weights for DNA from 1874 to 1962 (Fig. 6.6) shows how rapidly the estimates changed between 1935 and 1940. DNA and RNA now became macromolecular species, but the Protein Version of the Central Dogma remained essentially unshaken, for as Gulland remarked, the polytetranucleotide "would limit the potential number of isomers and hence diminish the possibilities of biological specificity" (Gulland, 1947, 12). In 1944 this great nucleic acid chemist had favoured the polytetranucleotide "as a practical working hypothesis which is helpful in the interpretation of experimental results, provided its limitations are fully realized" (Gulland, 1944, 216). Sound evidence, on the hypothesis, he said, was nonexistent, and a year later he felt the necessity of introducing the term "statistical tetranucleotide", to describe a polynucleotide with base composition equal to that for a tetranucleotide, but with a "random" base sequence (Gulland, Barker and Jordan, 1945). By 1946, when he addressed the symposium on nucleic acids organized by the Society for Experimental Biology, he reckoned he had titration data which could be interpreted as conflicting with the assumption of molar equality of the bases in DNA.

The conclusion I suggest, is justified, that independent of the work on bacterial transformation (see Chapter 12) the revolution in chemistry which led to the recognition of macromolecules, and was in turn to focus attention on chemical sequence, would in time have undermined the Protein Version of the Central Dogma. But the chemists were not left alone to achieve this process single handed. The cytochemists played a decisive part, and as we shall see in the next chapter, the effect of their work was to achieve only a *partial recognition* of the biological importance of DNA, and to protect the Protein Version of the Central Dogma from complete overthrow until the work of Avery stimulated them to reassess their data.

CHAPTER SEVEN

The Nucleoprotein Theory of the Gene

Each chromosome has a definite and constant morphology and is made of segments, each of which has a characteristic pattern of chromatic lines or broader bands . . . This discovery places in our hands, for the first time, a qualitative method of chromosome analysis and once the normal morphology of any given element is known, by studying chromosome rearrangements of known genetic character, we can give morphological positions to gene loci and construct chromosome maps with far greater exactness than has been heretofore possible.

(Painter, 1933, 585)

Cell genetics led us to investigate cell mechanics. Cell mechanics now compels us to infer the structure underlying it. In seeking the mechanism of heredity and variation we are thus discovering the molecular basis of growth and reproduction. The theory of the cell revealed the unity of living processes; the study of the cell is beginning to reveal their physical foundation.

(Darlington, 1937, 562)

The obvious procedure of applying to chromosomes the methods which have yielded information about other biologically less important molecules, and so attacking the problem from the start on fundamental lines, must wait upon development of technique which may easily take decades. If, therefore, our colleagues in Stockholm are able to give us information as to the varying distributions of proteins and nucleic acid during the mitotic cycle, we should welcome it with open arms as giving data which are very urgently required.

(Wrinch, 1935, 389)

The Copenhagen Conference [1938] was a great success. I even did not expect such a success from the *first* meeting. The whole group is very nice and we really worked hard. Everybody, including the physicists, was very enthusiastic. What the results of the discussions were you will see soon from our report, which is being written. I don't doubt that the next meeting will be still more interesting and efficient.

(Ephrussi, 1938)

Attempts to found the specificity of nuclear and cytoplasmic stains on a chemical foundation continued long after the death of Miescher, but fell short of success. Miescher, who had attached so much importance to this task, not only failed to disperse the clouds of uncertainty but further confused the issue. He claimed that the clear specificities obtained by such men as Ballowitz, Schweigger-Seidel and La Valette St. George with spermatozoa were not observed in his own work. He was suspicious of the "guild of dyers" as he called the cytologists (Miescher, 1897, i, 107–108), and would surely have applauded Alfred Fischer for his criticism of the chemical theory of staining specificity (1899) and Sir William Hardy for his exposure of protoplasmic structures as artefacts of fixation (1899) had he lived to see their work. It is not surprising, therefore, that cytochemistry failed to make a substantial contribution to knowledge of the biological function of DNA in the cell until a new technique was introduced, the specificity of which did indeed have a chemical basis, one which had to do with the oxygen atom at position C_2 of the sugar ring in DNA. This was the nucleal reaction which Robert Feulgen had devised for extracted nucleic acid in 1914 and applied to tissues nine years later. The demonstration of this new technique in Tübingen in 1923 greatly impressed the veteran nucleic acid chemist, Albrecht Kossel (see Felix, 1957), but does not appear to have been taken up by cytologists on the chemical side when it was published (Feulgen and Rossenbeck, 1924).

Already, in 1914, Feulgen had concluded that his nucleal reaction involved exposure of aldehyde groups which he located in the sugar rings. He thus paved the way for the work leading to the identification of the sugar in DNA as of the desoxyribose type. In 1923 he opened a way to establish the distribution of DNA and RNA in the cell. But he did more, for many years later Caspersson succeeded in making the Feulgen reaction quantitative so that the DNA content of nuclei could be followed during the life of the cell. The result was the confirmation of the long held belief in a cycle of synthesis followed by loss of DNA, as Heidenhain had pictured it in 1907. Because the synthesis of DNA was associated with the onset of cell division and chromosome duplication the Protein Version of the Central Dogma was modified to give an important role to DNA. This substance was now held to be necessary for the synthesis of new protein molecules of the gene. This involvement of DNA was pictured by Astbury in terms of a stretcher frame upon which the existing polypeptide chain of a gene was held in its extended form, in which position it could act as a template for the laying down upon it of an identical chain. In 1931 he perceived that this identity could be achieved provided mother and daughter chains ran in opposite directions. The resulting ordered structure would possess C_2 symmetry* (Astbury, 1931). Later he was to suggest that DNA acted as a "negative" between two "positives"—the polypeptide chain of the gene and that of the daughter gene (Astbury, 1939b). Strangely enough the latter idea did not also raise the question of a nucleotide sequence to match the amino acid sequence of the genic polypeptide.

In a similar way the independent work of Caspersson and of Brachet on the role of RNA in the synthesis of proteins in the cytoplasm assumed that here again the nucleic acid acted as a "midwife" for the reproduction of the all

* This is line symmetry, not space symmetry.

important proteins. Fundamentally, therefore, chromosomal and cytoplasmic reproduction were the same—they both concerned the duplication of polypeptide chains. This hypothesis for the role of DNA and protein we may express in the term "Nucleoprotein Theory of the Gene". Its strength lay in the fact that both DNA and protein had important roles, and the genetic specificity of the amino acid sequence in the polypeptide chains was retained. The thinking of biologists and chemists was much influenced by this theory when Avery's work was announced. In this chapter we shall survey the evidence which lay behind it.

Hammarsten and Caspersson

The most convincing evidence for a cyclical process of synthesis and loss of nucleic acid in the nucleus was provided from Stockholm. There the Hammarsten family had been involved in physiological–chemical research since the time of Olof Hammarsten. Only two of Olof's papers were on nucleoproteins, but his nephew, Einar Hammarsten, aided by his brother, the Gymnasium teacher Harald, devoted many years to the physical chemistry of thymonucleic acid, its viscosity, sensitivity to electrolytes and pH, affinity for proteins, and its osmotic pressure. In the course of this work Einar Hammarsten improved on Ivar Bang's extractive procedure, and by 1924 could describe his product as "snow-white and of a peculiar consistency like gun-cotton" (Hammarsten, 1924, 386). Since at no time during extraction did he allow the temperature to rise above 1 °C, and the pH to depart from neutrality; since he avoided the use of alkalis and acids, his precipitation of the sodium salt of thymonucleic acid in the form of white fibres was to be expected. Here then was work of a character radically different from Levene's. Whereas Levene unravelled the structure of the tetranucleotide, Hammarsten revealed the physical properties of the less degraded molecule, thereby rediscovering the extreme lability which Miescher had observed in salmon nuclein fifty years before. Levene worked within the discipline of organic chemistry and asked questions appropriate to that discipline, Hammarsten in that of physiological chemistry, where a tradition had been set by Miescher, Kossel, Neumann and Hammarsten's uncle, Olof. Unfortunately Einar's work was also in the discipline of colloid science. He did not, therefore, attribute low osmotic pressure to high molecular weight alone, but also to the effect of high molecular volume on cations of low molecular volume, these being as it were engulfed by the large molecules and thereby shielded from exerting their influence on the surrounding solution. And when he spoke of high molecular volume he had in mind a molecular weight of 1545—little larger than that of the tetranucleotide.

As a physiological chemist, Hammarsten wanted to relate his findings to the biological functions of the nucleus. Here he saw the relationship of nucleic acid with water, pH and proteins as significant. Perhaps it would account for the varied degrees of acidity of the nucleus in contrast to the fairly constant neutral reaction of the cytoplasm. He seems to have belonged to the "mechanical" school of Jaques Loeb and Sir William Hardy who saw in electrolytes, water and colloids the key to vital phenomena. The extreme viscosity of DNA therefore intrigued him, just as it had intrigued Miescher in the nineteenth century, and he looked for "a direct parallel between the role of lactic acid in muscles and that of nucleic acids in the cell nuclei, with respect to the very great influence of both upon the lability of water fluctuations" (Hammarsten, 1924, 385).

Hammarsten did indeed observe large changes in viscosity when the pH of DNA solutions was altered, and this he attributed to alterations in the degree of "hydration", the greater the hydration the greater was the influence of the nucleic acid molecules upon their aqueous surroundings, and thus the greater was the viscosity of their solutions. We have seen that in 1938 the nucleic acid prepared by the Hammarsten–Bang procedure in the Karolinska Institute was recognized as a macromolecular compound and not a mere aggregate or micelle of tetranucleotides. But this conclusion was virtually *forced* upon the Stockholm scientists. Hammarsten had begun by considering the "tremendous viscosity" of DNA as due to some "restrictions" on molecular motions, such as an oblong asymmetric molecule would yield. Runnström was able to demonstrate such asymmetry to Hammarsten from birefringence studies in 1927 and six years later Pederson used The Svedberg's ultracentrifuge to re-examine the question. Although the DNA behaved as a polydisperse system, the molecular weights obtained between pH 4 and 10 gave a value somewhere between a half and one million. (Many years later the figure was reduced to 250 000 (Svedberg and Pederson, 1940).) Yet Hammarsten, as late as 1939 vacillated between according to these results the significance of a particle, micelle, or molecule (Hammarsten, 1939).

Meanwhile Caspersson was trying to remove all traces of turbidity in his nucleic acid solutions in order to measure their opacity after treatment with Feulgen stain. Filtration, instead of retaining protein and leaving pure nucleic acid in the filtrate, retained the nucleic acid and allowed the lower molecular weight protein to pass through. In puzzled tones Caspersson wrote that "the complexes of the nucleic acid must be larger than the protein molecules . . . the ratio between the radii of the dispersed protein particles and of the nucleic acid molecules must be of a quite different order of size from that expected at the outset" (Caspersson, 1934, 162). Whilst he wrote of "Complexes", privately he began to entertain the idea that he might be dealing with a high polymer. Meanwhile, his quantification of the nucleal reaction by optical means had led him in another important direction.

The Nucleic Acid Cycle

About the year 1932, Caspersson tried illuminating a nucleic acid solution with ultraviolet light and he recorded the transmitted spectrum on a photographic plate. This, when developed, showed a pronounced absorption peak in the region 2600 Å (see Figs. 7.1 and 7.2). Unaware of previous reports,

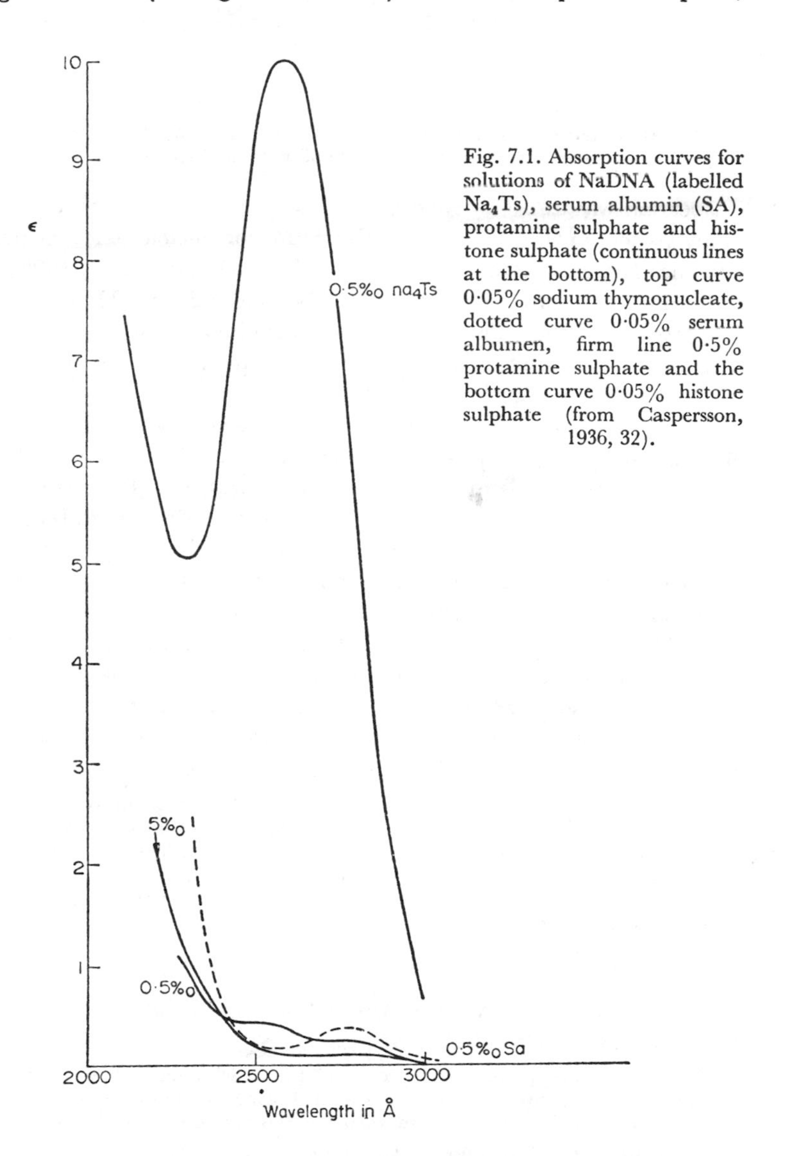

Fig. 7.1. Absorption curves for solutions of NaDNA (labelled Na_4Ts), serum albumin (SA), protamine sulphate and histone sulphate (continuous lines at the bottom), top curve 0·05% sodium thymonucleate, dotted curve 0·05% serum albumen, firm line 0·5% protamine sulphate and the bottom curve 0·05% histone sulphate (from Caspersson, 1936, 32).

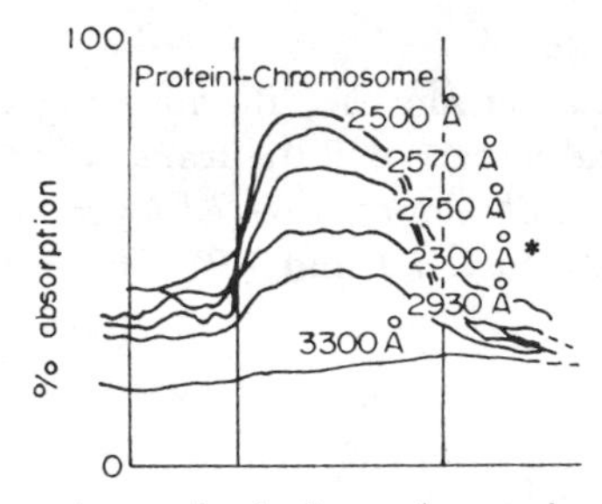

Fig. 7.2. Absorption curves at six wavelenths for a microscopic section containing a chromosome with protein on either side of it (from Caspersson, 1936, 94).

Caspersson was taken by surprise, but on searching the literature he found many papers on the ultraviolet absorption of nucleic acid, nucleoprotein, adenine, guanine, uracil and cytosine (Soret, 1883; Dhéré, 1906; Holiday, 1930; Heyroth and Loofbourow, 1931, and others). Here, then, was a property of nucleic acids which, taken in conjunction with the nucleal reaction of Feulgen, offered a means for the quantitative estimation of thymonucleic and ribonucleic acids in the cell.

The Swedish group had good reasons for exploiting this discovery because of their interest in the physiological significance of changes in the balance between nucleic acids and proteins. There was, too, the vexed chromatin problem: chromosomes could only be perceived as discreet rods of chromatin when the cell began to divide. In the so-called resting stage their form fragmented into a mass of granular matter, the stainability of which was slight. Boveri, who had established the individuality of the chromosomes through successive resting stages, had suggested that the permanent structure of the chromosome loses its basophyllic substance (chromatin) in the resting stage, to become recharged with it at the onset of division (Boveri, 1888). This "achromatin" hypothesis was espoused by Haecker, Montgomery, Mathews and Strasburger. The latter said of chromatin:

> This substance collects in the chromomeres and may form the nutritive material for the carriers of hereditary units which we now believe to be enclosed in them. The chromatin cannot itself be the hereditary substance, as it afterwards leaves the chromosomes, and the amount of it is subject to considerable variation in the nucleus, according to its stage of development. Conjointly with the materials which take part in the formation of the nuclear spindle and other processes in the cell, the chromatin accumulates in the resting nuclei to form the nucleoli.
>
> (Strasburger, 1909, 108; cited in Mirsky, 1968, 88)

E. B. Wilson also sided with Strasburger and in his influential book *The Cell in Development and Heredity*; he wrote:

> That the continued presence of "chromatin" (i.e. basichromatin) is essential to the genetic continuity of the chromosome has, however, become an antiquated notion . . . So far as the staining-reactions show, it is not the basophilic component (nucleic acid) that persists,

* Caspersson's 2300Å is almost certainly a misprint for 2800Å.

but the so-called "achromatic" or oxyphilic substance. The nucleic acid component comes and goes in different phases of cell-activity. . .

(Wilson, 1925, 895, 653)

When T. H. Morgan gave his Nobel lecture in Stockholm, in June 1934, he posed the question: "What are genes?"

Now that we locate them in the chromosomes are we justified in regarding them as material units; as chemical bodies of a higher order than molecules? Frankly, these are questions with which the working geneticist has not much concerned himself, except now and then to speculate as to the nature of the postulated elements. There is no consensus of opinion amongst geneticists as to what the genes are—whether they are real or purely fictitious—because at the level at which the genetic experiments lie, it does not make the slightest difference whether the gene is a hypothetical unit, or whether the gene is a material particle. In either case the unit is associated with a specific chromosome, and can be localized there by purely genetic analysis. Hence, if the gene is a material unit, it is a piece of a chromosome; if it is a fictitious unit, it must be referred to a definite location in a chromosome—the same place as on the other hypothesis. Therefore, it makes no difference in the actual work in genetics which point of view is taken.

Between the characters that are used by the geneticist and the genes that his theory postulates lies the whole field of embryonic development . . .

(Morgan, 1934, 315)

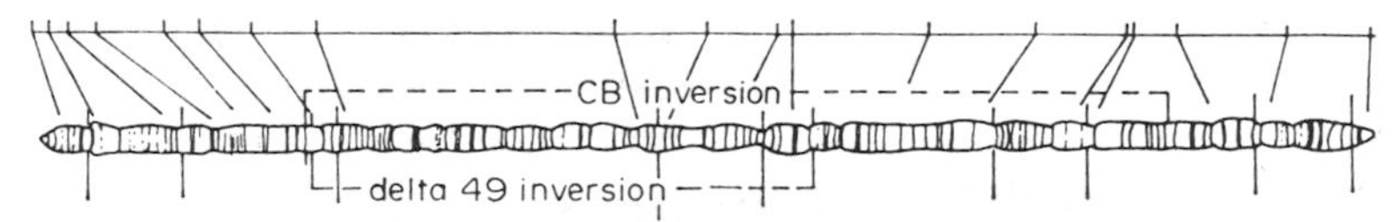

Fig. 7.3. Painter's Camera Lucida sketch of the X chromosome of *Drosophila* as seen in the salivary gland after treatment with acetocarmine (from Painter, 1933, 586).

One must remember that at this time little evidence of chemical differentiation along the chromosome was available. To be sure, the chromosome had been recognized as a row of deeply-staining chromomeres held together by a faintly-staining thread composed of a substance called "linin". But geneticists were justly unenthusiastic about the chemical basis for such vague distinctions. Their attitude began to change when E. Heitz advanced a new differentiation of the chromatin of the liverwort *Pellia* into two kinds of chromatin, morphologically sharply different—heterochromatin and euchromatin (Heitz, 1928), the former being stainable long after the latter had lost its stainability at the close of mitosis. Subsequently, he extended this discovery to *Drosophila* (Heitz, 1933). Even more compelling of the geneticists' attention was T. S. Painter's disclosure that the banding on salivary gland chromosomes of *Drosophila* was a constant and characteristic feature of them.*

In the euchromatic region of the X chromosome (see Fig. 7.3) he pictured more than 150 of the now well known bands, and gave the positions of breakage points and gene loci alongside them (Painter, 1933). Of this work Sturtevant remarked:

* Of course Painter was not the only contributor, the first to associate the banding with the genetic map being Kostoff (see Darlington, 1937, 175ff).

> Here at last was a detailed correspondence in sequence between the crossover map and cytologically visible landmarks, and a technique that was capable of refinement to give the precise loci of genes in terms of recognizable bands. Instead of two or three landmarks per chromosome (the ends and centromeres), there were now hundreds, and there soon came to be thousands for the whole complex.
>
> (Sturtevant, 1965, 76)

Now if the genes could be located on specific sites of the chromosome the next step would be to analyse the chemical constitution of such chromosomal segments. The task of linking the geneticists' abstract units with the chemists' organic compounds now seemed worth attempting. This was not just another discovery. It was a landmark in bringing together the discoveries of the geneticists and those of the cytologists. And more, it opened up the way to the chemical exploration of the fine structure of chromosomes.

In his Nobel lecture, delivered in June 1934, Morgan showed a slide of the famous salivary gland chromosomes of *Drosophila* in which he demonstrated the transverse banding in material treated with strong acetic acid. One can well imagine how interested Hammarsten and Caspersson were in Morgan's slides. The group in Stockholm had developed a reagent which digested the protein of the chromosome and retained the nucleic acid fixed as a lanthanum salt. This enzyme-lanthanum reagent they used to search for a "protein structure in cell nuclei, especially in chromosomes" (Caspersson, Hammarsten and Hammarsten, 1935, 367). Their plan was to show by ultraviolet microphotometry a change in the pattern of ultraviolet absorption before and after treatment with the reagent. The absence of any conspicuous differences led them to doubt that a block-like packing of protein between nucleic acid bands existed in their material (*Stenobotrus* spp.) but they expected such an arrangement would be demonstrable in Morgan's material (*Ibid.* 369). The strong acetic acid, they thought, had concealed a "negative" of protein concentration along the chromosome, the dark bands representing low protein content but high nucleic acid content.

When Hammarsten and Caspersson presented their report to the Faraday Society in London in 1934, their work was still in a preliminary stage. The ultraviolet microscope at their disposal was very imperfect and they used the double line of magnesium at 280 mμ for their light source. Later on, Caspersson greatly improved the quality of these microphotographs by using the cadmium light source which he passed through a monochromator to give 257 mμ—the portion of the ultraviolet spectrum most intensely absorbed by nucleic acids. The designer of the Zeiss ultraviolet microscope, August Köhler, had avocated the use of cadmium in 1904 when he published pictures he had obtained with it, the most striking of these were the larval gill bud cells of *Salamandra maculosa.* He saw the technique of ultraviolet microphotometry as a control on the results achieved by traditional staining techniques

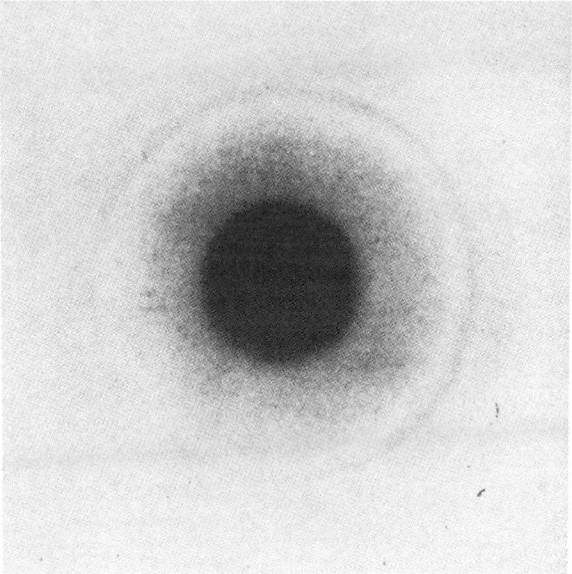

2 *Diffraction pattern from paraffin wax (from Friedrich, 1913).*

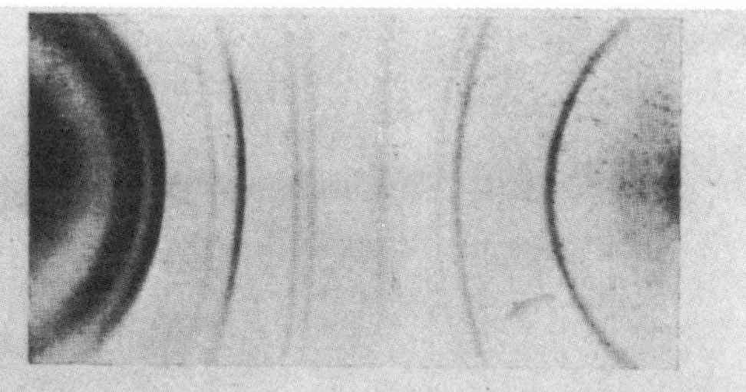

Fig. 1.

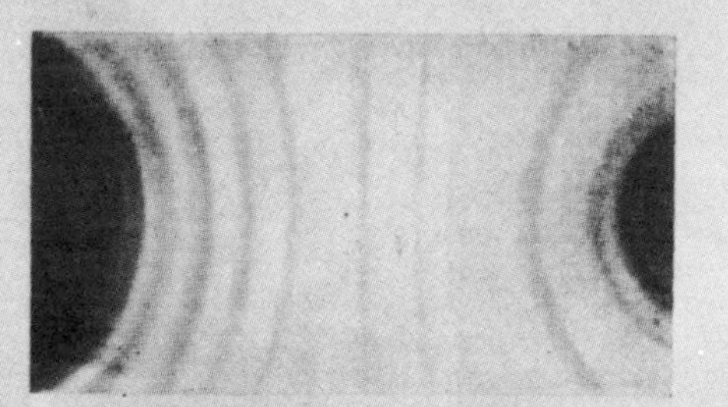

Fig. 2.

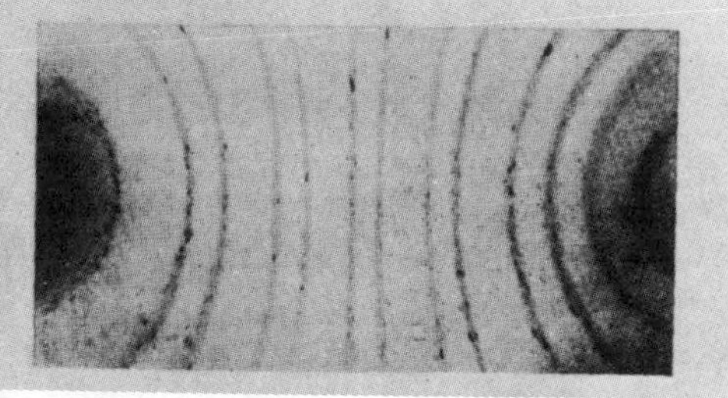

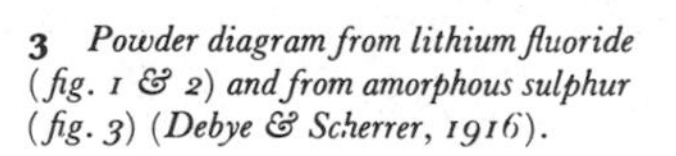

3 *Powder diagram from lithium fluoride (fig. 1 & 2) and from amorphous sulphur (fig. 3) (Debye & Scherrer, 1916).*

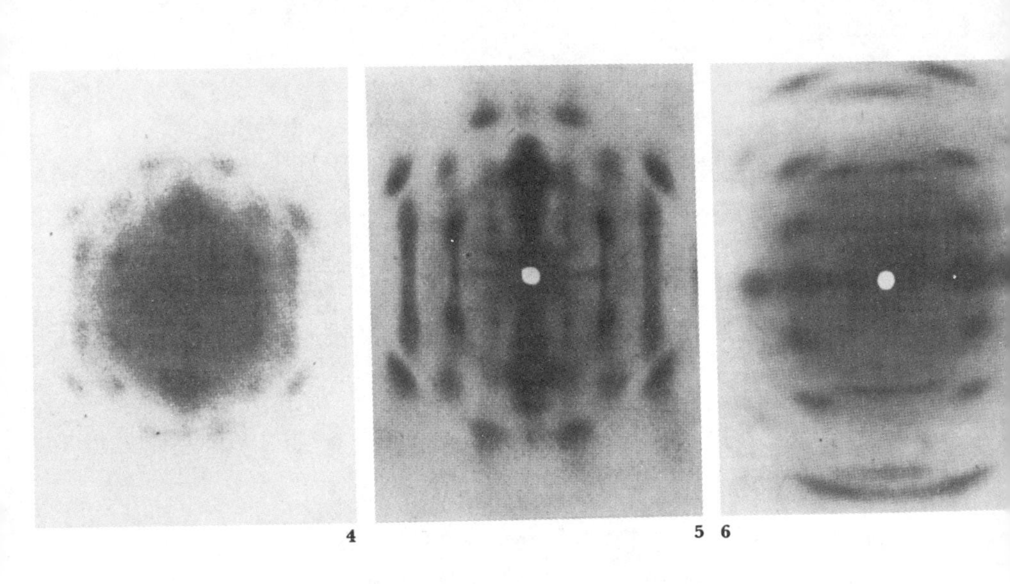

4 5 6

7

4 *Fibre diagram for cellulose (from Polanyi, 1921, p.339).*
5 *Fibre diagram for cellulose (hemp) (from Astbury, 1940, p.26).*
6 *Fibre diagram for silk (from Astbury, 1940, p.22: also Astbury, 1933. p.79).*
7 *Astbury, the 'John Bull' of British X-ray crystallography (left), with H. J. Woods.*

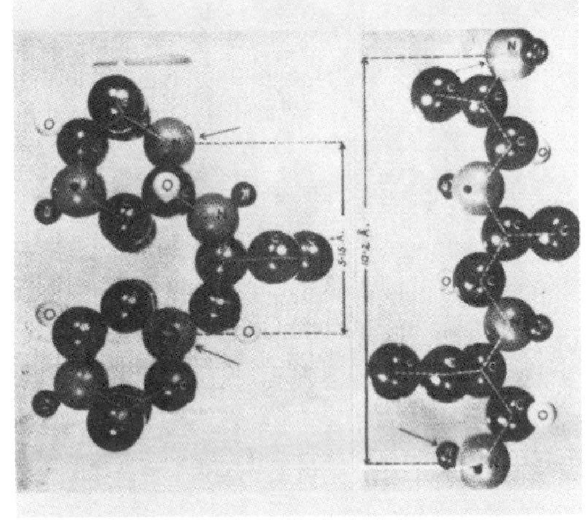

8 9

10

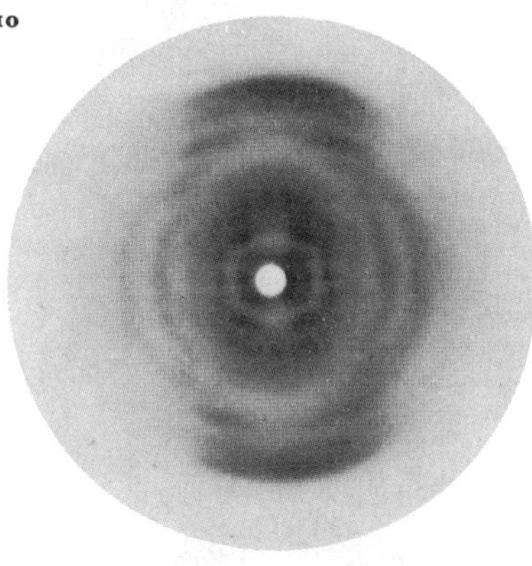

11

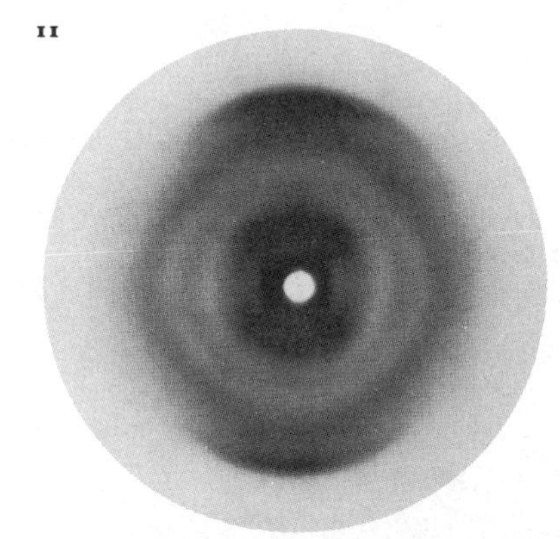

8 *X-ray diagrams for α and β keratin (from Astbury, 1930, Figs. 1a, 1b).*
9 *Astbury's billiard ball models for the molecular chains in α and β keratin (from Astbury, 1930, Figs. 2a, 2b).*
10 *X-ray diagram of thymonucleic acid – opalescent film stretched to 1.67 times its original length. (Most of the pattern is characteristic of the A form, but the marked meridional reflexion at 3.34Å shows the presence of the B form (from Bell, 1939, p.46).*
11 *X-ray diagram of the dried, rolled membrane formed at the interface between solutions of NaDNA and clupein. (The cross-ways feature and meridional reflexion of the B form can be seen clearly. One half of this pattern was reproduced in 1947, and studied by Watson and Crick.) (From Bell, 1939, p.66.)*

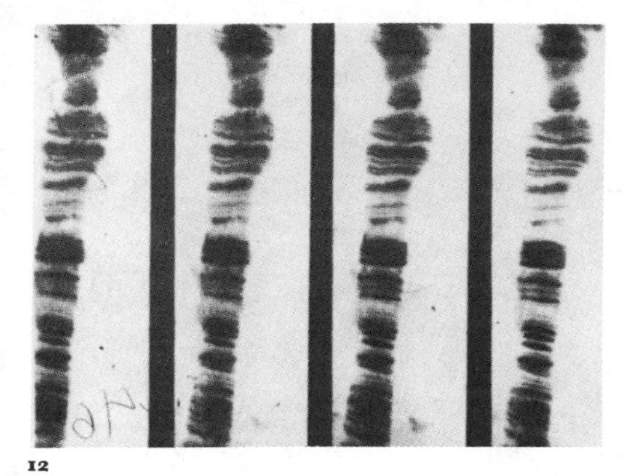

12

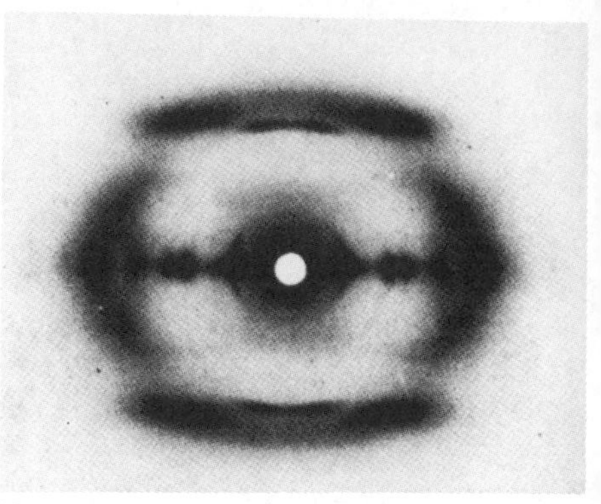

13

14

15

16

12 *The banding of* chironomus *chromosomes revealed by ultraviolet absorption photography (from Caspersson, 1936, Fig. 92).*
13 *X-ray diagram of stretched film of poly Benzyl-L-glutamate (from Bamford* et al, *1951, Plate 7, fig. 3).*
14 *J. D. Watson with Tracy Sonneborn.*
15 *X-ray diagram, the top half of NaDNA, the bottom half of clupein thymonucleate (Watson and Crick paid special attention to the latter) (Astbury, 1947, Plate 1, fig. 2b).*
16 *X-ray diagram of thymus DNA (supplied by Chargaff, photographed by Beighton in May 1951). The film was stretched to 3 times its original length at R.H. = 98%. This is a mixture of A and B forms.*
17 *X-ray diagram of the same specimen after relaxing at 98% R.H. for several days and stretching again. This is a remarkably fine B pattern, hitherto unpublished, and therefore unknown to Watson, Crick and Pauling.*

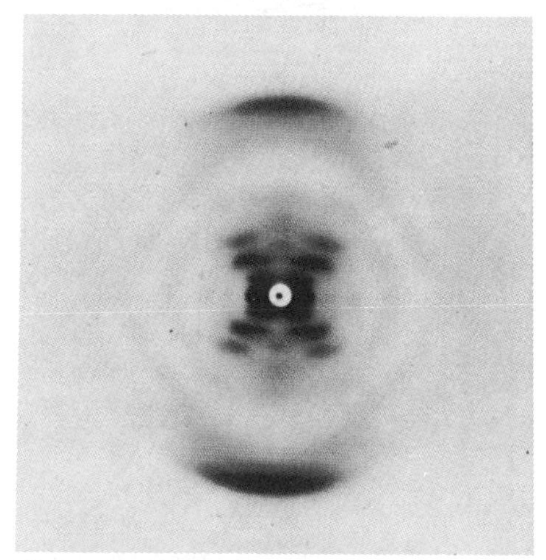

17

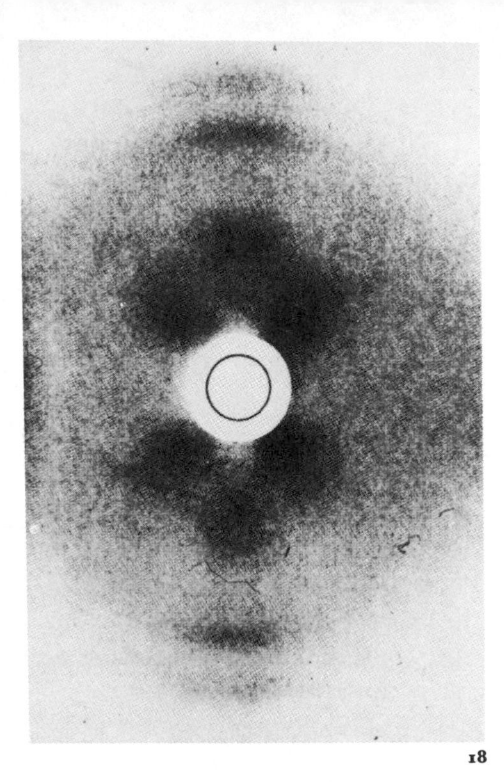

18

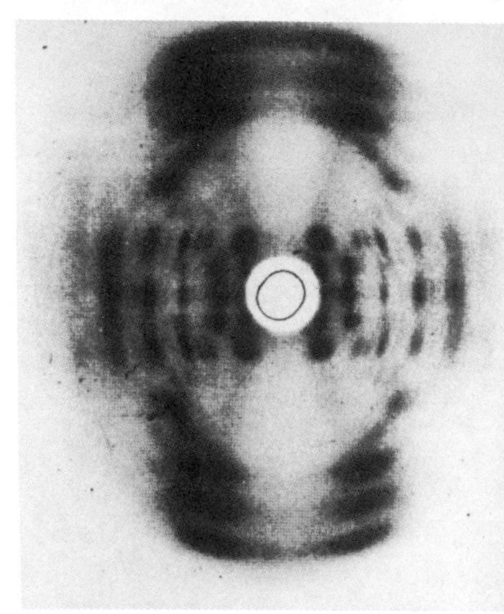

19

18 *"Sheet" pattern from calf thymus DNA made at King's College, London. (Material prepared by M. J. Fraser and Photographed by Gosling.) It shows the "Cross-Ways", meridional spot at 3.4 Å and spurious intensity along the meridian inside this arc.*
19 *X-ray diagram of "crystalline" DNA taken by Gosling in June 1950. This was the first example of a single-phase diagram, and shows the wealth of data which the A form yields.*
20 *X-ray diagram of* sepia *sperm, showing features of the B pattern. Photograph taken by Wilkins in the Winter of 1951/2.*
21 *X-ray diagram of* E.coli *DNA, taken by Wilkins about the same time as (20) – a good B pattern.*

20

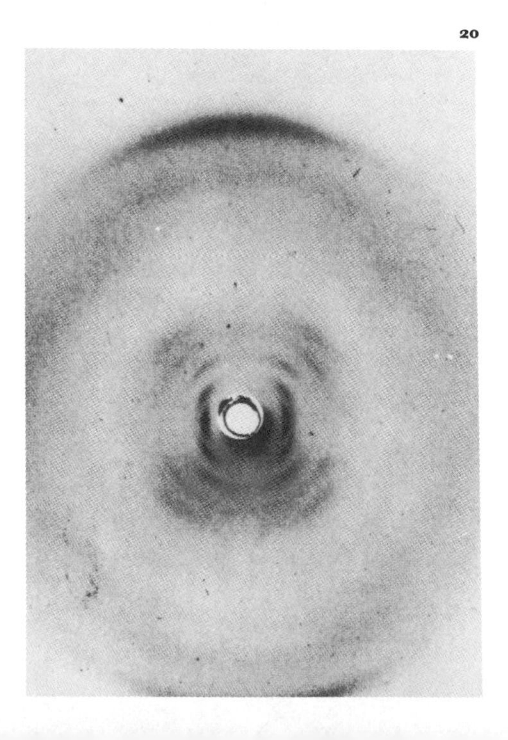

21

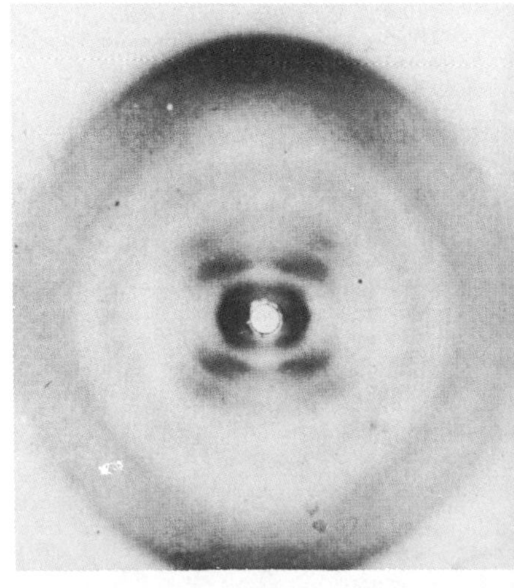

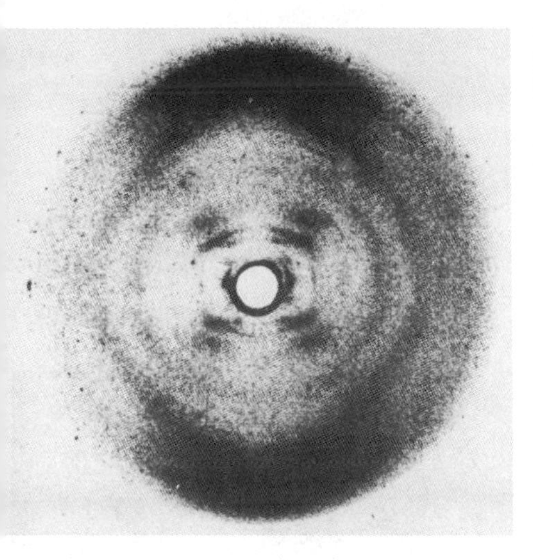

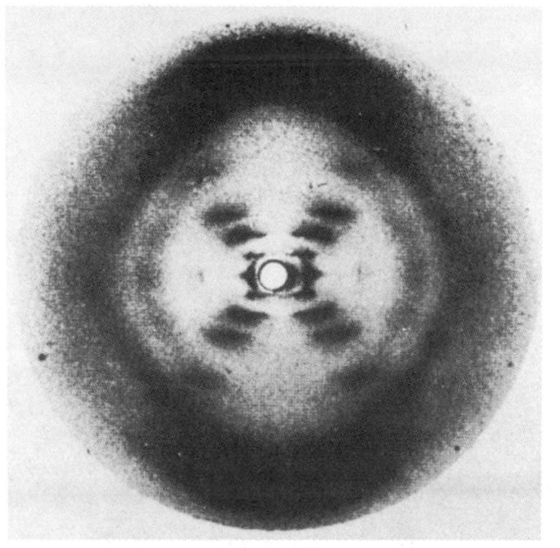

22 *Franklin's and Gosling's first B pattern (R.H. = 92%, 6-hour exposure of several fine fibres).*

23 *Franklin's remarkable B pattern of May 1952 (R.H. = 75%, 62-hour exposure of a single fibre, diameter = 50μ).*

24 *The pattern which showed "double-orientation" (R.H. = 75%, 100-hour exposure of a single fibre, diameter = 40μ).*

25 *The A pattern, showing the meridional arc on the eleventh layer line. (Photograph taken with the tilting camera.)*

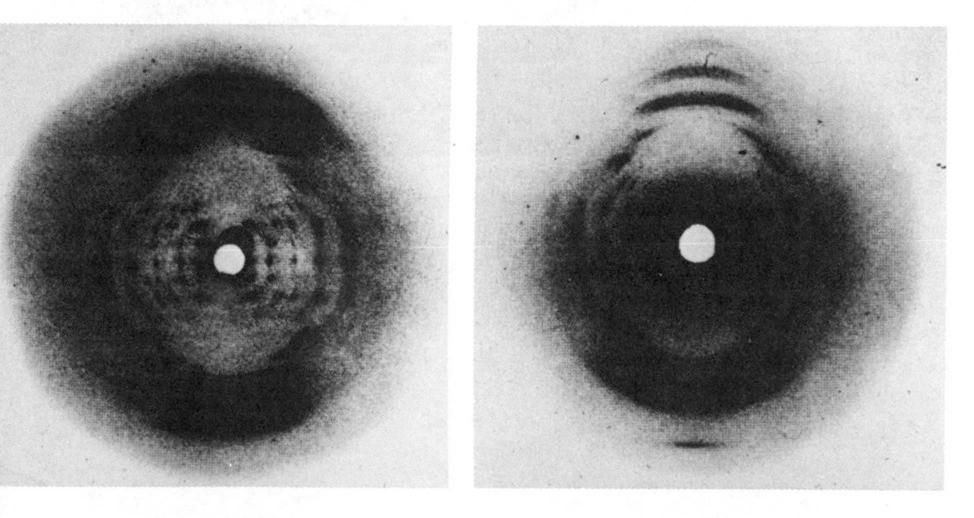

26 *Watson & Crick at the Cavendish Laboratory beside their model, May 1953.*

and as opening up a new field of research (Köhler, 1904, 302). Caspersson had taken slides to Köhler in Jena, probably in 1933, and had seen just what could be achieved with the Zeiss equipment there.

At long last, Köhler's toy was to prove its worth. Admittedly the hoped for duplication of resolving power at 275 mμ had not been achieved. An early attempt to use the microscope in the search for bacterial nuclei had brought disappointment (Schrötter, 1906), and Carl Zeiss only once demonstrated the instrument (1905). But when Caspersson used it on salivary gland chromosomes with Köhler's improved monochromator, what a wealth of detail was revealed! A new dimension in cytochemistry opened up which the Swedish scientists exploited to the full.

It was not that Caspersson was the first or the only one to use the ultraviolet microscope in the 1930's. There was Ralph Wyckoff at the Rockefeller and Muller and Prokofyeva in Russia, but the all important link between the absorption of ultraviolet light by the chromosomes and by nucleic acids was first seen by Caspersson. So here were these wonderful observations of chromosomes in the *living* cell as well as in the fixed cell, photographed under ultraviolet light, showing Painter's characteristic banded patterns. Was there after all a chemical morphology of which the genetic map was an expression? It seemed possible.

Having said all this, having urged the importance of the link between the genetic map and the distribution of nucleic acid, reluctantly we have to admit that the Hammarstens and Caspersson continued with the traditional assumption that the genes must be proteins. Referring to the proteins and nucleic acids as the "known substances" they wrote: "there are only the proteins to be considered, because they are the only known substances which are specific for the individual. *On that assumption a protein structure in chromosomes takes on a very great interest*" (Caspersson, Hammarsten and Hammarsten, 1935, 369). Consequently, their subsequent discovery of vigorous nucleic acid synthesis when chromosomes were thought to duplicate had to be turned on its head in order to preserve the protein version of the central dogma. What resulted we may term "The Nucleoprotein Theory of the Gene".

DNA Synthesis at the Onset of Cell Division

Caspersson observed such a marked increase in ultraviolet absorption during meiosis that there was no shred of doubt in his mind that: "In the time between early leptotene, where scarcely any absorbing substance can be detected in the nuclei, and the development of the complete nucleic acid-rich tetrads, a genuine fresh synthesis must evidently have occurred" (Caspersson, 1936, 120). Interphase and ensuing prophase were correlated with loss and renewal of absorbing substances. The absorbing substance was nucleic acid:

therefore, reasoned Caspersson, this substance is essential for chromosome duplication and hence for gene reproduction.

When Caspersson was joined by Jack Schultz in Stockholm, a correlation was found between nucleic acid content of certain bands in *Drosophila* salivary gland chromosomes and the variegated eye colour mutants discovered by Muller and attributed to "chromosomal displacements" (Muller, 1930, 299). Ultraviolet measurements in Stockholm showed that "the closer a given band is to the heterochromatic region, the greater the augmentation of its nucleic acid content" (Caspersson and Schultz, 1938, 295). A parallel increase in nucleic acid was noted when going from the eggs of XX females to those of XXY females. From their study of the salivary gland chromosomes, Caspersson and Schultz rightly concluded that the heterochromatin was not an inert region. Such regions of the chromosome, they argued, perform:

> . . . a function also performed by all other genes—namely, the synthesis of nucleic acid. The relation of these regions to the variegation, taken together with the local appearance of nucleic acid in the chromosomes, suggests that the synthesis of nucleic acid is closely connected with gene reproduction. The structure-forming properties of thymonucleic acid, and its ability to form high molecular weight polymers, as well as the correspondence of its X-ray diffraction pattern with that of the proteins, suggest a basis for this function.
>
> These considerations have an especial interest in the case of the other self-reproducing molecules—the viruses and the bacteriophage—all of which have been shown to contain nucleic acid. Moreover, in the inactivation of the bacteriophage by ultraviolet light, the curve for the efficiencies of the different wavelengths does not agree completely with that of the bacteriophage, but does with the nucleic acid absorption spectrum. It seems hence that the unique structure conditioning activity and self-reproduction, possibly by successive polymerization and depolymerization, may depend on the nucleic acid portion of the molecule. It may be that the property of a protein which allows it to reproduce itself is its ability to synthesize nucleic acid.
>
> (Caspersson and Schultz, 1938, 295)

As we read this passage our excitement rises. Surely they will conclude that the genes *are* nucleic acids! But no—they turn the evidence on its head and in the last sentence we get a view into their inner thoughts. Sure enough, within two years Caspersson had made this conclusion the basis of a comprehensive scheme for the protein metabolism of the cell. Here protein synthesis in the cytoplasm was the equivalent of chromosome duplication in the nucleus. Both events involved the self-duplication of protein molecules in the presence of nucleic acids. Caspersson did not stop here. He sketched out the phylogenetic development of the nucleus, starting with simple RNA-containing plant viruses, going on to DNA-containing yeast and bacteria, and finally to the chromosomal nuclei of higher organisms. Here, the more primitive genetic material was located in the heterochromatin, where it produced histones. These accumulated in the nucleolus and were ultimately transferred to the cytoplasmic proteins; such a scheme appeared appropriate to Caspersson's observations of the nucleic acid content of actively secreting cells.

One experiences difficulty in comprehending Caspersson's scheme if only because of what we now know, but we can see that in distinguishing between the functions of euchromatin and heterochromatin he was not making *all* protein synthesis dependent on a transfer of specific substances from nucleus to cytoplasm. It was as if major structural genes synthesized their protein products within the nucleus in the absence of RNA, but minor genes did so only by a more complicated pathway involving RNA in the cytoplasm. Nonetheless, one fact stands out—the genes were regarded as proteins, whether one had to do with viruses, bacteria, fungi, higher plants or animals.

Evidence for the Protein Nature of the Gene

At no time, as far as I am aware, did Caspersson state that *all* nucleic acid is removed from the chromosomes during interphase. The inability to demonstrate more than slight traces may have encouraged Caspersson to rule out nucleic acids as the genetic material, but more influential throughout the 1930's and 1940's was the belief that diversity of structure in nucleic acids was severely limited. One observation which may have encouraged his orthodoxy here was his success in demonstrating the presence of protein all along the salivary gland chromosome and his failure to detect nucleic acids in the inter-band regions. In his famous paper of 1936 he wrote:

> The protein component which forms [the parts of] the chromosomes between the large bands or band complexes, is easily digestible and absorbs extremely little ultraviolet light. Here the continuity of the chromosome is easily broken by mechanical agents and mechanical treatment. No trace of a structure can be revealed with ultraviolet light.
>
> Intimately mixed with the nucleic acids in the transverse bands there are also protein components. These are dissolved out after long continued digestion.
>
> If the genes are conceived as chemical substances, only one class of compounds need be given to which they can be reckoned as belonging, and that is the proteins in the wider sense, on account of the inexhaustible possibilities for variation which they offer. It is shown that both types of segment contain protein.
>
> On the localization of the genes in the chromosomes it seems more likely that just those bands which are highly organized because of a nucleic acid structure would be the carriers of the genes than that the less highly differentiated non-absorptive segments between them should be.
>
> Such being the case, the most likely role for the nucleic acids seems to be that of the structure-determining supporting substance [*Stützsubstanz*].
>
> (Caspersson, 1936, 138)

It is widely accepted now that data tend to be interpreted in such a way as to accommodate knowledge from as many related fields as possible, which usually involves fitting it into an established framework, or paradigm. Caspersson's scheme, in which nucleic acids served a supervisory and non-specific role in gene duplication, fitted admirably into the picture which was emerging of the way chromosomes duplicate as revealed by studies of chromosome mechanics. Here all discusions were based on the assumption that *the attractive forces by which a chromosome collects the sub-units out of which it fashions a replica of itself are the same as those by which like chromosomes—homologues—are drawn together in meiosis.* A study of chromosome mechanics, it was felt, should therefore serve to unravel the complex process by which the gene reproduces.

Chromosome Mechanics

The strangest paradox revealed by cytology was the contrast between the behaviour of the chromosomes in meiosis and mitosis. In meiosis entire chromosomes attracted each other and came together to form pairs, or bivalents, before they split into chromatids, but in mitosis the chromosomes repelled each other, and as soon as their form became clear they were observed to be split longitudinally into chromatids. On the whole, cytologists had been content to account for this paradox in a teleological manner, the pairing, or synapsis, of chromosomes in meiosis they regarded as the final step in the sexual process by which at last maternal and paternal chromosomes came together, effected an exchange of material, and segregated from one another. The fact that meiosis involved the formation of four cells instead of two was attributed to the need for a halving of the chromosome number before fertilization.

Now it is most important for the historian to distinguish between a correct representation of events in mitosis and a correct understanding of how the chromosomes seen in mitosis have arisen from the interphase nucleus. All the work of Boveri on the individuality of the chromosomes did not suffice to dispel the fallacious belief in a continuous spireme which by fragmentation gives rise to the chromosomes of prophase nuclei. In an otherwise excellent book on the history of cytology, this point is completely ignored (Hughes, 1959). Had it not been for belief in the spireme, the events of meiosis might have been understood more quickly. Those who accepted it accounted for the halving of the chromosome number in meiosis and its equality in mitosis (reductional and equational division) by simply postulating the fragmentation of the spireme in meiosis into half the number of threads formed in mitosis. As more details of meiosis were revealed the spireme theory ran into difficulties, and additional erroneous processes had to be put forward to save the theory, such as end-to-end pairing or "telosynapsis". Further confusion resulted from the study of polyploids and the ring-chromosome associations of *Oenothera*, which invited a telosynaptic interpretation. It was the young cytologist, C. D. Darlington, who cut through this confusion and offered a simple mechanistic scheme which incorporated meiosis and mitosis, diploids and polyploids, ring chromosomes, normal chromosomes and belatedly-, or non-pairing, chromosomes.

When nineteen years old, Darlington went to Merton on the outskirts of London, a degree in agriculture as his credential, and asked William Bateson the first Director of the John Innes Institute, if he had a place for him. Although Bateson did not take to the idea of employing a man without an Oxbridge degree, he was persuaded to let this London graduate work in his institute for three months without pay, for another three without pay, and finally appointed him a minor student in 1924. Darlington resisted Bateson's suggestion that he go and work for the Empire Cotton Breeding Research Association, and when the great prophet of Mendelism died, in 1926, Darlington stayed on under Daniel Hall, and finally succeeded him in 1939.

The Precocity Theory of Meiosis

In 1931 Darlington advanced the view that meiosis was an aberrant form of mitosis. In the prophase of mitosis each chromosome was already split into two chromatids. The attractive forces which he believed to emanate from the chromosomes at this time were therefore satisfied by the adherence of sister chromatids. They shortened, coiled around each other, and became more stainable, but *whole* chromosomes did not attract one another; indeed it seemed as if there were repulsive forces between them, so evenly did they space themselves out on the metaphase plate. In meiosis, on the other hand, prophase of division I began *before* the chromosomes had divided into chromatids: it was *precocious*, and the attractive forces could only be satisfied by chromosome–chromosome pairing. "It is therefore possible to regard the pairing of undivided homologous chromosomes at the prophase of meiosis as a re-establishment of conditions upset by their precocious contraction" (Darlington, 1931, 234). This became known as the Precocity Theory. On this argument one would expect homologous chromosomes to separate once they split into chromatids. The fact that they remained in synapsis Darlington had already attributed to the exchange of chromatid segments, or crossing-over.

The Molecular Spiral

The precocity theory naturally stimulated much controversy, and Darlington's book *Recent Advances in Cytology* (1932) was, to quote Haldane, "the object of numerous attacks. In fact, one of the sessions of the sixth International Congress of Genetics in 1932 was mainly occupied in disproving Dr Darlington's conclusions" (J. B. S. Haldane, 1937a, p. vi). Our concern here is not with the evidence for the precocity theory, but for the impetus it gave to a reductionist approach in cytology.

Consciously following the example of Newtonian mechanics, Darlington interpreted the movements of chromosomes in terms of a specific attractive force between paired chromosomes before they split, between chromatids after they split, and a non-specific repulsive force between different paired chromatids. Quoting Newton to the effect that the precise nature and causes of the forces he had postulated need not be defined, Darlington ended his analysis of chromosome movement with the words:

"Thus the author of the law of gravity: need we be more confident?" (1932, 353). But Darlington did pursue the nature of these forces in so far as the coiling and uncoiling of chromosomes allowed. The strains involved, he noted, were not due to lengthwise bending as T. H. Morgan believed, but to crosswise twisting (1932, 289). Now why, asked Darlington, do chromosomes coil? He began by distinguishing between the coiling of one chromatid around another (the relational spiral) from the coiling of a chromatid around its own axis (the internal spiral) as in Figure 7.4. He observed the belated end

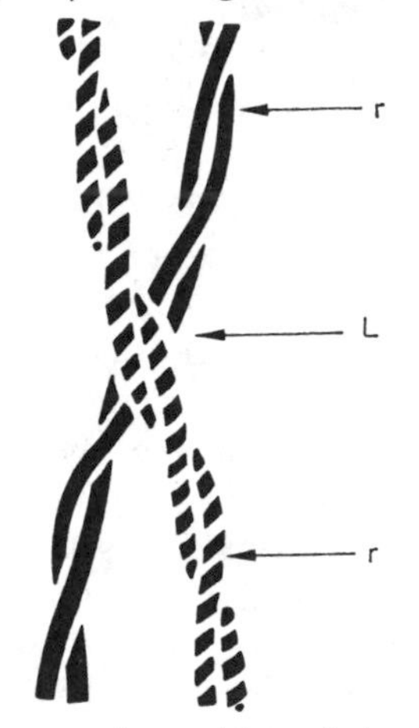

Fig. 7.4. Diagram of two chromosomes in a pairing relationship showing the internal spiral of the chromatid, the relational spiral of sister chromatids, and the relational spiral of homologous chromosomes (from Darlington, 1937, 549).

to the unwinding of these spirals as the cell began its next division (the relic spiral), and fresh coiling occurred. In the last analysis, he concluded, the torsions which brought about visible coiling must have been induced by the formation of an unseen *molecular* spiral. He saw the successive levels of coiling in a chromosome as the device by which torsion, arising from chromatid formation, was overcome; for if successive levels of coiling were in opposite directions, equilibrium would be achieved. "The production of the equilibrium of pachytene", he wrote, "is the object of all spinning operations" (Darlington, 1937, 549). It is interesting to see this idea developed to account for the "unwinding problem" in DNA synthesis (Person and Suzuki, 1968).

The Chemical Basis of Spiralization

It is noteworthy that Darlington did not discuss the chemistry of the chromosomes until 1939. To be sure, he noted the parallel between the degree of coiling of chromosomes and their affinity for nuclear stains (1937, 307) and shortly thereafter the effect of low temperature on the staining intensity of certain regions of *Paris* chromosomes (Darlington and La Cour, 1938), but it was not until December 1939 that he wrote to Astbury ascribing this discovery to "differential nucleic acid production" (Darlington, 1939b), and with La Cour wrote the paper on "Nucleic Acid Starvation of Chromosomes in *Trillium*" in which he launched into a discussion of the nucleic acid cycle in the section entitled, "Supply and Demand of Nucleic Acid" (Darlington and La Cour, 1940, 201 ff.). What had happened between 1938 and 1940? He had been to the Seventh International Congress on Genetics in Edinburgh in August 1939 and although Caspersson had not been able to attend, his big 1936 paper "Ueber den chemischen Aufbau der Strukturen des Zellkernes", reached Darlington soon afterwards, and had a profound impact:

> At last, I have been able to see your great 1936 paper. I borrowed it from Koller who borrowed it from Demerec at the Congress. Not having seen it before, its full significance had escaped me. La Cour and I have now made some discoveries which touch you very closely. We find that the differential segments in *Paris* . . . are super-charged with nucleic acid in resting stages, just as they are undercharged at metaphase.
>
> (Darlington, 1940a)

The result of Darlington's enthusiasm for Caspersson's work was a steady flow of papers between 1940 and 1947 on the relation between nucleic acid metabolism and cytogenetics, beginning with the following statement:

> During the resting stage it [nucleic acid] is present in the nucleus in smaller quantity and largely unattached to the chromosomes. Nucleic acids, having in their simplest units a molecular weight of at least 1300 (Gulland, 1938), presumably exist inside the nucleus as polymers arising from such units. During mitosis the nucleic acid increases in quantity and is attached to the protein framework of the chromosomes (Caspersson, 1937). It may be digested with nuclease to leave the framework intact (Mazia and Jaeger, 1939). Nevertheless, the nucleic acid does not lie on the surface of the spiralized chromosome but is bound up with the coiled thread itself . . .
>
> (Darlington and La Cour, 1940, 201–202)

In Edinburgh the previous August he had heard Astbury give his fine paper, "Protein and Virus Studies in Relation to the Problem of the Gene" in which Astbury stressed the equality of separation of nucleotides and amino acids at 3.3_4 Å as significant for an answer to the question: " . . . why, with the proteins, the nucleic acids are the chosen reactants in nuclear division . . ." (Astbury, 1939, 50). Darlington went a step further and suggested that "nucleic acid is formed through the agency of the genes arranged on the polypeptide chain and, being attached to them, serves as an agent in their reproduction, in their separation and even perhaps in their spiralization (Darlington and La Cour, 1940, 202).

The aim of giving this extended treatment of Caspersson's and Darlington's work has been to reveal the path by which chemistry made an entry into cytogenetics. From the late 1930's onwards cytochemistry enjoyed an ever wider popularity. Though some still denied the evidence from the Feulgen stain for the localization of nucleic acid in the chromosomes (Stedman and Stedman, 1947), though the dispute over the chemical specificity of this reaction dragged on, the link established between heterochromatic regions and nucleic acid content on the one side, and between gene action and the siting of the gene with respect to heterochromatin on the other, brought chemistry and cytology together. The most effective tools in this development were chromosome mapping, the Feulgen stain and ultraviolet microscopy. The most suggestive material was the chromosome complement of dipteran salivary gland cells. Unfortunately, Feulgen's nucleal reaction did not satisfy those who wanted firmer ground for chemical characterizations than was offered by a test of disputed chemical specificity. Thus one of Darlington's correspondents wrote urging him to be non-commital about the chemical identity of eu- and heterochromatin, and to state that Caspersson lacked definite proof that he was, in fact, working with nucleic acid (Carr, 1947?). Fortunately, such reservations did not prevail; the link with nucleic acid was forged.

Replication of the Gene

Darlington's precocity theory at once appealed to the American geneticist, H. J. Muller, and for obvious reasons. Way back in 1917, Muller had pondered the nature of gene duplication when he read L. T. Troland's remarkable paper "Biological Enigmas and the Theory of Enzyme Action" which we shall discuss in Chapter 9. Troland's concept of gene duplication as a form of growth in which sub-units of the gene were attracted to it by the similarity of their "force field patterns" led Muller four years later to consider the forces involved as identical with those by which homologous chromosomes appear to be drawn together at the beginning of meiosis: "it is evident that the very same forces which cause the genes to grow should also cause like genes to attract each other, but more strongly, since here all the individual attractive forces of the different parts of the gene are summated" (Muller, 1922, 38). Here then, in this Toronto address of 1921, we find his *influential but erroneous analogy between autocatalytic and synaptic attraction* clearly expressed: "If the two phenomena are thus really dependent on a common principle in the make-up of the gene, progress made in the study of one of them should help in the solution of the other" (*Ibid.*). Evidence for the polar nature of synaptic attraction he found in the suggestion that pairing of homologous chromosomes in triploids excludes the third member. Darlington's extension and revision of Belling's work reinforced Muller's views and the British cytologist's assumption of forces of attraction in pairs could not have been more in tune with Muller's enthusiasm for attack on the gene by the physicists. Muller maintained that these forces acted "over visible microscopic distances" (1936, 211), "distances much larger than in the case of the ordinary forces of so-called cohesion, adhesion and adsorption known to physical science" (1922, 37). He urged the physicists "to search the possibilities of their science and tell us what kind of forces these could be, and how produced. . . ." (1936, 211). As he concluded this visionary address, the enigmas revealed by cytogenetics—position effect, synapsis, chromosome repulsions, coiling and uncoiling—came before him and he admitted: "The geneticist himself is helpless to analyse these properties further. Here the physicist, as well as the chemist, must step in. Who will volunteer to do so"? (Muller, 1936, 214).

Whilst Muller invited the physicists, J. B. S. Haldane invited the chemists. He was imbued with Sir Archibald Garrod's conception of biochemical individuality, which he may well have learnt from F. G. Hopkins; and when he contributed an essay to the Hopkins Festschrift his subject was the biochemistry of the individual. Here he discussed the duplication of the gene. It could not be merely a case of growth followed by division, he argued.

> It must, on the contrary, be a process of copying. The gene, considered as a molecule, must be spread out in a layer one *Baustein* deep. Otherwise it could not be copied. The most likely method of copying is by a process analogous to crystallization, a second similar layer of *Baustein* being laid down on the first. But we could conceive of a process analogous to copying of a gramophone record by the intermediation of a "negative" perhaps related to the original as an antibody to an antigen . . .
>
> The whole problem of protein synthesis is an almost virgin field in biochemistry. And it is absolutely fundamental. If genetics had done no more than pose the question in its sharpest form it would still be a valuable stimulant to biochemical research.
>
> (Haldane, 1937b, 8–9)

The Klampenborg Meeting

In its policy of support for the application of physics to biology the Rockefeller Foundation funded three meetings attended by physicists and biologists at which problems of gene and chromosome structure were discussed. The first took place in Copenhagen, in 1936, on X-ray mutation and the target theory, the second at Klampenborg (near Copenhagen) in April 1938, on chromosome structure, division and conjugation, the third—a follow-up of the second—at Spa, Belgium, in October 1938. The ideas discussed at the Klampenborg meeting have been preserved in the reports of Darlington and Bauer and in the summary prepared by Waddington, extracts from which appeared in *Nature* recently (Waddington, 1969). These ideas can also be found in Waddington's book *Introduction to Modern Genetics* (1939), Darlington's *Evolution of Genetic Systems* (1939), Astbury's paper "Protein and Virus Studies in Relation to the Problem of the Gene", at the Edinburgh congress (1939), and in Bernal's book *The Cell and Protoplasm* (1940).

As Waddington has pointed out, the importance of this meeting lay in the fact that for the first time geneticists and crystallographers were brought together. Darlington, Bernal and Astbury, it is true, had already met at the British Association meeting in Blackpool in 1935 (Darlington, 1970), and had discussed chromosome pairing, but Bernal's enthusiasm for mitosis and meiosis clearly dates from the Klampenborg meeting. Waddington recalled the rough passage the British group had from Harwich to Esbjerg:

> We were travelling second class—this was before the days of reasonable travel grants. Most of us tried to sleep on the benches in the general saloon, but Darlington and Bernal kept sea-sickness at bay by the former teaching the latter "all the genetics and cytology anyone needs to know" throughout the course of the night. Before dawn Bernal had already decided that the mitotic spindle must be a positive tactoid, and that the separation of chromosomes probably resulted from the production by the centromeres of some fluid which formed negative tactoids within the larger positive ones.
>
> (Waddington, 1969, 318)

We shall refer later (p. 259) to Bernal's view on the forces operating in mitosis. All we shall note here is that like Muller, who was not present at Klampenborg, Bernal pursued the unfruitful path of long-range forces, and did *correctly* demonstrate their presence in nucleoprotein gels (Bernal and Fankuchen, 1941). Astbury, on the other hand, was chiefly concerned to promote his idea for the role of nucleic acids in gene replication. He argued that since the amino acids in proteins are all left-handed, there must be some device to prevent the production of the mirror image of the parent molecule when the gene duplicates. This might be achieved either by an "external molecular 'template' " (Astbury, 1939, 50) or by the intermediary link of a polynucleotide chain. This has:

> . . . two sides, one with purine residue and one with phosphoric acid. One of these would be attached to the original protein chain and the other would build an amino acid into a new chain, the interposition of the nucleic acid ensuring that the new chain would have the same configuration as the old and not be a mirror image of it.
>
> (Waddington's Report, 1938, 10; Waddington, 1969, 320)

Astbury returned to this idea again at the Edinburgh congress a year later when he speculated that "the functions of a column of nucleotides is that of bridging the enantiomorphous stage" (Astbury, 1939, 50). Many years later, pondering the reason for the association of proteins with nucleic acids he wrote: "Surely it is not mere 'numerology' to suppose that the answer rests in part on the close dimensional correspondence between polypeptide chains and polynucleotide columns; for whatever else is required, at least there must be a 'fitting' of molecules one against another . . ." (Astbury, 1948?, 21).

An Astbury-like conception of gene duplication was arrived at by the biophysicist, Friedrich-Freksa who suggested electrostatic attraction between the basic histones and acidic phosphate groups of nucleoproteins.

Friedrich-Freksa's conception of gene duplication was not independent of the discussions at the Klampenborg meeting for it was an address by Bernal in Copenhagen in 1938 (surely at the Klampenborg meeting) that stimulated the physicist, Pascual Jordan who in turn stimulated Friedrich-Freksa. Bernal had attempted to do away with specific forces of attraction to account for chromosome pairing, postulating instead long-range nonspecific forces. The physicist Pascual Jordan rejected Bernal's suggestion and put in its place specific forces of attraction based on resonance. He suggested that if the same sites on two identical or near-identical molecules differ in their energy states, the one being in an excited state, the other in the ground state, there would result a quantum-mechanical stabilizing interaction between them, sufficient to account for chromosome pairing and possibly for chromosome synthesis. He doubted that normal valency forces would be adequate, as had been suggested for less complex synthetic processes, but he was left with the problem of how the rigidity of a molecule can be disturbed so as to raise parts of it to excited states. This he overcame by rejecting the wholly molecular model of the chromosome and introducing a "quasi-liquid state of aggregation" of the atoms so that they are only loosely bound and can therefore undergo positional changes (Jordan, 1938, 714). Two years later, his paper was seen by Linus Pauling, who promptly wrote a refutation of Jordan's theory which Delbrück and he published in *Science*. They believed the solution to questions of biological specificity would be found in the well-known covalent and hydrogen bonds, van de Waals, and electrostatic interactions (Pauling and Delbrück, 1940). Independent of them, Friedrich-Freksa suggested electrostatic attractions between the basic histones and acidic phosphate groups of nucleoproteins in place of Jordan's resonance attraction (Friedrich-Freksa, 1940).

At that time all the staff of the Kaiser Wilhelm-Institutes in Berlin had the use of the Society's one and only swimming pool (Melchers, 1971). There, Friedrich-Freksa met Jordan and they discussed the subject of gene duplication. The result was the paper "Bei der Chromosomenkonjugation wirksame Kräfte und ihre Bedeutung für die identische Verdopplung von Nucleoproteinen". One may justly ask how Friedrich-Freksa expected the sequence of the polypeptide chain of the existing gene to be mediated to the new gene. Beyond speaking of local concentrations of acidic or of basic amino acids he did not consider this question.

The Chemical Sequence of the Gene

There may well be significance in the fact that no organic chemist or biochemist was present at Klampenborg in 1938, only the histochemist Albert Fischer. In the 1930's the link between physicists and geneticists had been established through X-ray mutagenesis, and with cytologists through chromosome mechanics. Caspersson's cytochemistry established a link between the cytologists and chemists mainly through the studies of nucleohistones, but not until the end of the thirties, and even then the state of nucleohistone chemistry appeared too unreliable for most biologists to place much reliance upon it. Consequently *it was not an obvious step to make the connexion between the linear sequence of the genes and that of the amino acids in a polypeptide chain.* One person in the thirties who did just this was Dorothy Wrinch, the British-trained South American mathematician. At an informal meeting of the Theoretical Biology Club in Cambridge in 1934, she expressed the view that the specificity of genes resides in the specificity of their amino acid sequences. This informal contribution led to a short note in *Nature* (Wrinch, 1934). In Joseph Needham's notes on her talk to the Club, the building blocks of the gene are represented as T-pieces 3.5 Å in length, so that a gene 100 Å long would contain twenty-eight such T-pieces. The stems of the Ts represented the transverse "woof" of nucleic acid, the tops of the Ts the longitudinal sub-units of the polypeptide chromosome thread. This protein "warp" contained many parallel and identical (for any one gene) polypeptide chains. She repeated this scheme at a lecture in the physics department of Manchester University in 1935 and published it in *Protoplasma* a year later. By 1935 Linderstrøm-Lange and Rasmussen had carried out analyses of the amino acid content of clupein and had found that for every two of the following amino acids: alanine, valine, serine, hydroxyvaline, proline and hydroxyproline, there were four to five arginine residues. Using the symbol A for arginine and M for the six monoaminoacids they constructed sequences to fit their data on amino acid proportions such as: M A A M A A A A M A A M A A M A A A. Dorothy Wrinch calculated that even if five sevenths of the polypeptide consisted of arginine residues, and the remainder of the six monoamino acids, the number of possible "patterns" in a 40 micron stretch was $10^{50\,000}$ (Wrinch, 1936, p. 558).

More might have been made of Wrinch's sequence hypothesis had not her molecular model of the chromosome come under attack. She had assumed cyclic tetranucleotides at right angles to the chromosome axis, so that the chromosome should have been positively, not negatively, birefringent. Schmidt, in Giessen, lost no time in denouncing the model because it conflicted with his observation of the negative birefringence of DNA fibres and chromosomes (Schmidt, 1932). Evidence of positive birefringence during mitosis (Caspersson, 1940) did not help Wrinch's model, since it was attributed to supercoiling at metaphase. Fortunately, hers was not the only statement of the sequence hypothesis. There were at least three others: J. B. Leathes in his address to the physiological section of the British Association in 1926, N. K. Koltzoff in his address to a congress in Leningrad in 1927, and Max Bergmann and Carl Niemann in *Science* in 1937. The first two merely pointed out the variety that varied amino acid sequences could yield.

Leathes declared it probable "that the order as well as the proportion in which each amino acid occurs in the molecule is fixed, and it is this specific order and proportion that accounts for the specific character and properties of the protein" (Leathes, 1926, 388). He went on to calculate that the variety of sequences possible in a polypeptide chain with fifty links and nineteen different amino acids was about 10^{48}. Extracts from Leathes address were reprinted by Gortner in his influential textbook, *Outlines of Biochemistry* (1929, 358) and by Gulick in his important paper of 1937 in the *Quarterly Reviews of Biology*. It was Gortner who brought Leathes remarks to the attention of McCarty (see p. 187).

Koltzoff related his discussion of the specificity of different amino acid sequences to chromosome structure, where the sequence was as unchanging as the sequence of letters in successive sheets printed from the same block. As all new chromosomes arose from pre-existing chromosomes, as cells were formed from cells, "we now add a new thesis: every albumin molecule originates in nature from an existing albumin molecule by the crystallization of the amino acids and other albuminous fragments in the solution around it. *Omnis molecula ex molecula*" (Koltzoff, 1928, 368).

Bergmann and Niemann published an extraordinary paper in which they sought to demonstrate certain compositional regularities in widely differing proteins. Their theory of amino acid sequence then followed from the assumption that the total contribution of an amino acid to the protein molecule *was distributed evenly in that proportion along the entire length of the polypeptide chain.* Svedberg had already produced evidence that proteins consist of sub-units of the same size (mol. wt. 17 600), now Bergmann and Niemann suggested that the sequences within these sub-units were subject to a general restriction—the composition of any fragment of the chain had to give the amino acid composition which had been obtained from the whole molecule. They argued that without some such restriction on the sequence it would be impossible ever to discover it. Given this restriction it was then possible to gain a clue about sequence from overall composition. For instance, glycine made up 50 per cent of the amino acids in silk fibroin. Therefore this amino acid was assumed to represent every other residue along the chain. In like manner alanine represented every fourth, tyrosine every sixteenth as in Table 7.1. It was the similar data of Kossel and Dakin on the protamines and later work by Linderstrøm-Lange and Rasmussen in Copenhagen that gave Wrinch support for her hypothesis of the amino acid sequence of the gene.

Having described these early attempts to devise a plausible sequence hypothesis, what do they amount to? Very little, perhaps. Yet their existence surely served to raise the question of sequence, which had fallen into neglect since the time of Emil Fischer. Bergmann's and Niemann's idea tended to be accepted by those who speculated on molecular structure in the early days.

TABLE 7.1

The Calculation of Amino Acid Sequence from Amino Acid Composition
(Adapted from Bergmann, 1936, 55)

Protein	Amino acid	Weight (%)	Wt./Mol. Wt. (%)	Ratio	Frequency
Gelatin	Glycine	25.5	0.34	6	1 every 3
	Proline	19.7	0.17	3	1 every 6
	Hydroxyproline	14.4	0.11	2	1 every 9
	Arginine	9.1	0.052	1	1 every 18
Fibroin	Glycine	40.5	0.54	8	1 every 2
	Alanine	25.0	0.28	4	1 every 4
	Tyrosine	11.0	0.07	1	1 every 16
Protamine	Arginine	87.0	0.50	2	2 every 3
	Monoamino acids	13.0	0.25	1	1 every 3

Its influence in Florence Bell's thesis of 1939 is very evident. Astbury certainly began by assuming that some such regularities must exist. On the other hand Pauling and Niemann expressed themselves cautiously on the subject in 1939.

The International Genetical Congress

A number of special sessions on the problems of the gene were held at this, the Seventh International Congress on Genetics, in Edinburgh in August, 1939. In the absence of Vavilov, F. A. E. Crew presided. Reports from those unable to attend were read and questions on gene and virus structure were debated in a discussion described by Muller as "animated and searching" (Muller, 1939, 816). In the combined gene and chromosome session opened by Astbury, Darlington noted "a practical convergence of biology and the physical sciences, a convergence that will probably mark the beginning of a new epoch" (Darlington, 1939, 817).

On the opening day 600 geneticists from 55 countries were present; enthusiasm for the success of the conference was widespread, but by the second day "the murmurings outside grew in intensity, . . . the air became disturbed" (Crew, 1939, 496). First to leave the congress was the 34-strong German delegation, followed by the 17-strong Dutch contingent when their army was mobilized. "From that moment onwards the citizens of Edinburgh looked upon the congress as a trustworthy barometer [of the international scene]. Hourly enquiries as to whether or not the Italian, the French and the Poles were still with us were received . . ." (*Ibid.* 497).

No more dramatic turn of events could have served to terminate international enthusiasm for the great problem of gene duplication. Maybe activity in such places as the Kaiser Wilhelm-Institut for biology in Berlin–Dahlem was not seriously curtailed until 1942 (Melchers, 1971), and until about the same time in the United States; in Sweden Caspersson's work continued unimpeded from 1944 with the help of a new building (Caspersson, 1944), but in the British Isles the impact of the Second World War was felt more rapidly by the scientific community. When Darlington was flown over to Sweden in December 1942, Astbury asked him to tell Caspersson of his difficulties, "a maximum of work with a minimum of assistants, besides being a navigation officer in the University Air Squadron" (Astbury, 1942b). Early that year he had written to Fankuchen in America: "there are far too many damned things to do, that's all, and I have no secretary, no personal assistant, and not even a lab. boy now! What a war!" (Astbury, 1942a).

The Cold Spring Harbor Symposium

In the peaceful environment of Cold Spring Harbor on the northern shores of Long Island, the ideas on the duplication of the gene which we have described were once again aired in 1941, with two of the Copenhagen conference participants—Muller and Delbrück—present.

On this occasion Muller asked Wrinch whether she still adhered to the molecular model of the chromosome which she had described at the Edinburgh congress: the gene and chromosome duplication she had pictured in terms of the opening out of the tetranucleotide rings around the polypeptide chains constituting the gene. She replied cautiously:

> I think this view might be worth considering. However, I have since been visualizing the possibility that protein synthesis occurs at the surface of native proteins, possibly in the type of structure I have now proposed for protoplasm.
>
> (Wrinch, 1941, 234)

In her account of this fabric structure for the cytoplasm there is, needless to say, no mention of amino acid sequences, and I am not aware that she ever referred to such sequences after giving her fullest account of the molecular model of the chromosome in 1936.

We come to a natural stopping point in the early years of the last war, by which time the one who had enunciated the Protein Version of the Central Dogma most precisely had fallen silent about it. As for nucleic acids, evidence which we can *now* see as indicative of their genetic specificity, was treated in a diametrically opposite manner. In his summing-up of the 1941 Cold Spring Harbor meeting, Muller coolly reviewed the facts of ultraviolet mutagenesis in the context of the protein nature of the gene and concluded by questioning Delbrück's contribution—an amino acid resonance model for gene duplication in which the well-tried idea of stereochemical fit between like amino acids was yet again assumed. In that Delbrück's position did not involve specific long range forces it was an advance on that of Muller. But as to the stereochemical fit, Muller could with justice complain that "it is not evident either on ordinary chemical or on geometrical considerations why this should be the case . . ." (Muller, 1941, 305). As for the nucleic acids, both men merely expressed concern that in the Delbrück model their function was left unexplained.

Again Muller could not resist returning to his long-treasured belief in the presence of special long-range forces operative in chromosome pairing and in gene duplication. He wondered "whether we can yet be sure of the negative proposition, that forces of specific attraction, at distances greater than the ordinary atomic ones, cannot exist between genes" (*Ibid.*, 306). He reviewed Pascual Jordan's model and even pleaded for its retention in a modified form such as would overcome the criticism which Linus Pauling and Max Delbrück had made of it in 1940. Little did Muller realize that this paper which Pauling had written and persuaded Delbrück to co-author contained a foreshadowing of the symmetry principle by which DNA duplicates:

> When speculating about possible mechanisms of autocatalysis, it would therefore seem to be most rational from the point of view of the structural chemist to analyse the conditions under which complementariness and identity might coincide.
>
> (Pauling and Delbrück, 1940, 78)

Eight years later, Pauling stated what these conditions might be:

> If the structure that serves as a template (the gene or virus molecule) consists of, say, two parts, which are themselves complementary in structure, then each of these parts can serve as the mould for the production of a replica of the other part, and the complex of two complementary parts thus can serve as the mould for the production of duplicates of itself.
>
> (Pauling, 1948, 10)

Nor was Pauling the only one to perceive how complementariness and identity could be achieved in one and the same structure. Astbury had envisaged complementary replication in terms of anti-parallel polypeptide chains as in Figure 7.5.

Astbury produced this scheme under the influence of Svedberg's discovery of a pattern of molecular weights in proteins which showed integral multiples of a basic number. Astbury interpreted this in terms of a basic polypeptide chain which by successive duplication generated the higher molecular weights. "The essence of the argument," he declared,

> is that protein chains are laid down side by side to form definite patterns, in which the chains are never related to each other as an object is to its mirror image. The latest developments in the theory of genetics require such a mechanism in order that the characteristics associated with each "gene" may be reproduced, and it is now admitted that the chain systems proposed may possibly supply the necessary basis.
>
> (Astbury, 1931, 21)

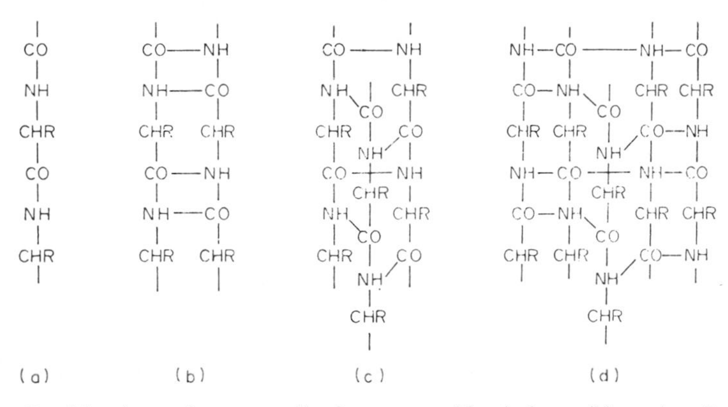

Fig. 7.5. Possible schemes for gene replication suggested by Astbury, (b) consists of complementary polypeptide chains (from Astbury and Woods, 1931, 664).

We have seen that by 1939 Astbury had gone over to nucleic acid as the intermediary between parent and nascent polypeptide chains. Direct complementary replication was sacrificed in the cause of the nucleoprotein theory of the gene, in which DNA had a role.

Conclusion

There is no escaping the fact that all attempts to establish the chemical identity of the genetic material by direct and indirect methods had failed. Chemistry had again been pressed into service in the new cytochemistry of Caspersson, but it had disappointed those who were awaiting the great event of the fusion of the chemical and the cytological approaches. We have seen Muller fascinated with the study of chromosome mechanics as a means for revealing the forces of gene duplication, the attempt of Wrinch to argue from the behaviour of whole chromosomes *down* to the structure of their constituent genes. Both failed, and as has been hinted, the fruitful approach involved starting right back at the beginning and working upwards from the principles of quantum mechanics, the structure of simple organic compounds, then the structure of more complex organic compounds extracted from the cell, finally returning to the cell and the chromosome. Chromosomes did not give the clue to the mechanism of gene duplication, physical chemistry and structural crystallography did.

CHAPTER EIGHT

The Physiology of the Gene

It has often been remarked that until the emergence of biochemical genetics from obscurity in the 1940's geneticists worked in a conceptual vacuum. To be sure they had the character, the gene, the allel, the locus and the chromosome map, but what was on the map? Loci? Genes? What *are* genes, and what is the causal sequence connecting the gene with the character it determines? With some justice, the foremost critic of the gene concept, Richard Goldschmidt, could castigate those who carried on correlating the statistics of character transmission with chromosome maps whilst they conveniently excluded discussion of the physiological problem as to how genes act. Just as F. G. Hopkins had distinguished between static and dynamic biochemistry (Hopkins, 1913; also Garrod, 1914) so Goldschmidt contrasted static and dynamic genetics (1938, 1; 1954, 85). Now how could the dynamic approach be achieved in the laboratory?

By the 1930's, viruses could be studied chemically and the results applied *by analogy* to genes; but oddly enough this approach did not lead to the correct conclusion. As early as 1902, known Mendelian character differences could be examined chemically so that such differences could be accounted for in chemical terms. In this way it was hoped that a knowledge of the nature and action of the gene would be won *indirectly* by way of its end product. The essence of the approach here was, then, *to start with the end product of the gene and to work backwards,* a promising approach, but one which in the long run submerged its advocates in the technical difficulties involved in characterizing unknown compounds and in establishing what appeared to be improbable metabolic pathways whose enzymes could not be isolated.

Nor did the *direct* approach, starting at the other end, and analysing the chemical constitution of the chromosomes fare much better. We have already seen just how inconclusive were such attempts in the nineteenth century. All that could be said at the close of the first three decades of Mendelian research was that many well known allelic genes controlled character differences whose basis was the presence or absence of a specific chemical reaction, this reaction presumably having been brought about by an enzyme, which in turn was the product of the gene, unless of course enzyme and gene were one and the same thing.

As far as the nature of the gene is concerned, it was the inconclusiveness of the indirect and analogical approaches which allowed the bizarre phenomenon of bacterial transformation to assume a central role in the demonstration that the gene is neither a protein, nor a nucleoprotein, but a nucleic acid. In Chapter 12 we shall examine this successful case of applying the direct approach. What concerns us now is the *inconclusiveness* of the other approaches. Indeed, both virus research and biochemical genetics appear to have led geneticists in the *wrong* direction, from which they were rescued by Avery and Chargaff.

Mendelism and Chemistry

The one gene–one enzyme hypothesis was not explicitly formulated until the early 1940's, but attempts to link Mendelian genetics with biochemistry date from the very early days of Mendelism, to be precise, from the year 1902, when William Bateson's book *Mendel's Principles of Heredity—a Defence* and his first report to the Evolution Committee of the Royal Society appeared. Already this Cambridge biologist had learnt of the work on alkaptonuria which the London physician Garrod, later Sir Archibald, was so actively pursuing. Fifty years later the Nobel laureate, George Beadle, was to draw attention to this important pioneer study and honour its author with the title "father of chemical genetics" (1950, 222).

What has puzzled both Beadle, and many other writers, is the neglect of Garrod's work; a neglect so great as to leave those like Beadle, who came to the subject of biochemical genetics in the 1930's, completely oblivious of it. It is worth examining this case of neglect if only to throw light on the limitations and difficulties inherent in Garrod's approach.

Attempts to account for the neglect of Garrod have, of course, been made before (Dunn, 1962; Glass, 1965; Strauss, 1960), but one point has been overlooked: *an association between the facts of Mendelism and those of biochemical individuality and metabolism was seen by the early Mendelians as natural.* In Bateson's mind such an association was justified, whereas the association between Mendelism and the chromosomes was not; and was this attitude not understandable? When one thinks of the various "Mendelizing characters" of those days, one class stands out prominently: colour differences. These were known to be due to chemical compounds whose nature was partly understood *before* the rediscovery of Mendelian ratios in 1900. Thus the anthocyanins of plants were widely believed to arise from the oxidation of a "chromogen" related to tannin—(Wigand, 1862). The black colour of Japanese lacquer was held to arise from the oxidation of a substance in the latex with the aid of an enzyme (Yoshida, 1883). This enzyme was later described as an oxidase and named "laccase". Certain white and yellow patches on the snake had been attributed to a combination of guanine and a "proteid base" (Leydig, 1888). The remarkable association of guanine with lipochromes and melanins in fish scales first observed by Barreswil in 1861 had been extended by Cunningham and MacMunn in 1893.

Finally, if the reader is still not convinced that specific differences were thought to have a chemical basis before 1900 let him consult Gowland Hopkins' paper on butterfly pigments of 1896. While Gull Research Student at Guy's hospital Hopkins had thought to show that the opaque white colour of the wings of a Cabbage White butterfly was due to uric acid (later shown to be a pteridine) excreted in the pupal stage, and that the yellow colour in the Brimstone's wings was that of a derivative of uric acid which he called "lepidoporphyrin". But when he tried to isolate similar compounds from butterflies outside the family Pieridae he failed. The yellow pigment of the Swallowtail's wing was not the same as that of the Brimstone. Such differences were also seen in cases of mimicry. " . . . the mimicking Pierid retains the characteristic pigments of its group, while those of the mimicked Héliconid are quite distinct" (Hopkins, 1896, 680). Hopkins found in the strict confinement of these special pigments to the Pieridae "interesting evidence justifying the customary classification of these insects as a natural group" (*Ibid*, 680).

Colour varieties and their corresponding albino forms cried out for an explanatory mechanism of gene action in terms of the presence or absence of an enzyme specific to a reaction in the biosynthesis of the colour pigment. Genes either produced enzymes, or were the enzymes themselves. The same went for the melanins of human skin in normal and albino individuals and the chromogens responsible for coat colour in mammals.

It was surely such a view of gene action which led three of the foremost Mendelians: Lucien Cuénot, Carl Correns and William Bateson to advocate the "Presence and Absence" theory according to which the dominance of one variety over another in the hybrid formed between them was due not to the relative strengths of the genetic units involved (*A* and *a*) but the simple fact that the recessive form lacked the gene which the dominant form possessed.

It would be a grave error to imagine that this hypothesis was the atypical product of Bateson's dogmatic school, for when he advocated it in 1906 he was only lending his support to the view already advanced by Cuénot, and by Correns in 1903. Even Morgan accepted it in 1910 only to reject it when he established the chromosome theory of heredity, for then he had to ask how the dominant gene (presence) could pair in meiosis with its recessive partner (absence)?

It should be clear, therefore, that there did exist an early tradition of biochemical genetics, which was not continued after the establishment of the chromosome theory of heredity. The period of neglect lasted from about 1912 to 1935. The presence and absence hypothesis, once relegated to oblivion by the demands of the Morganists (and the facts of multiple allelism) the links with chemistry loosened and geneticists forged ahead with chromosome mapping, analysis of position effect, the introduction of modifier genes; in short they fashioned an even more complex association of genes. But the question which Goldschmidt asked was: do all these discrete units really exist? Perhaps the existence of numerous expressions of one character (multiple allelism) had a physiological explanation. Whilst there was a natural concern by the 1930s to unravel the sequence of events by which the gene determines its character, this concern was deepened by the attacks of Goldschmidt on the Morganist school, and when two cases of multiple allelism—fruit fly and meal moth eye colour—appeared to offer a way into the mechnaism of gene action, active research programmes in biochemical genetics were once more pursued. But what were the conditions for the earlier phase of such work? Three factors stand out: Garrod's work, especially his study of alkaptonuria, the facts of plant and animal pigments, and the conception of gene function as auto- and hetero-catalytic.

Alkaptonuria

This name was given by Bödeker to the disease in which urine darkened as it took up oxygen (καπτεη = to suck up greedily, and the arabic word alcali). This very rare condition—in 3000 cases of urine abnormalities presented to F. G. Hopkins at Guy's he found only one clear case—owed its importance in genetics to the way it "advertised its presence", as Garrod put it. The parents of alkaptonuric babies were naturally worried that the dark staining on nappies left exposed to the air indicated some sinister malady. So they brought their babies to the attention of medical men and these cases got recorded.

As long ago as 1859 the substance in the urine with this darkening property had been isolated, and in 1891 its chemical structure was discovered through the classic research of Wolkow and Baumann. Their conclusions were supported subsequently by synthesis according to three different methods (Baumann and Fränkel, 1895; Osborne, 1903; Neubauer and Flatow, 1907). Nor was this all. The darkening substance, alkapton, was excreted in greatly increased quantities if the amino acid tyrosine was fed to the alkaptonuric patient. Wolkow and Baumann therefore concluded that alkapton was produced by the oxidation of tyrosine. How on earth did they jump to this remarkable and correct diagnosis? Eugene Knox has given us a lucid answer to this question (Knox, 1958). They had established its empirical formula as $C_8H_8O_4$, but none of the sixteen known hydroxy-acids corresponded to this formula. Nearest to it was gentisic acid, in which case alkapton represented the next, as yet unknown, homologue in the series, so they renamed alkapton "homogentisic acid".

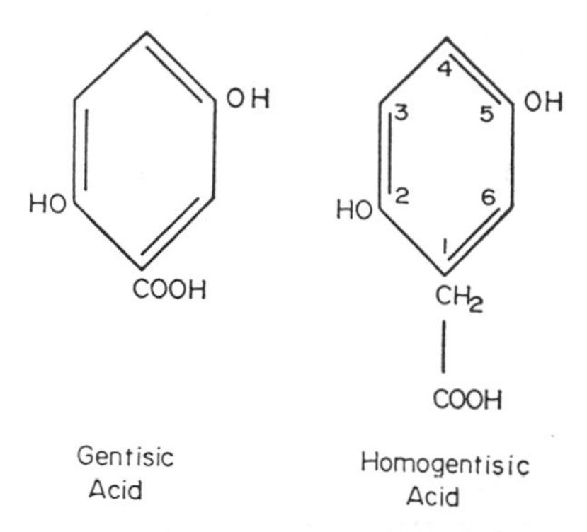

Gentisic Acid

Homogentisic Acid

Now Baumann was sure that animals cannot make aromatic compounds from aliphatic ones—they cannot synthesize the aromatic benzene ring—so where, in their diet, could the aromatic ring have come from? Assuming it to be a protein, he knew of but two possibilities: tyrosine and phenylalanine. As Wolkow and Baumann had only sufficient quantities of the former, this was used in the feeding trials which led them to the correct metabolic reaction:

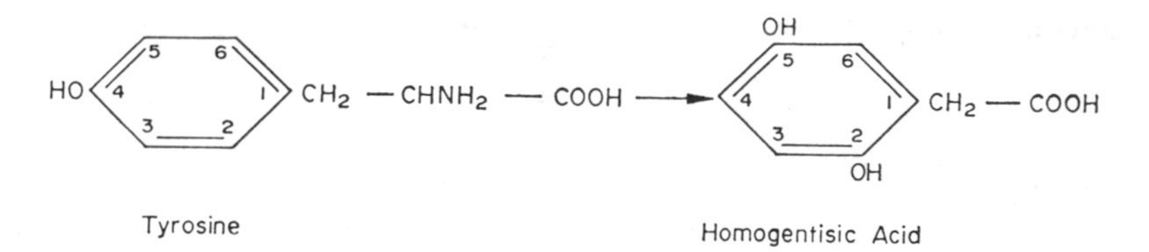

Tyrosine

Homogentisic Acid

Unfortunately this reaction involved the migration of a hydroxyl group. If it really does occur, declared the German authors, "then here we have an organism doing in metabolism what is *not already known to chemistry.* The certainty built up by the study of the metabolism of hundreds of compounds over the past decade would be lost. Metabolism might do anything, and to investigate the metabolism of any substance would be useless" (Wolkow and Baumann, 1891, 274; trans. in Knox, 1958, 97). The authors therefore felt compelled to attribute this extraordinary metabolic transformation to microorganisms of the gut, as if it did not matter for physiological chemistry so long as atypical reactions were confined to the metabolism of these bizarre creatures!

This objection to the tyrosine–homogentisic acid path as a *part of the normal* metabolic oxidation of tyrosine in the human body was temporarily overcome by the discovery of a side chain migration reaction in chemistry (*p*-cresol to toluhydroquinone; Bamberger, 1903). Unfortunately the eminent biochemist, Dakin, supplied evidence seven years later which was reckoned to rule out the hypothetical quinone pathway between tyrosine and homogentisic acid which Bamberger's reaction suggested, and it was not until 1949 that radioactive tracers showed up side chain migration in an appropriate metabolic oxidation (Schepartz and Gurin, 1949).

In the case of alkaptonuria, then, we have initial enthusiasm and subsequent rejection of the belief that in such sports of nature we gain an insight into the links in the normal metabolic pathways under the control of enzymes. If homogentisic acid represented *a member of a quite different pathway*, that of the alkaptonuric, one could learn nothing about the normal pathway from studying it. As if this was not enough to discourage the biochemist, no enzymes responsible for the formation of homogentisic acid were isolated until 1951, and the enzyme thought to be missing from alkaptonurics, which in normal individuals cleaves the benzene ring, was not isolated until 1958 (La Du).

This failure provides yet a further reason for the neglect of Garrod's work and one is left wondering how Garrod could have been so sure of his view that alkaptonuria *was* caused by a block in the normal metabolic pathway due to the absence of an enzyme.

Garrod's Conception of a Metabolic Block

Long before he was presented with an alkaptonuric patient Garrod had been directing his attention to the chemistry of urine. In 1895/96, when he was at St. Batholomew's Hospital, he had begun collaborative work in this area with F. G. Hopkins (later Sir Frederick), then at Guy's Hospital. Both Bateson and Garrod helped Hopkins in his work on the wing colours of butterflies. Garrod was conversant with Continental work, having paid a visit to Vienna after qualifying at Barts in 1884 (Rolleston, 1949). From 1902 onwards he repeatedly referred to the "most suggestive" rectoral address of Karl H. H. Huppert of Prague, "On the Maintenance of Species Characteristics" in 1895, which supported Garrod's belief that subtle differences in the chemistry of different individuals exist alongside more patent differences in external characteristics (Garrod, 1902, 118–119). Huppert, it should be noted, worked on alkaptonuria and confirmed the identity of alkapton as homogentisic acid in 1899.

Garrod is reported to have said that "patients did not really interest him and the complex problem presented by an individual who is ill did not really appeal to him for solution. In a patient he saw one symptom only, one abnormality, and on this he would concentrate . . ." (Frazer, 1936, 1808). All morning he would puzzle over a urine specimen, he would pop over to Dr Hurtley in the Chemistry Department, or consult later with his Oxford and Cambridge friends—no doubt Hopkins was often the one to help him. How natural, then, that Garrod should have shown an interest in the alkaptonuric baby under Dr Voelcker's care in the Hospital for Sick Children, Great Ormond Street, London. Garrod was an honorary consultant physician to this and two other hospitals. The boy was transferred to Garrod's care, probably in 1898. Within a year Garrod was able to support the conclusion of Wolkow and Baumann that the cause of alkaptonuria was homogentisic acid (Garrod, 1899). Two years later he presented fresh evidence allowing him to oppose the Germans' microbial theory of the origin of this acid.

In the spring of 1901, to Garrod's delight, this boy was no longer the only alkaptonuric in his family. A fifth child was born; both mother and nurse "were fully alive to the possibility that the child might show the same peculiarity as its elder brother, and were on the look-out for any indication that this was the case" (Garrod, 1902a, 72). Sure enough, by the second day the baby's nappy showed a slight staining, just as his elder brother's had done in 1897. Now the family allowed Garrod to have their elder son, by now four years old, under observation. So Garrod could study his output of homogentisic acid in relation to meal times. The maximum output came four to six hours after the main meal. During this period, Garrod pointed out, tyrosine could be absorbed and broken down in *the tissues* before being carried to the bladder. Moreover, no homogentisic acid could be discovered in the faeces. Confidently, Garrod declared to the Medico-Chirurgical Society in the Autumn of 1901:

> The facts here brought forward lend support to the view that alkaptonuria is what may be described as a "freak" of metabolism, a chemical abnormality more or less analogous to structural malformations. They can hardly be reconciled with the theory that it results from a special form of infection of the alimentary canal. There is here no question of the intensification of family tendencies by intermarriage, for in no instance were the parents themselves alkaptonuric, and . . . there is, up to now, no recorded instance of alkaptonuria in two generations of a family.
>
> (1902a, 71)

But more exciting still was the discovery that the parents of Garrod's alkaptonuric family were first cousins. From this time onwards Garrod enquired in every possible instance about the relationships in alkaptonuric families. It could not, he argued, be mere chance that three out of the four alkaptonuric families he had described in 1901 were first cousin marriages, when one recalled the low proportion of such marriages in the population—George Darwin had estimated the *upper limit* for the population at large at 3 per cent, in London the proportion being lower (Darwin, 1875).

The next year, 1902, saw the publication of Bateson's first Report (with Miss Saunders) to the Evolution Committee of the Royal Society in which Garrod's correlation between consanguinity and alkaptonuria was accounted for on the assumption that the condition was due to a recessive Mendelian factor. With evident delight Garrod quoted Bateson's words as follows:

"Now there may be other accounts possible, but we note that the mating of first cousins gives exactly the conditions most likely to enable a rare, and usually recessive, character to show itself. If the bearer of such a gamete mate with individuals not bearing it the character will hardly ever be seen; but first cousins will frequently be the bearers of similar gametes, which may in such unions meet each other and thus lead to the manifestation of the peculiar recessive characters in the zygote." Such an explanation removes the question altogether out of the range of prejudice, for if it be the true account of the matter it is not the mating of first cousins in general but of those who come of particular stocks that tends to induce the development of alkaptonuria in the offspring. Whether the Mendelian explanation be the true one or no there seems to be little room for doubt that the peculiarities of the incidence of alkaptonuria and of conditions which appear in a similar way are best explained by supposing that, leaving aside exceptional cases in which the character, usually recessive, assumes dominance, a peculiarity of the gametes of *both* parents is necessary for its production.

(Bateson and Saunders, 1902, 133 footnote; Garrod 1902b, 116-117)

In his classic book *The Inborn Errors of Metabolism* (1909), Garrod was able to go much further and state that the error in alkaptonurics involved a failure to break up the benzene ring in all acids like homogentisic acid in which the hydroxyl groups are in positions 2 and 5. The error was hence located "in the penultimate stage of the catabolism of the aromatic protein fractions . . ." This led to the final inference:

We may further conceive that the splitting of the benzene ring in normal metabolism is the work of a special enzyme, that in congenital alkaptonuria this enzyme is wanting, whilst in disease its working may be partially or completely inhibited.

(Garrod, 1909, 50)

This remarkable grasp of the nature of biochemical abnormalities was based upon his underlying picture of metabolism as a continuous movement from one intermediary product to another, the latter having but a momentary existence save when, in disease, one step in the process failed and an intermediate accumulated. In such cases wrote Garrod:

All that is known of the course of catabolism tends to show that . . . the intermediate product is being excreted as such, rather than that it is further dealt with along abnormal lines. . . . if the conception of metabolism in compartments, under the influence of enzymes, be a correct one, it is unlikely, *a priori*, that alternative paths are provided which may be followed when for any reason the normal paths are blocked.

(Garrod, 1909, 8)

These so-called diseases were then "errors of metabolism" which were congenital (existing from birth) and inborn (transmitted through the gametes). Should the reader regard this conclusion as over speculative in the context of knowledge in 1909, let it be emphasized that Garrod was able to argue cogently for the view that the normal body was equipped to deal with only certain of the aromatic compounds given to it. The majority escaped with the benzene ring intact and these were excreted either in combination with sulphuric acid (aromatic sulphates) or in combination with glycine (hippuric acid group). Even in the case of gentisic acid, a portion of the intake appeared in the urine in the form of aromatic sulphate. Not so tyrosine, phenylalanine and homogentisic acid. Clearly the body was equipped to deal with these. Was this not because they belonged to the *normal* series of reaction leading to their complete oxidation? And if the alkaptonuric individual could not complete this oxidation, then Garrod could point to evidence that this lack was total, not just impaired function but *no* function, from the second day of life onwards into old age. This was an all-or-nothing phenomenon like the clear discontinuous "contrasted characters" of Mendelian genetics. How strongly, as we may now see, did alkaptonuria and its like—albinism, cystinuria, and pentosuria—support the presence and absence hypothesis, and a direct link between Mendelian factors and enzymes.

Garrod's Influence

This appears to have been greater among biochemists than among geneticists, and least among the medical profession, with the exception of paediatricians. In England Garrod's influence was most clearly seen in the writings of Hopkins and Haldane. Now clearly, genetical experimentation with human abnormalities was out of the question, therefore the data was to remain totally inadequate as evidence for Mendelian heredity in man. Certainly, examples like alkaptonuria could be accounted for in Mendelian terms more satisfactorily than by any other theory, but in those early days of Mendelism more was required than a good explanation. Mendelism was under severe attack and convincing ammunition was sorely needed by the Bateson school with which to fight the school of biometricians under Karl Pearson's determined leadership.

In 1902 Bateson and Cuénot independently extended Mendel's laws to the animal kingdom. But as for man, Garrod's cases were so rare and seemed so bizarre as to be irrelevant to the work of the medical profession. Naturally, many physicians felt that efforts to show the inheritance and congenital nature of a disease were of little use to them in their work of curing diseases. Nor did alkaptonuria help the Mendelians at a time when the Hardy-Weinberg law governing the distribution of dominant and recessive genes in a randomly breeding population was unknown. Hence, in a large sample of the population a 3:1 ratio between normal and alkaptonuric individuals was expected on the Mendelian theory.

The influence of Garrod's work was undermined by the very incomplete nature of Mendelism as expounded from Mendel's paper in the first few years of this century. Now the biometricians were anchored in the discipline of human biology on the basis of Francis Galton's pioneer anthropometric measurements in the laboratory which he established in University College London. When R. C. Punnet addressed the Royal Medical Society on

Mendelian heredity in 1908 he was vigorously attacked by the biometricians. The report of this meeting presents a remarkably clear picture of the nature of this opposition. It was after this encounter with the statistician, Udne Yule, that Punnett asked the Cambridge mathematician, Sir G. H. Hardy, to answer Yule's question why Mendelizing characters are not distributed in human populations in the ratio of three dominants to one recessive. The result was the famous equation known as the Hardy Weinberg law of 1908.

Apart from this shortcoming, L. C. Dunn has drawn attention to the naive and misplaced aims of eugenicists in the early days of Mendelism, and the attendant lowering of standards in research on human genetics. Garrod's work, he said, "was caught up in the cross-currents to which all studies of man are exposed." Progress in human genetics was impeded less by "lack of means than by lack of a clear scientific goal . . ." (Dunn, 1962, 1).

When we turn to biochemistry, we find that Garrod's work stimulated research into the chemistry of metabolism. Admittedly there existed in the previous century a tradition of such work within the discipline of physiological chemistry. Even if Hopkins could deplore the dominant position of organic chemistry, which he saw as static, the references in Garrod's early papers to work on the chemistry of metabolism should suffice to show that Garrod was not so much creating a new tradition as making explicit a vague and insecure tradition. This he did by denying that metabolism in pathological states is radically different from the normal state. From this the concept of the metabolic pathway followed, a concept richly suggestive of experimental tests. True, Garnier and Voirin (1892) first suggested that homogentisic acid was a product of normal metabolism, yet it is to Garrod that we owe the concept of alkaptonuria as the result of *complete absence* of an enzyme capable of cleaving the benzene ring hydroxylated in positions two and five.

What we must not forget is that Garrod did not express this view in 1902 but in 1909 after a large number of papers by Abderhalden, Embden, Salomon, Schmidt, Falta, Neubauer, Langstein, Mittelbach and Meyer had been published. A study of the literature after 1909 would show whether Garrod's synthesis of the work up to 1909 stimulated the field. The impression one has is that his contribution was well known and well received in biochemical circles before the First World War, but even then it was not prized for its biological significance. Fölling's work apart, there seem to have been few important developments until after the last war, by which time there were available the new techniques of chromatography, autoradiography and much else that was new in enzymology. Both Continental and British biochemists of Garrod's day were not very excited by the hereditary aspects of the inborn errors, nor did they shout about the glimpses which these metabolic disorders offered of the pathways by which compounds are broken down in the body—always excepting Hopkins. The early years of this century saw the blossoming forth of biochemistry as a channelling of the older discipline of physiological chemistry into the study of enzyme controlled metabolic pathways. But it was not Garrod's approach which offered the best tool for attacking this programme. Instead, enthusiasm centred on the new enzymology which Buchner's successful extraction of zymase (1897) had opened up. The great tradition established by Harden, Young, Warburg, and Meyerhof was not concerned with such questions as how the body disposes of tyrosine but with how sugar is broken down to liberate energy. This new world of the young subject of biochemistry, it appears, had little contact with the *older* world of Garrod's physiological chemistry. And as for Garrod's work in forging links between genetics and biochemistry we might well have had to wait until the 1950's before anything substantial could have been achieved in view of the difficulty of isolating the enzymes involved. Fortunately, there existed another group of phenomena which from the early years of Mendelism invited collaboration between biochemists and geneticists—the genetics and biochemistry of plant and animal pigments.

Anthocyanins

Work on the genetics of anthocyanins was pursued both in Germany, Holland and England. In Cambridge, Bateson had not only seen the significance of Garrod's work but also of R. P. Gregory's study of the shape of starch grains in round and wrinkled seeds. Gregory had shown that round seeds of the edible pea contained simple elongated starch grains, whereas wrinkled seeds contained small irregularly shaped grains often compounded into clusters (Gregory 1903). Later, Bateson urged that this characteristic be studied further:

> The interrelations of round and wrinkled seeds are to be recommended as offering perhaps the most favourable example for an investigation of the chemical nature of a genetic factor. The wrinkling is evidently the consequence of a particular method of drying, and this must depend on the nature of the reserve-materials. A first step would be to determine the relative amounts of sugar and starch in the two chief types. It is natural to suppose that the wrinkled peas are those in which the transformation of sugar into starch has gone less far than in the round peas; but, as much starch is formed in the wrinkles, one ferment having this transformative power must be present in them. Hence we are led to suppose that in the round pea a second ferment is present which can carry the process further. As offering an attractive problem in physiological chemistry the phenomena are recommended to those who have the requisite skill to investigate them.
>
> (Bateson, 1909, 29–30)

A student of biochemistry, Muriel Wheldale (later Mrs. Onslow), began helping Bateson in 1903. Later she worked under Bateson on the inheritance of flower colour in snapdragons. Her work began in Cambridge, was continued at the new John Innes Institute in Merton on the outskirts of London, where she developed it in the direction of biochemistry. By the time of her return to Cambridge as a fellow of Newnham College, she was firmly in that tradition. Her famous book *The Anthocyanin Pigments in Plants* of 1916 is divided into two parts, one devoted to biochemical aspects, the other to genetical aspects, but in the latter there are no references to papers by Wheldale on the genetics of anthocyanins after 1910. Indeed there appears to have been a failure to carry this work much further on the genetical side until about two decades later. Thus, Rose Scott-Moncrieff could write that Wheldale's lead had only recently been "followed up by an attempt to link up this new knowledge of pigment chemistry with the fast accumulating information on the genetical factors controlling colour variation in plants". Hitherto it had only been by the "fortuitous coincidence" of chemical and genetic study by independent workers that any links were made whatever. But in the mid-thirties "a unique opportunity was offered by this institution [John Innes] for a combined attack on this interesting problem" (Scott-Moncrieff, 1936, 118). This opportunity was provided by the Robinsons' work on the anthocyanins which had yielded both an insight into their structure and the quick qualitative tests requiring only small quantities of flowers. These proved admirable for the genetic studies by W. J. C. Lawrence and Scott-Moncrieff. Muriel Onslow urged that the work be taken up again in 1928. In 1929 J. B. S. Haldane suggested contact be made with the John Innes, and the result was "the transference of the biochemical work back to Merton as in the pioneer days" (Scott-Moncrieff, 1937, 230).

Wheldale had found that an "ivory" snapdragon (very pale yellow) crossed with a pure white variety gave in F_1 the dull red flower colour typical of the common snapdragon. From the flowers of the ivory variety she extracted an anthocyanin, so she inferred that the pure white variety had introduced a "factor" responsible for the oxidation of the ivory anthocyanin to the dull red form (Wheldale, 1907; Bateson, 1909, 280).

Bateson was highly delighted with this result as he had been under attack from his opponents for the gratuitous way in which he had accounted for distant reversion by attributing to the wild-type, long-lost (distant) character, a compound nature lost by selective breeding and restored by reconstitution of its components in the act of crossing appropriate varieties. He wrote:

> Not often can we hope to be able to specify the complementary elements which must meet each other in order that a certain compound character may be produced. Nevertheless, by the co-operation of physiological chemistry with genetics there is every hope that in favourable cases of a simple order actual demonstrations of these elements may be carried out. Perhaps the nearest approach to such an achievement is that made by Miss Wheldale in her experiments on Antirrhinum (Snapdragon).
>
> (Bateson, 1909, 279)

The old tendency to confuse the gene with its product in describing such work (see Darlington, 1964, 94) was if anything encouraged by the association of Mendelian factors with enzymes. Indeed it led many, though not Bateson, to regard the Mendelian factors as enzymes. Consider the following statement by Wheldale:

> The foregoing representation of Mendelian factors in terms of oxydases is not purely speculative. Experimentally the petals of Sweet Pea and Stock plants of known pedigree have been found to vary with respect to their oxidative powers on guaiacum. It is evident that the loss of either element, oxygenase or peroxydase, from the initial reddening factor will give rise to albinism even though the albino may carry blueing or other modifying factors. Hence in any genus having a complex series of colour varieties, there are a number of possible albinos, both genetically and physiologically different from each other. By selection we are able to procure an albino carrying any particular factor and it is to a large extent the behaviour of these extracted albinos toward Guaiacum which had led me to infer that there is a definite connection between Mendelian colour factors and oxidizing enzymes.
>
> (Wheldale, 1910, 470)

This early work on anthocyanins lent valuable support to the idea that Mendelian differences have a chemical basis, and that genes act as enzymes. Admittedly the enzymes involved could not be isolated and it turned out later that the metabolic links which Wheldale regarded as oxidations were in fact reductions. Nonetheless, her work gave a glimpse of a metabolic pathway—what she called "a highly complex process involving a series of progressive stages" (1910, 467)—leading to the production of a Mendelian character. This picture served to keep alive the desire to push such enquiries further, and it formed the starting point for many speculative discussions of the chemical basis of heredity.

Haldane's Influence

Early in his life Haldane had taken part in physiological experiments with his father, and later he came under the influence of F. G. Hopkins when he was reader in biochemistry at Cambridge (1922–33). The pain caused by his dismissal, appeal and subsequent reinstatement (1925–26) at Cambridge probably caused him to look elsewhere for the realization of his ambitions. In 1927, whilst still reader in biochemistry he became officer in charge of genetical investigations at the John Innes Institute under the 63-year-old director, Sir Daniel Hall. The latter was expected to retire soon, so Haldane anticipated promotion. What better omen could this be for the re-establishment of biochemical genetics at the John Innes? In the biochemistry department at Cambridge were Muriel Wheldale, her husband Huia Onslow and Haldane's research student Murray, all interested in the physiology of gene action. Murray was working on the influence of the pH of plant sap upon flower colour (Haldane, 1929, 6). In 1928, Muriel Onslow urged that her earlier work at the John Innes be taken up again. Haldane was able to carry out her suggestion. He obtained support from the D.S.I.R. for Rose Scott-Moncrieff to work on the anthocyanins. Later he was to view with pride this association with her work. Her study of thirty-five genes controlling anthocyanin production he called "a model for future researches". Perhaps, he suggested, his initiation of her work was his most important contribution to biochemistry (Haldane, 1937b, 4).

It was Julian Huxley who advised Hall to call upon Haldane's services. He came for brief visits during term-time and a month at a stretch in vacations. This association was in Darlington's view the salvation of the John Innes. For Haldane it was the means of uniting his knowledge of enzymology with his interest in theoretical genetics. His book *Enzymes* (1930) shows no concern whatever for the relation between genes and enzymes, but in *New Paths in Genetics* eleven years later, he placed special emphasis on this very subject.

Scott-Moncrieff became a member of the permanent staff and continued her very fine study of plant pigments until she left in 1936. This extension of the early work gave added weight to the importance of genetical research for horticulture, plant systematics and organic chemistry. Here was evidence of the fruit to be reaped from collaboration between geneticists and organic chemists—the Robinsons at Oxford. What did Haldane contribute to this programme after he had set it up? It is difficult to find anything positive and original. Indeed the work seems to have continued in spite of him. His feud with Hall, the director, was making life difficult at the Institute. His approach to the organization of research was fragmented. In 1936 a committee was set up to inquire into the organization and staffing of the Institute. It considered dismissing Haldane, but was saved this embarrassment by the action of University College London where he was appointed to the full-time chair of biometry (Clark, 1968, 105).

When Haldane resigned from the John Innes the programme of biochemical genetics was still going strong. J. R. Price, Scott-Moncrieff's successor, saw the potential significance of the work when he wrote as follows in his first report:

> For the first time, then, gene action can be examined in its most fundamental sense, namely as governing simple (but essential) chemical changes. From this stand point alone, the potentialities of such work are very great. But this is only one aspect of the genetics of flower colour.
>
> (Price, 1938, 15)

Price went on to emphasize the biosynthetic aspect. A year before, Haldane had written his last report for the Institute. It was a melancholy swan song: "It will be seen from the reports of individual workers that, while a number of points of detail have been elucidated, no genetical work has been accomplished which is likely to be of fundamental scientific importance" (Haldane, 1937c, 8; cited in Darlington, 1968, 934). No doubt Haldane merely meant that no "new" genetical work had been accomplished . . ., but this only serves to underline his failure to give the work a fresh and deeper direction in contrast to Beadle and Tatum in their parallel and independent studies of eye colour in insects. Haldane could be speculative, imaginative, stimulating, a link-man between disciplines, but he could neither conceive nor direct the sort of experimental research programme required to explore the question of the modes of gene action. This demanded a different approach and different research material. Neither Haldane nor Darlington seemed to have grasped the limitations of the work with anthocyanins. They saw its successes in the context of the John Innes, an institute founded for the benefit of horticulture and therefore centred upon flowering plants. In Germany a similar association of early Mendelian genetics with higher plants played a part in the failure of German scientists to break fresh ground in their studies of the physiology of the gene.

The School of Correns

It was chiefly Correns who established a tradition of physiological genetics on the Continent. This retiring and reticent man had a remarkable understanding of the subtleties of his subject (see Olby, 1971c). His concern to understand the dominance–recessive relationship led to his hybridization of the anthocyanin-containing henbane (*Hyoscyamus niger*) with the anthocyanin-free form (*H. pallidus*). Since the F_1 hybrid flowers showed an intermediate coloration he was led to suggest the presence and absence theory (1903). In his paper on hybrids between annual and biennial henbane, a year later, he concluded that segregating colour characters of flower, fruit, cotyledon and foliage were to be traced back directly to chemical processes (1904, 517).

When at the last minute Boveri decided not to come to direct the new Kaiser Wilhelm-Institut für Biologie in Berlin-Dahlem, Correns was appointed instead. He was assisted in Berlin-Dahlem by the young Fritz Wettstein, later von Wettstein (1895–1945), who spoke of his master's continuing concern for the physiological aspects of genetics. There is scarcely a paper by Correns, declared von Wettstein, which does not repeatedly touch on such questions and contribute to their solution. In him von Wettstein saw the "general physiologist", not a man restricted in his view to the mathematical problems of the distribution of the *Anlagen*, but one who included in his purview the problems of their expression in the entire organism (Wettstein, 1938, 149).

Von Wettstein returned to the Kaiser Wilhelm-Institut to succeed his master after a period of ten years spent at various universities (1924–34). In those ten years he established his reputation on the basis of his research into the physiological genetics of mosses.* He it was who introduced the technique

* Though published in this period, these researches occupied him from 1919 in Berlin under Correns (D. von Wettstein, 1974).

of tetrad analysis, using chloral hydrate to retain the products of meiosis in their original positions in the tetrad, thus allowing distinction between the products of the first and second division to be made. It was he who used mosses to achieve a polyploid series and to express the varying degrees of dominance in the members of such a series in terms of the quantity of active gene material introduced into them. It was under von Wettstein in Munich that Maria Schönfeld and Heinz Wülker studied the genetics of the bread mold (*Neurospora sitophila* and *tetrasperma*) from 1932–34 (Schönfeld, 1935; Wülker, 1935). Again it was under von Wettstein that Melchers carried the *Hyoscyamus* studies of Correns to a deeper level of analysis in the search for the "reaction chains which lead in the course of ontogeny from the gene to the future character" (Melchers, 1939, 213).

Ephestia Studies

Before his appointment to Munich von Wettstein had nearly six years in Göttingen as professor of Botany. Here he had found in Alfred Kühn, a professor of Zoology, a man who thought like him about genetics. Together, they worked to break down the traditional barriers between zoology and botany. They held joint seminars and in their researches concentrated on basic physiological questions. At this time Kühn had a group working on the production of eye colour in the meal moth *Ephestia Kuhniella.* Ernst Caspari in Göttingen had found a form of this moth with pale eyes and testes. He had used the technique of organ transplantation to study the action of the diffusible substance or "hormone" which he believed to control pigment production in these organs. This technique, first developed for other purposes by Meisenheimer (1909), was used for genetic purposes by the Göttingen group, then adapted by Boris Ephrussi and Beadle in Paris.

It is remarkable that the German physiological geneticists appear to have possessed all that they needed for attacking the genetics of metabolism in the manner subsequently chosen by Beadle and Edward Tatum. In Göttingen there was the work on eye colour rivalling that of Beadle and Ephrussi. In Munich University there was the work on *Neurospora.* True, political difficulties in Göttingen during the Nazi period forced so firm a critic of this regime as Kühn to resign in 1937, and von Wettstein had left Munich for Berlin-Dahlem in 1934. But through the influence of Max Planck and von Wettstein, Kühn was able to join his Göttingen friend at the Kaiser Wilhelm-Institut where he took over the division formerly directed by Richard Goldschmidt. Here, in this island of freedom surrounded by an "ocean of loud propaganda" (Melchers, 1953, 14). Kühn continued his work, and in 1940 Butenandt, Wolfhard Weidel, and Erich Becker succeeded in identifying the eye pigment studied by Beadle and Ephrussi, to which they gave the name "kynurenine" and recognized it as a derivative of tryptophane.

Apart from the fact that war conditions were soon to make experimental work in Germany well nigh impossible, Kühn's school under A. Butenandt failed to ask themselves *whether it would not be better to turn their programme on its head* and instead of taking *known genetic features* and working out their chemistry, they should take *known chemical features* and work out the genetics. It was as if their success in achieving the goal of identifying kynurenine—Tatum and Haagen Smith succeeded a year later—prevented the German workers from re-examining their whole approach. Even had they done so, they surely lacked one essential piece of expertise—experience with the minimal media culture of the bread mold. The work on *Neurospora* in Munich had been carried out with either beer wort or malt extract agar. The cytogenetics of characters such as sex determination was being explored by means of tetrad analysis. Such a programme of research did not connect with the physiollogical studies in Göttingen and Berlin-Dahlem.

When Georg Melchers thought of all the researches going on under von Wettstein, Erwin Baur and Alfred Kühn he felt that von Wettstein's division came very close to making *Neurospora* the object of their developmental-physiological programme. But he concluded that: "We hung on too closely in those years to the higher plants to take this step as well" (Melchers, 1962, 23). Weidel himself, according to Melchers, always felt badly about his failure to perceive the possibility of using *Neurospora* in the way taken by Beadle and Tatum (Melchers, 1971).

After the last war, Butenandt, Weidel and Schlossberger returned to the eye colour pigments and were able to show that the cinnabar mutant eye colour in *Drosophila* was due to the oxidation product of kynurenine—hydroxykynurenine (1949). But by this time the key to such an identification had already been supplied by work on *Neurospora* (Mitchell and Nye, 1948).

Drosophila Studies

We have seen that the John Innes Institute was heavily committed to flowering plants. In Cambridge, Bateson's successor, R. C. Punnett, was still an enthusiastic tennis player but not a magnetic force for genetics. At Columbia University, T. H. Morgan (1866–1945) represented the tradition of close association between cytology and genetics, but with embryology in the background ready to be exploited when the right opportunity presented itself. But with such a rich harvest to gather in from the cytogenetics of *Drosophila* it is hardly surprising that Sturtevant's discovery of examples of non-autonomous gene action lay unexploited for seventeen years. Not until Ephrussi came to Caltech in 1933 and met the maize geneticist, George Beadle, who was currently investigating crossing-over in *Drosophila,* was the non-autonomous gene for eye pigment seen to offer a way in to the mystery of gene action.

It was, it seems, Ephrussi who stimulated Beadle's interest in gene action and exposed his latent unease with Morganist "formal genetics":

> . . . he and I had many long discussions in which we deplored the lack of information about the manner in which genes act on development. This we ascribed to the fact that the classical organisms of experimental embryology did not lend themselves readily to genetic investigation. Contrariwise, those animals and plants about which most was known genetically had been little used in studies of development.
>
> (Beadle, 1958, 590)

The obvious plan was to try to learn more about the embryology of *Drosophila* and with this in mind Beadle joined Ephrussi at the Institut de Biologie physicochimique in Paris where they began to attempt tissue culturing *Drosophila* cells.

Meanwhile, Caspari in Kühn's zoological institute in Göttingen, had utilized the technique of organ transplantation for genetic study in *Ephestia*. Ephrussi's suggestion to Beadle that they adapt this technique to *Drosophila* succeeded despite the pessimism of one noted Sorbonne authority. Now they could investigate the expression of genes of one type against a background of another and register the modification in the character produced. The obvious choice was eye colour where twenty-six genes had been recognized.

As we have seen, Sturtevant had found an exception to the general rule that the sex-linked genes in *Drosophila* were autonomous, that is to say, in an organism of sexually mixed constitution (gynandromorph) the genes in the male segment were not affected by their allels in the female segment. This exception was vermilion eye colour. This bright red eye colour was rendered brown (wild-type) "not by the genetic constitution of the eye pigment itself, but by that of some other portion of the body" (Sturtevant, 1920, 71). It was, to quote Beadle, as if some substance had diffused from wild-type tissue to the eye and caused it to become normally pigmented. With Ephrussi, Beadle transplanted vermilion-type eye tissue into wild-type caterpillars. Sturtevant's result was verified. The two collaborators found only one other eye colour that behaved likewise—was "reparable"—and that was "cinnabar".

Clearly both mutants possessed the bright red eye colour component, but lacked the brown component. Was the diffusible substance which caused the production of brown pigment the same in both mutants or different? If the same, then reciprocal mutant transplants would yield mutant-type eye colours, if different, the mutant genes brought together by recombination would complement each other and wild-type eyes would be produced. Hence at this stage (in 1935) Beadle and Ephrussi *had not the concept of sequential gene-controlled stages in mind*. Great was their surprise, therefore, when neither prediction was fulfilled. Instead, the vermilion eye had no influence on the cinnabar transplant, but the cinnabar host converted the vermilion transplant to wild-type. To account for this unexpected result Beadle and Ephrussi "formulated the hypothesis that there must be two diffusible substances involved, one formed from the other according to the scheme: Precursor → v^+ substance → Cn^+ substance → Pigment" (Beadle, 1958, 591). In the vermilion mutant the first reaction was blocked; the vermilion gene lacked the power to bring about this reaction. The cinnabar mutant allowed formation of the v^+ substance but blocked the second reaction leading to the formation of the Cn^+ substance. Only in the presence of this substance and not in that of v^+ could the cinnabar mutant produce the wild-type eye colour. Here then, were two "diffusible substances" which were members of a sequence of reactions leading to the production of wild-type pigment. They were described as hormones and in 1939 (but as far as I am aware not before) Beadle and Ephrussi introduced the hypothesis that where hormone production was blocked it was "due to a defect in the enzyme system involved in the synthesis of the hormone" (Beadle, 1939, 60). Beadle added that despite the absence of "direct evidence for this intervention of enzymes in the system, they are assumed throughout for the reason that for the moment they appear to provide a simple mechanism by which genes might control reactions" (*Ibid*, 61). This working hypothesis was later to harden into the one gene—one enzyme hypothesis.

The story of the long hard grind which finally led to the identification of the v^+ substance as kynurenine by the group under Butenandt in the Kaiser Wilhelm-Institut in 1940 and by Tatum and Haagen Smit a year later is well known. In the five years which passed before Beadle and Ephrussi succeeded there were many unexpected twists and turns in the work which has been so admirably described by Ephrussi in the *Quarterly Review of Biology* (1942). The last straw for Beadle was when Tatum's crystalline produce could not be identified. Later it transpired that the kynurine had esterified with sucrose!

Life would clearly be much easier if an organism could be used with very simple nutrient requirements. Even then the identification of the precise chemical disturbance involved in a given mutant character, they argued, was bound to be slow and discouraging. "Our idea—to reverse the procedure and look for gene mutations that influence known chemical reactions—was an obvious one," Beadle declared (1958, 594). *It does not appear to have been obvious to anyone else*. It was a reversal of the Bateson–Wheldale–Scott-Moncrieff approach.

CHAPTER NINE

The Enzyme Theory of Life

When Beadle was appointed professor of biology at Stanford University he was able to invite a biochemist to work on the identification of the v^+ substance. His choice of Tatum could hardly have been more appropriate. At Wisconsin Tatum had worked in the young and exciting field of growth factors of micro-organisms. A year in Europe had brought Tatum into contact with F. Kögl, who only shortly before had discovered biotin; it also introduced him to Niels Fries, the great Swedish mycologist and originator of the "Fries medium" used for *Neurospora* for many years.

Neurospora Studies

The *Drosophila* studies described in the previous chapter led Beadle to the concept of a sequence of gene controlled reactions leading to the synthesis of the wild-type product. They had also led him to assume *for each such reaction a single controlling gene*. In retrospect he said: "Thus it seemed reasonable to assume that the total specificity of a particular enzyme might somehow be derived from a single gene" (1958, 597). Only with this belief, did Beadle and Tatum have the confidence that they could reverse the analysis, select known metabolic requirements and investigate their genetic determination. Just as Mendel's success depended on his deliberate choice of an organism suited to answer a specific question (Olby, 1966, 115) so did Beadle's and Tatum's. All higher plants and animals were excluded because their nutrient requirements were not "completely under the control of the investigator", green algae because the breeding techniques involved were complex, bacteria and blue green algae because of no known sexual reproduction in them, protozoa because of their only partially understood nutrient requirement.

> Fungi, on the other hand, seemed more promising. The ascomycete *Neurospora* has a life cycle almost ideally suited to genetic studies, can be grown under aseptic conditions, is not unreasonable in its space requirements, and has been studied rather extensively from a genetic standpoint.
>
> (Beadle, 1945, 644)

Beadle described how *Neurospora* came to mind in his Nobel lecture thus:

> As a graduate student at Cornell, I had heard Dr B. O. Dodge of the New York Botanical Garden give a seminar on inheritance in the bread mold *Neurospora*. So-called second-division segregation of mating types and of albinism were a puzzle to him. Several of us

who had just been reviewing the evidence for 4-strand crossing-over in *Drosophila* suggested that crossing-over between the centromere and the segregating gene could well explain the result.

Dodge was an enthusiastic supporter of *Neorospora* as an organism for genetic work. "It's even better than *Drosophila*", he insisted to Thomas Hunt Morgan, whose laboratory he often visited. He finally persuaded Morgan to take a collection of *Neurospora* cultures with him from Columbia to the new Biology Division of California Institute of Technology, which he established in 1928.

Shortly thereafter when Carl C. Lindegren came to Morgan's laboratory to become a graduate student, it was suggested that he should work on the genetics of *Neurospora* as a basis for his thesis. This was a fortunate choice, for Lindegren had an abundance of imagination, enthusiasm and energy and at the same time had the advice of E. G. Anderson, C. B. Bridges, S. Emerson, A. H. Sturtevant and others at the Institute who at that time were actively interested in problems of crossing-over as a part of the mechanism of meiosis. In this favourable setting, Lindegren soon worked out much of the basic genetics of *Neurospora*. New characters were found and a good start was made toward mapping the chromosomes.

Thus, Tatum and I realized that *Neurospora* was genetically an almost ideal organism for use in our new approach.

(Beadle, 1958, 594–595)

Although they did not know the nutrient requirements of the bread mold, they had Niels Fries' assurance that these were simple. Soon Tatum identified the only growth factor required as biotin, and as luck would have it this hormone had just then become available in the sort of quantity required for these experiments.

It remained only to irradiate asexual spores, cross them with a strain of the opposite mating type, allow sexual spores to be produced, isolate them, grow them on a suitably supplemented medium and test them on the unsupplemented medium. We believed so thoroughly that the gene–enzyme reaction relation was a general one that there was no doubt in our minds that we would find the mutants we wanted. The only worry we had was that their frequency might be so low that we would get discouraged and give up before finding one.

We were so concerned about the possible discouragement of a long series of negative results that we prepared more than a thousand single-spore cultures on supplemented medium before we tested them. The 299th spore isolated gave a mutant strain requiring vitamin B_6 and the 1085th one required B_1. We made a vow to keep going until we had 10 mutants. We soon had dozens.

Because of the ease of recovery of all the products of a single meiotic process in *Neurospora*, it was a simple matter to determine whether our newly induced nutritional deficiencies were the result of mutations in single genes. If they were, crosses with the original should yield four mutant and four non-mutant spores in each spore sac. They did.

In this long, roundabout way, first in *Drosophila* and then in *Neurospora*, we had rediscovered what Garrod had seen so clearly so many years before. By now we knew of his work and were aware that we had added little if anything new in principle. We were working with a more favourable organism and were able to produce, almost at will, inborn errors of metabolism for almost any chemical reaction whose product we could supply through the medium. Thus we were able to demonstrate that what Garrod had shown for a few genes and a few chemical reactions in man was true for many genes and many reactions in *Neurospora*.

(Beadle, 1958, 595–596)

In addition to the fruitful results in the B-vitamin group Beadle and Tatum were able to establish the sequence of gene-controlled reactions leading to tryptophane via anthranilic acid and indole. When David Bonner and Norman Horowitz joined them in 1942 the range of gene-controlled biochemical reactions exposed to view was increased. By 1945 some 60 000 single spore strains had been exposed to X-rays, neutrons and ultraviolet light.

The One Gene–One Enzyme Hypothesis

The phrase "one gene–one enzyme" was *not* stated in any of the major papers by Beadle and Tatum between 1941 and 1945. With reason, they were more cautious and spoke of gene and enzyme specificities as being "of the same order" (1941, 500), of a "one-to-one relationship between gene and specific reaction" (1945, 643), of further research being needed to show whether "only one gene is ordinarily concerned with the immediate regulation of a given specific chemical reaction" (1941, 505). But in 1945 Beadle launched into a discussion of the relation between gene and enzyme. When, in conformity with his accustomed style, he spoke of every chemical reaction being controlled primarily by a single gene he qualified himself with the words: "or perhaps more precisely every enzymatically catalysed reaction" (1945, 643). Clearly, the relation between gene and enzyme, however implicit in the earlier work, only became explicit when the relation between gene and reaction in *Neurospora* was pondered.

Beadle's reflections on this relation offer an intriguing view into his conception of protein synthesis and will be described shortly. All we need add here is that there was a conflict in Beadle's mind between the presence or absence of a functional enzyme being dependent on a *single* gene, and the concept of stepwise synthesis of end products, including enzymes, each step being determined by a separate gene, the entire synthesis therefore involving *many* genes. This led him to distinguish these two aspects by what he called "primary" and "secondary" genic control.

Theories of Gene Expression

Long ago it had been suggested that the action of the nucleus on the cytoplasm was mediated by "ferments" (Driesch, 1894), and by 1902 hereditary factors were associated with ferments in the minds of several early Mendelians. As such an association became popular the earlier craze for self-reproducing particles died away. This craze of the nineteenth century had resulted from the inadequacy of knowledge about such complex molecules as the proteins, and from the recognition that self-reproduction called for some higher level of organization than the chemical molecule could provide. Therefore Herbert Spencer, Charles Darwin, Hugo de Vries and August Weismann invoked self-reproducing physiological and morphological units composed of many chemical molecules.

The twentieth century saw the discussion of gene duplication as an autocatalytic process analogical to crystal growth (Ostwald, 1908; Hagedoorn, 1911). Then the necessity for the old morphological units no longer pressed so strongly. In his very thoughtful essay "Autocatalytic substances: the Determinants for Inheritable Characters, a Biomechanical Theory of Inheritance and Evolution", the Dutch geneticist, Avend Hagedoorn, attacked the old tradition in which all the sub-units of living organisms were assumed to possess vitality. Like Wilhelm Roux before him, Hagedoorn argued that life was to be identified with the *association* of these subunits but not with the subunits themselves.

> The hypothesis that these hereditary things are vital units, composed of protoplasm and capable of assimilation and growth, certainly fits the facts. But we ask more of a theory of heredity and evolution. A working-hypothesis, to be of any use as an instrument of research, must explain the facts in terms of what is already known. It is inadmissable to try to explain the facts of evolution and inheritance by the behaviour of living particles which have been invented simply to admit of this explanation.
>
> (Hagedoorn, 1911, 23)

A like view was later expressed by C. M. Child who declared that the nineteenth century corpuscular theories do not "help us in any way to solve any of the fundamental problems of biology; they merely serve to place these problems beyond the reach of scientific investigation" (1915, 11–12). Troland, who echoed Child's view, greatly influenced H. J. Muller. Many years later this great American geneticist wrote:

> Those who postulated "pangenes", "determinants", or other self-reproducing particles in the old days did not seem to realize what a monster they had by the tail. They were still subconsciously, so close to the ancient lore of animism . . . that to attribute reproduction to a particle, especially when this formed a component of a larger thing, itself living anyway, seemed to present no problem.
>
> (Muller, 1951, 95)

Hagedoorn saw his role as a modernizer of the views expressed by Roux and Loeb. The mechanism of inheritance, according to the Dutch scientist's hypothesis now involved "numerous independently transmitted substances, each having autocatalytical properties" (Hagedoorn, 1911, 28). Troland embraced this hypothesis gladly for he saw how neatly it dispensed with vital particles and avoided the problem of an infinite regress. Driesch, he remarked:

> . . . argues that to explain the reproduction of a nuclear "machine" which determines development, we must postulate another machine to carry out the operation, and so on *ad infinitum*. The nature of the autocatalytic process, however, shows that this conclusion is in error, since pure autocatalysis would tend to bring about an exact qualitative reproduction of any given plane or linear mosaic of specific units.
>
> (Troland, 1917, 343)

There is no doubt that this catalytic theory of gene replication was in tune with the climate of opinion of the day. For this was the golden era of enzymology. By 1916 we find Troland writing a paper with the title, "The

Enzyme Theory of Life". A year later he spoke of "what we call life", as fundamentally a product of catalytic laws acting in colloidal systems of matter throughout the long geological periods of time (Troland, 1917, 327). To him "intra-vital or 'hereditary' determination is, in the last analysis catalytic" (*Ibid.*), and Bateson's opinion that the hereditary factors "themselves can scarcely be genetic factors, but consequences of their existence" (Bateson, 1909, 86) was in Troland's eyes nothing less than "intellectual blindness" (1917, 328). Bateson was still left with the question: "What are the factors themselves? Whence do they come?" But for Troland it was simple: "On the supposition that the actual Mendelian factors are enzymes, nearly all these general difficulties instantly vanish" (*Ibid.*).

Troland distinguished between the catalytic action of the gene whereby its like is synthesized (autocatalysis) from that in which a different substance is synthesized (heterocatalysis). The former was the more fundamental of the two, but both were akin to crystal growth in which the fields of force around the crystal attract appropriately shaped molecules from the solvent. This would tend to be a specific process, and the more complex the molecule the greater would be the specificity. The attraction would be like Fischer's lock and key relation between enzyme and substrate (*Ibid.*, 337).

Such ideas greatly stimulated Muller, as he admitted, at the end of his life:

> Troland's remarkable paper, which I read when it appeared in 1917, startled me, for it expressed a number of ideas very similar in essentials to some I had already formulated. However, there were several crucial differences in our conceptions. For one thing, although believing in the importance and multiplicity of specific enzymes in protoplasmic reactions and the dependence of their presence on genes, I believed it illegitimate to *identify* genes with enzymes and too early to decide in what then-known category of chemical substances (if any) genes belonged. Even in sperm there was still an unclassified residuum of material, and known enzymes were widely thought to be proteins.
>
> (Muller, 1967, 423–424)

Troland had in fact identified the substance of the genes with nucleic acid and had looked for the required subtlety of composition to chemical sequences which chemical techniques of the day were inadequate to reveal! These gave instead a statistical average composition (Troland, 1917, 342).

Unfortunately Troland's ideas did not offer suggestions for experimental testing. They served to keep alive the hope that the physiological action of the gene and its replication might one day be accounted for in terms of physical chemistry. They pin-pointed the autocatalytic function of the gene as *the great enigma of biology*. Although Hagedoorn's paper was very widely quoted, Troland's was not, and it was Muller who repeatedly reminded geneticists of this great enigma and who played a part in attracting physicists into biology to solve it.

Beadle's Conception of Genetic Determination

There is little indication in Beadle's papers that he was influenced by the speculative ideas we have described. He was far more cautious in his statements than any of the advocates of the autocatalytic and heterocatalytic functions of the gene. Instead, it was Scott-Moncrieff, Kühn, Melchers, Wheldale and, by 1939, Garrod who attracted his respect. (It was probably J. B. S. Haldane who drew Beadle's attention to Garrod's work by his essay of 1937.) Beadle's approach is well illustrated by the conclusion to his paper, "Genetic Control to the Production and Utilization of Hormones", read at the Edinburgh genetical Congress in 1939. If the work he had described there was important, he said, it was not "because it tells us much specifically about what genes do, but rather because it may indicate a method of attack which we may hope will become increasingly useful to both geneticists and biochemists" (Beadle, 1939, 61).

Six years later Beadle had behind him the substantial achievement of the *Neurospora* work. This in Bernard Strauss' words: "showed the closeness of the relationship between genes and metabolism" (1960, 6). Now Beadle felt he could speculate on the manner in which genes determine their end products. The immediate gene product was a hormone, enzyme, or just a protein. How did the gene determine this product? Here the model of antigen specificity as a *particular molecular conformation* influenced him and he wrote:

> Since many cases are known in which the specificities of antigens and enzymes appear to bear a direct relation to gene specificities, it seems reasonable to suppose that the gene's primary and possibly sole function is in directing the final configurations of protein molecules.
>
> Assuming that each specific protein of the organism has its unique configuration copied from that of a gene, it should follow that every enzyme whose specificity depends on a protein should be subject to modification or inactivation through gene mutation. This would, of course, mean that the reaction normally catalysed by the enzyme in question would either have its rate or products modified or be blocked entirely.
>
> Such a view does not mean that genes directly "make" proteins. Regardless of precisely how proteins are synthesized, and from what component parts, these parts must themselves be synthesized by reactions which are enzymatically catalysed and which in turn depend on the functioning of many genes. Thus, in the synthesis of a single protein molecule, probably at least several hundred different genes contribute. But the final molecule corresponds to only one of them and this is the gene we visualize as being in primary control.
>
> (Beadle, 1945, 660)

His collaborator, David Bonner, expressed himself in a similar vein at the Cold Spring Harbor meeting in 1945 when he said:

> There is quite general agreement at present that genes contain nucleoprotein as an essential component of their structure. One should expect, therefore, that genes, like other proteins, have specific configurations, the configuration of a single gene being characteristic of itself alone. These various considerations suggest the view that the gene controls biochemical reactions by imposing, directly or indirectly, a specific configuration on the enzymes essential for the specific reactions.
>
> (Bonner, 1946, 21)

It would be most unjust to state boldly that this conception of gene function led in the wrong direction. It pictured genes acting specifically by determining the conformation of particular proteins, a conception which still finds a place in the theory of protein synthesis, for the gene does determine amino acid sequence and this sequence, under given conditions, determines conformation. But there was no mention of amino acid sequence in Beadle's discussion of 1945, and the distinction between subunit assembly and final conformation shows surely that he did not have the modern picture in his mind. Where this work may be said to have led in the wrong direction was in the conclusion it invited: the gene is a special kind of enzyme, and therefore its specific action resides in its protein component. Consider the following passage from Beadle's 1945 paper:

> Genes are thought from various lines of evidence to be composed of nucleoproteins or at least to contain nucleoproteins as essential components. They have the ability to duplicate themselves which of course they do once every cell division. The manner in which this self-duplication is brought about is one of biology's unsolved problems, but it is thought to involve a kind of model-copy mechanism by which the gene directs the putting together of the component parts of daughter genes. If this is the mechanism and genes contain protein components, gene reproduction is a special case of protein synthesis.
>
> (Beadle, 1945, 660)

The Analogy with Viruses

Those who have read Gunta Stent's fine historical introduction to *The Molecular Biology of Bacterial Viruses* will know that the discovery of bacteriophages sparked off a debate as to whether the causal agents were organisms or enzymes. On the one side was Felix d'Hérelle director of the Pasteur Institute in Paris, who maintained that the phage was an "invisible microbe" (1917, 373); on the other side were André Gratia and Jules Bordet, the latter being director of the Pasteur Institute in Brussels. They supported the alternative view that the agent of bacterial lysis was an enzyme produced by the bacterium itself. The British microbiologist, Frederick Twort had considered this among the three possible explanations of bacterial lysis (1915, 1242). This enzyme theory was already orthodox for plant and animal viruses. It was A. F. Woods' theory for tobacco mosaic virus (TMV) in 1899 and 1900. In the necrotic spots chlorophyll was changed to xanthophyll, owing to local accumulation of oxidizing enzymes. Basically the same view was taken by F. W. T. Hunger in 1905. In Hagedoorn's essay, which we have discussed, genes were likened to the "filterable viruses" which caused rabies and poliomyelitis. The latter were simply "chemical substances with autocatalytic properties"; they lacked the "structural relationships" of bacteria and yeast cells and therefore did not lose the power of autocatalysis on passage through the finest filters (Hagedoorn, 1911, 26–27). *The discovery of the filterable viruses and bacteriophages thus offered valuable support to the enzyme theory of life and to the conception of autocatalysis as the fundamental and primitive characteristic of life.* Hagedoorn did not exclude the possibility "of creating 'living' organisms by the combination of non-living things like the 'filterable viruses' and other autocatalytic substances . . ." (1911, 27).

Northrop's Autocatalytic Theory of Phage Reproduction

This theory was enunciated in a remarkable paper, "Concentration and Purification of Bacteriophage", in 1937. Its author, Northrop, saw himself in the same role as Buchner, who some forty years earlier had demonstrated that a reaction long regarded as "vital" could be brought about by a cell extract. The fermentation of sugar then became a chemical reaction only requiring the presence of the enzyme, zymase. But then along came the bacteriophage. It increased dramatically in the host cell, thus appearing to be a living organism. A fresh controversy resulted, and Northrop sought to lay this new ghost of vitalism. He rightly argued that the bacteriophage had no perceivable organized structure and metabolism of its own (as then known). He found no proof of its independent autocatalytic increase.

> . . . since the increase occurs only in the presence of living cells, and since under these conditions, the normal proteins and enzymes of the cells are also increasing. In fact, it has been recently found by Kunitz that an extracellular proteolytic enzyme produced by a mold (*Penicillium*) increases in an analogous manner. The only difference between the phage and "normal" enzymes or proteins, then, is the fact that not all cultures contain it. If all cultures did contain it, it would simply be regarded as a normal enzyme and the culture would be said to autolyse readily.
>
> (Northrop, 1938, 359)

Was this autocatalytic increase of phage so special, asked Northrop? Surely there are other examples of substances which, when added to living cells, increase. In 1922 Gratia had drawn an analogy between the increase of thrombin during clotting and the production of phage by a bacterium. Six years later Griffith discovered the transformation of pneumococcal types. Was this not a case of the production of a new compound—Type III coat—under the stimulus of dead Type III cells? And did not the inactive compounds pepsinogen and trypsinogen change into active pepsin and trypsin *in vitro* when presented with the active forms?

> A similar mechanism will evidently account for the increase of bacteriophage and other viruses in the presence of living cells. The cells synthesize a "normal" inactive protein. When the active virus or bacteriophage is added, this inactive protein or "prophage" is transformed by an autocatalytic reaction into more active phage.
>
> (Northrop, 1938, 361)

This was the theory already advocated for TMV by Stanley in 1936. Well, there were some problems with it! If, as Northrop admitted, the conversion of an inactive to an active protein involved hydrolysis and therefore cleavage of the molecule, why was the molecular weight of TMV obtained from ultracentrifugation so high (3×10^8)? And as F. C. Bawden (later Sir Frederick) was to ask, of what benefit was it to the host bacterium to manufacture these virus precursors? Where, too, was the analogy between the quantitative *in vitro* conversion of pepsinogen to pepsin and of phage precursor to phage? There just was no evidence of such large quantities of Northrop's "pro-phage" in the host cell or of its equivalent in the tobacco plant. There was no evidence of it at all. (Bawden, 1939, 280). In 1937 Stanley had to admit that his TMV had a molecular weight in excess of 10 000 000, yet he could find no particles in healthy tobacco sap with molecular weights above 30 000 (Wyckoff, Biscoe, and Stanley, 1937).

Here, then, was the all-embracing enzyme theory of life being employed to attack the vitalism of phage lore. As Northrop admitted, the mechanism he invoked was like that proposed by Bordet in 1931, and by Stanley in 1936. This "enzyme camp" in the Rockefeller Institute appears not to have been influenced by Muller's advocacy of the virus–gene analogy. Muller expressed this analogy first at the AAAS Meeting in Toronto in 1921, and again in 1929 and 1936. On the first occasion he argued that

> . . . if these d'Hérelle bodies were really genes fundamentally like our chromosome genes, they would give us an utterly new angle from which to attack the gene problem. They are filterable, to some extent isolable, can be handled in test tubes, and their properties, as shown by their effects on the bacteria, can then be studied after treatment. It would be very rash to call these bodies genes, and yet at present we must confess that there is no distinction known between the genes and them. Hence we cannot categorically deny that perhaps we may be able to grind genes in a mortar and cook them in a beaker after all. Must we geneticists become bacteriologists, physiological chemists and physicists, simultaneously with being zoologists and botanists? Let us hope so.
>
> (Muller, 1922, 48; quoted in Carlson, 1971, 162)

In 1936 he referred to Stanley's crystalline TMV product, which was "apparently" a pure protein.

> We judge that this material has the properties of a gene, in as much as it can reproduce itself, i.e., it can undergo autosynthesis when present in a cell and it is probably mutable, since different species of it are known. We may provisionally assume, then, that it represents a certain kind of gene. The weight of its giant molecule is of the order of several million, and this agrees as well as could be expected with the very approximate estimates of size hitherto made for the genes of flies.
>
> (Muller, 1936, 213)

By the time of the Genetical Congress in Edinburgh we find Astbury reading a paper with the title, "Protein and Virus Studies in Relation to the Problem of the Gene", in which he declared: "The two simplest reproductive systems that we know—'simplest' in the sense of being nearest the beginning of things—are the viruses and the chromosomes" (Astbury, 1939, 49). On this occasion, too, H. H. McKinney and J. W. Gowen spelt out the analogies between virus particles and genes. The former entitled his contribution to the congress "Virus Genes". TMV, he asserted,

> . . . possesses the basic functions assigned to genes, i.e. it serves as a determiner of the specific reactions which make for its duplication (autosynthesis) and it mutates, forming a series resembling in certain particulars a series of allelomorphs derived from a wild-type. These views seem not to impair the conclusions drawn from the chemical evidence that the virus is a parasitic nucleoprotein.
>
> (McKinney, 1939, 202)

The recognition of the nucleoprotein constitution of viruses and of the virus–gene analogy did not lead to the destruction of the enzyme theory of life nor to the overthrow of the Protein Version of the Central Dogma. Instead, what we can now see as conflicting evidence was accommodated very conveniently by a hypothesis which lasted for a whole decade.

The Gene as a Conjugated Protein

Perhaps, after all, it was irrelevant to ask in which portion of the molecule the specificity and replicating power resided. This was Jack Schultz' conclusion in his paper, "The Evidence of the Nucleo-protein Nature of the Gene", at Cold Spring Harbor in 1941. He likened the functioning of the conjugated protein of the gene to that of the nucleoprotein enzymes of respiration described by Otto Warburg in 1938. Here, said Schultz, "the active group is the conjugated protein; neither component without the other is effective . . . the specificities of the gene reside in the nucleoprotein, and the continuous structure of the chromosome is a protein fibre" (1941, 62).

By the 1940's, then, there was no longer a straight confrontation between protein or nucleic acid as the genetic material, but rather between protein and nucleoprotein. *If genetic specificity could not be attributed to the protein alone, nor could it be attributed to the nucleic acid alone.* On the whole, though, the *storage* of genetic information was still associated with the protein part, the *expression* and *duplication* of this material requiring the nucleic acid part. This view of the matter persisted long after 1953 and is well illustrated from the writings of Richard Goldschmidt. One might have expected this great advocate of biochemical and dynamic genetics to embrace the Watson–Crick model gladly in 1953. That he did not do so is evidence of the "paradigm shift" required. He was clearly rooted in what we have called the "Protein Version of the Central Dogma". In 1955 he described the Watson–Crick model as "ingenious" and "experimentally well founded" (1955, 37), but he assured his readers that the geneticist:

> . . . will not be content to accept a fine explanation of self-duplication as proof that the substance in question, DNA, is the genic material. Such a proof requires agreement with the entire body of biochemical, genetical, and cytological facts . . ., and the exclusion of alternative interpretations.
>
> It is at once clear that there should be no difficulty in reversing the theory and concluding that the DNA molecule is the scaffold which keeps the proteinic genic material in place for its reduplication . . . Such a function of the stiff DNA fibre was frequently assumed for former models of this material.
>
> (Goldschmidt, 1955, 40–41)

CHAPTER TEN

The Chemistry of Virus–Genes

> Our results for the investigated wavelengths yield a maximum of activity at 265 mμ, both for reduction of sporogonial extension and for frequency of mutations. This is just the region of wavelengths in which the absorption maximum was found for the nucleic acids . . . in the chromosomes; whereas no maximum is known in this region for the proteins concerned. . . . The elucidation of the detailed relationship between thymonucleic acid and the actual genetic substance must await further research.
>
> (Knapp, *et al.* 1939, 304)

> The original structure of the protein [of TMV] thus appears [after removal of nucleic acid] to be fully preserved. Against this the biological examination in the individual test upon *Nicotiana glutinosa* . . . showed that after separation of the nucleic acid the biological activity has practically completely vanished. Hence the necessity of nucleic acid for the process of multiplication is established for TMV.
>
> (Schramm, 1941, 536)

> The report of Schramm that the [nucleophosphatase] enzyme effects the complete removal of nucleic acid from the virus and yields an inactive phosphorus-free protein which is the same as the intact virus with respect to size, homogeneity in the ultracentrifuge, and electrophoretic mobility has not been confirmed.
>
> (Cohen and Stanley, 1942, 870)

> . . . the production of radiation mutations of the tobacco mosaic virus molecule is to be traced back to quantitative and qualitative changes in the nucleic acid portion of the virus molecule, especially for the reason that here the highest radiation absorption is to be expected on account of the accumulation of atoms of higher atomic number—phosphorus.
>
> (Pfankuch, Kausche and Stubbe, 1940, 255)

> Two proteins have been isolated from plants infected with this [turnip yellow mosaic virus] disease and both have been crystallized. One is a nucleoprotein . . . and the other appears to be the same protein without the nucleic acid. . . . Only the nucleoprotein is infectious to plants, and the presence of combined nucleic acid would appear to be necessary for virus multiplication.
>
> (Markham, Matthews and Smith, 1948, 90)

The Chemistry of Virus–Genes

At first sight it must surely seem strange that the identity of the hereditary substance was first demonstrated in research arising out of the study of epidemics. Of course this was only so because the direct approach proved so difficult. The nucleic acid in the chromosomes of higher organisms is bound to histones and associated with globular proteins in a complex manner. Attempts to free the chromosome of its protein or nucleic acid component and to show that it did or did not fall apart served only to establish the one or the other component as the structural framework of the chromosome thread. At the Edinburgh Congress in 1939 Daniel Mazia and L. Jaeger described their enzymatic studies of salivary gland chromosomes which led them to assert the presence of a "continuous protein framework composed of protamines and histones" and to deny that nucleic acid molecules were necessary for the structural integrity of the chromosome (Mazia and Jaeger, 1939, 213).

Seven years later, at the symposium on nucleic acids held by the Society for Experimental Biology the same subject cropped up and the Stedmans denied the classical view of the sperm nucleus as being made up almost exclusively of nucleoprotein. It will be recalled that Miescher had found 81.6 per cent of the sperm head accounted for as nucleoprotein. Realizing that Miescher's extractive procedure (with dilute HCL) had not been exhaustive, Schmiedeberg had computed the nucleoprotein left behind and when he added this to Miescher's figure of 81.6 per cent he arrived at 96 per cent. This classical view was supported by Mirsky and Pollister when they introduced their molar salt extraction technique (1942), which, when applied to trout sperm, gave them a nucleoprotein content of 91 per cent (Pollister and Mirsky, 1946). Nevertheless, they felt that the remaining 9 per cent was significant (*Ibid.*, 114). It contained, in their view, another nuclear component which gave occasion to a further paper entitled: "Chromosin, a Desoxyribose Nucleoprotein Complex of the Cell Nucleus". If the reader turns to Taylor's *Classic Papers in Molecular Genetics* he will find the first of these two papers, but not the second. In the latter, Mirsky and Pollister described non-histone protein which contained tryptophane and which amounted to 27 per cent of the nuclear protein. Clearly, they argued, it was misleading to describe the material extracted from the nucleus by their molar NaCl technique as nucleoprotein, for it was a complex of nucleoprotein plus this "residual protein". And since this complex contributed almost all of the substance of the chromatin it was best known by the name "Chromosin" (Mirsky and Pollister, 1946).

These authors could point out that the classical picture of the constitution of thymus nuclei had already been opposed by Edgar and Ellen Stedman, who had used Miescher's extractive technique and had found only 70 per cent of the sperm head in the form of nucleoprotein; the remaining 30 per cent they attributed to proteins higher than the protamines and histones, which they called "Chromosomin" (1943; 1947, 239). This, they claimed, was a group of proteins, the amino acid content of which differed from species to species. In contrast to the histones the proteins of chromosomin were acidic, like those proteins whose absorption spectra Caspersson had noted in the interbands of salivary gland chromosomes. "Far from being a synthetic product characteristic only of the metaphase nucleus", as Caspersson believed, "The protein of the interbands evidently consists of part of the continuous thread or threads of chromosomin which run throughout the length of the chromosomes . . ." (Stedman and Stedman, 1947, 250). Now the Stedmans could overcome the failure of chemistry to provide the required subtlety of genetic diversity, for cell nuclei did not after all possess a similar simple chemical composition but a varied one dependent on the constitution of the chromosomin component of the chromosomes. Nucleoprotamine, on the other hand, was really the material constituent of nuclcar sap! To be sure it was condensed *on* the chromosomes during certain phases in the nuclear cycle—as in the sperm head—but this was a transitory state of affairs (*Ibid*, p. 247).

By 1947 when there was a symposium on nucleic acids and nucleoproteins at Cold Spring Harbor, the picture of the chemical constitution of the nucleus had become highly confused. On that occasion Mirsky gave a paper describing the great variation in the proportion of the residual tryptophane-containing protein in chromosomes—5 per cent in fish erythrocytes and about 50 per cent in liver chromosomes—and in the discussion which followed Seymour Cohen asked if the term "chromosin" could be given up. Mirsky replied that:

> In the present paper the term chromosin was not used, and this obviously indicates that we have already abandoned the term as we have come to a better understanding of the nature of our "chromosin" preparations . . .
>
> (Mirsky, 1947b, 146)

Margaret McDonald expressed her perplexity that

> Dr Mirsky, using apparently identical procedures, has prepared materials containing (a) only nucleic acid and histones, and (b) nucleic acids, histones and tryptophane-containing proteins. Would he tell us the technical reasons for not finding the tryptophane-containing proteins in his first preparations . . .

Mirsky replied:

> In our first work the tryptophane-containing protein was in large part removed by repeated precipitation, solution and centrifugation. In the work described in the paper that has just been presented, the tryptophane-containing protein fraction was clearly identified just because the procedures followed were quite different.
>
> (*Ibid.*)

Biologists had called for chemical facts on extracted chromosomes. Mirsky and Pollister supplied them with a wealth of technical detail demonstrating the demands which such a request made upon the resources of the biochemist, and the pitfalls which beset the evaluation of the data obtained. The simple answer which the biologist wanted, the chromosomes of higher organisms did not yield.

But what of viruses? Surely the analogy between viruses and genes was appreciated from the late 1930's onwards, and crystalline virus material was used to try to solve this question of the chemistry of the gene. The tradition of virus research was international, being solidly based on the great economic importance of both plant and animal viruses. Bacterial viruses, too, had become widely popular as research topics on account of the promise they appeared to hold out for the control of bacterial diseases. The obvious candidates for solving the question were the crystalline plant viruses, the pathogens in tobacco, tomato and potato plants, and the best of all was tobacco mosaic virus or TMV.

The Chemistry of TMV

This is an extraordinary story. The view that viruses were proteins was suggested by the successful precipitation of TMV using protein precipitants (Mulvania, 1926). A year later C. G. Vinson, working at the Boyce Thompson Institute for Plant Research, New York, was able to assert the presence of nitrogen in the product he had thrown down with lead acetate. This success was followed by the claim of Barton-Wright and McBain to have produced nitrogen-free crystals of TMV (1933). The pathology division of the Rothamsted Experimental Station was not slow to examine this achievement, with the result that the Barton-Wright–McBain crystals were identified as potassium dihydrogen phosphate, derived from the salting-out agent and with TMV as an impurity! (Caldwell, 1934).

The following year brought Stanley's dramatic announcement "Isolation of a Crystalline Protein Possessing the properties of TMV", in *Science*. From the juice of infected leaves of the Turkish tobacco plant the Rockefeller staff had extracted a product one thousand times more active than the raw juice. Stanley declared he had "strong evidence that the crystalline protein herein described is either pure or is a solid solution of proteins" (1935, 644). TMV was therefore "an autocatalytic protein which, for the present, may be assumed to require the presence of living cells for multiplication" (*Ibid.*).

In England virus research was first directed to the economically important potato virus X, tobacco not being a commercial crop in this country. Long ago, under the influence of Bateson, R. N. Salaman had initiated research into the genetics of the potato at his home in Barley, Herts, which he continued from 1906 to 1925. He wrote up the results of this fine work in a little book in 1926 and was then able to influence Sir Daniel Hall at the Ministry of Agriculture to establish a Potato Virus Research Unit in the Agriculture Department of Cambridge University. There, Kenneth Smith started work on what he later described as Potato Virus X. Because this work was in progress, Norman Pirie in the biochemistry department was able to start extracting the active agent in the form of suspensions. In this work he collaborated from Cambridge with Frederick Bawden of Rothamsted Experimental Station. They showed that potato virus X had a high nitrogen content and serological activity, and when they applied the same technique to TMV they confirmed most of Stanley's findings. In their first publication on TMV (November 1936) they noted the presence of 2.5 per cent carbohydrate and 0.5 per cent phosphorus. Five months later they described TMV as a nucleoprotein (Bawden and Pirie, 1937). Markham, who joined Pirie in 1938 and took the work up again after the war, commented that it is difficult to miss the phosphorus. The Rockefeller analyst must have left the nucleic acid behind in the supernatant from the alcohol precipitation of the virus. So Stanley had effectively thrown away the major part of the nucleic acid. When he admitted the presence of phosphorus in aucuba mosaic virus, a year later, he attributed it to nucleic acid which was not, in his opinion, part of the infective particle! "It was found possible", he declared, "to remove the nucleic acid and to obtain phosphorus-free protein possessing virus activity by dialysis at pH 8 or 9 . . ." (Stanley, 1937, 329).

Stanley's work was the continuation of the programme begun by Vinson under Louis Kunkel at the Boyce Thompson Institute. The latter had left that Institute in 1931 to direct the newly formed division of plant pathology at the Rockefeller Institute in Princeton. From the beginning he had Stanley on the staff, and it was this young man's combination of Vinson's lead acetate procedure for TMV with Northrop's procedure for the crystallization of proteolytic enzymes that yielded a liquid–crystalline preparation for TMV. Both Stanley's extractive procedure and his extensive study of the inactivating agents of TMV pointed to the protein nature of the material. Other workers (Dvorak, 1927; Purdy, 1929) had shown the serological activity of TMV, a clear evidence of its protein nature.

In the winter of 1936/37 Stanley had every reason for playing down Bawden's and Pirie's isolation of phosphorus and carbohydrate—only 0.5 per cent phosphorus anyway. Even if nucleic acid was a genuine constituent of TMV it could hardly have an important function, since RNA was still thought to be a tetranucleotide (Myrbäck and Jorpes, 1935), and Levene's authority was still behind this opinion. Moreover, the staff of the Princeton branch of the Rockefeller Institute were just as bound to the Protein Version of the Central Dogma as were the staff of the New York Laboratories. The suggestion that viral activity and specificity should be associated with a nucleoprotein, rather than simply with a protein, was not to be entertained without strong evidence, in the intellectual milieu of the Rockefeller, the home of Levene, Landsteiner and Mirsky, the scene of the crystallization of enzymes by Sumner, Northrop and Kunitz. TMV must be a protein, and an autocatalytic one at that.

A further year passed before Stanley came out in support of the nucleoprotein character of TMV. At a symposium in Indianapolis in December 1937, he commented that "virus proteins isolated so far have been found to be nucleoproteins" (Stanley, 1938, 119). But this statement was made casually after he had repeatedly referred to virus as a protein. Earlier in 1937 he had described the crystalline protein of aucuba mosaic virus, which differed from the protein of TMV as demonstrating "that two different strains of a virus give rise to two different proteins" (Stanley, 1937, 340). This adherence to the protein character of viruses, very understandable as it was, did do harm, as it conditioned the thinking of such influental scientists as Muller and Delbrück in their earlier discussions of the identity of the genetic material.

Curiously enough, the analogy between viruses and chromosomes, which their chemical constitution supported was not pursued by Stanley. Instead he suggested that in "this combination of a nucleic acid and high molecular weight protein we have sufficient organization within a single molecule to endow it with the life-like properties that characterize it" (Stanley, 1938, 119). For him the mechanism of viral reproduction was the mechanism for the growth of protoplasm. The important event in growth was not cell division but the growth which led up to it. "I think we may dismiss at once the mechanism by means of which cells divide and thus reproduce, for we are interested in a fundamental mechanism, the one by means of which the cell grows until it is in a position for the second mechanism to operate" (Stanley, 1938, 120). The growth of protoplasm was, of course, a case of protein synthesis mediated by intracellular enzymes.

When, with the aid of hindsight, we look back at these early studies of the action of proteolytic enzymes, pH, heat, and nitrous acid, how well they fit into present views of the overriding importance of the nucleic acid. Was some of this evidence too good? What should we make of Pfankuch's detection of a difference in electrophoretic mobility of the nucleic acids but not of the proteins from two related strains of virus, from which he was led to suggest that their biological differences were due to the nucleic acid portion of the virus (Pfankuch, 1940)? Or consider the case of the different electrophoretic mobilities of the nucleoproteins belonging to five related strains of virus. The mutations by which these strains had arisen, the authors believed, were to be attributed to "quantitative and qualitative changes in the nucleic acid portion of the virus molecule, especially as the highest absorption of radiation is to be expected here, owing to the accumulation of atoms with a higher atomic number—phosphorus" (Pfankuch, Kausche, Stubbe, 1940). Or we may turn to the repetition of this type of experiment at the Kaiser Wilhelm-Institut in Tübingen six years later, which yielded characteristic curves for the pH-dependent electrophoretic mobility of each virus nucleoprotein studied (Schramm and Rebensburg, 1942). But as Seymour Cohen remarked, these "differences in mobilities of the mutants do not reflect differences in the mobilities of the RNAs of the mutants. They possibly reflect the extent to which some phosphates do poke through the protein coats and the exposure of carboxyls in the variously folded protein subunits" (Cohen, 1973).

It would be quite wrong to give the impression that the German workers were right and the American workers wrong. In 1948 Melchers recalled that Schramm was looking for the chemical basis of strain differences in the protein portion of virus nucleoprotein, not in the nucleic acid portion (Melchers, 1948, 116). When these electrophoretic studies were repeated in America using the ordinary and the rib strain TMV only, a single boundary was observed (Knight and Lauffer, 1942). Schramm and Müller tried "covering" the amino acid groups of TMV with acetyl or phenylureido groups, but although they achieved 70 per cent coverage, the infectivity of the virus was no less than that of the untreated virus. The same experiment repeated in America resulted in diminished infectivity before complete coverage (Miller and Stanley, 1941). Nor was the German work purporting to show that a phosphatase could inactivate TMV (Pfankuch and Kausche, 1938) verified by Loring or by Cohen at the Rockefeller. Small wonder, then, that in England and America there was scepticism concerning the German work. Yet, in spite of the war, the further biophysical studies of viruses by Schramm and his colleagues was followed by Stanley's group in Princeton with just as much interest as was the work at Rothamsted. The Americans had access to all the papers of the Germans and *vice versa*. Cohen recalled that "the Rothamsted and Princeton labs disagreed with each other at least as much as either [of them] did with the German labs. I never heard (that I remember) criticism of Schramm's work for any political reason. We repeated everything of everyone's work, just to be sure" (Cohen, 1973).

In 1943 Schramm observed the fragmentation of TMV at pH 9 into two types of subunits, all of uniform size. One type contained nucleic acid, the other lacked it, but both had a particle weight of 360 000, in this they closely approached the molecular weight derived from the unit cell (370 000—by Bernal and Fankuchen, 1941). But what amazed Schramm was the fact that when he returned these subunits to pH 5 he could no longer detect them in the ultracentrifuge. Instead, rods had been formed just like those of the original TMV! This find led Schramm to conclude that the nucleic acid did *not* determine the characteristic shape of the TMV particle, and it did not appear to hold the protein subunits together. Neither the subunits nor the reconstituted nucleic acid-free particles were infective. After the war Schramm carried this work a stage further when he preserved the nucleic acid of TMV intact, after stripping off the protein, but this RNA core did not prove to be infectious (Schramm, 1948).

Another eight long years passed by before the infectivity of the RNA core of TMV was demonstrated (Gierer and Schramm, 1956; Fraenkel Conrat, 1956), and even then available techniques did not prevent depolymerization by RNase to such an extent as to render the RNA barely infectious. The idea that removal of the protein coat was simply exposing the RNA core to enzyme action "was not at all obvious" (Melchers, 1965, 120), and the restoration of infectivity by reconstitution of viral subunits was thought to signify "a 'resynthesis' of elementary systems of life" (Stanley, 1955), in which the protein of course had a vital role. These virologists, it seems, were trapped in a Kuhnian box. Crick recalled that he "tried to get Markham and Smith [in Cambridge] to do the reconstruction experiments—later done by Fraenkel-Conrat—but they refused to take my suggestion seriously" (Crick, 1973). Nor were the virologists in California any more enlightened than their Cambridge counterparts. As late as 1942 they were trying to show that a dethreonized virus represented a laboratory mutant. Such a virus with 7 per cent of its threonine removed, was infective and yielded the normal threonine-containing virus particles. But they did not appear to be perturbed! (Harris and Knight, 1952).

What, then, was achieved by these studies of TMV? They ultimately, if belatedly (1956), brought conviction that viral RNA, not viral protein, carried the genetic specificity of the virus. But in the late 1930s they showed what the study of the sperm head had failed to do, that the infective particle consisted of nucleoprotein and nothing else.

Viral Inactivation and Mutation

To take the analysis a stage further—to decide between nucleic acid and protein as the bearer of genetic specificity and the agent of infection and replication—called for a discussion of the meaning of the results obtained from the exposure of viruses to chemical and physical mutagens. Here Stanley ran into difficulty, for nitrous acid seemed not to damage the protein of the aucuba mosaic virus, yet it inactivated the virus. As he assumed all the vital functions to reside in the protein he found this result "strange" (Stanley, 1938, 331). On the other hand TMV subjected to proteolytic enzymes lost infectivity, possibly due to the presence of RNase as a contaminant. But what of the action of ultraviolet light? As early as 1928 (see Table 10.1) the bactericidal action of this radiation had been associated with the absorption maxima for nucleic acid constituents (Gates, 1928). By the time of the Cold Spring Harbor symposium of 1941 Stadler, Hollaender and Emmons were well aware of the link between ultraviolet mutagenesis and nucleic acid absorption maxima. But Stadler favoured the view that the observed loss of genetic markers was due to chromosome breakage brought about by damage to DNA as the structural substance of the chromosome (Stadler, 1939). He was in search of a mutagen which would cause transmutation of the gene

TABLE 10.1

Ultraviolet inactivation and mutation

Date	Authors*	Material	Irradiation	Acting upon	Effect
1917	Harris and Hoyt	bacteria	ultraviolet	aromatic amino acids	death
1928	Gates	bacteria	ultraviolet	cytosine thymine uracil	death
1933	Altenburg	Drosophila	ultraviolet		mutation
1934	Noethling and Stubbe	Antirrhenium	2967 Å		mutation
1936	Stanley	TMV	ultraviolet	protein	inactivation
1939	Knapp *et al.*	liverwort	2650 Å	thymonucleic acid	mutation
1936	Stadler and Sprague	maize pollen	2537 Å		mutation
1939	Hollaender and Emmons	ringworm	2650 Å	nucleic acid	mutation
1945	Beadle and Tatum	Neurospora	2537 Å	nucleic acid	mutation
1948	Latarjet	bacteriophage	ultraviolet		mutation

*See Bibliography.

and not mere rearrangements of the genetic material along the chromosome. His problem was to distinguish between these two processes. Hollaender and Emmons, after assembling a remarkably strong case for the involvement of nucleic acid in ultraviolet mutation (see Fig. 10.1) went on to caution their audience:

> It is probably somewhat dangerous to overemphasize the importance of nucleic acid in the study of radiation effects on living cells. It is very well possible that in radiation produced mutations, the nucleic acid is only the "absorbent" agent, then transfers the absorbed energy to the protein closely associated with it.
>
> (Hollaender and Emmons, 1941, 185)

This enunciation of the Energy Transfer Hypothesis enabled them to drop their earlier circumventions of the ultraviolet data: that "a living cell may contain essential compounds in very low concentration with high sensitivity at 2650 Å or that morphological structures of the cell may protect the essential features at shorter wavelengths" (Hollaender and Emmons, 1939, 400). Beadle considered the Energy Transfer Hypothesis as an alternative to the simpler hypothesis that "nucleic acid is a component of genes . . . it is possible that the nucleic acid is extragenic and that it transfers the mutation-producing energy to the gene" (Beadle, 1946, 51). Nor did the discovery of the ultraviolet inactivation of bacteriophage help matters, for soon the

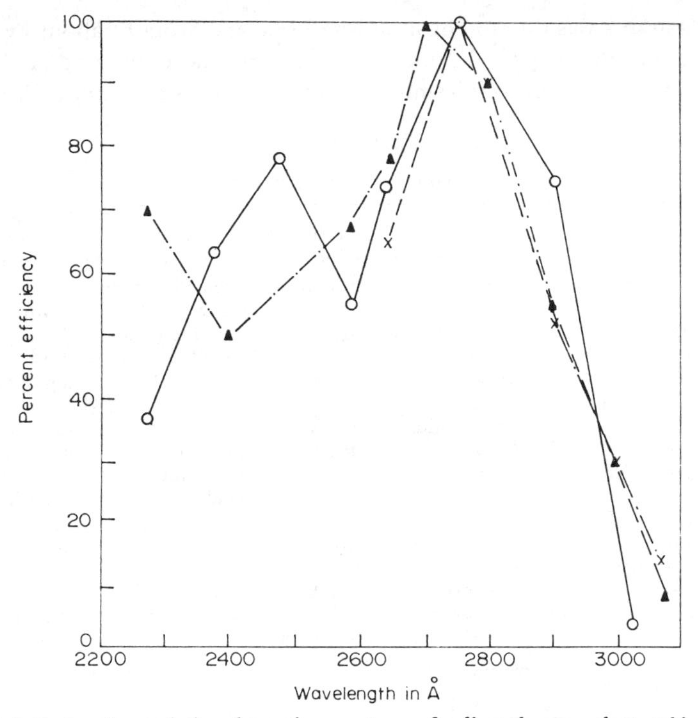

Fig. 10.1. Broken line: relative absorption spectrum of sodium thymonucleate taking absorption at 2600 Å as 100 per cent. Solid line: relative effectivity for mutation production in fungi. Dotted line: relative effectivity of mutation production for liverwort spores (Knapp, Reuss, Risse and Schrieber, 1939) taking the effectivity at 2650 Å as 100 per cent (from Hollaender and Emmons, 1941, 183).

phenomenon of photoreactivation was discovered (Kelner, 1949; Dulbecco, 1949). Not only did it take all the phage workers by surprise—Delbrück described it as a "shocker" of a discovery (Delbrück, 1948)—but it made the whole subject of ultraviolet mutagenesis look complex. And again, the ultraviolet inactivation spectrum for TMV (see Fig. 10.2) was misleading, for it seemed to support Stanley's belief in the protein nature of the viral genetic material, maximum absorption being well below that for nucleic acid. In TMV, too, there were special experimental difficulties in studying its mutation, due to the effects of competition being confused with those of mutation. With justice, Melchers later underlined the importance of Mundry's and Gierer's work showing that by causing partial inactivation a mutagen diluted the inoculum of TMV, thereby reducing the competition between mutant and normal infective particles. To show, therefore, that a mutagen had increased the mutation rate of TMV and not merely to have allowed spontaneous mutants to infect the host, it was necessary to carry out

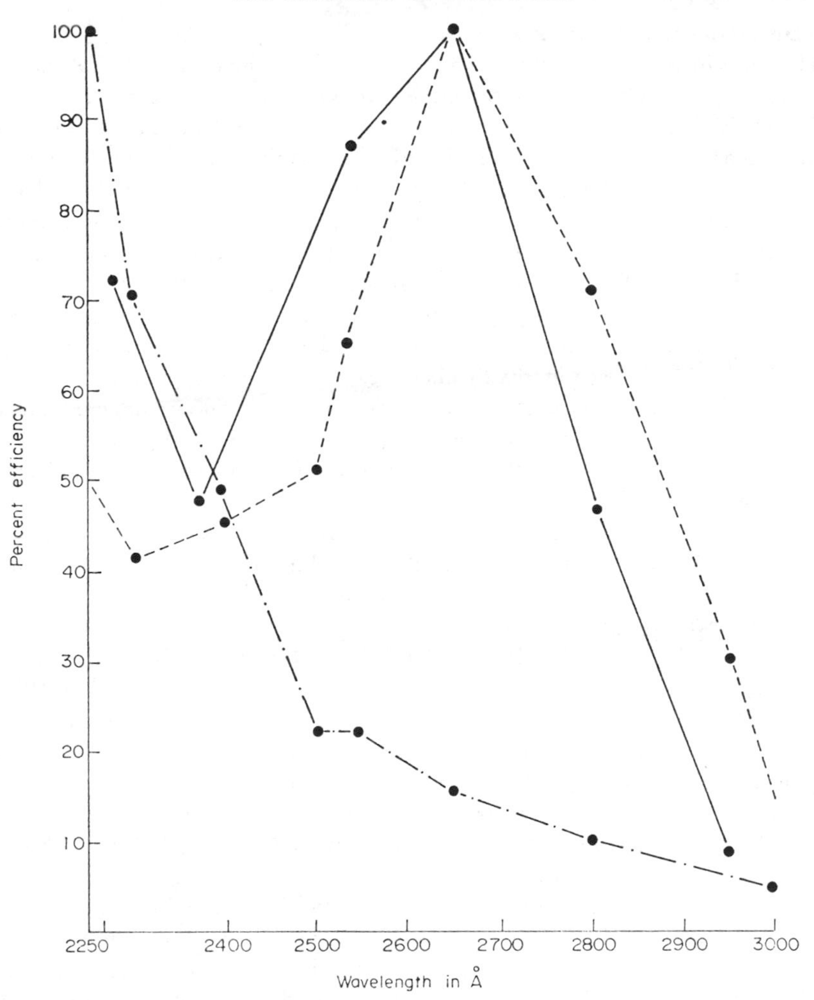

Fig. 10.2. Relative efficiency curve for fungicidal action (solid line); bactericidal action (dotted); and for the inactivation of the virus of tobacco mosaic (broken) (from Hollaender and Emmons, 1939, 400).

controls in which the inoculum was diluted with water. This was "not 'trimming on the cake' but a *conditio sine qua non*" (Melchers, 1965, 124). On the average tobacco leaf there were only a few hundred points of entry for the virus, but millions of virus particles in a typical inoculum. Nitrous acid mutagenesis of TMV did in the end (1958) give vital clues to the genetic code, but what a long grind and how many disappointments lay behind this final vindication of work on TMV.

The Structure of Viruses

Here was international activity aimed at clarifying the nature of viral replication by an attack on its mutagenesis, chemistry and structure. There was Louis Kunkel's school at the Princeton branch of the Rockefeller, Bawden's at Rothamsted in collaboration with Pirie at Cambridge, Melcher's school in Berlin-Dahlem and later at Tübingen, and much later Roy Markham's group at the Molteno Institute, Cambridge. There were also lesser known workers like Pfankuch in Freiburg and subsequently at the Biochemical Institute in Berlin-Dahlem, Erwin Weineck in the Institute for Hygiene in Leipzig, and Takahashi in Japan, not to mention the physicists who worked on viruses under the stimulus of the plant virologists—Bernal and Fankuchen in Cambridge, subsequently Birkbeck College, London, and Astbury and Bell in Leeds. Well might Astbury exclaim in a little known paper: "To the molecular biologist there can be no question but that the most thrilling discovery of the century is that of the nature of the tobacco mosaic virus: it is but a nucleo-protein" (Astbury, 1939b, 125).

In the early days of virus research the active particles were assumed to be spherical (Bechhold, 1902; Bechhold and Schlesinger, 1933; K. M. Smith, 1935, 55–57). The first hint that this might not hold true for some viruses came, as we saw in Chapter 1, from birefringence studies (Takahashi and Rawlins, 1932; 1933; Bawden *et al.*, 1936) and by estimation of the length–breadth ratio of TMV particles from X-ray data (Bawden *et al.*, 1936, 1052). At the same time sedimentation velocity and electrophoresis had been used in Uppsala to yield a molecular weight of 15 to 20 thousand (Eriksson-Quensel and Svedberg, 1936). This was followed by the work of Max Lauffer in Stanley's laboratory. From viscosimetry he calculated the length and breadth to be 430 mμ and 12.3 mμ respectively (Lauffer, 1938). With such dissymmetry the molecular weight could not be in the region of 20 000 but far higher—Svedberg had wrongly assumed for TMV the dissymmetry of an average protein. With Lauffer's value the molecular weight came to 42.5×10^6. Only a year passed before the long thin particles inferred from the physical chemistry of TMV were photographed under the electron microscope (Kausche, Pfankuch and Ruska, 1939). How amenable viruses were to the requirements of the biophysicist, and what a contrast with those awkward things called chromosomes! In the next chapter we shall see how great was the impact of the plant viruses on physicists. How promising they looked and yet how difficult they proved to be!

Should we laud Astbury for his simple enthusiasm and castigate Bernal and Fankuchen for their caution? So simple a reading of the past serves no useful purpose. Bernal and his collaborators early perceived the subunit construction of TMV particles from their X-ray diffraction pictures. By 1941 they estimated the size of the larger subunits as 44 Å × 44 Å × 22 Å, and that of the smaller as 11 Å × 11 Å × 11 Å. Astbury preferred their earlier estimate of the subunits at 22 Å × 20 Å × 20 Å (Bernal and Fankuchen, 1937, 924) for this was equivalent to one nucleotide and 54 or $2^1 \times 3^3$ amino acids (Astbury, 1939b, 126). But Bernal and Fankuchen were not to be drawn on to speculate in this way. Moreover it was, they felt, much too early to make any pronouncement on the biological significance of this subunit construction:

> . . . the crystalline nature of the particles that we have studied cannot in itself be given a biological significance nor [can it] give an answer to the question as to whether they are or are not the infective agents, nor to the far more metaphysical question as to whether they are to be considered living organisms.
>
> (Bernal and Fankuchen, 1941, 161)

Although they noted "marked resemblances with the structure of both crystalline and fibrous protein", they held that the virus structure "does not belong to any of the known classes hitherto studied. There are indications that the inner structure is of a simpler character than that of the molecules of crystalline proteins" (Bernal and Fankuchen, 1941, 164).

We know that Astbury, with typical enthusiasm, discussed the significance of the TMV work with Bernal, so some of the caution in Bernal's paper may well represent a reaction against Astbury. Unfortunate, however, was the complete absence of any reference to the gene–virus analogy in the 54 page article of Bernal and Fankuchen in the *Journal of General Physiology* (1941). In fact, the team at Rothamsted, like Stanley's in the United States, was not enthusiastic for so simple an analogy; doubtless this attitude influenced Bernal. Instead, Fankuchen and he were enthusiastic about the way in which their measurements of narrow-angle spacings revealed a structural organization intermediate between that of the molecules themselves and that observed under the light microscope (see Chapter 16).

In truth TMV, which had seemed so promising to the biophysicists and biochemists, proved elusive. It did not, after all, reveal the chemical identity of self-reproducing units. It failed them partly because of experimental difficulties, partly because no one tried to infect tobacco plants with protein-free RNA from TMV. Seymour Cohen is of the opinion that the RNA which he prepared in Stanley's laboratory by heat denaturation of the virus (Cohen and Stanley, 1942) was probably infectious (Cohen, 1973). What prevented him? Surely the Protein Version of the Central Dogma. Was this not true also for Markham, and for Schramm? Thus did the walls of the plant virologists' paradigm prevent them from showing the way to the chemical compounds which contain the genetic information of genes and viruses. This task was, instead, left to a group of outsiders, in that "other world" of medical research.

SECTION III

BACTERIAL TRANSFORMATION, ITS NATURE AND IMPLICATIONS

GRIFFITH, AVERY, BOIVIN, VENDRELY, CHARGAFF, WYATT

CHAPTER ELEVEN

Bacterial Transformation

> It may very properly be asked whether the attempt to define distinct species, of a more or less permanent nature, such as we are accustomed to deal with amongst the higher plants and animals, is not altogether illusory amongst such lowly organized forms of life as the bacteria. No biologist now believes in the absolute fixity of species. . . . But there are two circumstances which here render the problem of specificity even more difficult of solution. The bacteriologist is deprived of the test of mutual fertility or sterility, so valuable in determining specific limits amongst organisms in which sexual reproduction prevails. Further, the extreme rapidity with which generation succeeds generation amongst bacteria offers to the forces of variation and natural selection a field for their operation wholly unparalleled amongst higher forms of life.
>
> (Andrewes, 1906, 14)

> When confronted on every hand with such pictures of bacterial instability . . . it is logical, first to inquire whether the confusion we observe is pure chaos, or whether there exists any trace of orderliness amidst the general disorder. . . . Cocci become rods and rods cocci or spirals; forms of growth change overnight; motility is lost or regained; fermentation reactions are modified by time and opportunity; spore formers become sporeless; haemolytic activity comes and goes; capsulated bacteria lose their capsules, and capsules are gained by noncapsulated forms; antigenic power vanishes and reappears; cultures become spontaneously agglutinative or fail of agglutination; virulent cultures become harmless and harmless cultures virulent.
>
> (Hadley, 1927, 5)

In the 1870s there was considerable scepticism on the subject of bacterial species. The botanists, Carl Nägeli and Ferdinand Cohn denied their existence (Cohn, 1875; Nägeli, 1877) but Robert Koch affirmed it and he won the day. His doctrine of constant bacterial species was first undermined by studies of a variable strain of the colon bacillus known as *mutabile* (Neisser, 1906; Massini, 1907), and their studies were soon supported on the basis of the single-cell culture technique. This work was rapidly followed up by other workers. "The result" wrote Hadley, "was to bring into the field of study of bacterial variation the de Vriesian term 'mutation', imported from the botanical literature" (Hadley, 1927, 10). Inevitably, the whole question of bacterial variability was open once more, and evidence was not lacking. Thus, changes in the ability to ferment specific media and attenuation of virulence were observed at the end of the nineteenth century. Roger, at the Pasteur Institute, observed what was probably a spurious case of attenuation in pneumococcal cultures in 1891, later, more definite evidence was obtained by Neufeld at the Robert Koch Institut, Berlin, in 1902. But what was the significance of these changes? Were they lasting changes like de Vries' mutations, or were they equivalent to his "fluctuating" variations? An understanding of such changes, it was hoped, would help to explain the fluctuations in the severity and incidence of diseases. Such was the background to Griffith's discovery of bacterial transformation in pneumococcus.

The Discovery of Rough and Smooth Forms

The British bacteriologist, J. A. Arkwright, at the Lister Institute, London, studied the characteristics of the virulent and attenuated strains of several bacteria, chiefly Shiga's bacillus. In 1921 he gave a clear description of their colonies, the virulent ones being smooth, dome-shaped, and regular; the attenuated ones being granular, flat and irregular. He introduced the terms rough R and smooth S, and described them as persistent variations or "mutants" (Arkwright, 1921, 55). Since the R forms were only observed under artificial conditions, Arkwright saw that the observed uniform character of bacteria like *B. typhosus* and *B. dysenteriae in vivo* was due to selection. "The human body infected with dysentery may be considered a selective environment which keeps such pathogenic bacteria to the forms in which they are usually encountered" (*Ibid.*).

Arkwright had picked out an important example of microbial variation. He had shown that, once produced, the R-form was reproduced faithfully in the subsequent colonies subcultured weekly, but above all he had given very clear and simple criteria for distinguishing the two forms visually. Small wonder then that his work was "seized upon at once, first and foremost by the English school" (Hadley, 1927, 12). The R and S forms were described in streptococci in 1922, pneumococci in 1923, *B. enteritidis* in 1924 and Salmonella in 1925. At the Rockefeller, Paul de Kruif studied the S and R forms in the bacillus of rat septicaemia. In London, Frederick Griffith demonstrated reversion of R to S in pneumococcus, by animal passage and in plate culture (1923), and this was confirmed by Levinthal working in Berlin under Neufeld (1926).

The R and S Forms of Pneumococcae

The extension of Arkwright's R and S forms to pneumococcus was achieved by Griffith, a quiet and retiring medical officer of health at the Ministry's pathology laboratory in Endell Street, London. He was the most English of Englishmen. "He was a civil servant and proud of it. He had that kind of a mind and the integrity that often goes with it. He did not allow his fancy to roam . . . and being employed by the Ministry of Health to do a specific job, he believed in fulfilling his contract however frustrating that might be" (Elliot, 1970). Allison has described just how retiring Griffith was: vain were all attempts to persuade him to attend the meetings of the Pathology Society and the Royal Medical Society, or to read a paper to the Medical Research Club. To get him to the International Congress for Microbiology to hear Rebecca Lancefield talk on streptococcal types Allison and Scott had to order a taxi and bundle Griffith into it (Allison, 1969; Pollock, 1970, 7). No wonder that the great immunochemist, Avery, never came to know Griffith personally and never corresponded with him (Pollock, 1970, 10).

The problem that the Rockefeller Hospital scientists were trying to solve was the production of an effective immune serum for treatment of patients suffering from acute lobar pneumonia. Long ago Neufeld, at the Robert Koch Institute, had classified pneumococci serologically into three types, and had thus laid the basis for the recognition of Types I, II and III. The Rockefeller groups added a fourth Type which, because of its heterogeneity, became known as the "American Scrap Heap". Griffith renamed this collection "Group IV".

Neufeld observed that Type I pneumococcae were the commonest to be found in cases of pneumonia brought to his attention in Berlin. Therefore he called this type "typical" in contrast to the other two types which he called "atypical". He suggested that in order to produce a successful immune serum an attempt should first be made to prepare one against Type I pneumococcae, rather than against the "atypical" type. The Rockefeller group and Griffith also found Type I most common. They set to work and prepared an immune serum, but as Griffith's colleague, Dr Arthur Eastwood wrote, it was far from an unqualified success.

> Whilst appreciating the value of the progress which has been made, it must be admitted that the present position is unsatisfactory. The Rockefeller investigators can only offer serum therapy if the case is found to be due to Type I; in that event, their experience is that large and repeated intravenous injections of specific serum will bring down the mortality due to this type from about 25 to about 10 per cent. Diagnosis of the type should be made at an early stage of the disease and treatment should follow immediately. But, in hospitals, patients are often in an advanced state on arrival; and, owing to the special skill and care which are required and to the very large quantities of serum needed, this treatment is not likely to be readily adopted by the general practitioner.
>
> If the precise antigenic characters of the infecting strain of pneumococci are all important, one can understand that it would be difficult, if not impossible, to provide therapeutic sera which would be useful for infections with "atypical" strains; but, even on this assumption, there is no generally accepted explanation why immunization of horses with the second and third of the "fixed" types has failed to produce a good therapeutic serum, when immunization with the first has succeeded.
>
> (Eastwood, 1922, 18–19)

What a contrast was pneumococcus with the straightforward behaviour of diphtheria and smallpox! It threw up problems which called for a very thorough knowledge of its antigenic properties. Had it not proved so difficult the science of immunochemistry would surely not have been so strongly promoted, and the transformation of pneumococcal types might never have come to occupy so important a role in the identification of the hereditary material! One thing the experiences at the Rockefeller Hospital taught Avery and Dochez was just how distinct were the various types of pneumococcus. Their study of the immunity reaction, on the other hand, showed them that, like diphtheria, live pneumococci do secrete soluble substances into the host organism, and that the quantity liberated seemed to be related to the amount of capsular material around the pneumococcus—much in Type III and least in Type I. Although the evidence pointed in another direction they considered the soluble substance to be "of protein nature or to be associated with protein" (Dochez and Avery, 1917, 493).

Five years later Avery was given something of a jolt when an extract like his specific soluble substance was obtained from a number of bacterial species, including pneumococcus, and shown to be protein-free (Zinsser and Parker, 1925). This work had been carried out at a rival institute—the College of Physicians and Surgeons in New York—and its authors suggested that they were dealing with an antigen composed of two parts, the one a nucleoprotein which stimulated antibody production, the other a non-protein "residual antigen" which reacted with the antibody. The latter they speculated might be an example of the hypothetical hapten molecules of which Landsteiner had written (Landsteiner, 1919).

Then it was Avery's turn to show that the specific soluble substances in pneumococcus were unusual polysaccharides and that the capsular carbohydrate, long known, was in fact involved in the determination of serological activity (many papers starting with Heidelberger and Avery, 1923; Avery and Heidelberger, 1923).

What had been a series of serological types now became in addition a set of chemically distinct types.

Change of Pneumococcal Types and the Onset of Acute Pneumonia

When in 1923 Griffith discovered the S and R forms in pneumococcus and their interconversion *in vivo* and *in vitro*, the Rockefeller concept of type specificity was not *directly* challenged, for reversion of R to S forms always led to the production of a specific soluble substance identical with the S type from which the R form had originated. But Griffith was exploring a more daring suggestion. He knew that associated with the R and S forms was the property of non virulence and virulence respectively. He knew furthermore that the Rockefeller scientists had shown Types I and III to be associated with acute lobar pneumonia whereas members of their "Scrap Heap"—Group IV—were found in the sputum from healthy individuals and in patients recovering from acute lobar pneumonia. Whereas the Americans simply concluded that Types I and II died out during convalescence and were replaced by Group IV, which was not invasive and therefore remained in the mouth, Griffith as early as 1922 wrote:

An alternative theory is that the virulence of Types I and II becomes attenuated during convalescence, and this change is accompanied by mutation of type characters, which now become degraded into those of the heterogeneous and less virulent group termed IV. There are two experimental difficulties about this view. . . . Still it is theoretically possible that mutation may occur in nature, though it cannot be reproduced *in vitro*.

(Griffith, 1922, 35–36)

To test for such an event he sought for different serological types from one and the same patient during the course of the disease and he found them.

If mutation occurred, one might expect some regularity in the serological characters of the strain which replaced the Types I and II. . . . Although many of the infections have been apparently pure, it is very striking how often a typical bile soluble diplococci can be found in sputum, even during the acute stage, together with the Types I and II. . . .

(Griffith, 1922, 36)

From the beginning, it was Griffith's desire to show affinities rather than differences between the pneumococcal types.

The various races of pneumococci resemble each other so closely in appearance of colonies and in the characteristic of bile solubility that there can be no doubt that they belong to one species.

(*Ibid.*)

When he introduced his discovery of S and R forms in pneumococcus he wrote: "The conception of a 'pure culture' of a bacterium as a number of absolutely identical individuals is no longer tenable" (1923, 1). Griffith seems to have felt that mutation *within the limits of the species* was acceptable since it was a device by which the species adjusted itself to changes in the environment. It was a "natural tendency with many species of bacteria" (Griffith, 1923, 11). The S to R change was one example; it was "attributed to degenerative changes" and was "associated with the loss of certain antigenic qualities" (*Ibid.*). Because it was achieved regularly when immune serum had been added to the medium Griffith suggested that the serum not only sensitized the bacteria in preparation for phagocytosis (Neufeld's bacteriotropic theory) but it caused those which escaped this fate to become rough and therefore non-virulent (Griffith, 1923, 12–13).

We can, I believe, conclude that Griffith would have denied the transformation of one bacterial species into another, but for him the pneumococcal types and R and S forms were mutable characteristics *within* the species. Such "mutations" were distinct from the important mutations discovered by de Vries. In this Griffith sided with P. Hadley who felt it was not in accord with biological principles

that variations of hereditary significance (that is, true mutations) would be formed as easily or as commonly as we observe to be the case . . . or that micro-organisms in general are addicted to discarding permanently their ancient hereditary characters with such apparent nonchalence . . .

(Hadley, 1927, 224)

Griffith's Discovery of Transformation of Types

From Avery's point of view the story took an unexpected and regrettable turn when in 1928 Griffith reported the transformation of the pneumococcal types which Avery and Heidelberger had so firmly established as invariable. The story seemed improbable therefore from the beginning, but even more so when the method by which Griffith had achieved this transformation was considered. He had injected the living cells of the R form of Type I pneumococcus into a mouse, together with heat-killed cells of the S form of Type II. The mice succumbed to the infection and died. From their blood Griffith isolated colonies of the S form of Type II! Not only had there been a reversion from R to S but a *change of type*. The capsular substance of the colonies isolated was identical with that of the dead Type II cells, and not of the capsular type from which the living R cells had been obtained by serial sub-culture.

If this phenomenon could have been simply a question of the dead polysaccharide coats being used by the living cells Griffith's result would have been considered more plausible, but not only did the polysaccharide belong to a different pneumococcal type from that of the living R cells but it was resistant to steaming, yet the dead S cells lost the power to transform the living R cells if they were heated above 80°C. For Griffith then, the transforming substance had to be thermolabile; the polysaccharide was thermostable. He called it S substance and wrote: "By S substance I mean that specific protein structure of the virulent pneumococcus which enables it to manufacture a specific soluble carbohydrate" (1928, 151).

Now, he argued, how could the Type I cell utilize the S substance of a Type II cell unless it already contained some of the S antigen of Type II. Here he seemed to slip from S substance to S antigen and thus made his meaning ambiguous. But for Griffith, as for Avery, the S antigen was a compound structure *not* equivalent to the soluble specific substance. Otherwise the latter would be found capable of stimulating antibody production, but it was not (Zinsser and Parker, 1923; Avery and Heidelberger, 1923). It was therefore quite consistent for Griffith to speak of utilizing the S antigen from dead cells. He wrote:

> Since virulence and the capacity to form soluble substances are attributes of the S strain, their possession may for convenience be ascribed to a special antigen which may be termed the S antigen.
>
> (Griffith, 1928, 149)

His argument was, then, that an R form of Type I may possess a rudiment of the antigenic protein required to make the specific soluble substance of a Type II S form. All it required was more of this protein in order to function effectively. Now the transformation of RI to SII became in Griffith's eyes not so very different from ordinary reversion of RI to SI. In the latter process the S antigen "remaining in an R strain may be regenerated and reach its original abundance under suitable conditions" (*Ibid.*, 152), as when he innoculated the R form with heat killed dead cells of the S form of the *same* type, or when the R form multiplied in a protective nidus under the skin and the cells which survived the host's reaction utilized the remnant of the S antigenic protein from those pneumococci which had not survived attack. Now suppose such an R cell "may contain in addition to its major antigen a remnant of the other type antigen" (*Ibid.*). What then was the difference between the R form of Types I, II and III and Group IV? For Griffith there was little if any difference.

> When pneumococci of Types I and II are reduced to their respective R forms by growth in homologous immune sera, they lose nearly all their major S antigen though they may retain their minor S antigens which are presumably not affected by the heterologous immune substance. But the major S antigen apparently still preponderates, since an R strain on reversion to the S form regains its original type characters.
>
> (*Ibid.*)

Griffith went on to point out the fact that in those R strains of Type I which did not revert spontaneously to their S forms there was no such preponderance of the residual Type I S antigen and hence there was no difference between the R form of these strains and that of a Type II pneumococcus! Then he waxed bolder and suggested that between Types I and II "there is no essential distinction".

> In fact, there are certain indications that the R pneumococcus in its ultimate form is the same, no matter from what type it is derived; it possesses both Type I and Type II antigens in a rudimentary form or, as it may be differently expressed, it is able to develop either S form according to the material available.
>
> (Griffith, 1922, 153)

If by "is able to develop either S form . . ." Griffith meant that the R form had the protein required to synthesize either type, then the following oft-quoted passage makes more sense:

> When the R form of either type is furnished under suitable experimental conditions with a mass of the S form of the other type, it appears to utilize that antigen as a pabulum from which to build up a similar antigen and thus to develop into an S strain of that type.
>
> (*Ibid.*)

This idea of the pabulum jars on our ears, accustomed as we are to far more explicit statements of the relation between genetic constitution and its expression in the visible characteristics of the organism. It may help, therefore, to understand Griffith's thinking if we turn to the report which his much more speculative colleague, A. Eastwood, wrote for the Ministry of Health in 1923. At that time transformation of pneumococcal types had not yet been discovered, but bacterial variation was well known, especially Arkwright's S to R change, and Griffiths' observation of it in pneumococcus, together with reversion of R to S. Eastwood belonged to the orthodox school who, as we saw in the last chapter, adopted the enzyme theory of life. For him there were two phases in the life cycle of a bacterium: (1) Catalytic, or as we would say catabolic; (2) Synthetic, that is, the building up of new protoplasm. In the former the union of enzyme with substrate was transient, after which the digested substrate was liberated, in the latter the union of enzyme with substrate was stabilized and the resulting complex yielded protoplasm with all its properties, such as ability to manufacture the specific soluble substance. ". . . bacterial protoplasm may be regarded as a complex of enzymes and the products of enzyme action, a complex which involves the synthesis of these products (e.g., amino acids) into proteins" (Eastwood, 1923, 19). He argued further that there existed a "critical phase" in the life of the growing bacterial cell when a transition from catalysis to synthesis occurred. This was a delicate time when external influences could easily upset the balance and initiate fresh formation of protoplasm before all the ingredients from the environment were available. Defective protein would result. Or the catalytic stage might be prolonged and bacterial lysis would result. (Eastwood belonged to the Bordet/Gratia School and denied the existence of d'Hérelle's bacteriophages.) And what happened when the S form of pneumococci grew in a culture containing immune serum?

> In this case it is reasonable to postulate that, when digested food particles are being synthesized into protoplasm, the antibodies find in some of those particles their appropriate antigen and "pick them off"; the new bacterium is synthesized, but it is an impoverished bacterium (a variant), because it has been robbed of some of its antigenic components.
>
> (Eastwood, 1923, 21)

Eastwood's conception of metabolic activities was clearly cast in the mold of nineteenth century chemical vitalism. We find references to the "side chains" and "active groups" of protoplasm which remind us of Ehrlich and of the complex protoplasmic molecule. Finally, in the foreword to Griffith's and Eastwood's reports we find George Newman using the word "pabulum" (Newman, 1923, iv). This takes us back to the days when enzymes were chiefly known for their catabolic roles and when the distinction between growth and replication was not recognized. It shows us too, how long-lasting in some quarters was the vitalistic conception of protoplasmic synthesis as originally detailed by Pflüger in 1875.

Griffith's Interpretation of his Results

It has been stated recently that Griffith's demonstration of type transformation "must surely have been made almost *despite* his own emotional inclinations, rather than, as is so often the case, because of them" (Pollock, 1970, 7). We have seen that there are good reasons for believing that almost the converse of this was true, and that in 1922 and 1923 Griffith was toying with the idea that within the "species" there are characteristics like the type of polysaccharide capsule *subject to mutation in response to environmental conditions.* Admittedly, he had no evidence of type mutability from plate cultures, but he did have the observation of more than one type in the same patient. Was this due to multiple infection or to mutation? The work described in his 1928 paper was aimed at deciding between these two explanations. Having shown that mutation can be produced experimentally Griffith opted for this alternative, and explained it in nineteenth century Darwinian terms.

Like Darwin and the nineteenth century animal breeders, Griffith pictured the progressive "fixing" of characteristics and he asked whether, if transformation of type "is a question of altered environment, are the influences which initiated the divergence of type still at work, i.e., are the type characters still in a state of flux, or have the different varieties become stabilized?" (Griffith, 1928, 148). Well, he showed that they were indeed still in a state of flux. The S form of Type II stimulated the production of a "specific immune substance" by the host which caused clumping together of the pneumococci when they became susceptible to phagocytosis.

> By assuming the R form the pneumococcus has admitted defeat, but has made such efforts as are possible to retain the potentiality to develop afresh into a virulent organism. The immune substances do not apparently continue to act on the pneumococcus after it has reached the R stage, and it is thus able to preserve remnants of its important S antigens and with them the capacity to revert to the virulent form.
>
> (1928, 156)

The bacterium did not necessarily play a purely passive part in the host–pathogen relationship, "the various forms and types may be assumed by it to meet alterations in its environment" (*Ibid.*).

In contrast to Griffith, Arkwright had presented a view of bacterial mutation which was modelled on de Vries' mutation theory, but even Arkwright was worried by the widespread occurrence of the same mutation S → R. And after Griffith had introduced the idea of major and minor antigens, Neufeld and Levinthal happily repeated it, and Dawson and Alloway explicitely supported it.

The Significance of Transformation for Epidemiology

We have already seen that the Rockefeller Hospital staff found Group IV pneumococci rarely in association with acute lobar pneumonia in contrast to Types I, II and III. Griffith considered it likely that Group IV represented the non-invasive type which could survive in the upper respiratory tracts without causing the onset of disease. From this situation it could spread by aerial infection to other hosts. To become invasive it "evolved" into Type I, II or III, reached the lower tracts of the lungs and there brought about acute lobar pneumonia. Should the host recover, the pathogen simply changed back into Group IV and survived in the upper respiratory tracts of the convalescing patient.

But what of epidemics of pneumonia? Could these be accounted for in terms of transformation of type? It is clear that Griffith had hoped from the outset that some such mutation did lie behind the spread of pneumonia through a population. He had been impressed by the fall in the frequency in Smethwick of Type II from 32.6 per cent of the cases in 1920 to 7.4 per cent in 1927. This was paralleled by a rise from 30 per cent to 53 per cent in the frequency of Group IV. Surely this change had resulted from transformation of type? We can now see Griffith's work in its context. To learn how to control the incidence and spread of chronic lobar pneumonia called for a detailed knowledge of the host–bacterium relationship. Griffith believed he had shown there was more to this relationship than was generally believed. When bacteria were killed by the host their presence along with living bacteria did not merely negate the action of leucocytes by the "aggressin" popularly held to be liberated from them. The living R cells "actually make use of the products of the dead culture for the synthesis of their antigen" (Griffith, 1928, 150). And as his experiments showed, this could lead either to reversion of R to S of the same type or to that of another, since all pneumococcal types retained a rudiment of the protein structure necessary for making *any* of the various polysaccharide coats.

What an unfortunate turn events were taking in the world of microbiology! The extreme position of Robert Koch and F. Cohn, who believed in the fixity of bacterial types, had already been undermined. Now Avery's and his colleagues' demonstration of the constancy of pneumococcal types appeared to be going too. The opposite position from Koch and Cohn, represented by Carl Nägeli, who asserted the interconvertibility of bacterial species (1877), seemed in danger of coming back into fashion. Griffith's work was not just an oddity that could be shrugged off; it was a bombshell which fell into a fused situation, and Avery had every reason for *not* accepting it. Griffith's reputation was high, but his 'Lamarckian' ideas and his vague talk of a "pabulum" could hardly have appealed to Avery, who, though cautious to the point of conservatism, was at least committed to strictly chemical explanations. Small wonder, then, that he was not the prime mover behind the repetition of Griffith's experiments at the Rockefeller.

The Confirmation of Griffith's Results

At the Robert Koch Institute in Berlin, Neufeld and his assistant Levinthal were so quick to repeat Griffith's work that their confirmation of his results appeared in the same year, 1928, as Griffith's own paper. This had been possible because Levinthal had already been working on bacterial variation, having achieved reversion of R to S pneumococci without change of type in

1926, and also because Neufeld had visited Griffith's laboratory while the transformation studies were in progress and therefore knew the details (McCarty, 1968; Neufeld and Levinthal, 1928, 324). In 1929, in far-off Peking, H. A. Reimann also confirmed Griffith's work. But it was the fact that so reliable a bacteriologist as Neufeld has been able to reproduce bacterial transformation that made urgent the repetition of Griffith's work in the Rockefeller itself. This was done not by Avery but by the strongly pro-British Canadian, Henry Dawson, who "took advantage of the fact Avery had to be away for more than six months (because of hyperthyroidism) to repeat and confirm Griffith's experiments" (Dubos, 1972; Dawson, 1929). In his biographical memoirs of Avery, Dubos wrote: "For many months, Avery refused to accept the validity of this claim and was inclined to regard the finding as due to inadequate experimental controls. This scepticism was understandable in one who had devoted so much effort and skill to the doctrine of immunological specificity" (Dubos, 1956, 41). It was as if history was repeating itself, just as the firm ground won by Robert Koch was undermined, so was that of Avery!

According to George Corner, Avery had "asked Dawson to look into Griffith's transformation" (Corner, 1964, 461). This seems unlikely since it was never Avery's custom to ask anyone to undertake a specific piece of research. Had Avery played any part in the confirmatory work one would expect there to be some reference to him in Dawson's two papers of 1929, but there is none. Avery's name is on neither paper, yet when Dubos isolated a substance capable of digesting the capsular polysaccharide of pneumococcus while Avery was on holiday in the summer of 1930, Avery's name went first on the paper published by *Science* in August. Corner gives no source for his information. Dubos was there at the time. Dawson's confirmation of transformation by subcutaneous injection was reported in his second paper, received by the *Journal of Experimental Medicine* in July 1929. Avery now had no alternative but to face Griffith's discovery.

CHAPTER TWELVE

The Identity of the Transforming Substance

The story of the advances made by Dawson and Sia (*in vitro* transformation, 1933), Avery, MacLeod and McCarty (identification of the transforming substance, 1944) has been told many times. I shall deal here with only one aspect of the work—the progressive characterization of the transforming agent—since it was this aspect which led to the first evidence for the genetic role of DNA. It also serves to throw into relief the very different approaches of the Lamarckian Griffith together with those who confirmed his work and adopted his interpretation on the one hand, and on the other the group under Avery who demanded and obtained a precise chemical account of bacterial transformation.

Naturally the early workers thought the agent of transformation must in some way be dependent upon a protein. They had Griffith's evidence that it was thermolabile. Perhaps, wrote Dawson, "the S vaccine disintegrating in the animal tissues supplies a suitable pabulum . . ." (1928, 121). But as it could not be the specific soluble substance itself, perhaps it was a co-ferment which, together with the synthesizing enzymes of the receptor cell, made possible the production of the carbohydrate capsular substance. Neufeld and Levinthal thought: "Possibly a precursor of the specific carbohydrate or a carbohydrate bound to a protein" was necessary (1929, 340). When Dawson and Sia had succeeded in producing transformation *in vitro* they realized that the transforming substance bore a "striking resemblance to the 'antigenic specific substance described by Day" (1931, 709). H. B. Day at the Institute of Pathology, St. Mary's Hospital, London, had shown that the antigenic specific substance must consist of two portions, one thermostable which reacted with immune serum, the other thermolabile and susceptible to the attack of bacterial enzymes. This latter portion provoked the production of antibodies (Day, 1930). Evidence for such a twofold character to the bacterial antigen of course goes back to earlier work (Zinsser and Parker, 1923, Avery and Heidelberger, 1923) when the theromolabile portion was described as a nucleoprotein. Griffith's immediate successors, however, did not pursue this identification. They had enough to do trying to achieve an experimental situation in which transformation could be reproduced consistently and *in vitro*.

When Wilhelm Baurhenn in Heidelberg confirmed the *in vitro* transformation achieved by Dawson and Sia he supported Dawson's "co-ferment" idea for the agent responsible (1932, 91). Lionel Alloway, who prepared the way for Avery, MacLeod and McCarty by his achievement of cell-free *in vitro* transformation, avoided committing himself on the chemical identity of the agent. It could not, he reasoned, be the polysaccharide alone unless this specific soluble substance was present "in a different physical state, or in combination with some other substance . . ." (1933, 276). Later, we are told, he favoured the protein-containing full capsular antigen (Hotchkiss, 1966, 184).

Alloway had used sodium desoxycholate (bile salt) to liberate the contents of the pneumococcal cells and these, when extracted with salt solution, could be passed through Berkefeld filters and yet remain as active in transformation as the intact pneumococcal cells. Now when Alloway slowly added this bacterial salt solution to 500 cc of chilled absolute alcohol "a thick syrupy precipitate formed . . ." (1933, 266). This must surely have consisted of fibrous, biologically active DNA (Hotchkiss, 1965, 5), but "it was *not* implicit in general experience, in Alloway's time, that a thick, stringy alcohol precipitate meant DNA: some mucus, linear polymers and what was called 'renosin' (not *known* to contain DNA) behaved like that" (Hotchkiss, 1972). At the time, then, this biologically active stringy precipitate would most likely have been considered a protein or nucleoprotein but definitely not a polysaccharide, since the latter was not precipitated by alcohol.

Three years after Alloway described this alcohol-precipitated fibrous substance Avery is reported to have said that "the transforming agent could hardly be carbohydrate, did not match very well with protein, and wistfully suggested that it might be a nucleic acid!" (Hotchkiss, 1965, 5). Certainly this solubility in salt solution and precipitation by alcohol looked more like that of a nucleic acid than of a protein, and then there were the repeated reports of the isolation of an antigenic, though not type specific, nucleoprotein, by salt extraction and precipitation with the dropwise addition of acetic acid (Avery and Morgan, 1925). And had not Dubos shown that RNase destroyed the antigenicity of dead encapsulated pneumococci? (see Chapter 6). A biological function in the host-bacterium relationships was thus associated with a nucleic acid of the yeast type. Hence, in 1936 Avery either did not concern himself with the question of the type of nucleic acid involved or he would perhaps have favoured the yeast-type. But this is just speculation and: "In fact, it was not like him to settle on any single substance until there was strong evidence for it" (McCarty, 1972). We must remember, too, that by the time Hotchkiss arrived at the Rockefeller (1935) Avery was telling his intimates of the separation of the protein-containing antigen from the transforming activity (Hotchkiss, 1966, 184). Even so, it is doubtful that thymus-type nucleic acid was seriously considered, especially since it had not been extracted from pneumococci, whereas nucleic acid susceptible to RNase was extracted not long thereafter (Thompson and Dubos, 1938).

Characterization by Enzymology

The Rockefeller led the world in enzymology, so what more obvious and appropriate strategy could there be than to pursue the identity of the transforming agent by studying the enzymatic destruction of its activity? In 1946 McCarty and Avery had this to say of their approach:

> The enzymatic analysis was begun early in the course of the attempts to determine the nature of the transforming substance. Relatively unpurified pneumococcal extracts were subjected to enzymatic activity in the hope that by this approach some clue might be obtained as to the identity of the biologically active constituent. Crystalline trypsin, chymotrypsin, and ribonuclease had no effect on the transforming substance, but it was found that certain crude enzyme preparations were able to bring about complete loss of transforming activity. When the possible importance of DNA was suggested by chemical fractionation, the experiments with crude enzyme preparations were extended to determine whether their ability to destroy the activity of the transforming principle could be correlated with any enzymatic action on authentic samples of DNA of non-bacterial origin.
>
> (McCarty and Avery, 1946a, 89)

From this passage, written only three years after the identification of the transforming substance as DNA, it is clear that the crystalline enzymes trypsin, chymotrypsin and ribonuclease were used early on. Their failure to inactivate the substance threw doubt on the possibility that it could be a protein. Such doubt would have been strengthened had they been able to use crystalline pepsin, but at the optimum pH for this general proteolytic enzyme the activity of the transforming agent was destroyed. By 1940 Moses Kunitz (at the Rockefeller) had prepared crystalline RNase and this, when used on the transforming substance, failed to inactivate it. By 1940, therefore, Avery and Colin MacLeod knew that their active substance was not RNA. At this point in the story Levene's contribution to nucleic acid enzymology came in handy. As we saw in Chapter 6, he and Schmidt had detected the presence of a DNA depolymerase in the secretion of the intestinal mucosa of a dog. Avery and MacLeod therefore used Levene's intestinal extract on the transforming substance. It did inactivate it. Perhaps, incredible though it might seem, the substance was thymonucleic acid!

Enzymatic studies were not alone in pointing in this direction. Avery and MacLeod, like Alloway before them, had precipitated the active substance using absolute ethyl alcohol. This was a well-known precipitant in several methods for the preparation of thymonucleic acid. Avery and MacLeod therefore tried the technique of fractionation to increase the purity of their substance. Alcohol was added dropwise until at a critical concentration "varying from 0.8 to 1.0 volume of alcohol the active material separates out in the form of fibrous strands that wind themselves around the stirring rod" (Avery, MacLeod and McCarty, 1944, 143). When this fibrous substance was analysed it revealed the presence of phosphorus, it absorbed ultraviolet light at a maximum in the region 2600 Å, and it had a molecular weight of at least half a million. It gave a strong Dische reaction for DNA, but also a weak Bial reaction for RNA. When compared with the theoretical value for sodium thymonucleate the elementary analysis of the transforming substance agreed fairly well. By the time Maclyn McCarty joined in the work in 1942 the possibility that they were dealing with DNA was well founded, but by no means certain.

The sequel to the use of crude enzyme extracts on the transforming substance was their use on non-bacterial samples of DNA. Mirsky supplied them with such samples obtained from mammalian tissues and from fish sperm. Both were attacked.

Still not content, Avery encouraged McCarty to pursue the evidence yet further. He used crystalline RNase to remove RNA from the transforming substance and similarly the Dubos SIII enzyme to remove polysaccharide. After the publication of the great 1944 paper McCarty succeeded in purifying DNase to the point where it contained only traces of a proteolytic enzyme and no ribonuclease activity (McCarty, 1946a). This led in turn to an improved method for the preparation of the transforming substance (McCarty and Avery, 1946b), and to a striking demonstration of the power of this enzyme, even when present in minute quantities, to destroy the activity of the transforming substance permanently (McCarty and Avery, 1946a).

From this account it seems that the identity of the transforming substance was revealed slowly step by step. Just as there were no bold *a priori* ideas which experiment later validated, so there were no unfortunate and lengthy false trails which led nowhere. Illustrative of the way in which the problem of identity was solved, was the procedure of isolation. This came to incorporate stages for eliminating all the components whose inactivation or removal had early been shown to have no effect upon the activity of the transforming substance:

> the protein by the chloroform method, the capsular polysaccharide by digestion with a specific bacterial enzyme which hydrolyses it, the somatic polysaccharide by fractional alcohol precipitation, and ribonucleic acid either by enzymatic digestion with ribonuclease or by alcohol fractionation.
>
> (McCarty, 1946a)

All that has been said so far is in accord with the late Colin MacLeod's recollection:

By the time McCarty joined us we were virtually certain of what we were dealing with, both on the basis of the methods of preparation, the physical–chemical properties, and the elementary analysis. Moreover, we had pretty good evidence that the enzyme which destroyed activity was DNase from a variety of lines of approach . . . Maclyn McCarty was a great help in tying things down and in getting further evidence that the enzyme was indeed DNase through the purification of that enzyme from pancreas.

(MacLeod, 1967)

The Interpretation of Bacterial Transformation

All the writers before Avery, MacLeod and McCarty were inclined to accept Griffith's interpretation of transformation according to which the recipient cells had retained the power to elaborate the capsular polysaccharide of *several* serological types and needed only the specific *stimulus* of the transforming principle in order to produce any one of them. The serological types which resulted were of course determined by the type of the donor bacteria. Surprising as this may seem to us today, the truth of this assertion is based upon good evidence, as the following quotations show:

The R form, therefore, probably results from attempts of S bacteria to adapt themselves to unfavourable environmental conditions. Once reduced to the S state the organisms potentially have the capacity to develop the S structure of any of the various specific S types.

(Dawson, 1930, 143)

The exact nature of the active material in these extracts still remains to be determined. That it acts as a specific stimulus to the R cells which have potentially the capacity of elaborating the capsular polysaccharides of any one of the several types of pneumococci seems clear.

(Alloway, 1933, 277)

But the decisive cause of the behaviour of the R form is present here . . . in the character of the variants themselves, for in them the process of degradation S to R has evidently not yet worked deeply, so that each specific stimulus to the formation of an heterologous S structure brings about its full regeneration as in the case of the residual *Anlage* of the type specific structure of the receptor R form.

(Baurhenn, 1932, 84)

The 1944 Interpretation

Whatever has been said about the conservative stand taken by Avery and his colleagues in 1944 when they discussed the significance of their results (Pollock, 1970, 14; H. V. Wyatt, 1972, 87) it cannot be said that they followed the "Griffith line" described above. Certainly, they belonged to the empirical tradition. They set themselves a narrowly defined goal and pursued it single mindedly between 1940 and 1943. Experiments suggested ideas and these were in turn "controlled", to use a famous term of Claude Bernard, by further experiments, and they wrote:

The major interest has centred on attempts to isolate the active principle from crude bacterial extracts and to identify if possible its chemical nature or at least to characterize it sufficiently to place it in a general group of known chemical substances.

(Avery, MacLeod and McCarty, 1944, 138)

Now Dawson had left the Rockefeller in 1929 after confirming Griffith's work and had done the experiments with the Chinese scientist Richard Sia at Columbia University nearby. Alloway also left the Rockefeller after achieving transformation with cell-free extracts in 1932. When Colin MacLeod came two years later, one of his aims was to work on bacterial transformation—he had read Griffith's paper as a medical student, but no one was working on it at that time. Avery himself was away at a sanatorium for the second time, receiving treatment for what was later identified as hyperthyroidism. Meanwhile, MacLeod taught himself to isolate R forms of Type II pneumococcus and when Avery returned in the fall of 1935 they set to work using Alloway's *in vitro* cell-free system. From this point on it is clear that Avery was very much involved in the work. He was a bachelor and the Rockefeller was his second home. Like MacLeod, therefore, he would come in at week-ends to continue the work which involved regular 7 a.m. to mid-night sessions when transforming extracts were being prepared (MacLeod, 1968).

Now what can we learn about Avery? Was he the original genius who stimulated all those around him and without whom immunochemistry at the Rockefeller would be unthinkable? Dubos wrote that he "was not as broadly informed a scholar as one would assume from his achievements and fame" (Dubos, 1956, 42). On the other hand he was convinced that biological specificity was determined by chemical specificity and he focused his attention on this to the exclusion of lesser matters. His greatest contribution, said Dubos, "was not so much a vision as a thread along which all the observa-tions were organized. Avery was the person who went to the clinician, Dochez, and to the chemist, Heidelberger, . . . and crystallized all these fragments together" (Personal communication). "Nothing was of use to him unless it could be integrated into an intellectual picture" (Dubos, 1957). Those who came to work with him contributed their piece, and the picture took shape. This was the sense in which Avery acted as the leader. He did not organize and direct a team in the modern sense. He never suggested a topic to a young scientist, but all those who came to the Rockefeller hospital, especially those in the Pneumonia Service, "ended up in his office which was the ward kitchen up on the sixth floor, and Dr Avery would sit and tell them the laws of pneumococcus" (McCarty, 1968).

These talks were known as the "Red Seal Records"; they started off with background history, the long debate over the chemical basis of biological specificity. Could the specificity of an enzyme like Sumner's urease, or that of Northrop's pepsin, trypsin and chymotrypsin, really reside in the crystal-line protein molecule? Similarly, could the reaction of pneumococcus with immune serum really depend upon the chemical composition of its poly-saccharide coat? His account was interlaced with little aphorisms for the experimenter such as: "It is lots of fun to blow bubbles, but it is wiser to prick them yourself before someone else tries to" (Avery, 1943). These talks led to wide reading by the young scientist and further discussions with Avery, out of which would emerge an idea for a research programme.

There is no escaping the conclusion that Avery had a profound influence on these young men. The admiration and loyalty expressed by those who worked with him is testimony to this. Just how much of the laboratory work in the transformation story he did himself does not therefore seem important. Nor should we be misled by his public caution into assuming that he was not privately confident about the success and significance of the work. Avery and MacLeod found the transformation system extremely difficult for the first three or four years, but despite temporary setbacks they "never felt they were not going to pull it off" (MacLeod, 1968). By the winter of 1941/42 they were "quite confident" that they were on the right track. It looked like a nucleic acid of the thymus type, but they had not narrowed down their focus to this class of compounds alone. Protein was still felt a possible claimant since some proteins were known to resist the action of trypsin and chymotrypsin and not to be destroyed by chloroform (McCarty, 1968).

In March 1943 Avery had reported at a meeting of the Trustees of the Rockefeller on the chemical identity of the transforming substance (Coburn, 1969, 628). That summer saw lengthy and agonizing discussions of the text of the famous paper which was eventually submitted to the *Journal of experimental medicine* in November 1943. According to McCarty there were no referees' reports on it. Peyton Rous "was carrying the complete editorial load himself, as he did for many years. (The names of Simon Flexner and Herbert Gasser appeared on the journal as Editors, but they delegated all authority to Rous)" (McCarty, 1970). Avery delivered it to Rous personally,

> and told his friend and colleague of some thirty years that he wanted him to review it just as he would a manuscript submitted by an unknown outsider. Two to three weeks later, I was present in Dr Avery's office when Dr Rous presented his comments and corrections to us verbally. He began by reminding us that Dr Avery had asked for a truly editorial review and stated that he had taken him at his word. The manuscript was covered with the pencilled notations and comments that were so characteristic of Dr Rous's editorial method. Some of these were small matters of wording or presentation, but there were also some more substantial suggestions. For example, we had included a quotation from J. B. Leathes in the discussion which was concerned with the speculation that nucleic acids might some day be found to surpass the proteins in importance. Rous pointed out that this, being merely a speculation, added little to the argument. It was deleted.
>
> (*Ibid.*)

Here, then, was one influence which served to dilute the impact of the discovery. We have already seen how many biologists were impressed with Leathes' address in the late 1930s (see p. 117). So with this unfortunate piece of prudery, the link between the old tradition of chemical individuality and the new discovery of the specificity of DNA was surpressed!

What did come out clearly in the discussion at the end of the '44 paper was that DNA was much more than a mere "midwife molecule", it was not just a structural frame, for it was "functionally active in determining the biochemical activities and specific characteristics of pneumococcal cells" (Avery, Macleod and McCarty, 1944, 155). Just how the determination was brought about was, of course, unknown, but Avery, MacLeod and McCarty suggested that the transforming principle "interacts with the R cell giving rise to a co-ordinated series of enzymatic reactions that culminate in the synthesis of the Type III capsular antigen" (*Ibid.*, 154).

But how was the DNA acting? Was it behaving as a gene as Burnet (1944) had suggested, or as a mutagen which caused mutation of the genetic material in the recipient cell? (Gortner, 1938, 548; Dobzhansky, 1951, 48; Beadle, 1948, 71). Or was it behaving as a virus? (Stanley, 1938, 491). On these alternatives our three authors would not be drawn. All they would say was:

> If the results of the present study on the chemical nature of the transforming principle are confirmed, then nucleic acids must be regarded as possessing biological specificity the chemical basis of which is as yet undetermined.
>
> (Avery, Macleod and McCarty, 1944, 155)

This was surely a case of sitting on the fence. Did Avery privately believe the transforming substance to be a gene? And if so why did he not say so? Hotchkiss has testified that Avery "was well aware of the implications of DNA transforming agents for genetics and infections" (1965, 6). We know that in May 1943 Avery wrote his brother a famous letter in which we find almost the identical phrase which cropped up in the '44 paper: "nucleic acids are not merely structurally important but functionally active substances in determining the biochemical activities and specific characteristics of cells," and he went on, "Sounds like a virus—may be a gene." Avery then added, as if hastily: "But with mechanisms I am not now concerned. One step at a time and the first step is, what is the chemical nature of the transforming principle? Someone else can work out the rest. Of course the problem bristles with implications." He went on to assure his brother that a lot of well documented evidence was needed before anyone could be convinced that protein-free DNA had the properties he claimed. In other words Avery deliberately concentrated his attention upon the chemical identity of the transforming substance and excluded other aspects.

One would still like to know just what Avery's private attitude was to the idea that transformation involved the incorporation of a gene into the recipient cell. MacLeod said: "He was not so much resistant to the idea as cautious. He was almost neurotic about overstating the case" (1968). It would be useless to expect that geneticists at the Rockefeller would have pushed this suggestion, for there were no geneticists at the laboratories and hospital in New York. At the labs in Princeton there had been John Gowen, a pupil of T. H. Morgan, whose study of TMV mutagenesis we have noted, but he left Princeton in 1937 to become professor of genetics at Iowa State College and

> basic genetic investigations did not gain a solid foothold in either the New York or Princeton laboratories. Such studies, developed chiefly in zoological and botanical laboratories, apparently did not appeal to the administration as part of a programme then largely oriented toward pathology and physiology. Only years later, when gene action began to come within the grasp of biochemistry, was basic genetics to return to the Institute, under the leadership of Rollin D. Hotchkiss.
>
> (Corner, 1964, 309)

MacLeod recalled that since none of them was a geneticist "we all found ourselves reading genetic texts avidly" (MacLeod, 1968).

Dobzhansky's Interpretation

Sometime between 1940 and early 1942 Dobzhansky visited Avery's laboratory "and tried to argue that what were being observed were mutations like the mutations in *Drosophila.* Avery was slightly sceptical about it but said, 'I will look into the matter' " (Dobzhansky, 1968). About two weeks later Avery telephoned Dobzhansky and thanked him "for making an interesting suggestion that is probably what is taking place" (Dobzhansky, 1968). In the second edition of Dobzhansky's *Genetics and the Origin of Species,* the introduction to which is dated March 1941, there is the following statement:

> If this transformation is described as a genetic mutation—and it is difficult to avoid so describing it—we are dealing with authentic cases of induction of specific mutations by specific treatments—a feat which geneticists have vainly tried to accomplish in higher organisms.
>
> (Dobzhansky, 1941, 49)

Whether Avery really held this as the most likely explanation seems doubtful, but clearly he entertained the possibility seriously. Not until characteristics other than capsular polysaccharide were transferred in this way could the concept of directed mutagenesis be discarded. Harriet Taylor was later to be particularly forthright in rejecting the mutagenesis idea.

Dobzhansky was perhaps behaving as one would expect a geneticist to behave. Here, it seemed, was a mechanism for producing mutations, not just at random, but in a *predetermined* direction. He admitted that as a geneticist what interested him was that transformation involved a mutation; what it was produced by interested him much less (Dobzhansky, 1968). Likewise, Beadle in his Silliman lecture of 1948 put pneumococcal transformation alongside Auerbach's chemical mutagenesis and said:

> As a matter of fact, Pneumococcus type transformations, which appear to be guided in specific ways by highly polymerized nucleic acids, may well represent the first success in transmuting genes in predetermined ways.
>
> (Beadle, 1948, 71)

The three Rockefeller scientists therefore had good grounds for remaining non-committal, and when asked about this in 1966 MacLeod said: "at that time [1940s] all the genetic interpretations (plasmagenes, directed mutation, conversion, or Muller's pairing and crossing over) seemed plausible and they did not particularly favour one view" (Carlson, 1972).

Muller's Interpretation

At a conference held in New York in January 1946 Muller heard Delbrück report results for phages that seemed in principle like Avery's for pneumococcus, in that there was an apparent "conversion" of a viable type into a non-viable type of the sort with which it had been mixed. Now in Delbrück's experiments the rate of "conversion waș so high as to discount the possibility of random mutation and selection, which Avery's results were really open to," and instead, Muller put forward the following idea:

> To my mind this suggests strongly that in both Delbrück's and Avery's cases what really happens is a kind of crossing over between chromosomes or protochromosomes of the inducer strain and those of the viable strain.
>
> (Muller, 1946)

It is well known that this suggestion has since been widely accepted. At the time of the conference it is said to have created quite an impression.

At the same conference Mirsky gave reasons for believing that the transforming substance contained chromosome material to which proteins were bound as in "chromosin". McCarty had tested the transforming activity of chromosin extracted from pneumococci by Mirsky and found it effective. Mirsky believed the thread-like bodies in his chromosin to be chromosomes or fragments of chromosomes. Muller was delighted with this suggestion because it brought the transformation process into the realm of the chromosome theory of the gene. No doubt it was Mirsky's paper at this New York meeting which suggested to Muller this chromosomal interpretation and he wrote:

> . . . we should have to suppose that these chromosomes can survive "extraction", that is, that they float more or less freely in the medium and can nevertheless, on coming into contact with the bacterial cell, enter into synapsis with homologous chromosome parts already there.
>
> (Muller, 1946)

Mirsky's Criticism

It is very understandable that Mirsky, who had devoted much of his time since 1942 to the chemistry of the nucleus, was critical of the efforts of three Doctors of Medicine to characterize the transforming substance of pneumococcus. Whilst the biochemists were showing how complex was the chemical constitution of chromosomes, Avery, MacLeod and McCarty were suggesting that a single substance could alone transfer biological specificity from one cell to another. If they had identified this substance as a protein it would not have been so bad. But they made the revolutionary claim that it was a nucleic acid of the thymus type! They had, moreover, used a complex system which was inefficient (or so it appeared to be), unreliable and dependent upon competence factors whose chemical basis was unclear. As for the evidence from enzymology, Mirsky could point out that trypsin and chymotrypsin were inadequate as agents to destroy *all* types of protein. And in any case they only acted on proteins which had been partly or wholly denatured. Now pepsin is a general proteolytic enzyme, but this could not be used owing to the destructive effect of working at the required pH. (Here, pronase, had it been known at the time, could have filled the gap).

Apart from the evidence from enzymology there was the question of the purity of the DNA. Mirsky claimed that "pure, protein-free" nucleic acid could contain as much as 1 or 2 per cent of protein which histochemical tests would fail to reveal.

> One of the most sensitive direct tests for protein is the Millon reaction, but in our experience a nucleic acid preparation containing as much as 5 per cent of protein would give a negative Millon test. At present the best criterion for the purity of a nucleic acid preparation is its elementary composition and especially the nitrogen: phosphorus ratio. Presence of 2 per cent of protein would increase this ratio, but only by an amount that is well within the range of variation found for the purest nucleic acid preparations.
>
> (Mirsky and Pollister, 1946, 135)

Mirsky passed no comment on the evidence for purity of DNA from immunological tests, which Avery, MacLeod and McCarty claimed was capable of showing protein at a dilution of 1:50 000 (1944, 150). And even if this test was valid one could still claim that very little "genetic protein" was needed when so few cells in the recipient strain were transformed. On the other hand the transforming principle was active at a dilution of 1 in 6×10^8!

Nor was the purity of the transforming principle Mirsky's only concern. Following Alloway, Avery, MacLeod and McCarty had used sodium desoxycholate to liberate the nucleoprotein. But this substance acted as a detergent and was therefore likely to denature proteins. Mirsky, as we have seen, prepared "chromosin" from pneumococcal cells by his technique with molar NaCl. This procedure was less likely to denature the proteins and was therefore, he implied, preferable. Is this not carping over details? For as Mirsky himself admitted, the desoxycholate technique gave a higher yield of nucleic acid than the sodium chloride technique. And if residual protein was denatured by the detergent, then the conclusion that the transforming activity was associated with the nucleic acid and not with the protein was surely all the stronger. And was it not inconsistent for Mirsky to suggest in 1946 that desoxycholate denatures the protein in chromosin and in 1947 to discredit the evidence from the use of trypsin and chymotrypsin on the grounds that these enzymes only act on denatured proteins? This looks suspiciously like a rearguard action fought by one who had backed the wrong horse!

Now it may be argued with some justice that Mirsky, in common with other biochemists, was merely adopting an empiricist stand, as when he said: "... it is not yet known which the transforming agent is—a nucleic acid or a nucleoprotein. To claim more, would be going beyond the experimental evidence" (1946, 135). In fact, it is doubtful that there is any such thing as a purely empiricist stand. Those like Mirsky, who preferred to withhold their support from Avery's conclusion, had very good reasons. These arose out of their knowledge of protein specificity and its chemical basis on the one hand, and the supposed lack of any chemical basis to specificity in the nucleic acids on the other. It can hardly be called empirical to go on *assuming* that nucleic acids lacked the required chemical sophistication for biological specificity, especially when evidence of immunological specificity had been forthcoming (pneumococcal antigenic RNA was species specific, as was TMV RNA). When Mirsky attended the Cold Spring Harbor Symposium on "Nucleic Acids and Nucleo-proteins" in June 1947 he came out with the same empiricist declaration: "In the present state of knowledge it would be going beyond the experimental facts to assert that the specific agent in transforming bacterial types is a desoxyribonucleic acid" (1947, 16). The French microbiologist, André Boivin, whose confirmation of Avery's work was the subject of Mirsky's criticism, conceded to Mirsky his empiricist statement but insisted that "the burden of the proof rests upon those who would postulate the existence of an active protein lodged in an inactive nucleic acid" (Boivin, 1947, 16). Looking back on those days of protein conservatism, Hotchkiss said he often wondered "which of our ideas take root merely because it becomes impracticable and then impolitic to take up the effort of questioning them!" (1966, 191). We have seen, also, that geneticists were chiefly interested in transformation as a case of mutation. Furthermore, when Muller gave the Messenger lectures in 1945 it was not evident that the acceptance of genetic specificity on the part of DNA *would deny it to proteins* and he said: "it may even be suspected ... that for each different gene protein there is a special form of polymerized nucleic acid to match it" (1947a, 6). Before these lectures were published he had time to add a footnote about Mirsky's opinion that nucleic acids "do not constitute the main seat of the specificities" (*Ibid.*). After hearing Mirsky at New York in January 1946 he wrote to Darlington:

> Mirsky gave reasons for believing that Avery's so-called nucleic acid is probably nucleoprotein after all, with the protein too tightly bound to be detected by ordinary methods, and that what he had was free chromosomes, or pieces of chromosomes. This protein is a higher protein, like Stedman's chromosomin, in which Mirsky now believes, the histone and protamine being easily removed while the higher protein is left attached to the

> nucleic acid. All this too seems to me to fit into the same picture, and I am trying to induce Mirsky to bring along some of his and of Avery's extracted material to be looked at by Mrs. Baylor under the electron microscope at the University of Illinois.
>
> (Muller, 1946)

Muller inserted a very similar passage in a footnote to his Pilgrim Trust Lecture of 1946 before its publication (1947b, 23), and he ended the address as follows:

> Thus it may be that nucleic acid in polymerized form provides a way of directing such a flow of energy into specific complex patterns of gene building or for gene reactions upon the cell. But to what extent the given specificity depends on the nucleic acid polymer itself, rather than upon the protein with which it is ordinarily bound, must as yet be regarded as an open question.
>
> (Muller, 1947b, 24)

It may well have been Muller's acceptance of Stanley's initial identification of crystalline TMV as a protein which had already set his view on the chemistry of the gene (Carlson, 1966, 128). Whilst this may be true of Muller in the late '30s, I doubt it can be held responsible for his view in 1946, when Mirsky's authority seems to have been the major influence. Through Muller's widely read Pilgrim Lecture, this influence was spread to a wide audience. How should we evaluate this influence today? Was it beneficial or harmful? Clearly it was positive in the sense that more evidence was called for. On the other hand it protected the Protein Version of the Central Dogma on the grounds that the task of demonstrating the specificity of DNA had yet to be achieved.

CHAPTER THIRTEEN

Support for Avery

Recent investigations of the role of nucleic acids in biology have verified the opinion that they are comparable in importance to the proteins, especially with respect to the problem of the structure of the gene. The work of Avery on the relation of nucleic acids to the change in type of pneumococci provides a further illustration of the fundamental significance of these substances.

(Beadle, Pauling and Sturtevant, 1946, 30)

Here surely is a change to which, if we were dealing with higher organisms, we should accord the status of a genetic variation; and the substance inducing it—the gene in solution, one is tempted to call it—appears to be a nucleic acid of the desoxyribose type. Whatever it be, it is something which should be capable of complete description in terms of structural chemistry.

It has been a matter for rejoicing to his many admirers, friends and followers in many countries that Avery, a veteran now among investigators, should thus, on the eve of his retirement, have attained this new peak of discovery—a fitting climax to a devoted career of such wide influence on the progress of science.

(Sir Henry Dale, 1946, 128)

None of the experiments or facts, from the very beautiful biochemical research on transforming principle to the possibly equally informative work of cytologists and geneticists, leads directly and unambiguously to the conclusion that transforming material or genes are nucleic acids, or largely composed of nucleic acids. I would appreciate learning whether or not the decision, so widespread today, that nucleic acid is a transforming principle has in fact been decided by an unequivocal experiment, or whether it is no more than a voted agreement at the present time.

(Cooper, 1955, 19–20)

When a discovery is made that calls in question an established paradigm like the Protein Version of the Central Dogma one might expect the community of scientists directly involved to reject its claims and to fight a rearguard action against it on the grounds that the evidence was inadequate and could not bear an interpretation in harmony with established thought. One would then witness the familiar sequence of neglect, rediscovery and final recognition. Although it is possible to find striking examples of individuals who resisted the implications of Avery's work (Cooper, 1955; Sevag, 1952) this was far from being the only reaction to the discovery of the chemical identity of the transforming principle. This discovery was not neglected, not rejected, and not rediscovered. The Kuhnian blindness which we observed in the work on TMV in Chapter 10 was not a general feature of the discussions of Avery's work. In this and the subsequent chapter we shall explore the several ways in which Avery's evidence was made more convincing and ways in which the implications of his discovery were followed up.

Strengthening the Evidence

In 1943 the Rockefeller group had used the reaction with appropriate anti-serum as a sensitive test (1:50 000) for the presence of protein in the transforming principle. This was far more sensitive than the histochemical and analytic (N/P ratio) evidence to which Mirsky objected. But one could still assert that during extraction and purification protein in the transforming principle was altered in such a way as to prevent it from reacting with anti-serum.

MacLeod had left the Rockefeller in 1941 to take the chair of medicine at New York University's College of Medicine. There, with Austrian, he continued his study of transformation in which he demonstrated the transfer not of just one hereditary character but of three, each of which behaved independently of the others. Directed mutation, therefore, seemed not to be the mechanism behind transformation and they concluded that there must exist a "multiplicity" of DNA molecules, each "specificity" being determined by a different one (Austrian and MacLeod, 1948, 458).

Avery, though he officially retired in 1943, continued to work at the Rockefeller for several years, and collaborated with McCarty in devising an improved procedure for extracting the transforming principle. This gave five times the yield of the 1943 technique and could be used on Types II and VI as well as on Type III (McCarty and Avery, 1946b). The pneumococcal cells were allowed to autolyse, but in the presence of citrate. McCarty had discovered the dependence of DNase upon magnesium ions (McCarty, 1946a). Citrate removed these and thus prevented depolymerization of DNA. McCarty himself had been trying since 1943 to strengthen the evidence for the identity of the transforming principle by purifying DNase. This had been available only as a crude extract of intestinal mucosa in 1943. By 1946 he was able to demonstrate activity of DNase at very low concentrations, ten thousand times weaker than the concentration at which proteolytic activity could be demonstrated! The biochemist, Rollin Hotchkiss, who had come to the Rockefeller in 1935 to work with Walther Goebel and Charles Hoagland, took up the question of the chemical identity of the transforming principle in 1947. A year later he was able to report to a conference in Paris, base compositions for the transforming principle which differed from those for thymus nucleic acid and for a tetranucleotide. He also reported inactivation of the transforming principle by the crystalline DNase which Kunitz had recently prepared (Hotchkiss, 1949; Kunitz, 1948). Next he showed that the small quantity of amino acid that could be obtained from the transforming principle was all accountable as glycine, which could be traced to the decomposition of adenine. This important find allowed him to conclude that the maximum contamination of the transforming DNA with protein was 0.02 per cent.

Hotchkiss also sought for other markers which could, like the power to produce a capsule, be transferred to recipient cells. If what was transferred was genetic material then markers should behave independently. In the case of penicillin resistance and capsule formation he was able to demonstrate this (Hotchkiss, 1951). Later Julius Marmur worked under Hotchkiss on a strain of pneumococcus which possessed an adaptive enzyme for utilization of the sugar, mannitol. This gave the first case of linkage (with penicillin resistance) (Hotchkiss and Marmur, 1954).

These findings allowed the Rockefeller scientists to express themselves less cautiously. When McCarty addressed a symposium of the American Chemical Society on "Biochemical and Biophysical Studies on Viruses" he concluded with the words:

> It will be observed from the foregoing discussion that while the pneumococcal transforming substance is virus-like in certain of its properties, there is some evidence inconsistent with its classification with the viruses, despite the diversity of this group of agents. However, if one accepts the validity of the view that the biological specificity of the transforming substance is the property of a desoxyribonucleic acid, the results of the present study serve to focus attention on the nucleic acid component of virus nucleoproteins. In addition to its probable role in the self-reproduction of the virus molecule, the nucleic acid moiety may carry a specificity which is a determining factor in the ultimate structure of the virus.
>
> (McCarty, 1946b)

A month later (May) McCarty gave the Eli Lilly Award lecture to the Society of American Microbiologists. By this time he felt justified in concluding "that the accumulated evidence has established beyond reasonable doubt that the active substance responsible for transformation is a specific nucleic acid of the desoxyribose type." And in the body of the lecture he called for a "reconsideration of the possible role of nucleic acids in vital phenomena . . ." (1946c, 48), in the light of the "two cardinal effects" associated with DNA, namely the induction of "predictable and heritable modifications and the self-reproduction of the active agent in transformed cells" (*Ibid.*). He went on to draw analogies between the biological properties of the transforming substance and those of genes and viruses: transmissibility, recovery in quantities far exceeding the original inoculum, and unitary behaviour. "Although the validity of these analogies may be questioned, they serve to underline the possible implications of the phenomenon of transformation in the field of genetics and in virus and cancer research" (*Ibid.*).

These were all carefully measured words, and Hotchkiss was equally careful. He shied away from such crude statements as, the gene *is* DNA, or, the transforming agent is DNA. There was always the possibility lurking in the background that the agent was DNA-dependent but not itself composed of DNA.

> We in the Avery laboratory were concerned throughout with the possibility that traces of very active protein might account for transformation. My own respect for proteins owed very much to long hours of fascinating learning from Alfred Mirsky during the thirties. Quite on my own, then, I felt the same doubts he did: that the nitrogen–phosphorus atom ratios of nucleic acid and protein could vary only as much as the phosphorus—that DNase, purified, in fact all but discovered by McCarty out of a proteinase-rich pancreas fraction, might still have mild proteinase action. Mirsky spoke about these objections, but not very much to Avery's group or he would have learned as I did how eager they were to see the search for traces of protein continued.
>
> (Hotchkiss, 1966, 189)

And so the "ifs" and "seems probables" remained: "If this be true, it is of especial interest that these determinants are available for chemical, physical, and biological study in the form of the isolated, purified transforming desoxyribonucleates of bacteria" (Hotchkiss, 1952, 436). "These analyses seem to support the earlier inferences that the determinants being transferred in the DNA transformation are the bacterial genes themselves" (Hotchkiss, 1955b, 5). Much later Hotchkiss wrote:

> . . . people *engaged in the serious analysis* of genetic mechanism were not ready themselves to be stampeded into public generalization, or beguiled entirely by visions not necessarily prophetic. The historian can later see where more emphasis, exaggeration, exposure, boldness or cajolery, would have been "justified". But do not historians also sometimes observe the danger of the "bad guesses", the places where overemphasis or persuasiveness have given a generation viewpoints that had tediously to be unlearned?
>
> (Hotchkiss, 1972)

Confirmation from Paris

Results identical with those obtained by Avery, MacLeod and McCarty were achieved with the colon bacillus *Escherichia coli* by André Boivin and his collaborators Roger Vendrely and Yvonne Lehoult. Their work is of special interest because, in addition to numbering among the first cases of transformation outside pneumococcus, its significance was hailed by Boivin in terms which, by comparison with Avery, McCarty and Hotchkiss, were recklessly speculative.

Boivin's contact with nucleic acids went back to the late 1920s when he studied among other things the metabolism of purines and pyrimidines. In the early 30s, as professor of medical chemistry in Bucharest, he collaborated with the Mesrobeanus and the Magherus. At first he studied the nucleic acid constituents of bacteria, then turned to immunochemistry and isolated the important O antigen. 1936 saw him installed in the annex of the Pasteur Institute in Garches where work continued throughout the war. There with A. Delaunay and Miss Corre, Boivin did for the antigens of the colon bacillus what Avery and Heidelberger had done for the pneumococcus. They found:

evidence of the extraordinary multiplicity of antigenic types among the colon bacilli, each type possessing its own polysaccharide, characterized by a special chemical constitution and by a particular serological specificity. Each type remains stable through successive cultures; like the pneumococcus types, it can undergo antigenic degradation leading from form S (smooth) to form R (rough) by losing its polysaccharide, and, like the pneumococcus types again, it has the value of a true elementary species within the immense species of *Escherichia coli.*

(Boivin, 1947, 7)

In 1942, after ten months spent in captivity in Germany, Roger Vendrely returned to France and joined Boivin's group. He found Boivin anxious to establish the chemical basis of transformation which he believed must involve nucleic acids. Already in 1941 Boivin had tried a variety of *in vitro* arrangements to "discover whether, like the pneumococci, the colon bacilli might not give way, by controlled mutation, to the process of type transformation" (*Ibid.*). He had read the work of Griffith, Dawson, Sia and Alloway and in 1941 he seemed to be thinking of type transformation as a phasic development (Boivin, 1941, 799). Vendrely, as the biochemist, was put on to the extraction of the transforming substance. The donor cells were killed with chloroform and allowed to autolyse for two days at 37°C. Nucleoprotein was then precipitated from the autolysate by acetic acid. The Paris workers were then told about Avery's 1944 paper revealing the role of DNA. "Inspired by this work," said Boivin, "we too have obtained evidence of the intervention of desoxyribonucleic acid in directed mutations in bacteria. We take pleasure in acknowledging the priority of the American authors in this field" (1947, 9). All that Boivin and his colleagues had to do, it appeared, was to treat their nucleoprotein extract with pepsin at pH 2, or better, to use the Sevag-chloroform technique, to strip off the protein. At a meeting of the Académie des Sciences, in November 1945, Boivin announced his success in achieving transformation in *E. coli.* From the R form of Type S, they produced the S form of Type S_2 when cultured with a nucleic acid extract of the latter. Likewise they achieved transformation of S_2 to S_1. In December 1945 *Experientia* published these findings. The title of the paper contains the phrase "Significance for the biochemistry of heredity". The conclusion, in which the donor type is referred to as C_1 and the recipient type as C_2, translates as follows:

It seems well established now that the bacterial cell possesses a small nucleus of thymonucleic acid immersed in a cytoplasm of ribonucleic acid. Surely the principle derived from C_1, which demonstrated its ability to impose on C_2 a new molecular constitution for its polysaccharide and a novel enzymatic equipment, results from a simple "solubilization" of the rudimentary chromosomal apparatus of C_1? The hypothesis seems likely. If it corresponds to reality, it opens altogether novel horizons, and how promising these are for the biochemistry of heredity. In particular, it is on the side of the nucleic acid and not at all on that of the protein of the nucleoprotein macromolecule constituting a gene that one must find the basis for the inductive properties belonging to the gene. That would lead one to envisage the possibility of a "primary" or more likely "secondary" structure able to differentiate between the various desoxyribonucleic acids within their natural state of polymerization.

(Boivin, Delaunay, Vendrely and Lehoult, 1945, 335)

When Boivin attended the Cold Spring Harbor Symposium in June 1947, he gave a remarkable paper in which he related the work on bacterial transformation to Beadle and Tatum's work on biochemical genetics, described Tulasne's confirmation of Robinow's work on the bacterial nucleus (1947), and the chemical mechanism involved (Vendrely and Lipardy, 1946), and gave Tulasne's and Vendrely's cytochemical evidence, using RNase and DNase, for the localization of RNA in the bacterial cytoplasm and DNA in the nucleus of *E. coli* (1947).

When we look back over the mass of literature in the 1940s, it seems scarcely possible that André Boivin could have so accurately predicted the structure which the nascent subject of molecular genetics was to take. Consider the following statements:

We may, at the most, catch a glimpse of a series of catalytic actions which set out from primary directing centres (the desoxyribonucleic genes) proceed through secondary directing centres (the ribonucleic microsomes–plasma–genes) and thence through tertiary directing centres (the enzymes), to determine finally the nature of the metabolic chains involved, and to condition by this very means, all the characters of the cell in consideration.

Thus, this amazing fact of the organization of an infinite variety of cellular types and living species is reduced, *in the last analysis*, to innumerable modifications within the molecular structure of one single chemical substance, nucleic acid, substratum of heredity as well as of acquired characters. This is the "working hypothesis" quite logically suggested by our actual knowledge of the remarkable phenomenon of directed mutation in bacteria.

Thus there exist in the bacterial nucleus, as in the cell nucleus of higher organisms, desoxyribonucleoprotein genes which serve as a substratum for the characters of the species. It follows that whatever happens in the phenomenon of directed mutation can hardly be interpreted otherwise than as a result of solution of the bacterial chromosome apparatus without total destruction of its functional value.

In bacteria—and, in all likelihood, in higher organisms as well—each gene has as its specific constituent, not a protein but a particular desoxyribonucleic acid which, at least under certain conditions (directed mutations of bacteria), is capable of functioning *alone* as the carrier of hereditary character; therefore, in the last analysis, each gene can be traced back to a macromolecule of a special desoxyribonucleic acid. . . . This is a point of view which, in respect to the actual state of biochemistry appears to be frankly revolutionary.

(Boivin, 1947, 12–13)

In the ensuing discussion, Brachet expressed surprise that Boivin's autolysates were active in transformation. Surely DNase was present and therefore the activity must have been due to some substance other than DNA. All Boivin could do was to assure Brachet that whereas colon bacilli contained a very active RNase there was no evidence of a similarly active DNase, for Vendrely had found the proportion of RNA to DNA to fall very markedly during autolysis (Vendrely, 1947). Probably informally, Boivin was asked why he had not used sodium desoxycholate, like Avery, or molar NaCl like Mirsky.

So he added a footnote to his paper saying that colon bacilli resisted both reagents so strongly that only very poor yields resulted from these extractive procedures. Hotchkiss had this to say about Boivin and his evidence from *E. coli*:

> Boivin was respected for his identification of bacterial antigens far more complex than those Avery had identified a couple of decades before. Boivin was I think also widely considered an honest scientist, optimistic and given to simplistic logic. It is fitting that he should appear in your lights as an early molecular biologist! But he lacked the quantitative sense and self-critical attitude the best molecular biologists were to show. I doubt if we will ever know whether he and his coworkers ever really achieved a DNA-caused transformation since (I am told) his strains spontaneously go through the same change. It would have been outside his realm of inquiry to consider the role of *selection* in fostering the conversions—so with Avery's prior example just before him, I think he was overpersuaded by his own scant observations.
>
> Boivin called it "fifty per cent transformation" when (literally, when pursued) he meant that transformation *occurred in* one half of the *flasks* treated with DNA under best conditions. We asked, by 1951 and before, what per cent of *treated cells* are changed? His personal magnetism and enthusiasm were great, but these things are dangerous when matters are in a qualitative stage.
>
> (Hotchkiss, 1972)

Accordingly, we find another footnote in Boivin's Cold Spring Harbor paper which reads:

> Despite apparently identical experimental conditions, the transformation of R_2 into S_1 through the action of the desoxyribonucleic acid of S_1 is not regularly produced. In a dozen tubes, containing the same volume of medium and the same dosage of desoxyribonucleic acid, inoculated with the same number of bacteria, one frequently finds tubes giving rise to transformation side by side with others where no transformation occurs. The number of bacteria at the beginning and end of the culture and the concentration of the desoxyribonucleic principle do not allow an explanation on statistical grounds of the proportion of positive results obtained in the different experiments. All takes place as though a factor, still unknown, were able to facilitate or to prevent transformation.
>
> (Boivin, 1947, 8)

To make matters worse, other workers had difficulty in repeating Boivin's work, perhaps due to the difference in the competence of different strains (Ravin, 1969, 65). Had Boivin's strains 17 and 24 been available to other workers, confirmation might have resulted, but these strains "were lost when the tubes containing the parent strains were broken in a careless accident" (Vendrely, 1972). Boivin was at the time in hospital following his first serious attack of cancer, and Lederberg and Tatum, who received strains from Boivin in 1947, "never confirmed his finding. In correspondence with Tatum, Boivin admitted that these might have lost their competence in his own hands, and he stated he would try to recover others on which he could verify the transformation himself. His illness supervened" (Lederberg, 1972a).* At Columbia, where "Avery's work was very well known" (Lederberg, 1973). Ryan and Lederberg had in June 1945 tried "to emulate Avery by trans-

* Transformation has since been achieved in *E. coli* (see: Oishi and Cosloy, 1972; Wackernagel, 1973).

forming *Neurospora* mutants with DNA extracts. This was unsuccessful, and was therefore regarded as unworthy of report" (Lederberg, 1972b). It had been carried out precisely to clarify "whether 'transformation' was a typical gene transfer" (Lederberg, 1972a). Meanwhile, Boivin's work found its way into the literature as a confirmation of Avery's discovery, and Arthur Pollister, who had collaborated with Mirsky, was greatly impressed by the French work, especially that of Boivin's collaborator, Vendrely (see Chapter 14).

The Debate over Bacterial Transformation

It has been urged that the famous Avery, MacLeod, McCarty paper was not widely read because it was published in a journal normally found only in medical libraries (Wyatt, 1972). It is true that Avery made no attempt to get a short report of the work published in a widely circulated journal like *Science* or *Nature*. On the other hand news does not have to be published in order to travel! Quite apart from visits made by such scientists as Gulland (1946) and Macfarlane Burnet (1943), transformation was discussed at three unpublished symposia: The Mutation Conference, New York, January 1946. Biophysical and Biochemical Studies on Viruses, Atlantic City, April 1946. Conference with unknown title, Hershey (Penn.), October 1946.

The subject of the Cold Spring Harbor symposium of July 1946: "Heredity and Variation in Micro-organisms", had been chosen in the year Avery's paper appeared and the meeting would have been held in 1945 had not travel restrictions made it impossible. When the participants met in 1946 Avery, McCarty and Harriet Taylor were present and reported on the progress they had made in identifying the environmental factors essential to transformation. At yet another meeting in 1946—Society of American Microbiologists at Detroit—McCarty read almost the same fine paper that he had given in Atlantic City a month before. On this occasion he received the Eli Lilly Award in Bacteriology and Immunology. Nor did this flush of interest subside in 1947. The Cold Spring Harbor Symposium of June 1947 was devoted to nucleic acids and nucleoproteins. Boivin, Chargaff, Hotchkiss, Mirsky, Pollister and Harriet Taylor were present.

These meetings gave adequate opportunities for geneticists, virologists and biochemists to discuss transformation. In Europe the scene was less conducive to the publicizing of Avery's work. When the Society for Experimental Biology held a meeting on nucleic acids in Cambridge (1946), no one was invited to talk on transformation, and in his paper on bacterial nucleic acids and nucleoproteins M. Stacey succeeded in submerging his account of Avery's work in a list of what everyone else had done. Enthusiastic though he was, Stacey's own interpretation was clearly in the nucleoprotein camp (Stacey, 1947, 96).

There was one bright exception—the exciting colloquium held in Paris in 1948 by the Centre Nationale de la Recherche Scientifique (C.N.R.S.) with support from the Rockefeller Foundation. But even this meeting failed to have the wide impact which it most certainly deserved, perhaps because the proceedings were published in French in a limited number of copies of the C.N.R.S. colloquia.

This gathering had been planned by André Lwoff and Boris Ephrussi. They entitled it: "Biological Units Endowed with Genetic Continuity". "A year later", wrote Hotchkiss, "before I had realized that I almost never would find anyone who had read the symposium article, I was distressed to find that a tired abstractor for *Chemical Abstracts* had covered my own and also Ephrussi's conference papers in two short words, 'a review' " (Hotchkiss, 1966, 190). But what a grand colloquium it had been! Hotchkiss had reported his work with crystalline DNase and his quantitative chromatography of the transforming DNA. Boivin later referred to this as "the first direct argument of a chemical nature in support of the existence in nature of a multiplicity of nucleic acids" (Boivin, 1948, 1258). Harriet Taylor had described "intermediate" and "extreme" forms of pneumococci and had provided evidence for the presence of at least two functionally distinct DNAs in one and the same bacterial extract. Boivin, who by this time had moved with Vendrely to Strasbourg and had been joined by the cytochemist R. Tulasne, described their work on the localization of DNA and RNA in the bacterial cell, and their measurement of the DNA content of diploid and haploid cells.

These important contributions were summed up by André Lwoff in the following words:

> The transforming principle of pneumococcus is deprived of proteins and appears to consist exclusively of desoxyribonucleic acid. This is probably the case also for *Escherichia coli*. The importance of DNA is indicated by the fact that diploid nuclei have twice the DNA of haploid nuclei. The study of the transforming principle of pneumococcus has led to the conclusion that the purine and pyrimidine bases are not present in equimolar proportions. This gives an inkling of a possible explanation for the specificity of nucleic acids. Once the transforming principle of pneumococcus is introduced into a bacterium it confers on it permanently a given specificity. But this principle is susceptible of modification and even at the present time we know of two varieties of specific nucleic acid of type III pneumococcus. They have been compared to allelomorphic genes. In fact, they exclude each other reciprocally as if in competition for the same receptor.
>
> This fact thus throws light on the idea that the specific nucleic acids normally could and should be combined with another constituent, probably a protein.
>
> (Lwoff, 1948, 202)

The Influence of Studies in Transformation

There is no doubt whatever that the Avery paper of 1944 had a profound effect on biochemists. "These wonderful discoveries", said Mirsky, "have caused chemists to consider critically the evidence for uniformity among nucleic acids, and the generally accepted conclusion is that the available chemical evidence does not permit us to suppose that nucleic acids do not vary" (Mirsky, 1947, 15). I fear that his criticism of the evidence for DNA as the transforming substance did incline influential geneticists like Muller to retain the nucleoprotein conception of the gene rather than to go over to the DNA conception. But within the ranks of biochemistry Mirsky's criticism may well have served to stimulate further work. We have already referred to the genetic studies by Harriet Taylor and Rollin Hotchkiss. Two further lines of research naturally suggested themselves: studies of DNA content of cells and analysis of the base constitution of DNAs from different species. In short, Avery's work on bacterial transformation was not neglected. It did lead directly to further chemical and genetic studies, the outcome of which was crucial for Watson and Crick. This is not to say that those microbiologists and histochemists who supported the new view rapidly dominated the entire scene; conservatism lingered on as seen in Kenneth Cooper's outburst of 1955. Nor was the new view influential in England, where the old school of geneticists and plant virologists complacently carried on with the nucleoprotein gene. Perhaps it was Gulland's death in 1947 which left the British scientists to all intents and purposes blissfully unaware of the new developments in nucleic acid chemistry across the Atlantic and in Paris.

The Significance of Bacterial Transformation

With the passage of time the work of Avery, MacLeod and McCarty looks, if anything, more significant than in 1958; perhaps it was the most important event in undermining the pre-eminence of the protein, the culmination of a series of achievements which established the chemical basis to enzyme action, antigenicity and finally transformation. It marked the beginning of a new era in which a search for the chemical basis of nucleic acid "specificities" was undertaken. Muller admitted that if DNA was the transforming substance as Avery, MacLeod and McCarty concluded "their finding is revolutionary" (Muller, 1947a, 22). Boivin likewise felt that in the state of biochemistry at that time the postulation of many different DNAs "appears to be frankly revolutionary" (Boivin, 1947, 12). One does not find scientists describing a discovery as "revolutionary" every day of the week. This discovery was special. It demanded a re-examination of the Protein Version of the Central Dogma and of the tetranucleotide hypothesis. It suggested there must be hidden in the molecules of nucleic acids chemical specificities as rich as those of proteins. The debate over bacterial transformation therefore marks as Kuhnian a revolution as did the debate over macromolecules.

When we enquire into the reception Avery was given in the Rockefeller itself we find that on the surface at least there was enthusiasm. P. A. Levene, who died in 1940, was of course aware of the work on transformation in its earlier stages. "He was sceptical about the possible role of DNA in the transformation reactions and when Dr. Avery and I [MacLeod] described the system to him . . .—what the results were—and the properties we then knew about the active material, he was highly sceptical that it could be DNA" (MacLeod, 1968).

By 1943, when Avery read the great paper at the formal after-tea meeting in the Rockefeller, "there was next to no discussion . . . because it was a standing ovation afterwards. There was obvious recognition and a terrifically warm reception. Nobody mentioned any objection" (McCarty, 1968). The story that Avery was vehemently attacked in a discussion following the lecture and as a result dared not show his face in the Rockefeller for several weeks thereafter is clearly untrue.

There was, of course, opposition from Mirsky which was expressed publicly at meetings outside the Rockefeller. We have already examined the scientific grounds for this opposition. But such opposition can rarely be considered in isolation from other less objective grounds. We have noted that all three men —Avery, MacLeod and McCarty—were trained in medicine, not in biochemistry. True, they differed from men like Griffith, whose interests in epidemiology did not spill over into biochemistry. But medical research was messy, the systems used in experimentation were complex to the point where the results obtained from them were unreliable, and the transformation system was far from being an exception. No doubt, therefore, there was that feeling of professionalism on the part of biochemists in the Institute, like Mirsky, which predisposed them to question the contribution of "doctors", like Avery, MacLeod and McCarty, in the Institute's hospital. The same attitude exists to this day on the part of many a "pure" science department towards an "applied" science department.

Not only did the transformation story involve this conflict between biochemist and medic but in Sir Macfarlane Burnet's view it signalled a change from applied to pure research.

> Looking back I fancy that it was only in the 1930s that medical scientists began to be really interested in "pure" research . . .
>
> What swung microbiology perhaps for ever away from a primary desire to prevent and cure infectious disease to its current preoccupation with molecular biology, was probably Avery's discovery . . .
>
> (Burnet, 1968, 59)

Burnet visited Avery in 1943 and wrote home to his wife telling her that Avery:

> "has just made an extremely exciting discovery which, put rather crudely, is nothing less than the isolation of a pure gene in the form of desoxyribonucleic acid." I think that must be almost the last time I ever wrote DNA in full. Nothing since has diminished the significance or importance of Avery's work. Neither he nor I knew it at the time but in

> retrospect the discovery that DNA could transfer genetic information from one pneumococcus to another almost spelt the end of one field of scholarly investigation, medical bacteriology, and heralded the opening of the field of molecular biology which has dominated scholarly thought in biology ever since.
>
> (*Ibid.*, 81)

Parallel with this change, the introduction of new drugs like penicillin and the sulphonamides made further attempts to develop immunological aids of the sort the Rockefeller bacteriologists had been working on superfluous. Men like Hotchkiss who had started out contributing to a problem in immunochemistry found themselves drawn into the genetics of bacterial transformation.

Thus it came about that work begun by Griffith, a civil servant in the Ministry of Health, was taken up by a medical institute, and was there developed to answer the question: "On what compound does the specificity of the gene depend?" The Rockefeller's administrators, who had shown so little interest in genetics in the early days, now became very definitely committed to it.

CHAPTER FOURTEEN

Base Ratios

In the summer of 1950 a group of physiologists gathered in Woods Hole to pay tribute to Leonore Michaelis. Here at the Marine Biological Laboratory the sort of physiology which disgusted Delbrück had been carried on since Jacques Loeb set up the Physiology Course in 1898. We will pass over such contributions as dealt with the viscosity of the cytoplasm and the physiology of cell division, in neither of which nucleic acids found a mention; and will turn instead to the contribution from Daniel Mazia, of the University of California at Berkeley. In 1939 Mazia had concluded that the genetic substance was protein (see p. 154). Now he asked:

> If we suspect that DNA or any other constituent of the chromosome is the vehicle of heredity, what questions may we ask of it? If the cytological regularities of "chromatin" are a criterion for the behaviour of genetic material, the first expectation is that our chemical candidate will show the same regularities. It should be quantitatively the same in every diploid cell of a species, should be quantitatively proportional to the degree of haploidy, and should double exactly somewhere in the mitotic cycle. A second expectation is that it should be very stable, though it is difficult to set exact criteria of stability. Third, it should be capable of specificity, of existing in a large variety of individual configurations. Fourth, we may set as the ultimate test the possibility of transferring it from one cell to another and obtaining the same results as when genes are introduced by genetic techniques.
>
> (Mazia, 1952, 111)

I wish to suggest two reasons for Mazia's *volte face* over the significance of DNA. One concerns the metabolic studies of cellular nucleic acids, the other the significance of bacterial transformation. Both conspired to make histochemists ask questions which in the 1930's had been disposed of by superficial experimentation, or had been suppressed by slavish acceptance of the tetranucleotide hypothesis. The outcome of this questioning was that several histochemists and biochemists became convinced that DNA formed *a part* of the hereditary material, and they began seriously to doubt the supposed genetic role of the proteins. As Boivin had declared in 1947, the burden of proof no longer lay with those who believed that DNA was the hereditary substance, but with those who claimed this role for the proteins. At the same time, the historian must be careful to note that the proteins were still very understandably thought to contribute to the overall specificities of the gene. The burden of this chapter is to record the events around the year 1950 which brought about this transformation of attitudes and yielded a major clue to the structure of DNA—base ratios. These ratios resulted, then, from the biochemists' and histochemists' recognition of the implications of Avery's work. This progress did not demand acceptance of DNA as the gene, only that it could transfer some of the specificities of the gene. Even this limited interpretation of Avery's work, however, called for an attempt to discover the chemical basis of such a function. The long-accepted tetranucleotide hypothesis had to be examined afresh. New methods were called for. The result was, as Erwin Chargaff could claim in the summer of 1949, that there was "an enormous revival in interest for the chemical and biological properties of nucleic acids". What, he asked, provided the impulse for this sudden rebirth? Was it Hammarsten's work on highly polymerized DNA, the work of Caspersson and Brachet, "or was it the very important research of Avery and his collaborators . . . that started the avalanche?" (1950, 201).

I believe that Chargaff has identified the three most important influences; moreover his description of the rebirth of interest in nucleic acids as an "avalanche" at this time shows clearly that the revolution which was to bring the nucleic acids to the centre of the stage had begun before the Hershey–Chase experiment and the Watson–Crick model. It took place among students of cell chemistry, not among members of the phage group, not in the laboratories of structural chemists, biophysicists and geneticists. The first to explore the implications of bacterial transformation were André Boivin and Erwin Chargaff.

The Boivin–Vendrely Rule

Roger Vendrely recalled that it was the study of *E. coli* transformation that had convinced Boivin of the importance of DNA. So with his enthusiasm and "unyielding character" he decided "it was time to tackle the problem with full force and affirm the extremely important role of DNA. Everything, it seemed, was in favour of such an opinion" (Vendrely, 1972). Vendrely himself was more reserved in his opinion, and felt "it was still possible that proteic fractions intervene at some level in the functioning of the gene" (*Ibid.*).

Boivin devised a broad programme which included the demonstration of the presence and distribution of DNA and RNA in *E. coli*, the conditions for their liberation, the absolute quantity of DNA and RNA in mammalian nuclei and a comparison of the results with those from fish and birds. They isolated large quantities of nuclei, estimated the total number by a dilution technique, extracted and measured total nucleic acid in terms of purine nitrogen, and evaluated the DNA fraction either colorimetrically or in terms of the acid precipitable purine nitrogen. In 1948 they announced the diploid and haploid DNA cell contents as 6.5×10^{-6} and 3.4×10^{-6} γ (Boivin, Vendrely and Vendrely, 1948).

It was just as Boivin had hoped. The halving in the number of genes from somatic cells to germ cells inferred from the facts of genetics and cytology was paralleled by an approximate halving in the quantity of DNA per nucleus. Different individuals of the same species—calf, bullock and bull—yielded the same amount of DNA per cell. When Boivin announced this discovery to those attending the CNRS colloquium in Paris in 1948, Darlington objected that diploid cells should have more than twice the haploid quantity of DNA, since a large proportion of them would be in the act of duplicating their genetic substance (Vendrely and Vendrely, 1949). Vendrely's reply was that he had used tissues whose mitotic activity had come to an end. But when Mirsky and Ris published findings in support of Vendrely's a year later, their results for the nuclei of beef liver were too high (1949). Further work on rat liver nuclei led them to the discovery of a step-wise range of DNA values in the ratio of 1:2:4 associated with polyploidy (Ris and Mirsky 1949). Subsequent work in Chicago extended this discovery to a wider range of animal tissues and to plants (Swift, 1950a; 1950b). The constancy of DNA per cell and its halving in germ cell formation was then known as the Boivin–Vendrely rule.

Viewed in isolation this rule might have been taken merely to lend further support to the old Astbury–Caspersson conception of DNA as a framework upon which the genic proteins duplicated themselves. But when Mirsky and Ris viewed their data in relation to the chromosome theory of heredity they went further and concluded that DNA "is part of the genic material. This does not mean, however, that the gene consists of nothing but nucleic acid" (Mirsky and Ris, 1949, 667). Others went further and called into question any involvement of protein in the genic substance. There was a reason for this boldness. Since 1943 the studies of radioactive incorporation of ^{32}P by Hevesy and his colleagues in Sweden had made plain the contrast between DNA and other cellular compounds, including RNA (Hevesy and Ottesen, 1943; Hammarsten and Hevesy, 1946). Its turnover was but a fraction of the others. Also J. N. Davidson, at St. Thomas' Hospital, London, tried starving rats of protein for two days and observing the effect on the DNA and RNA levels in the liver. Whilst the former was unaffected, the latter dipped (Davidson, 1947a, 54). Boivin took up Davidson's idea and in 1948 a group in Paris presented more striking results which they claimed justified the following conclusions:

> ... the absolute quantity of desoxyribonucleic acid enclosed in *each nucleus* maintains an unchanging value in the course of the most severe fast ...
>
> The invariability of the DNA appears as a natural consequence of the special function which is now attributed to it, that of being the depository of the hereditary characters of the species. As for the variability of the RNA; this is easily explained by the very active role which one is inclined to give it in the process of cellular synthesis—(Brachet, Caspersson).
>
> (Mandel, Mandel and Jacob, 1948, 2020–2021)

At a time when isotopic studies were revealing the very rapid turnover of metabolites here was DNA maintaining a much slower turnover. When Mirsky and Ris confirmed the Boivin–Vendrely rule they also showed how very variable were the quantities of their "residual protein"—8.5 per cent to 29 per cent of the chromosomal mass, whereas DNA only varied between 26 per cent and 39 per cent. Now it became clear that the earlier reports of wide fluctuations in DNA percentage content were artefacts. They reflected changes in the absolute quantities of chromosomal proteins. Of these, histones were the predominent class, and recent work had shown that their composition in different tissues of the same organism varied immensely, the arginine from 19 per cent in red blood corpuscles to 87 per cent in mature sperm (Stedman and Stedman, 1947). Mazia pointed out that in the sperm cell of fish the histone type protein has deteriorated into a much simpler protamine type.

> The sperm cell has no function other than to serve as a genetic bridge between two generations; its physiological capacities are limited to activities that will facilitate the transmission of a nucleus into the egg. If the basic protein fraction actually degenerates in the one case where we can prove that a full genetic complement is transmitted by the nucleus, it is not tempting to think of this fraction as embodying the properties of genes.
>
> (Mazia, 1952, 114)

As far as concerned the apparent metabolic inactivity of DNA, Mazia saw this as agreeing with the "template" conception of the gene. "The logical implication is that the gene need not 'do' anything but that it merely provides a blueprint for syntheses" (Mazia, 1952, 115). In conclusion, he named DNA as the most likely candidate for the role of genetic material.

Mazia was not the only cell chemist to be impressed by the work of Avery, Boivin and Vendrely. Arthur Pollister at Columbia University, Hewson Swift at Chicago and Max Alfert at Berkeley were all stimulated by these pioneers to establish further parallels between the genes and DNA, and when a discussion meeting was held at Oak Ridge in 1950 on current problems in the biochemistry of nucleic acids, Pollister spoke warmly of Boivin and described the work of the Vendrelys. Nor was Sir Cyril Hinshelwood, the student of bacterial metabolism, unaware of the Boivin–Vendrely rule, and from his laboratory of physical chemistry in Oxford came evidence for the constancy of the DNA content and the variability of the RNA content of bacterial cells grown under a wide range of conditions (Caldwell and Hinshelwood, 1950a). After suggesting a nucleic acid code for amino acid sequencing (but not a one-way code) they concluded:

> The relative stability of the desoxy-component, and its constancy per cell, perhaps reflect its biological function, in so far as it is believed to be the principal seat of the unchanging hereditary characters, in contrast with the more changeable ribose nucleic acid component. This appears in part to be responsible for the bulk of the protein synthesis, and will presumably be the seat of the variable and adaptable characters.
>
> (Caldwell and Hinshelwood, 1950b, 3159)

Of course there were objections and difficulties—thus, ^{14}C labelled glycine appeared to go much more rapidly into purines than did ^{32}P into DNA (Le Page and Heidelberger, 1951), thus calling into question the accepted slow turnover rates for DNA. The Boivin–Vendrely rule was confused by the effects of polyteny and polyploidy, but these were not seen as serious objections. What puzzled men like Pollister was how a substance which did not vary from one tissue to another could yet be the cause of the development of cells so different as a tiny mammalian haemoblast on the one hand and the immense neuroblast on the other. He felt he was left on the "horns of the old differentiation dilemma" (Pollister *et al.*, 1951, 114). Others, like Boivin, saw RNA as responsible for tissue characters, and DNA for specific characters. But Boivin in his enthusiasm went further and gave to RNA molecules the rôle of transmitting acquired characters and of being themselves self-reproducing. As for Pollister, he felt like Mirsky that important genetic components other than DNA might yet be discovered. Meanwhile another line of evidence from the chemistry of DNA was being assembled by a group at the College of Physicians and Surgeons of Columbia University in New York which lent very powerful support to the Boivin–Vendrely rule.

Erwin Chargaff

Chargaff's interest in nucleic acids dated back to the war years. Born in Vienna, he came to the United States in 1928. During the war he had a contract at Columbia with the Army Medical Service which among other things included a study of the chemistry of the Rickettsias. What followed, Chargaff described in the following words:

> I induced Seymour Cohen, my first graduate student in the '30s here in Columbia, to come back from Princeton where he had worked with Stanley on viruses, to work with me on the chemistry of rickettsia, and in the course of this work we isolated the nucleic acid of rickettsiae. So I became aware of the presence of nucleic acids in these very small organisms and had decided from that time on that in addition to our work on lipids and lipoproteins I would try to become interested in the chemistry of nucleic acid. That is when I started to read up in detail what was known at the time.
>
> In 1944 Avery, MacLeod and McCarty published their famous paper on the transforming principle of pneumococci. This was really the decisive influence, as far as I was concerned, to devote our laboratory almost completely to the chemistry of nucleic acids and nucleoproteins. I read some more but it looked hopeless with the small amounts available that chemistry could be done on these very difficultly obtainable DNA's.
>
> (Chargaff, 1968b)

Hope then came in the form of a suggestion by Aaron Bendich that the qualitative paper chromatography developed by Martin and Synge in 1944 for the amino acids might be used to separate out the purines and pyrimidines in nucleic acids. Chargaff therefore formed a plan to attack this problem. While Bendich isolated calf thymus DNA and Stephen Zamenhof prepared yeast DNA (never done before), Chargaff searched for a good postdoctoral student, trained to the high standards of continental European organic chemistry, to join his team. So he wrote to a fellow organic chemist from his prewar days—Bernhardt—asking for a good student. Working under Bernhardt at that time was Ernst Vischer, on isotopic studies of fatty acid metabolism. He had come from Basle where, under Reichstein, he had learnt to use microchemical methods and had synthesized a desoxy sugar, d-Oleandrose.

When Vischer came as a Swiss–American exchange fellow to Columbia, Chargaff suggested to him a number of topics, including a problem in the chemistry of inositol. But Vischer chose the purine pyrimidine analysis of nucleic acids. Chargaff told him about the one and only good book on these compounds—by Levene—and spoke critically of the tetranucleotide theory there enshrined. He presented Vischer with a Beckmann photometer (one of the first ever sold in the United States), filter paper and nucleic acid. Vischer's brief was to re-examine the tetranucleotide hypothesis and search for unusual bases by developing a chromatographic technique for separating the bases, and estimating their proportions.

Looking back on those days, Vischer thought their approach very crude. He recalled how, when Martin came to New York and lectured at Columbia he brought with him an example of a chromatogram. "We tore off a little bit of the paper to check whether we had the right kind" (Vischer, 1970). Vischer tried to use an ultraviolet lamp to show up the purine and pyrimidine spots. Either the Hanovia lamp they used was too weak or its wavelength was too broad and was in need of filtration. This failure was tiresome, for it meant that they had to make a double run: show up the bases by conversion to their mercury salt, mark the spot on an untreated analogue and elute that. Nor was Vischer able to carry out his purine and pyrimidine analyses in one operation, as the behaviour of the two groups to hydrolysing agents was markedly different. In a single hydrolysis he could not prevent the deamination of cytosine to uracil. Despite these difficulties Vischer succeeded brilliantly. The technique was cumbersome, but it was a new venture in microchemical methods appropriate to Chargaff's quest. The result was the discovery of A:T and G:C ratios as well as the overthrow of the tetranucleotide hypothesis, the rejection of a claim for the presence of 5-methyl cytosine and the demonstration of phylogenetic differences in DNA composition.

In the Spring of 1947 he had made the chromatography of the purines quantitative (Vischer and Chargaff, 1948). That summer their first results on the molar proportions of the bases in calf thymus and spleen DNA were sent to the *Journal of Biological Chemistry* (Chargaff *et al.*, 1949). How far these departed from the expectations of the tetranucleotide hypothesis can be seen from Table 14.1. When rounded up to whole numbers these ratios gave 10 cytosine:16 adenine:13 guanine:15 (or 13) thymine. Now if DNA was the hereditary substance one would expect its base composition in different tissues and in individuals of the same species to be the same. The results obtained with calf thymus and beef spleen were not far from identical. The group at Columbia argued therefore that, if as Avery's work suggested, "certain nucleic acids are endowed with a specific biological activity, a search for chemical differences in nucleic acids derived from taxonomically different species should be conducted, and microorganisms would appear to be one of the most promising sources" (Vischer, Zamenhof and Chargaff, 1949, 429). Chargaff was also anxious to confirm the report of an unusual base—5-methyl cytosine—from the tubercle bacillus by Johnson and Coghill at Yale in 1925. He had his doubts, and so this microorganism was chosen. (He was also very familiar with tubercle bacilli from his work with them in Yale, Berlin and Paris (Chargaff, 1973).) Martin Pollock has recently pointed out how fortunate was this choice. *E. coli* would have given him the

TABLE 14.1

Base ratios obtained by Vischer, Zamenhof and Chargaff in 1949

DNA source	Adenine	Thymine	Guanine	Cytosine
Calf Thymus	1.7	1.6	1.2	1.0
Beef Spleen	1.6	1.5	1.3	1.0
Yeast	1.8	1.9	1.0	1.0
Tubercle Bacillus	1.1	1.0	2.6	2.4

(From Vischer, Zamenhof and Chargaff, 1949, p. 433, and Chargaff *et al.*, 1949, p. 413).

proportions predicted by the tetranucleotide hypothesis (1970, 16). Perhaps we have to thank Johnson and Coghill for their error, and the tubercle bacillus whose pathogenicity assured it the attention of physiological chemists from 1898 onwards!

In the summer of 1948 Vischer overcame the difficulties of matching the purine and pyrimidine analyses, with their different percentage yields, by estimating total phosphorus in *all* the hydrolysates. That July they had in their hands the first good base analyses. Unwittingly they had selected a high GC type bacterium in Helen Saidel's tubercle bacillus DNA, and a high AT type fungus, Zamenhof's yeast DNA. There was no escaping the remarkably close approach to equality between the now well known pairs of bases. Accordingly, we find in this paper of July 1948 the brief statement: "A comparison of the molar proportions reveals certain striking, but perhaps meaningless, regularities" (Vischer, Zamenhof and Chargaff, 1949, 436).

At this stage the regularities were more of an embarrassment to Chargaff than a pleasure. He had anticipated that perhaps total purine was equal to total pyrimidine, but the aim of the work was to show up differences, to establish a parallel between taxonomic relationship and DNA composition. This was very apparent in the yeast bacterium comparison. In the autumn of 1949 Chargaff extended the comparison to include human DNA (Chargaff, Zamenhof and Green, 1950). That summer he had undertaken a lecture tour in Europe. He could then claim the DNA composition to be "characteristic of the species, but not of the tissues" (Chargaff, 1950, 208). He could point to the marked difference between the composition of native, digested, and residual "core" DNA as an indication perhaps of the presence of "clusters of nucleotides (relatively richer in adenine and thymine) that were distinguished from the bulk of the molecule by greater resistence to enzymatic disintegration" (*Ibid.*, 208). Later he attributed this to the complicated structure of the polymer which "is composed of tracts of polynucleotides, differing in the proportions, and therefore in the sequence, of their components, and in their susceptibility to enzymatic attack" (Zamenhof and Chargaff, 1950, 13).

In turning back to these papers from Chargaff's group, papers which were written over twenty years ago, one feature stands out—the influence of Avery's work upon Chargaff's approach. His former collaborators, G. Brawerman and Vischer recalled the many times the work on bacterial transformation came up in conversation. It was, said Vischer, as if Avery was Chargaff's god. In his *Experientia* paper Chargaff referred to the possibility that "the disappearance of one guanine molecule out of a hundred, could produce far-reaching changes in the geometry of the conjugated nucleoprotein" which might be among the causes of mutations (Chargaff, 1950, 202). In his conclusion to this paper he affirmed the existence of "an enormous number of structurally different nucleic acids; a number, certainly much larger than the analytical methods available to us at present can reveal" (*Ibid.*, 209). He went on to calculate the number of possible sequences exhibiting the same base proportions as those of the ox. For a chain of 100 nucleotides it was 10^{56}, for 2500 the number was 10^{1500} (*Ibid.*). When he addressed a symposium on cytochemistry in 1950 he started by referring to Schrödinger's concept of the hereditary codescript and he went on to give evidence in support of the association of DNA with it. He drew attention to the extension of Avery's work to *E. coli* and *Hemophilus influenzae*. He described his own results, including the classification of DNA into AT and GC types (Chargaff, 1950, 655). These facts show that Avery's work did provide the direct stimulus for the most important development in the chemistry of the nucleic acids of that time. No one could claim that Chargaff was blind to the significance of his work, or that he did not foresee the broad direction in which it would be taken.

Base Ratios

We saw that the famous Chargaff base ratios were perceived in the summer of 1948 from a comparison of the data on yeast and tubercle bacillus DNA. Vischer remembered Chargaff sitting at the desk one evening and calling him over "saying: 'Look here, if we interpret the results like that we will get a constant sum [G/C = A/T = 1]'. That was his idea when he was reviewing the paper" (Vischer, 1970). This observation, said Chargaff, "was arrived at purely empirically and almost reluctantly" (Chargaff, 1968b) by inspection of the data. The testing of the tetranucleotide hypothesis called for a representation of molar ratios as close as possible to unity. As they improved the technique, so these base ratios stood out more clearly. C. E. Carter at the Oak Ridge National Laboratory introduced them to the "Mineralight" short wave ultraviolet lamp with which they could directly locate the bases (Chargaff, Magasanik, Doniger and Vischer, 1949). In the summer of 1950 Mrs C. Green and M. E. Hodes succeeded in developing a single hydrolysis procedure using formic acid and requiring only a single chromatogram (Zamenhof and Chargaff, 1950, 3). As the variety of sources for the DNA increased, so the striking constancy of the base ratios stood out in contrast to the wide variations in the (G+C)/(A+T) ratio.

The Significance of the Chargaff Ratios

These famous ratios were announced by Chargaff in the lectures he gave to the chemical societies of Zürich and Basle in June 1949, to the Société de chimie biologique in Paris, and to the universities of Uppsala, Stockholm and Milan. They appeared in print in the May 1950 issue of the Swiss journal, *Experientia* in a short paragraph which reads:

> The results serve to disprove the tetranucleotide hypothesis. It is, however, noteworthy—whether this is more than accidental cannot yet be said—that in all desoxypentose nucleic acids examined thus far the molar ratios of total purines to total pyrimidines, and also of adenine to thymine and of guanine to cytosine, were not far from one.
>
> (Chargaff, 1950, 206)

At the same time a short communication appeared in *Nature* in which Chargaff stated that it was "noteworthy, though possibly no more than accidental, that in all desoxypentose nucleic acids examined thus far the molar ratios of total purines to total pyrimidines were not far from 1. More should not be read into these figures" (Chargaff, Zamenhof and Green, 1950, 757). In retrospect Chargaff claimed more for these statements than was justified (Chargaff, 1963, p. vi). There was, I believe, a good reason both for this claim and for the extreme caution with which the actual discovery was stated. Chargaff was, he said,

> very aware of the fallacy of small irregularities in large molecules. Our work was the main proof that there were no single tetranucleotides or polymers of tetranucleotides in

nucleic acids. And I even said that "we ought to avoid falling into a streamlined version of the old trap which in the past tripped so many excellent workers in the field of nucleic acid chemistry" [Chargaff, 1951, 43].

(Chargaff, 1968b)

So we have on the one hand statements in Chargaff's papers like: "an ounce of proof still weighs more than a pound of prediction" (1950a, 201), "generalizations in science are both necessary and hazardous" (1950a, 208) and "it is difficult to say where the danger line lies beyond which over simplification will produce a dogmatic ignorance" (Chargaff, 1951, 659). All very true and salutary. One reason why Chargaff made them was perhaps that he was deliberately checking his own tendency to speculate. Vischer described him as "full of ideas, and he was thinking all the time—it could be this, it could be that. This could be done and that could be done." It was not possible to recall what explanation of the base ratios was favoured because so many possibilities were put forward, but a double helix was certainly not among them. Chargaff would come and discuss the work every day. "Each day he had a new idea . . . You had to defend yourself in some way to stick to the problem you were working on. He was very much striving to get new things done" (Vischer, 1970). Be that as it may, one can from Chargaff's publications, reconstruct a picture of the sort of structural explanations for the base ratios which he favoured. He became convinced "that there was something essentially different between nucleic acids as macromolecules and proteins." This conviction was all the stronger because it was arrived at reluctantly. How was this equipoise between the purines and pyrimidines achieved when, as the facts of genetics demanded and he believed, "the bases are distributed almost randomly along the chain? So I became simply aware that structurally there must be some special and fundamental difference between the nucleic acids and the proteins" (Chargaff, 1968b).

In 1948 Zamenhof, in an attempt to get an idea of the distribution of the bases in DNA, analysed the base composition of a nucleic acid at varying stages during its enzymatic breakdown. Chargaff reckoned that the results ruled out the existence of nucleotide sequences of the required frequency to yield the correct ratios, for as enzymatic breakdown proceeded, the ratios of purines to pyrimidines and of A:G and T:C rose significantly (Zamenhof and Chargaff, 1950).

Strange to relate, a scheme of hydrogen-bonded base pairing had in fact been put forward by the Brooklyn physical chemist, K. G. Stern, in lectures which he gave in Yale and at the Mount Sinai Hospital in New York. This was in 1946, before the base ratios had been discovered, and before he had read Avery's work. Hence, it was a nucleoprotein model of the gene, but unlike Wrinch's model the genetic specificities did reside in the nucleotide sequence along the DNA chain.

Stern had read Schrödinger on the size of the gene, Darlington on the molecular coil, and Pauling on hydrogen bonds. He chose the correct tautomeric forms of the bases and therefore noted "that the two purine and the two pyrimidine bases possess configurations at corresponding carbon atoms, namely, amino groups in adenine and cytidine [*sic*] and keto groups in guanine and thymine, which are known to give rise to strong hydrogen-bond formation" (Stern, 1947, 944). Now since Stern was visualizing hydrogen-bond formation between *adjacent* pairs of bases it had to be between either two purines or two pyrimidines, to get the keto and amino groups one directly over the other (see Fig. 14.1). Such hydrogen-bonding perpendicular

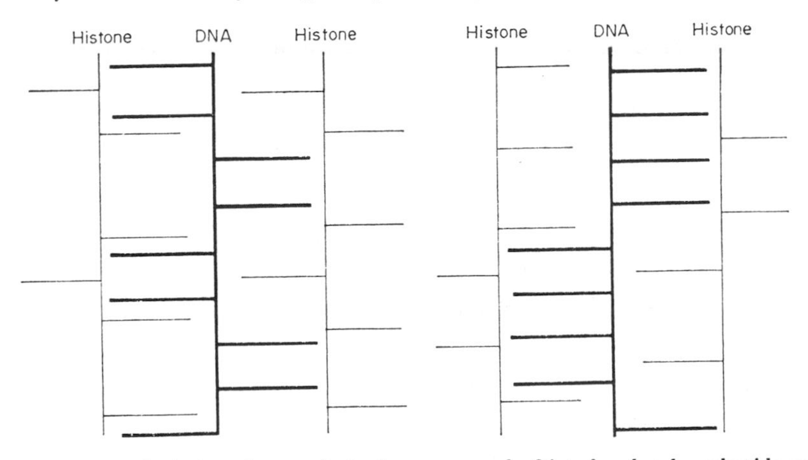

Fig. 14.1. Hypothetical nucleoprotein lattices composed of interlaced polynucleotide and polypeptide chains.

to the plane of the bases was, of course, impossible, but Stern like many of his contemporaries in the 1940's, had evidently not been convinced of the narrow range to the direction of these bonds. The models which he built were far too amateur to reveal the limitations which stereochemistry imposed. Such linear zig-zag chains as Stern visualized were just as impossible as the Astbury model upon which they were based. But Stern had a neat idea which was very appealing. He argued that the code resided not only in the linear sequence of bases, along the polynucleotide chain but in their *three-dimensional* configuration. By itself the polynucleotide chain would be genetically "neutral" since by the free rotation of bonds the orientation of the purines and pyrimidines could alter relative to the backbone. But if a polypeptide chain of the chromosomal protein lay parallel to two polynucleotide chains it could "lock" the latter into a given three-dimensional configuration so that base pairs projected from one side or the other of the chain in a "modulated" manner (Stern, 1947, 944) (see Fig. 14.1). He suggested that the phosphate groups of the polynucleotide chains were alternately associated with sodium ions and the arginine residues in the histone. Both portions of the nucleoprotein molecule were therefore implicated in the specificity of the gene, one was sequential, the other three-dimensional. It was all very neat in a crude sort of way. The only snag was that by the time Stern came to correct the paper for publication he had learnt of Avery's work which, as Stern admitted in a footnote, ran counter to the prediction from his model that "free nucleic acid chains could hardly retain their genic configuration" (Stern, 1947, 944).

When Chargaff discussed the question of gene specificities at Cold Spring Harbor in 1947 he put forward differences in the proportions and the sequences of the bases in nucleic acids as possible grounds for their specificities. For the 6-amino:6-keto = 1 regularity in RNA (still unexplained) he thought the most plausible hypothesis was "a geometrically unique combination between a specific nucleic acid and a specific protein". This hypothesis he also considered for other nucleic acids, including DNA.

> It is, for instance, conceivable that a particular sequence of the nucleotides in the nucleic acid could make for the accumulation of amino groups (from guanine, adenine, or cytosine) at certain spots in the long chain, thus creating centres that could serve to anchor the nucleic acid in space in a specific manner, once it is combined with a protein. The specific nature of a nucleoprotein would then be vouchsafed not only by the structure of the protein but also by that of the nucleic acid.
>
> (Chargaff, 1947, 32)

Since he expected nature to choose a three-dimensional shape to carry biological specificities he concluded his lecture by describing an experiment in topological replication—the Moebius strip. With this, he said, a child "can make many fascinating discoveries about the inheritance of geometrical peculiarities; and when it grows up and remembers them, they may help to take some of the terror from the seemingly automatic nature of the life processes" (Chargaff, 1947, 33).

Early in their work the group at Columbia had noticed the equality of bases carrying a 6-amino group and those carrying a 6-keto group (or hydroxyl as they termed it). Chargaff recalled that they "determined the ratio of hydroxyl groups to amino groups and found that regardless of the composition this ratio was the same in all DNA. It was almost the first ratio that we discovered" (Chargaff, 1968b). After the Watson–Crick model had been proposed Chargaff still inclined to the opinion that "the most satisfactory construction that could impose this uniformity 6-amino to 6-keto is one in which two polynucleotide chains are bonded to a polypeptide chain, so that each peptide carbonyl is linked by a hydrogen bond to the 6-amino group of adenine or cytosine and each peptide amino group to the 6-keto group of guanine or uracil (or thymine)" (Chargaff, 1957, 525). He felt that mere base sequence was not enough, and all the wisdom of the structural crystallographers seemed to him out of place when organic forms were being analysed. Nature, he quipped, "is not just an interior decorator." In the imaginary conversation between an old chemist and a young molecular biologist that he published in *Essays on Nucleic Acids*, Chargaff made the chemist warn his young disputant about the dangers of glibness.

> It was in 1889 that the great Swiss historian Jacob Burckhardt wrote a letter to a friend in which he warned of the oncoming of what he called "les terribles simplificateurs". Just as the locusts, once they are through with a field, have simplified it horribly, could we not say that this is also true of some of the great generalizations in biology?
>
> (Chargaff, 1963, 183)

We may well imagine that the man who later saw molecular biology as an inglorious piece of oversimplification had earlier considered some form of base pairing and rejected it as too simple. The evidence that we have, however, suggests rather that Chargaff was thinking in terms of some pattern of hydrogen bonding between the bases and the associated polypeptide chains of histone. Recently he had this to say about his attempts to arrive at a structural explanation of the ratios:

> We played with various schemes and models, in which some sort of 6-amino to 6-oxy pairing and purine to pyrimidine pairing played a role; but I do not believe we ever thought of a double stranded structure. We were, however, quite handicapped by not having accurate molecular models to fit together . . .
>
> The nearest I came to acknowledging our attempts at pairing models was in the following passage in a paper published in 1951 (*J. Biol. Chem.*, **192,** 223):
>
> As the number of examples of such regularity increases the question will become pertinent whether it is merely accidental or whether it is an expression of certain structural principles that are shared by many desoxypentose nucleic acids, despite far reaching differences in their composition and the absence of a recognizable periodicity in their nucleotide sequence.
>
> (Chargaff, 1970)

Stern's Explanation of the Chargaff Ratios

Stern worked at the Brooklyn Polytechnic Institute not far from Columbia where Chargaff worked. They were acquaintances but they never got together over the mystery of the base ratios. Not until August 1951 does it appear that Stern even knew about these ratios. The occasion was a symposium on the nucleus at Brookhaven National Laboratory. A Canadian, Jerry Wyatt, had read a paper describing his fine work on insect viruses, 5-methylcytosine and his results with the formic acid hydrolysis of a wide variety of DNAs (Table 14.2) which thoroughly confirmed the existence of Chargaff's ratios. Wyatt's study of the chemical taxonomy of DNAs led him to conclude that in higher organisms these molecules must have a very complex constitution, and he went on:

> It is difficult, however, to reconcile the complexity implied by this concept of nucleic acids with the apparent regularity connoted by the curiously constant ratios between

> certain of the bases. If the nucleic acids of genes do differ in composition, their variation must follow a fixed pattern by which these ratios are kept unchanged. We can scarcely even speculate upon how this could occur.
>
> (Wyatt, 1952a, 214)

When Wyatt read this paper the meeting had begun to thin out. It had been very hot, it was at the end of a long afternoon. So Wyatt decided to get through his paper as quickly as possible. The discussion which followed was brief but revealing. Gerald Oster suggested that the idea of irregular nucleotide sequences might not be consistent with the X-ray results which Denis Riley and he had got from DNA "powders". These indicated such a regular structure that variations in the bases could surely not exist. Wyatt suggested that "if you have a spiral structure" it was quite possible to have

TABLE 14.2

Base ratios and AT GC types according to Wyatt in 1952

Source of DNA	Adenine	Thymine	Guanine	Cytosine	5-methyl C.	$\frac{A+T}{G+C+MC}$
Calf thymus	1.13	1.11	0.86	0.85	0.052	1.27
Herring sperm	1.11	1.10	0.89	0.83	0.075	1.23
Wheat germ	1.06	1.08	0.94	0.69	0.23	1.15
Gypsy moth virus	0.845	0.80	1.225	1.13		0.70
Tent caterpillar virus	1.18	1.11	0.91	0.80		1.34
Pine sawfly virus	1.29	1.21	0.785	0.715		1.67
European Budworm virus	1.285	1.235	0.775	0.70		1.71
Rickettsia prowazeki	1.43	1.27	0.684	0.616		2.08

(From Wyatt, 1952a, 206–209, and Wyatt and Cohen, 1952, p. 846).

the bases "sticking out free so that they don't interfere with each other. Then you could have a regular spacing down the backbone of the chain, in spite of the differences in sequence" (Wyatt, 1952a, 216). Stern supported Wyatt and he went on to talk about his old idea of hydrogen-bonded base pairing. For a moment you hold your breath and ask—surely he will see the connexion between base pairing and Wyatt's ratios. But no! His pairing was 6-amino purine with 6-keto purine and 6-amino pyrimidine with 6-keto pyrimidine. They were the wrong base pairs! Two years passed before the explanation occurred to Wyatt. Then in March 1953 the *Biochemical Journal* received his paper with Cohen containing the following sentence: "One is tempted to speculate that regular structural association of nucleotides of adenine with those of thymine and of those of guanine with those of cytosine (or its derivatives) in the DNA molecule requires that they be equal in number" (Wyatt and Cohen, 1953, 780).

Wyatt returned to Canada in 1950 and it was three years before he had saved up enough leave to pay a return visit to Europe. In the spring of 1953 he came, and because his wife wanted to discuss insect tissue culture with André Lwoff they visited the Pasteur Institute. Watson was also visiting at the time. The conversation that ensued left a lasting impression on Wyatt.

> I recall this small group sitting round in the laboratory. Jim Watson was sitting on a high stool with his legs wrapped around it and was describing to us this wonderful idea that he and Crick had just had about the structure of DNA. It hit me immediately that it must be correct, that it explained the data that I had, in which I had not been able to see any real sense.
>
> (Wyatt, G. R., 1968)

SECTION IV

INTELLECTUAL MIGRATIONS

DELBRÜCK, SCHRÖDINGER, BERNAL, PERUTZ, PAULING, WATSON AND CRICK

CHAPTER FIFTEEN

Physicists in Biology: The Informational School

> The first stimulus for interesting myself in biological problems I derived from discussions with Bohr in 1931 on the bearing of quantum mechanics on biology. By devious ways of searching for a research project that held promise of throwing light on the fundamental problem in question I finally decided on this problem [phage replication] where we find the simplest case of duplication of highly complex molecules, under conditions that allow controlled quantitative experiments. Since I started this work I have become more and more convinced of its importance and its great experimental possibilities.
>
> (Delbrück, 1939)

In the opinion of one molecular biologist turned historian an important element in the "psychological infrastructure of the creators of molecular biology" was the romantic idea that "other laws of physics" might be discovered from the study of the gene (Stent, 1966, 4). This quote comes from the Delbrück *Festschrift*, *Phage and the Origins of Molecular Biology*. The uninformed reader going quickly through this book might come away with the idea that molecular biology came into existence because three physicists in search of paradoxes—Niels Bohr, Erwin Schrödinger and Max Delbrück—thought to find them in biology.

A number of biologists, chemists and physicists, some attracted by Bohr, some inspired by Schrödinger, attended the Phage Course given at Cold Spring Harbor and were thus initiated into the Phage Group whose self-appointed task it was to analyse a universal phenomenon of life—genetic replication—in its simplest possible form, and thus to discover the paradoxical mechanism by which individual molecules can reproduce their like with such remarkable fidelity, so reliable a copying process as to suggest that other laws of physics were involved.

Well, the paradoxes did not emerge. Instead, traditional physical laws and orthodox structural chemistry were shown to govern such replication, and the result was the establishment of a purely molecular description of this process. When John Kendrew reviewed the Delbrück *Festschrift* in the *Scientific American* he stated candidly that it was a little odd to find "in nearly every contribution to this book the explicit or implicit assumption that molecular biology had its only real beginnings in the Phage Group, as if the central theme of the subject is biological information." He could justly point to the work of the structural school of X-ray crystallographers, founded by Astbury, Bernal and their pupils. Were they not pursuing molecular biology? And did this structural school not forge links with the Phage Group whose approach differed from theirs and was best represented by the label Informational School?

Kendrew's review was the stimulus behind the famous lecture given by Stent at the Collège de France entitled "That was the molecular biology that was". Now Stent said it was his belief that the structural school's activity reflected a preoccupation with a down to earth view of the relation of physics to biology, "namely, that all biological phenomena, no matter what their complexity, can ultimately be accounted for in terms of conventional physical laws" (Stent, 1968, 392). The trouble with this approach, was that it was slow and it was not revolutionary. The informational school, on the other hand, "working on the physical basis of biological information must have seemed more like a pie-in-the-sky activity, for there was then hardly any common ground between genetics, on the one hand, and physics and chemistry, on the other" (*Ibid.*). In contrast to the structural school's very reasonable and down to earth view, the members of this informational school according to Stent "were motivated by the fantastic and wholly unconventional notion that biology might make significant contributions to physics" (*Ibid.*). For them orthodox physical laws were not going to be adequate to account for the facts of genetics, and hence new physical laws might well emerge from an analysis of genetic processes.

One cannot but have the uneasy feeling that Stent, who was himself motivated by the search for the paradox, has allowed his own enthusiasm for this romantic phase in his life to carry him away when he has attempted to reconstruct the early days of molecular biology. A glance at the table of contents to the Delbrück *Festschrift* will show that in the first section entitled "The Origins of Molecular Biology" there are two contributors who never attended a Phage Course and who certainly cannot be said to belong to the Phage Group. Although the fifth contributor attended the Phage Course he cannot be said to have belonged to the Group in any other sense, which leaves only two contributors—Stent, and Delbrück himself. This is not to criticize the way in which the *Festschrift* was organized, but simply to warn the sociologist against using it uncritically as a source for the influence and character of the Phage Group. All too easily one may assume that those who attended the Phage Course belonged equally to a close network of collaboration, correspondence and co-authorship, whereas in fact there was a tightly knit core and a much more diffuse penumbra. Nicholas Mullins has recently described very sensibly the various social features which marked the Phage

Group in its development, using the terms: Paradigm group, Network, Cluster and Speciality (Mullins, 1972, 53).

The motivation of those who turned to phage cannot be attributed simply to the "search for a paradox". Not even in Delbrück's case does this label do justice to history, for as he remarked: "The 'search for a paradox' was only a means of focusing on the 'nature of replication' " (Delbrück, 1972). In choosing an appropriate system in which to study replication, clearly the superiority of phage over *Neurospora* and *Drosophila* played a part. The virologist working with phage, was dealing with individual genetics particles, or so it seemed to him, and was free from the complications of an organized genetic system. Moreover, unlike the bacteriologist and cytogeneticist, he could manipulate this sub-cytological world. "He is not limited to dealing with integrated units of reproduction at the cellular level, but can control to a certain extent what goes into his cells. Because of this, virology's methods may lead more directly to solving the problem of the mode of replication of genetic material" (Luria, 1951, 463). Support for this demoting of the importance of the paradox hunt may be found in the fact, admitted by Stent, that none of his friends, save of course Delbrück, was primarily influenced by this aspect of the phage programme.

We shall now examine in some detail how the distinction between the informational and structural schools was due, as Stent has suggested, to two quite different influences which had in the first place brought physicists to address themselves to problems in biology. One such influence was the desire to apply orthodox physical principles to biology, the other to discover new principles. The thinking behind the latter school is represented by Bohr's application of his principle of complementarity to biology, whereas that behind the former school is seen in Schrödinger's book *What is Life?* At the time Stent, it appears, thought the message behind this book was like Bohr's, whereas in fact the new laws of physics which Schrödinger expected to emerge from the study of living organisms were not going to be as revolutionary as Bohr's new laws; physicists were not going to come up against an impenetrable barrier when they applied physical methods in biology, as Bohr at one time conjectured they might. On the contrary Schrödinger was on the side of the structural school. Let us begin our analysis with the informational school, the central figure in which was Max Delbrück.

Delbrück as a Physicist

Before the rise of National Socialism, Delbrück was attempting a career first in quantum chemistry and then in theoretical nuclear physics. He had a great reputation to live up to. The family name was legendary in the world of academia and government. His father Hans was Berlin University's professor of history, the liberal who with Harnack organized the moderate petition to counter the aggressive pan-German movement within the universities. As a student Delbrück moved from one university to another in typical German style finally settling in Göttingen, pursuing his boyhood interest in astronomy, which had taken him to the field of astrophysics. Failing in his attempt to produce a satisfactory thesis in this area he switched to theoretical physics, and prepared a thesis on the theory of homopolar bonding in the lithium molecule. This was an elaboration of the Heitler–London theory of the hydrogen molecule. Max Born and W. Heitler supervised this thesis. On this early phase of his scientific career Delbrück had this to say:

> My interest in astronomy dates back to boyhood. I would now say that I seized upon this subject as a means of finding my own identity in an environment of many strong personalities all of them senior to me, many of them with high accomplishments, but none a scientist, with one exception, the eldest boy in the Bonhoeffer family, Karl Friedrich, eight years my senior, a physical chemist of high distinction, who became my mentor and lifelong friend. The shift to theoretical physics during the latter part of my graduate studies was an easy one from astrophysics and a natural one in the late 20's, just after the breakthrough of quantum mechanics for which Göttingen was one of the centres.
>
> (Delbrück, 1967a)

The influence of such a variety of accomplishments and interests around Delbrück clearly left its mark on him. Here was no footslogger who would toil on year after year at some colossal but dull task. His interests were at first narrow, but gradually roved from one field to another, and towards the end of his Göttingen years they ranged deeply into social sciences and philosophy, but chemistry and biology were completely ignored. Bohr therefore found it somewhat surprising that he could interest this young man in what he saw as the great paradox of biology—his principle of complementarity applied to biology—when Delbrück came to Copenhagen on a Rockefeller fellowship in Physics in 1931.

Complementarity acccording to Bohr

This principle has been subjected to all sorts of misinterpretations, some more crass than others, which have been analysed by Philipp Frank in his book *Modern Science and its Philosophy* (1950). Frank gives the following definition:

> If in the description of an experimental arrangement the expression "position of a particle" can be used, then in the description of the same arrangement the expression "velocity of a particle" can *not* be used, and *vice versa*. Experimental arrangements, one of which can be described with the help of the expression "position of a particle" the other with the help of the expression "velocity", or, more exactly, "momentum", are called *complementary* arrangements and the descriptions are referred to as *complementary* descriptions.
>
> (Frank, 1950, 163)

It is possible to analyse light in its propagation in terms of waves and in its emission and absorption in terms of particles, but it is not possible to describe the propagation in terms of particles and the emission in terms of waves. And if we attempted to trace the path of a single quantum of light in its propagation, our measuring instruments would interact with the light quantum and thus vitiate the success of our observation. We shall return to this remarkable limitation on the power of physics to determine phenomena, in the old classical sense, when we come to examine Schrödinger's work. Whereas Schrödinger sought to overcome this limitation, indeed he turned to biology in the hope of discovering in organisms the conditions for a return to the strict determinacy of classical physics, Bohr saw no such possibility. Instead, he viewed physical indeterminacy as an example of the epistemological problems which had fascinated him as a precocious adolescent, when he attended Høffding's lectures on the history of philosophy and read P. M. Moller's *Tale of a Danish Student*. There, young Bohr found the problem of analysing thought humorously depicted as in the following passage:

> When you write a sentence, you must have it in your head before you write it; but before you have it in your head, you must have thought it, otherwise how can you know that a sentence can be produced? And before you think it, you must have had an idea of it, otherwise how could it have occurred to you to think it? And so it goes on to infinity, . . .
>
> (cited in Rosenfeld, 1963, 48)

When Bohr returned at the close of his life to the subject of complementarity in biology he emphasized that the use of complementary modes of expression —teleological and reductionist—did not "in itself imply any limitation in the application to biology of the well-established principles of atomic physics" (Bohr, 1963, 197). But he concluded his address by underlining the difficulties which remained in overcoming the complementary aspects of consciousness. The feeling of wonder which the physicists felt thirty years ago for the phenomena of life, he said, still remained, what had changed was "the courage to try to understand" (*Ibid.*).

Once quantum mechanics was established, Bohr set about exploring the precise nature of complementarity with reference to specific examples. Believing it to be a really fundamental principle, he sought to explore it in all possible directions and biology did not escape such an analysis. Just as atomic physics had "revealed the limitation in principle of the so-called mechanical conception of nature" so biology, aided by the microscope, had revealed "unsuspected fineness in organic structure and regulatory processes." As an adolescent, Bohr had listened to the discussions which took place in his house in the circle of physiologists centred around his father, Christian. Here, the insufficiency of the purely mechanical conception of nature was emphasized, but there was no question of a return to "primitive ideas of a life force"

(Bohr, 1958, 95). This is clear from the passage from Christian Bohr's paper "On Pathological Lung Expansion" which his son, Niels, included in his essay, "Physical Science and the Problem of Life". Christian had been justly impressed by the suitability of the haemoglobin molecule for its function as oxygen transporter, as seen in the S-shaped curve of its binding with oxygen at different oxygen tensions (Bohr, Hasselbalch and Krogh, 1904). Such fine details in physiology impressed him with the "purposiveness of organic functions. His anti-reductionist stand was echoed by Niels Bohr in the famous address "Light and Life" of 1932:

> On the one hand, the wonderful features *which are constantly revealed in physiological investigations and differ so strikingly from what is known of inorganic matter*, have led many biologists to doubt that a real understanding of the nature of life is possible on a purely physical basis. On the other hand, this view, often known as vitalism, scarcely finds its proper expression in the old supposition that a peculiar vital force, quite unknown to physics, governs all organic life. I think we all agree with Newton that the real basis of science is the conviction that Nature under the same conditions will always exhibit the same regularities. Therefore, if we were able to push the analysis of the mechanism of living organisms as far as that of atomic phenomena, we should scarcely expect to find any features differing from the properties of inorganic matter.
>
> (Bohr, 1933, 457–458)

But to push it so far, to analyse it so far, as to "describe the role played by single atoms in vital functions" would bring about the death of the animal, and he concluded with the famous statement:

> In every experiment on living organisms, there must remain an uncertainty as regards the physical conditions to which they are subjected, and the idea suggests itself that the minimal freedom we must allow the organism in this respect is just large enough to permit it, so to say, to hide its ultimate secrets from us. On this view, the existence of life must be considered as an elementary fact that cannot be explained, but must be taken as a starting point in biology, in a similar way as the quantum of action, which appears as an irrational element from the point of view of classical mechanical physics taken together with the existence of the elementary particles, forms the foundation of atomic physics.
>
> (*Ibid.*)

As he went on, Bohr revealed that what would be destroyed by physico–chemical analysis was the organization by which the animal preserved and propagated itself. This teleological will to survive was foreign to mechanical analysis just as the quantum of action was foreign to classical mechanics. The physical and the physiological analysis of the organism involved, then, what Bohr called "complementary" experimental arrangements. In 1950 Frank described this approach as possible and perhaps even desirable, whereas in the transition from classical to quantum mechanics the complementarity mode of expression was necessary (Frank, 1950, 170). Bohr, on the contrary, was convinced that it was just as necessary in biology as it was in atomic physics.

Delbrück's Entry into Biology

Delbrück had entered atomic physics when the quantum of action was at last finding its place in the subject. He had come to Copenhagen in the spring of 1931 before Bohr gave his famous address "Light and Life" to the International Congress on Light Therapy in August 1932. Delbrück had been in Bristol in the summer of 1932, but on the day of Bohr's lecture he returned, and Rosenfeld met him at the station, and took him straight to the lecture. There in the public gallery, all alone, Delbrück and Rosenfeld listened to Bohr.

> It would be a romantic exaggeration to say that we were fascinated by the lecture, but it is a fact that when Delbrück afterwards read the text and pondered over it, he was so enthusiastic about the prospects it opened up in the vast field of biology that he there and then decided to take up the challenge.
>
> (Rosenfeld, 1967, 134)

Shortly thereafter Delbrück became assistant to Lise Meitner in Otto Hahn's Kaiser Wilhelm-Institut für Chemie. This move to Berlin, wrote Delbrück, "was largely determined by the hope that the proximity of the various Kaiser Wilhelm-Institutes to each other would facilitate a beginning of an acquaintance with the problems of biology" (Delbrück, 1967). There is no doubt that the Kaiser Wilhelm-Gesellschaft not only intended but succeeded in bringing about such a cross-fertilization between disciplines, and this character of the work carried on in its institutes proved crucial to Delbrück. Although he was in the radioactivity division of Hahn's institute in Berlin-Dahlem, he was able to collaborate with N. W. Timoféeff-Ressovsky in the genetic division of the institute for brain research, in Berlin-Buch, and with K. G. Zimmer in the radiation division of a hospital, the Cecilienhaus, in Berlin-Charlottenburg.

Because of the repressive state of Nazi Germany at that time

> . . . official seminars became dull. Many people emigrated, others did not leave but were not being permitted to come to official seminars. We had a little private club which I had organized and which met once a week, mostly at my mother's house. First just theoretical physicists (I was at that time a theoretical physicist), and then theoretical physicists and biologists. The discussions we had at that time have had a remarkable long-range effect, an effect which astonished us all.
>
> (Delbrück, 1971a, 4)

Now according to George Gamow there was the feeling at that time that physics would not make rapid progress until the high energy accelerators necessary for analysing nuclear structure had been built. Gamow may have influenced Delbrück to move into biology, but what proved crucial for Delbrück was his encounter with the brilliant and lively Russian geneticist, N. W. Timoféeff-Ressovsky (referred to in future as Timoféeff).

Among those who belonged to Delbrück's circle in Berlin were Hans Gaffron, the expert on photosynthesis and Kurt Wohl, who instructed the others on the principles of photochemistry and reaction kinetics. As a result of long discussions with Timoféeff and Zimmer, Delbrück came to write a theoretical section for the paper, describing the careful and thorough experiments carried out by Timoféeff and Zimmer on the X-ray mutagenesis of *Drosophila*. Because it took so many authors to write this paper it became known jokingly (perhaps derisively) as the "Three-Man-Work", or more kindly "The Green Paper" on account of the colour of the reprint covers. This paper, entitled "On the nature of Gene Mutation and Gene Structure", is of fundamental importance in the history of molecular biology. Unfortunately its publication in the *Nachrichten der Gesellschaft der Wissenschaften in Göttingen* meant that it was virtually inaccessible save through reprints. As a result few who refer to it have read it.

The Three-Man-Work

The three collaborators approached the subject of X-ray mutagenesis from different points of view. Timoféeff saw it as the culmination of a long series of attempts to bring about artificial mutation of the gene. In 1926 Muller had succeeded in transmuting the gene. Eight years later Timoféeff could say: "The treatment with short-wave radiations and high speed electrons . . . is so far the only effective method of inducing mutations, giving constant and measurable results" (1934, 451). He looked forward to the day when by "methods of treatment which would work differentially" we will be able to "induce at will certain types or groups of mutation" and produce new genotypes—"genetic engineering" (*Ibid.*). Here then was a biologist whose long term aim it was to develop a more speedy procedure for altering the genetic constitution of animals and plants in the service of man.

Zimmer, on the other hand, as a physicist, was fascinated by the problem of what physico-chemical changes were produced by ionizing radiations. Why, for instance, was it that even very small quantities of energy when administered in the form of X-rays were mutagenic, but when given in some other form, such as heat, were without any effect whatsoever? And why did the dose-effect curve for X-ray mutagenesis show no threshold value below which there was no effect? It was, recalled Zimmer, "the discovery of the form of the dose-effect curve for which a plausible explanation did not seem available that led to an entirely new line of thought: the application of concepts of quantum physics to biological problems. These concepts in a generalized form have been well justified as a working hypothesis, there is no doubt but that in this way modern physical concepts came into contact with biology and that the synthesis of the specialities so initiated has been remarkably fruitful" (Zimmer, 1966, 35).

Timoféeff had already come to the conclusion that X-rays acted directly and simply (1934, 439). When Zimmer joined him they analysed the various mechanisms that had been put forward to account for the mutagenic action of short wave radiations. Timoféeff did the genetic analysis, Zimmer the dosimetry, "at that time a tricky piece of experimental physics", commented Delbrück, "and absolutely crucial for the work, you might say Zimmer did the abscissa, Timoféeff the ordinates of the data graphs" (Delbrück, 1972).

In the Three-Man-Work they came to the conclusion that they had to do with an elementary process, in the quantum mechanical sense. This event was a "single hit", which represented either the formation of an "ionic pair" or an "excitation" within a minimum volume of genetic material known as the "sensitive volume". In this they came out in support of the "target theory" of F. Dessauer (1922) and J. A. Crowther (1926/27). The "Three-Man-Work", then, was more than a paper developing the target theory—the parallels between spontaneous and artificial mutations was examined, the significance of back-mutation rates for the structure of the gene was emphasized—and an estimate for the minimum size of the gene was actually given.

When Timoféeff and Zimmer joined Delbrück's circle their work became a central topic for discussion, in the course of which Delbrück rephrased their conclusions in the language of quantum mechanics. His contribution to the paper was entitled: "Physical–atomic Model of Gene Mutation", which became known as the quantum mechanical model of the gene.

Delbrück's Model

Delbrück's interest in mutagenesis differed from both Timoféeff's and Zimmer's. What intrigued Delbrück was the autonomous character of genetics. It was a quantitative science, to be sure, but the units which had to be scored were not masses, charges or velocities, but individual organisms showing certain characteristics, and this made genetics independent of the measurements of physics. How, then, was genetics to be united with physics? Not, he argued, through chemistry, for one could never hope to have enough material in exactly the same molecular state to characterize it accurately by chemical analysis, nor was there sufficient knowledge of how the gene acted chemically in its catalysis of development. What we had to do, then, was to grasp the problem "from a more primitive side" (Timoféeff-Ressovsky, Zimmer and Delbrück, 1935, 226). We had to investigate "the *nature of the limits of the stability* of the gene and see whether it agreed with what we know about well defined atomic associations on the basis of the atomic theory" (*Ibid.*, 226).

Delbrück went straight to the heart of the problem. How was physics to account for the two aspects of mutation—change and constancy? Judged by the characteristics it determined, the gene was remarkably stable. On those rare occasions when it mutated, giving rise to a new characteristic, the latter was transmitted again with remarkable fidelity, indicating that the new state of the gene was as stable as the old. The recognition that only certain forms of energy could bring about such a change artificially suggested to the physicist that here one had to do with the stable configuration which atoms adopt when held together in the molecule of the chemist. This stable configuration was only disturbed when energy in an intense form—short wave radiations—was applied to it. Such radiation increased the amplitude of the vibration of one or more electrons to the point where it overstepped the limits of the existing configuration and "jumped" to a fresh orbit, thus achieving a fresh stable configuration. The result was a rearrangement of the atoms within the molecule. The activation energy required to bring about such a change would vary depending upon the original arrangement of atoms in the gene. Furthermore it would be possible for the change to be reversed by further irradiation. The activation energy necessary for this back-mutation would not necessarily be the same as that for the forward mutation. The reaction was temperature dependent, but time independent. Finally, all three authors were agreed that the atomic configuration of the gene must be a special one and not merely a long chain of similar sub-units as in a polymer (Goldschmidt's quantitative model), for if the latter were the case it was inconceivable that back mutation rates could ever approach forward mutation rates. Mutation, involved an alteration in the arrangement of atoms in a certain minimum, "sensitive" volume. It was possible by a statistical method to arrive at an estimate of this volume. According to Delbrück this volume contained approximately one thousand atoms, well below the maximum size arrived at from cytological observations, which was of the order of a cube with sides 300 Å long.

To some scientists it was very attractive to find that this minimum volume was equivalent to the molecular weights then recognized for some of the proteins. But to Delbrück the importance of this work lay in demonstrating that the facts of genetics *could* be related to fundamental physical theory. He does not seem to have been as interested as were his co-authors in demonstrating that the polymer sub-unit theory of the gene was untrue, nor even in attributing the stability of the gene to a *molecular* configuration. He was far more cautious and spoke only of an "atomic association" (*Atomverband*) rather than of a molecule. This did not prevent the fundamental message of the "Three-Man-Work" from coming over: the stability of the gene was due to the strength of inter-atomic forces, and its mutation was due to a quantum jump from one stable configuration over the energy "hump" which separates one configuration from another.

Today this conclusion may seem a truism, but in the mid-30s, according to Delbrück it was not. "Genes at that time were algebraic units of the combinatorial science of genetics, and it was anything but clear that these units were molecules analysable in terms of structural chemistry. They could have turned out to be submicroscopic steady state systems, or . . . something unanalysable in terms of chemistry as first suggested by Bohr . . ." (Delbrück, 1970, 1312). This opinion should be compared with the conclusions arrived at in Chapter 14, on the chemistry of the gene. Perhaps Delbrück's view reflected both his own and Bohr's scepticism about the structural chemical approach to the nature of the gene. What irony that, as Delbrück later admitted: "The road to success effectively by-passed radiation genetics" (*Ibid.*). Even more ironical was the fact that the single-hit form of the dose-effect curve later turned out to represent a multi-hit curve which, because of variation in target volume, simulated a single-hit curve! (Zimmer, 1941; 1950; 1961).

The Difficulty in Arriving at a Chemical Description

Delbrück admitted that the structural chemist's successful application of quantum mechanics to more and more complex molecules could not suffer any limitation in principle; in practice, however, Delbrück saw a difficulty in extending this approach to the genetic substance, for it presupposed the availability of "a practically infinite number of molecules of the same kind and of practically infinite stability . . ." (Delbrück, 1949, 20; Timoféeff-Ressovsky, Zimmer and Delbrück, 1935, 225). How could the chemist obtain such quantities when the inhomogeneities in the cell "go right down to the atomic scale?" How could he hope to extract sufficient quantities of one gene when this existed in the chromosome only in a single copy? By 1949 he knew of Avery's work and had discussed it with his friends (see Chapter 18). But his point remained a valid one until, by sequence analysis in combination with genetic analysis, the nucleotide sequences of the gene were established. Delbrück was not against a chemical analysis just for the sake of it! Just as Bernal saw limitations in the direct assault on protein structure by the techniques of X-ray crystallography of the 1930s, so Delbrück doubted seriously that the chemical techniques of that time could expose the atomic structure of the gene. *What was lacking at that time, but has since been provided, was a chemical, as opposed to a phenotypic, characterization of the gene.*

No wonder Delbrück's followers recalled that he "deprecated biochemistry" (Benzer, 1966, 158), and that in contrast to Seymour Cohen, who *wanted* biochemistry to explain genes, "Luria and Delbrück opted for a combination of genetics and physics" (Watson, 1966, 241). The key word here is "opted"; Watson has, I believe, exactly expressed the initial attitude to chemistry of these two founders of the Phage Group. Delbrück's message in the "Three-Man-Work" was that we had to account for the extraordinary stability of the gene and its power to duplicate itself in mitosis. This was the challenge to the physicist. When Delbrück began to work on phage he wrote: "Certain large protein molecules (viruses) possess the property of multiplying within living organisms. This process, which is at once so foreign to chemistry and so fundamental to biology . . ." (Ellis and Delbrück, 1939, 365)—the die was cast and the stage set for the quantitative study of this autocatalytic reproduction of protein molecules! For him the task was to get at the fundamentals and brush all else aside. Thus Herman Kalckar, who attended Delbrück's phage Course in 1945 said that Delbrück "wanted to bypass all the phenotypic diversities of the living cell and go straight to the problem of gene replication and gene action. Was it possible to find fundamental laws for gene action and gene replication?" (Kalckar, 1966, 46).

But what events led Delbrück to turn to phage at a time when *Drosophila* was still the favourite organism for the geneticist? We know that after attending a meeting organized by Niels Bohr and the Rockefeller in Copenhagen in September 1936 to which he travelled with Muller and Timoféeff from Berlin, Delbrück returned to that city. A year later he was stimulated by Stanley's earlier (1935) announcement of the crystallization of TMV protein to write a short memorandum to himself entitled "Riddle of Life" (Delbrück's suggestion that he wrote this as a summary of the Copenhagen meeting is clearly unlikely since it is dated a year later, August 1937 (Delbrück, 1970, 1313).) Now what attracted Delbrück to TMV was the fact that Stanley's preparations demonstrated great uniformity in their behaviour—they migrated with uniform velocity in the electrophoresis apparatus, their infectivity was not impaired by recrystallization, they gave reproducible chemical analyses—they were protein molecules which could reproduce themselves using the host cell as a nutrient medium. Here, then, was a close approximation to the real thing and only the real thing—replication unclouded by the complexities of recombination. "We want", he said, "to look upon the replication of viruses as a particular form of a primitive replication of genes, the segregation of which from the nourishment supplied by the host should in principle be possible" (Delbrück, 1970, 1315).

For the next event which brought Delbrück closer to biology, and bacteriophage in particular, we have to thank Warren Weaver, who played an important part in formulating a policy of support by the Rockefeller Foundation for the recruitment of chemists and physicists into biology (see Chapter 17). It was probably Weaver or Tisdale who visited Delbrück in Berlin in 1937 and suggested that he applied for a fellowship to work in biology. "I chose Caltech as the place to go to because of its strength in *Drosophila* genetics, its sympathetic attitude towards my scientific interests in general, and to some extent because of its great distance from the impending perils at home" (Delbrück, 1967).

As the shadows thickened over the once brilliant life of Berlin's Kaiser Wilhelm-Institutes, Delbrück left for the United States of America, the manuscript "Riddle of Life" in his baggage. At Caltech he met Emory Ellis who was working on phage. What followed has been described delightfully by Delbrück himself in his Harvey Lecture of 1946. He opened his presentation with a reference to the heroic age [1920s] of the study of bacteriophage. Now, he said, this age was a matter of the past, "the passions have subsided and minor men [the phage group] with different interest are beginning to settle on these grounds" (Delbrück, 1946, 161). These men came from far outlying fields such as physics and biochemistry. Like the physicists of fifty years ago, who saw exciting development coming to a focus on the constitution of atoms, so now these "outsiders" felt that many lines of biological research were converging on a central problem, the organization of the cell. "They feel that the field of bacterial viruses is a fine playground for serious children who ask ambitious questions" (*Ibid.*).

> You might wonder how such naïve outsiders get to know about the existence of bacterial viruses. Quite by accident, I assure you. Let me illustrate by reference to an imaginary theoretical physicist, who knew little about biology in general, and nothing about bacterial viruses in particular, and who accidentally was brought into contact with this field. Let us assume that this imaginery physicist was a student of Niels Bohr, a teacher deeply familiar with the fundamental problems of biology, through tradition, as it were, he being the son of a distinguished physiologist, Christian Bohr.
>
> Suppose now that our imaginary physicist, the student of Niels Bohr, is shown an experiment in which a virus particle enters a bacterial cell and 20 minutes later the bacterial cell is lysed and 100 virus particles are liberated. He will say: "How come, one particle has become 100 particles of the same kind in 20 minutes? That is very interesting. Let us find out how it happens! How does the particle get in to the bacterium? How does it multiply? Does it multiply like a bacterium, growing and dividing, or does it multiply by an entirely different mechanism? Does it have to be inside the bacterium to do this multiplying, or can we squash the bacterium and have the multiplication go on as before? Is this multiplying a trick of organic chemistry which the organic chemists have not yet discovered? Let us find out. This is so simple a phenomenon that the answers cannot be hard to find. In a few months we will know. All we have to do is to study how conditions will influence the multiplication. We will do a few experiments at different temperatures, in different media, with different viruses, and we will know. Perhaps we may have to break into the bacteria at intermediate stages between infection and lysis. Anyhow, the experiments only take a few hours each, so the whole problem can not take long to solve."
>
> Perhaps you would like to see this childish young man after eight years, and ask him, just offhand, whether he has solved the riddle of life yet? This will embarrass him, as he has not got anywhere in solving the problem he set out to solve. But being quick to rationalize his failure, this is what he may answer, if he is pressed for an answer: "Well, I made a slight mistake. I could not do it in a few months. Perhaps it will take a few decades, and perhaps it will take the help of a few dozen other people. But listen to what I have found, perhaps you will be interested to join me."
>
> (Delbrück, 1946, 161–162)

This paper is a piece of vintage Delbrück. He told his audience in no uncertain terms that he was not interested in the medical applications of the work (*Ibid.*, 163) but that he and his collaborators wanted "to get to the bottom of what goes on when more virus particles are produced upon the introduction of one virus particle into a bacterial cell. All our work has circled around this problem" (*Ibid.*, 162). He could hardly have been more inviting and welcoming to those who might turn to work on phage when he concluded by saying that the field was wide open and full of promise. "A strong feeling of adventure is animating those who are working on bacterial viruses, a feeling that they have a small part in the great drive towards a fundamental problem in biology" (*Ibid.*, 187). Small wonder that he attracted co-workers and the Phage Group of enthusiastic devotees was formed.

The Phage Group

Much has been written about this group associated with Delbrück and Salvador Luria (Fleming, 1968; Stent, 1966, 1968; Mullins, 1972). What emerges from these accounts is the role of Delbrück in co-ordinating the work on phage. This impression may have been unduly emphasized since both Fleming and Mullins have based their accounts on the contributions to a book whose *raison d'être* was to honour Delbrück's sixtieth birthday, a book in which the introduction was written by one who was captivated by the Bohr/Delbrück paradox. Even so, no one who has read the lectures Delbrück gave at conferences in the 1940s can doubt that his influence in the phage work was a powerful one. All who attended the Phage Course at Cold Spring Harbor beginning in 1945 came under his influence. But what, one wonders was special to the Group? Delbrück's answer was that it "was open and co-operative in strict imitation of the Copenhagen spirit in physics" (Delbrück, 1972). This meant that the work of the Group's members was criticized, the phage worker was made to defend his results and the interpretations he put upon them. What was the use of having a group in which work was passively accepted? It was necessary to say, as Delbrück so often did, "I don't believe a word of it!"

Perhaps those early years among so many bright stars of German physics had helped to create in Delbrück this desire to mediate and had accustomed him to an atmosphere in which criticism was meted out ruthlessly, no holds barred. But one suspects that his bite in those days was not as bad as his bark, and that behind his demanding and ruthless exterior, then as now, there lay "the conviction that the game of science is pointless if you do not put aside personal pride" (Delbrück, 1972). With Delbrück around, things moved, papers got written—even if he had to encourage his students to remove themselves to a secluded spot for a few days until they had written up their work—ideas were discussed, hypotheses formulated and tested. At the same time Delbrück's strategy—to use physics and genetics rather than chemistry—tended to be accepted within the group at least in the early years. Later, when Putnam, Kozloff and Cohen had studied phosphorus transfer in phage infection (all of them very much in the Group)—Delbrück's attitude began to change and by 1950 Luria was for sending Watson to Europe to learn nucleic acid chemistry.

The Initial Paradigm

When Delbrück came to Caltech in 1937 he met Emory Ellis and collaborated with him in his work on phage. As a physicist, Delbrück was searching for a simple system in which to observe replication. Higher organisms were too complex, their heredity was bi-parental, but phages, like the plant viruses which Stanley had crystallized, appeared simply to replicate using the host as a nutrient medium: no recombination; no segregation; just unadulterated replication of nucleoprotein molecules; pure autocatalysis. Maybe it was as the California scientist A. P. Krueger believed: phage was a self-reproducing enzyme-like protein, in which case "the mechanism of phage production can be studied like any other mechanism of enzyme formation under conditions which set it apart from the complexities of cellular growth" (Krueger, 1937, 380).

Such was Delbrück's enthusiasm for autocatalysis that he put forward a scheme to account for it in terms of short-range chemical interactions between the amino acids in a polypeptide chain and those free in the surrounding environment, which would lead to the synthesis of a replica polypeptide. The occasion was the 1941 Cold Spring Harbor meeting; a vigorous debate followed. At the same time he concluded a review article on phage with the words:

> It is likely that its solution [the problem of autocatalytic synthesis] will turn out to be simple, and essentially the same for all viruses as well as genes. The bacterial viruses should serve well to find this solution, because their growth can be studied with ease quantitatively and under controlled conditions. The study of the bacterial viruses may thus prove the key to basic problems in biology.
>
> (Delbrück, 1942, 30)

Developing an Assay System

Delbrück's first task was to improve on the disastrous assay systems used by Northrop and Krueger. This effort led to the development of "one-step" growth curves for phage which buried Northrop's and Krueger's continuous growth curves. Delbrück went on to question Krueger's evidence for the existence of a virus precursor, like the precursors of active enzymes. In this phase of the work Delbrück did for phage what the German workers did for TMV (see Chapter 14): he unscrambled a complex situation.

How different the system turned out to be from what Delbrück had initially supposed! When Luria and he tried mixed infections they expected the two types to replicate independently of each other and at their own rates in the host cell. But there was evidence of interference. One type excluded the other, and even more strange, the "excluded" phage depressed the yield of the replicating phage which had "penetrated" to the cell's metabolic machinery (Delbrück, 1945; 1946). Was there a limiting key enzyme or a limited number of "receptors" which, once occupied by one phage, could not accommodate the other. Or did entry by the first phage cause a "fertilization" membrane to form around the bacterium barring entry to the second? Then came the discovery of a transfer of genetic material from one phage to another. Such a surprise was this, and so high the frequency, that Delbrück took it as he did Avery's pneumococcal transformation for a case of directed mutation. When Hershey showed it to be a case of recombination Delbrück found the news "exciting principally by the blow it deals to our fond hopes of analysing a simple situation" (1949, 14). Nor did Doermann's ingenious technique for liberating the contents of the bacterial cell prematurely bring an essential simplification to the problem, for replication and recombination apparently occurred in the "dark period" when no active particles could be detected. What could Delbrück salvage from these disappointments? By December 1949 he was no longer putting his trust in a combination of genetics and physics. Instead, he was looking to the control and variation of the chemical conditions under which phage replication took place. The phage multiplied by commandeering the assimilatory activity of the host cell and diverting the host's products to the synthesis of fresh phage particles. Now one could use the techniques of biochemistry—and they were much more advanced in 1949 than they had been when Delbrück first began to work on phage—to analyse this shunting of host products from host functions to phage reproduction.

Delbrück could now (1950) sense that his hunt for physical paradoxes in phage reproduction had been illusory. Orthodox chemical techniques were now needed to pursue the mystery of biological replication and to open up the "black box" of the phage-infected bacterial cell. Just as he had left astrophysics for atomic physics, and atomic physics for genetics, he now left genetics for sensory physiology. By this time, too, another physicist, who in 1944 had addressed himself to biology in a famous little book, no longer seemed interested in the problems he had earlier pondered so deeply, and this was none other than the famous author of the Schrödinger wave equation.

Schrödinger's Book *What is Life?*

Schrödinger's interest in the problem of biological reproduction seems to have stemmed from his concern to discover in nature the strict determinism which had been banished from physics. He did *not* accept Bohr's belief in a bar to the possibility of explaining genetics in terms of physics (Schrödinger, 1951, 64). His search for other laws must, therefore, be distinguished from Bohr's.

Probably through his friendship with Delbrück he had heard of the famous "Three-Man-Paper" and came to accept the quantum mechanical theory there described. This theory figured prominently in the lectures he gave in Dublin in 1943, which were published a year later in his very influential book *What is Life?* (referred to in future as *Life*).

On the question of communications between Delbrück and Schrödinger, Delbrück had the following to say:

> You [Olby] are in error if you came to the conclusion that Schrödinger and I did not overlap in Berlin. He was still very much there when I came to work with Lise Meitner in the fall of 1932 and I saw him a number of times and was quite friendly with him. However this was before my collaboration with Timoféeff. I do not think that Schrödinger met either Timoféeff or Zimmer. I am not certain how he came to take an interest in our paper. I presume that I sent him a reprint when the paper appeared in 1935, I believe, to wherever he was, either at Oxford or at Dublin. But of course his book was ten years later and I was not in contact with him during these subsequent years, so I do not know at what time he chanced to pick it up and read it.
>
> The Schrödingers must still have been in Berlin *at least* till February, 1933, because I remember a costume party in their apartment about that time, for which I borrowed the uniform of the porter at the Harnackhaus and acted the part of their butler at their party. When the news of Schrödinger's Nobel prize came in October, 1933 I wrote a letter of congratulations to Oxford, pretending to be his old butler Schulze, and received a reply to "My dear Schulze . . . remembering your many years of faithful service I am giving you an annuity of one hundred pounds (of potatoes)."
>
> (Delbrück, 1971b)

Of the several problems which Schrödinger tackled we need only consider those relating to the hereditary substance: How does it encode so much information preserved against the onslaught of thermal "noise", and how is it reproduced with such fidelity every time a cell divides?

Order-from-Order Laws

The problem facing the physicist who tried to account for the facts of genetics may be stated as follows: conformity to the laws of physics is far from perfect. Laws governing molecular events display a relative order of inaccuracy of the order of $1/\sqrt{n}$ where n is the number of molecules involved. Thus the departure from the predictions of the gas laws would be 10 per cent for $n = 100$ and 0.1 per cent for $n = 1\ 000\ 000$. But according to the target theory of radiation mutagenesis, the volume of the gene within which a "hit" had to take place in order to overcome the stability of the gene was equivalent to only one thousand atoms. How could so few atoms retain so constant a configuration as the facts of hereditary transmission indicated? Inheritance, it appeared, was not based on the typical laws of physics, which were statistical and approximate. These were what Schrödinger called "order-from-disorder" laws, so well demonstrated in the law of diffusion, where a uniform, average distribution of molecules was achieved by the disordered Brownian motion of very large numbers of individual molecules.

When Schrödinger considered the Habsburg lip which had been reproduced faithfully in this famous family over a period of three centuries, the problem seemed baffling, for throughout that period the gene concerned had been kept at 98°F, far above the absolute zero.

> How are we to understand that it has remained unperturbed by the disordering tendency of the heat motion for centuries?
>
> A physicist at the end of the last century would have been at a loss to answer this question, if he was prepared to draw only on those laws of Nature which he could explain and which he really understood. Perhaps, indeed, after a short reflection on the statistical situation he would have answered (correctly, as we shall see): These material structures can only be molecules. Of the existence, and sometimes very high stability, of these associations of atoms, chemistry had already acquired a widespread knowledge at the time. But the knowledge was purely empirical. The nature of a molecule was not understood—the strong mutual bond of the atoms which keeps a molecule in shape was a complete conundrum to everybody. Actually, the answer proves to be correct. But it is of limited value as long as the enigmatic biological stability is traced back only to an equally enigmatic chemical stability. The evidence that two features, similar in appearance, are based on the same principle, is always precarious as long as the principle itself is unknown.
>
> (*Life*, 47)

Schrödinger went on to show how the quantum-mechanical theory as applied to the chemical bond by Heitler and London in 1926–27 justified and accounted for the existence of stable aggregates of atoms called molecules to the satisfaction of the physicist. The atoms in a molecule are in an energy well, and to rearrange or extricate them requires the supply of energy, not just any supply of energy, but a sufficient and intense quantity to raise the energy state of the atoms *above a threshold value* to get them "over the hump" which separates one stable configuration from another "isomeric with it."

Unlike Bernal and Astbury, Schrödinger had not noted the fact that viruses and chromosomes are composed of giant nucleoprotein molecules—surely a strange omission when he is trying to drive home the macromolecular nature of the genes (*Life*, 56, 69). But internal evidence suggests that Schrödinger deliberately avoided using chemical evidence. What he called a macromolecule was to Schrödinger, the physicist, ultimately indistinguishable from other aggregates in the solid state. For him, therefore, the problem remained, of the atoms in a gene being held together in a fixed sequence despite thermal agitation. Instead of accepting the solution he had already given, Schrödinger went on what now appears as a wild goose chase to find another solution. Thus he sets out the following table:

Molecule—solid—crystal
Gas—liquid—amorphous

which schematizes the view that all stable arrangements of atoms are confined to the crystalline state, and what makes for stability in solids is the fact of crystallinity. All true solids are crystalline; hence all molecules with stable configurations of their atoms must be in the form of crystalline solids, for in liquids and gases there is no crystallinity and the individual molecules are at the mercy of thermal agitation. For genes to have a permanent arrangement, therefore, they must be solids, that is, crystals. To the question, "Why do we wish a molecule to be regarded as a solid —a crystal?" he said:

> The reason for this is that the atoms forming a molecule, whether there be few or many of them, are united by forces of exactly the same nature as the numerous atoms which build up a true solid, a crystal. The molecule presents the same solidity of structure as a crystal. Remember that it is precisely this solidity on which we draw to account for the permanence of the gene!
>
> The distinction that is really important in the structure of matter is whether atoms are bound together by those "solidifying" Heitler–London forces or whether they are not. In a solid and in a molecule they all are. In a gas of single atoms (as e.g. mercury vapour) they are not. In a gas composed of molecules, only the atoms within every molecule are linked in this way.
>
> (*Life*, 60)

Now of course there is a sense in which all interatomic forces of affinity can be viewed as the same—they can all be described in terms of the Schrödinger wave equation. This was clearly Schrödinger's approach, for he denied any fundamental distinction between macromolecules, the aggregate of atoms in a coin, and a batch of crystals in a copper wire (*Life*, 58, 61). Here Schrödinger was doing what the opponents of the macromolecule had done earlier, refusing to recognize a fundamental distinction between a crystalline aggregate and a macromolecule bound only by covalent bonds. But Schrödinger's reason differed from that of the champions of the aggregate theory. Whereas their reason had to do with the properties of colloidal aggregates, Schrödinger's had to do with the unique character of Heitler–London forces.

From the theme running through *What is Life?* it is clear that the new physical laws would be what Schrödinger called order-from-order laws, in contrast to known order-from-disorder laws, and he termed the former "dynamical" and the latter "statistical", following Max Planck, who in his paper, "Dynamische und statistische Gesetzmässigkeiten" of 1914 (Planck, 1958, iii, 77–90) identified dynamical laws with the microscopic world of individual molecules, and statistical laws with the macroscopic world of large numbers of molecules. In physics, as in biology, there existed a problem of reducing one set of laws to another, the macroscopic to the microscopic. Opinion differed as to whether this would or could ever be achieved. Whereas Maxwell ruled out the possibility of reducing statistical laws to the effects of Newtonian mechanics operating on individual molecules (Maxwell, 1890, ii, 374; Heimann, 1970), Planck saw in such an attempt "one of the chief tasks of progressive science" at which, for example, V. Bjerknes was working in his attempt "to trace all meteorological statistics back to their simple elements, that is, to physical regularities" (Planck, 1958, iii, 87). From this paper of 1914 it is clear that the Planck of those days hoped to bridge the gap between the behaviour of the macroscopic and microscopic worlds by reducing statistical laws to dynamical laws. Planck, too, long treasured the classical conception of the role of probability theory in physics as a provisional device to be disposed of when the still deeply hidden causal factors are revealed (Planck, 1958, ii, 259, 289). Boltzmann had the same desire and by his writings and followers he influenced Schrödinger (Boltzmann, 1896). Maxwell's legacy—the problem of treating molecular collisions dynamically—was for Planck, Boltzmann, and Schrödinger a challenge and a sore in the body of physical theory.

The contrast between the microscopic and macroscopic worlds was sharpened by the discovery of the discontinuity implied by the quantum of action (h). How was the continuity of energy assumed by classical physical theory to accommodate it? From 1900 to 1911 the quantum of action posed an increasing threat to the analogy between the continuity of energy at the macro-level and the behaviour of energy at the micro-level. In 1911 Planck attempted first a classical interpretation of h (Planck, 1958, ii, 249–259), while at the Solvay Conference eight months later he argued that a physical interpretation of h could only be achieved by developing the quantum hypothesis, and that the laws of the theory of quanta must apply to all particles united by the molecular bond (Planck, 1958, ii, 285–286).

As if this were not enough, a further consequence of the quantum theory was pointed out by Werner Heisenberg when he enunciated the Uncertainty Principle which denies the determination of position and velocity of an electron, allowing only a probability estimate of the one parameter and a determination of the other (Heisenberg, 1927). Speaking of this in 1933, Schrödinger said:

> Will we have to be permanently satisfied with this? On principle yes. On principle there is nothing new in the postulate that in the end exact science should aim at nothing more than the description of what can really be observed. The question is only whether from now on we shall have to refrain from tying description to a clear hypothesis about the real nature of the world.
>
> (Schrödinger, 1933, 316)

This I believe to be the background to his book *What is Life?* In the living cell, where order is based on order, he hoped physical laws of a deterministic kind would be found. The organism is a macroscopic system which behaves in some of its aspects very like matter close to the absolute zero, where "molecular disorder is removed" (*Life*, 70). How does the hereditary substance achieve this? By being built like a clock of solids, which are kept in shape by Heitler–London forces, strong enough to elude the disorderly tendency of heat motion at ordinary temperatures (*Life*, 85). This is where the purpose of Schrödinger's crystalline-solid analogy comes in. A gene and a clock are similar in that they are held together by Heitler–London forces. Strictly speaking, a clock behaves statistically, but for practical purposes we may say it behaves dynamically, for in the solid state, matter at room temperature is equivalent to matter near the absolute zero where dynamical law reigns.

Now why did Schrödinger go to all this trouble to unfold before his audience Delbrück's quantum mechanical model of the gene? As Schrödinger himself remarked: "Was it absolutely essential for the biological question to dig up the deepest roots and found the picture on quantum mechanics? The conjecture that a gene is a molecule is today, I dare say, a commonplace. Few biologists, whether familiar with quantum theory, or not, would disagree with it" (*Life*, 57). Schrödinger's answer was, that quantum mechanics is the first theoretical approach in physics which accounts from first principles for all sorts of aggregates that one is likely to encounter in nature. "The Heitler–London bondage is a unique, singular feature of the theory, not invented for the purpose of explaining the chemical bond. It comes in quite by itself, in a highly interesting and puzzling manner, being forced upon us by entirely different considerations" (*Life*, 57). This theoretical approach was unique; it accounted for the facts of chemistry; so beautiful and so unique was this theory that the likelihood of another theory emerging to account for the facts of chemistry was most unlikely.

> Consequently, we may safely assert there is no alternative to the molecular explanation of the hereditary substance. The physical aspect leaves no other possibility to account for its permanence. If the Delbrück picture should fail, we would have to give up further attempts.
>
> (*Life*, 57)

To this great physicist, who had himself been so personally involved in the quantum revolution, is it so surprising that he should have been magnetized by the wonderful correspondence between genetic facts and the principles of quantum mechanics. He could demonstrate that even if genes were not chemical molecules the physicist could still allow them to have stability and mutability, because of quantum mechanics. Now a physicist reading this book could get excited about genetics. Schrödinger made the facts of genetics meaningful to the physicist. He did *not* offer his readers the bait of a fresh mutually exclusive complementarity relationship, as did Bohr. He offered them "other laws" to be sure, but not of the kind Bohr envisaged. They would be related to known physical laws, just as the laws of electrodynamics were related to the more general laws of physics. The cell was like the dynamo—"The difference in construction is enough to prepare him [the physicist] for an entirely different way of functioning" (*Life*, 77).

The Hereditary Codescript

I have dealt with Schrödinger's discussion of order-from-order laws first because I believe the quest for such laws was his motivation for writing the book. We come now to what we can see in retrospect as the most positive and influential aspect of this little book: the concept of an hereditary codescript. Schrödinger recognized that the replication of the gene was not just the "dull device of repetition" that one found in the growth of a crystal in three dimensions. On the contrary, the replicated structure was "aperiodic" as in complicated organic molecules, where every atom and every group of atoms plays a special role. "We might quite properly", he said, "call that an aperiodic crystal or solid and express our hypothesis by saying: we believe a gene—or perhaps the whole chromosome fibre—to be an aperiodic solid" (*Life*, 61). Schrödinger went on to calculate the amount of information that could be encoded in a small quantity of such aperiodic material as the chromosomes represented. His aim, he said, was to illustrate simply that "with the molecular picture of the gene it is no longer inconceivable that the miniature code should precisely correspond with a highly complicated and specified plan of development and should somehow contain the means to put it into operation" (*Life*, 62). The reader will recall that Miescher, Fischer, Kossel, Leathes and Wrinch all drew analogies between alphabetical codes and chemical codes in the form of isomeric arrangements of the sub-units in a large molecule, and we also saw that Leathes' discussion of this analogy was widely quoted. But it required a physicist, and an eminent one at that, to make physicists aware of the concept, and perhaps most important of all, to relate the codescript to crystallography in the conception of an aperiodic crystal.

Schrödinger's Influence

If the above analysis of *What is Life?* appears to be critical, the aim behind it was not. I have sought to show that Schrödinger hoped the organism would somehow supply those "ideal" conditions in which order reigned and the process of hereditary transmission obeyed "order-from-order" laws. This in no way denies to Schrödinger his important positive role in introducing physicists to the concept of the hereditary codescript and the aperiodic crystal.

Those who read *What is Life?* found in it what they were looking for. Few were interested in what Schrödinger had to say about the "order-from-order" laws, but the following were influenced by the "aperiodic crystal".

Luria: I remember finding Schrödinger's book exciting for the formulation of "aperiodic crystal", not for the speculations of new laws of physics.

(in Stent, 1968, 395)

Crick: On those who came into the subject just after the 1939–1945 war, Schrödinger's little book, *What is Life?* seems to have been peculiarly influential. Its main point—that biology needs the stability of chemical bonds and that only quantum mechanics can explain this—was one that only a physicist would feel it necessary

> to make, but the book was extremely well written and conveyed in an exciting way the idea that, in biology, molecular explanations would not only be extremely important but also that they were just around the corner. This had been said before, but Schrödinger's book was very timely and attracted people who might otherwise not have entered biology at all.
>
> (Crick, 1965, 184)

> I cannot recall any occasion when Jim Watson and I discussed the limitations of Schrödinger's book. I think the main reason for this is that we were strongly influenced by Linus Pauling, who had essentially the correct set of ideas. We therefore never wasted any time discussing whether we should think in the way Schrödinger did or the way Pauling did. It seemed quite obvious to us that we should follow Pauling.
>
> (Crick, 1970)

> Wilkins: Schrödinger's book had a very positive effect on me and got me, for the first time, interested in biological problems. I think it had the same effect on other physicists. I think one reason for this is that Schrödinger wrote as a physicist. If he had written as an informed macromolecular chemist it probably would not have had the same effect. The aperiodic crystal idea, although not so near the truth as the macromolecular idea, was something which appealed to physicists.
>
> (Wilkins, 1970)

Apparently both Seymour Benzer and J. D. Watson were also attracted to the molecular basis of genetics by reading *What is Life?* (Stent, 1968, 395). The enthusiasm of these six scientists for Schrödinger's "aperiodic crystal" is in marked contrast to their feelings for Bohr's principle of "complementarity", of which Luria was suspicious; Hershey described it as "double-talk", whilst in the intellectual development of Watson and Crick it played no part whatever. Only Delbrück, and after him Benzer and Stent, were also attracted by Bohr's idea. Having noted this contrast let us remember that the fundamental work upon which both Delbrück and Schrödinger built was the "Three-Man-Work". Perhaps molecular biology owes more to the geneticist who began that work—Timoféeff-Ressovsky—than has so far been admitted.

CHAPTER SIXTEEN

Physicists and Chemists in Biology: The Structural School

One day in Cambridge, in the year 1927, Arthur Hutchinson, Master of Pembroke and professor of mineralogy in the University, chaired a meeting called to appoint a lecturer in structural crystallography. This post was a revival of the lectureship which Hutchinson had filled for the five years before he became professor. Among those interviewed were W. A. Wooster, the new demonstrator in mineralogy, and Astbury. When the latter was asked his opinion on collaboration he replied very rudely: "I am not prepared to be anybody's lackey!" (Bernal, 1963, 27). A third candidate was the late J. D. Bernal. When he came into the room the committee saw before them a man with a "shock of fairish hair gone wild and humorous hazel eyes" (Snow, 1966, 19); he appeared shy and sullen as he sat down, and his head sank onto his chest. The committee plied him with this question and that, but to no avail. All he would say was "yes" and "no". Finally, the chairman, by this time in despair, asked him what he would do with the crystallographic laboratory were he appointed. At this, wrote Snow,

> Bernal threw his head back, hair streaming like an oriflamme, began with the word No (as he usually began his best speeches), and gave an address, eloquent, passionate, masterly, prophetic, which lasted forty-five minutes. "There was nothing for it but to elect him," said Hutchinson, who, incidentally, correct and courteous himself, had a touching respect for talent, and who had been responsible for getting the lectureship founded.
>
> (Snow, 1966, 24)

This appointment was undoubtedly the main reason why molecular biology developed at Cambridge despite the forces within the University which opposed it.

Hutchinson's Role

Men like Astbury, Wooster and Bernal were taught crystal physics as undergraduates by Hutchinson. The latter had served under the ageing W. J.

Lewis who had held the chair of mineralogy until the day of his death in 1926 at the ripe age of seventy-nine, thus establishing a record of forty-five years of tenure. Hutchinson had done most of the work and possibly all the teaching, serving as demonstrator for twenty-eight years and lecturer for five, on a salary of £100 p.a.

Since Lewis' death the statutes relating to the tenure of university chairs had been very rightly changed, but the alteration meant only five years in the chair for Hutchinson. In those years he continued the crystallographic tradition established by Miller (of the Miller indices) and pushed it in the direction of physics and chemistry rather than in that of geology and petrology. First, he appointed Wooster to the vacant demonstratorship in mineralogy, and asked him to take over the teaching of crystal physics and to introduce X-ray crystallography into the course. This consisted of twenty-four lectures and included demonstrations on the properties of crystals: refractive index, extinction angles, optical rotation, optical axial angles, thermal conductivity, thermo-electric and pyro-electric properties. Hutchinson claimed this course as "the most complete of its kind in existence" (Hutchinson, 1929, 392). He had a right to be proud for since 1927 it had included X-ray crystallography, probably the only course for undergraduates in this country that did so. Wooster had not been long appointed before Hutchinson succeeded in reviving the lapsed lectureship in crystallography, renamed it structural crystallography and appointed Bernal. He was responsible for the maintenance of the X-ray equipment, instructing advanced students in its use and how to interpret their results. He also gave an elementary and an advanced course.

These two newcomers had scarcely found their feet before a syndicate was appointed by Council to Senate to report on the position of mineralogy in the studies of the university. Hutchinson later complained that he only learnt of the constitution of this body "from the casual conversation of friends in London" (1930, 647). But he was invited to join several of the five meetings of the Syndicate. His excellent memorandum on the history and present state of the department of mineralogy was received and published in the *Cambridge University Reporter* (10.XII, 1929). The report resulting from these meetings was a compromise. It advised the establishment of two chairs, one for crystallography, and one for mineralogy and petrology; each was to be associated with a new department; mineralogy and petrology should have a new building and crystallography should have the existing buildings.

The Report on Mineralogy Discussed

When the report came up for discussion in 1930 Sir William Bragg, Hutchinson, Wooster and Bernal argued for the importance of developing the "new crystallography", as they called their subject. Their words fell on barren soil. To the men of geology the X-ray technique had yielded but one valuable adjunct to their science—the X-ray goniometer. X-ray determinations took too long; the superficial study of crystals by the optical methods sufficed in the majority of cases for the geologists. One never knew where fresh aspects of science would develop. X-ray crystallography had been with us for some time. It was unwise to associate future developments in crystallography so narrowly with the X-ray method. Bernal accused many of the participants in the debate of not appreciating the "novelty" of the methods of X-ray crystallography.

> The extent and the importance of the structure of crystals was much greater than could ever be the case with superficial Crystallography. Crystallography could not only be studied in connexion with Mineralogy and Geology on the one side, and Chemistry on the other; but it effected very strongly Physics and Technology . . . even the study of liquids was now really an essential part of common Crystallography.
>
> (Bernal, 1930a, 653)

Crystallography, he declared, had moved away from the geological point of view. Now it was becoming a "branch, or sub-division, of chemistry, and like it, followed directly from fundamental physics."

> And the new methods were synthetic: there were now means of saying *a priori* that such and such an arrangement of atoms was possible and would produce a crystal of such and such properties. That was very different from the old descriptive side of mineralogy. It required a complete remodelling of the teaching, and, if that teaching was to be remodelled in time for the demand for trained workers in a field which was getting wider and wider, the sooner it was begun the better.
>
> (Bernal, 1930a, 654)

Just as the petrologists and geologists feared, crystallography at Cambridge was moving away from them. They had every reason to act in time and rescue the chair of mineralogy so as to continue the promising development of petrology and to follow the example of their cousins in the United States and build up a research school along the lines of the Carnegie Institution, whose Geophysical laboratory they greatly envied. There the laboratory study of artificial synthesis and transformation of minerals was yielding quite new knowledge which the young lecturer in petrology, C. E. Tilley, greatly prized. Just as there was a new crystallography there was also a new mineralogy. It was experimental and genetic. It concerned the "mode of formation, stability relations, transformations, and genesis of minerals" (Tilley, 1930, 648). The era of description and classification was drawing to a close. Now the principles of physical chemistry could be applied to this mass of data.

At this debate there was the familiar pattern of Old Boys who came over to Cambridge, congratulating her on her past excellence in geology and calling to her for inspiration and leadership in the new development of the subject. Industry, the mines, oil exploration were all calling for young men,

Cambridge must provide them, men soundly trained in mineralogy and petrology.

The outcome of the Report was that the General Board simply stated the inability of the University to find the money for fresh chairs, and in 1930 a grace was passed separating the teaching of crystallography from that of mineralogy and petrology but not associating this division with any suggestions or assumptions about separate departmental structures to follow. Council to Senate issued a report the following year which was debated very vigorously in Senate. It proposed one chair of mineralogy and petrology with crystallography to be associated with that department. It also recommended the use of the Plummer Bequest to finance the chair, a bequest which was made for "chemistry, biochemistry, physical science, or such other allied subjects as John Humphrey Plummer trustees should think fit." Council thus proposed to cut down the University's commitments to the subject by utilizing this very handy bequest that had just been received. The outcome was the appointment of Tilley, lecturer in petrology, to the new chair with no change for Bernal, Wooster, or crystallography save the repair of the roof to their hut, which had leaked at every conceivable point, and the promise of a little more space when the mineralogical collection was moved to the new building for mineralogy and petrology. Meanwhile the summer heat in the hut drove Wooster and Bernal out. The winter cold, which froze the gas meter thus turning off the gas fires and freezing the solvent benzene, continued. The petrologists had won the day.

An Institute of Physico-Chemical Morphology

Further attempts to persuade the University to recognize the importance of X-ray crystallography failed, as when Joseph Needham urged that the Jacksonian Chair of Experimental Philosophy (established 1783) be "occupied by a crystal physicist . . . one deeply interested in the biological implications of his subject" (Needham, 1935). "Perhaps", wrote Needham,

> the biological importance of crystal physics is not yet as widely appreciated as would be desirable. In the first place, this study provides one of the few methods at our disposal for attacking a certain range of size. The gap between the smallest morphological structures is the key position in biology at the present time. Until fundamental progress has been made within this sphere, the organic chemistry of the materials of life will remain arbitrary and morphological generalizations will possess that air of ultimate irreducibility which has always hitherto appertained to organic forms. This has best been exemplified by the progress which has been made during the past few years in the X-ray analysis of semi-crystalline fibres. It may be truly said that vast realms of biological phenomena are, in effect, available in the study of fibre properties. It is becoming clearer and clearer that *the* one of the bases of morphological structure is the fibre, whether intracellular, extracellular, or incorporated in the cell wall. Such workers as Meyer and Mark have explored the phenomena of molecular contractility and made it possible to identify the linearity necessitated by the biological data (such as the linearity of the chromosome, the muscle fibre, or the mycelles of embryonic connective tissue) with the linearity or the contractile but "crystalline" protein molecules.

Next there is the profound importance of the paracrystalline state for biology. The "liquid crystal" is not merely a model for what goes on in the living cell; it is in point of fact a state of organization actually found in the living cell . . .

Finally, the value of the X-ray analysis of crystals in helping to settle the structural formulae of biologically important compounds, which can only be obtained in very small quantities or with regard to which purely chemical evidence gives equivocal results, need hardly be emphasized. Reference may be made to the impressive development of the biology of the sterol group which has followed upon that decision regarding the structure of the cholane ring in which X-ray analysis played so important a part.

(Needham, 1935)

At the same time Needham, advised by Dorothy Wrinch and J. D. Bernal, was exploring another possibility for establishing a research institute in Cambridge which would bridge physics, chemistry and biology. This idea was prompted by Tisdale's visit to Cambridge in 1934 on behalf of the Rockefeller Foundation. Warren Weaver, shortly after his appointment in 1932 as director for the Foundation's funding of the natural sciences had

urged the Trustees, with the full backing of the then president of the Rockefeller Foundation, Max Mason, that the science program of the foundation be shifted from its previous preoccupation with the physical sciences, to an interest in stimulating and aiding the application, to basic biological problems, of the techniques, experimental procedures, and methods of analysis so effectively developed in the physical sciences.

This proposal was accepted and approved by the Rockefeller Foundation Trustees, and progress in the program was sufficiently prompt and promising so that when I drafted the "natural science" section of the Annual Report of the Rockefeller Foundation for 1938 this section began with a sixteen-page portion, pages 203–219, which was headed in large type, MOLECULAR BIOLOGY, the first sentence being "Among the studies to which the Foundation is giving support is a series in a relatively new field, which may be called molecular biology, in which delicate modern techniques are being used to investigate ever more minute details of certain life processes."

(Weaver, 1970, 582)

After Tisdale's visit to Needham in 1934 there followed a further visit from both Weaver and Tisdale in May 1935. But the two Americans were not persuaded into supporting the setting up of an entire institute of experimental embryology as Needham had suggested. Nor were they prepared to support this scheme when Needham had modified it on the advice of Wrinch and Keilin and renamed it "Institute of Physico–chemical Morphology" (Wrinch's term). Then it was to implement "a vigorous attack by X-ray technique on morphological problems"; Bernal was suggested as the director of the Institute's crystal physics division, Crowfoot (Dorothy Hodgkin) as his research assistant. Weaver and Tisdale would go as far as funding the latter post but wisely avoided any commitment to the larger programme. At the same time they gave £10 000 to Astbury and a large sum to Linus Pauling.

In 1936 Needham made one further effort to gain the Foundation's support. The Rockefeller replied that if the University would support his scheme then the Rockefeller would too. The University did not. The bridge-building research school envisaged by Needham embodied many of the aspirations of the informal circle who met sometimes in Cambridge, and sometime in London, or at Woodger's house on Epsom Downs. Bernal was not a frequent participant, but he did contribute a memorandum to Needham on the importance of crystal physics in biology. This was written after he had made the wonderful discovery that pepsin crystals, bathed in their mother liquor, gave single crystal X-ray diffraction patterns. He wrote:

> Until 1934 our only knowledge of the molecular structure of the proteins apart from that of their chemical constituents was derived from the X-ray studies of Meyer, Herzog and Astbury on protein fibres such as silk, collagen and hair. All of these showed a polymerized chain structure similar to cellulose but presumably of a polypeptide character. Attempts had also been made to investigate crystalline proteins by X-ray methods but always without success. This we now see was due to the fact that the earlier investigators used crystal powders, a useless method with such large [unit] cells, and also worked with dry material in which the crystal structure of the protein had collapsed.
>
> In the spring of 1934 I was fortunate to obtain crystals of crystallized pepsin prepared by Philpot, by Northrop's method, and, foreseeing the above difficulties, I examined a single crystal bathed in its mother liquor. A good diffraction pattern was obtained and the main outlines of the crystal structure were made out. The unit cell is relatively enormous—461 × 67 × 67 Å . . . of molecular weight 36 000, independently confirming Svedberg's values . . . Later, insulin was examined by Miss Crowfoot at Oxford and was found to possess a similar structure with the same spheroidal shape of molecule but more closely packed, possibly owing to the binding effect of the essential metal atoms, zinc and cadmium.
>
> (Bernal, 1935a)

On the strength of this achievement Bernal proposed two lines of approach (1) "The systematic examination of a series of typical proteins: albumens, globulins, conjugated proteins etc., to discover the main lines of their architecture . . ." (*Ibid.*). (2) An attack on "the internal structure of the protein molecule based on intensity analysis. This in turn depends on (1) because it is first necessary to know from preliminary examinations which protein is most likely to yield results" (*Ibid.*). He went on to point out the difficulties: low angle diffraction spots, long exposure times—up to 200 hours or more for one photograph—but more serious was the problem of obtaining good crystals.

> In spite of a promising start, protein analysis in Cambridge is indefinitely held up, not for lack of X-ray apparatus and technique, but for suitable preparations of crystals, which can no longer be made here now that Miss Crowfoot (who was a trained chemist) and Dr Philpot are in Oxford . . .
>
> If the Rockefeller Foundation could see its way to providing this assistance they would be coming to the help of research in what I am convinced is a most promising junction between biology and physics at a critical stage in its development.
>
> (Bernal, 1935a)

Bernal's Goal

Bernal's interests were wide-ranging to say the least, but overriding them all was his vision of a biological science based on known molecular structures.

This vision is referred to in his address: "The Place of X-ray Crystallography in the Development of Modern Science" (Bernal, 1930b), in the opening remarks of his pioneer study of the crystallography of amino acids and peptides (1931), in the evidence he submitted to the University in 1928 and in his contribution to the debate of 1930 on the place of mineralogy in the university.

In contrast to Bohr he saw biophysical methods as especially appropriate to reveal the subtle details of living organisms. With such refined techniques it was "easier to approximate to the detailed study of the mechanism of an intact animal or plant" (Bernal, 1939a, 339). According to C. P. Snow, even before Bernal was appointed lecturer in structural crystallography at Cambridge, the mission of carrying physics and chemistry into biology had become the "cardinal theme of his scientific life."

> He had equipped himself, through crystallography, with a powerful technique for probing into the structure of materials: now he began to use that technique on materials of biological significance, in the first place amino acids, sterols, and vitamins. Then he went on to water, since most organisms are made of it . . . Then to proteins and viruses.
>
> (Snow, 1966, 25)

It was Bernal who popularized Ewald's conception of the reciprocal lattice and who drew attention to its heuristic value; it was Bernal who in the same paper gave a systematic account of rotation methods and printed the Bernal charts for plotting the layer lines (1926). Among his subsequent achievements may be mentioned the following:

1931 unit cells of 8 amino acids and of 2 oligopeptides
1932 carbon skeleton of the sterols
1933 structure of water
1934 first protein single crystal pictures and introduction of technique of bathing crystals in the mother liquor
1937 diffraction patterns of crystalline viruses

How did Bernal become so interested in compounds of biological importance?

Bragg's Influence

It is well known that the Braggs had come to a sensible arrangement whereby father worked on organic compounds and his son on inorganic compounds. At University College, Sir William began his famous study of naphthalene and its derivatives, and Müller and Shearer began their remarkable study of long chain hydrocarbons. When Bragg moved to the Royal Institution in 1923 he took Müller, Shearer, Lonsdale (*née* Yardley) and Astbury with him. Bernal, whose undergraduate prize essay had been brought to Bragg's attention, joined him in 1922. Not surprisingly, therefore, Bernal formed the ambition shared by almost all those in the Royal Institution to tackle organic compounds of biological and economic importance.

The Braggs had opened the way with the structure of diamond (1913), Bernal with the structure of graphite (1924), but optimism really stemmed from Sir William's successful though partially erroneous, attempt upon naphthalene and anthracene (1921). That year he addressed the Physical Society on "The Structure of Organic Crystals", a subject which had been neglected, he claimed, because of the complexity of such crystal structures. "Yet if a way could be found of making determinations of structure, in spite of the complexity, it seems likely that they would quickly be fruitful" (Bragg, 1921, 33). Let us suppose, he said, that the benzene ring or naphthalene double ring has a "definite size and form, preserved with little or perhaps no alteration from crystal to crystal . . ." (*Ibid*). And he went on to show that the diffraction pattern was in accord with the regular packing of such rings. Seven years later, Kathleen Lonsdale working in Leeds, gave definitive evidence for the existence of the benzene ring in tetramethylbenzene, an achievement made possible by the fact that all the atoms lay in the same crystallographic plane.

During the 1920s the significance of the carbon–carbon distances in diamond (1.54 Å) and in graphite (1.42 Å) were gradually perceived as prototypes for aliphatic and aromatic structures. Bragg, through his popular lectures at the Royal Institution, became increasingly enthusiastic for the application of his work to industry and biology (see Chapter 5). In X-ray crystallography he saw a powerful new tool. It cried out to be applied to metallurgy, textiles and organic chemistry.

How could anyone *not* be influenced in so lively and happy an environment under the genial, fatherly and unsnobbish Sir William? Those years, recalled Bernal, were "the most exciting and the most formative of my scientific life" (1962, 522). He regarded himself as exceptionally lucky to have been there in the 1920s when "a new field, the arrangement of atoms in crystals, was just being worked out. No one will ever have precisely that kind of excitement again now that the main types of crystal structures are known" (*Ibid.*).

Bernal's Approach

In contrast to Astbury, Bernal was only too aware of the difficulties of interpretation when dealing with complex molecules. His decision to look at amino acids was a wise one, later followed by Pauling and Corey. Unfortunately the crystals made for him by A. Leese in the biochemistry department were too small and imperfect and with so many parameters involved Bernal did not attempt to apply Fourier methods.

When Bernal's friend Glen Millikan brought back from Sweden the beautiful crystals of pepsin which John Philpot had made whilst in Uppsala, Bernal took the first ever X-ray pictures of a crystalline protein. The blurred spots on the film extended all over it to spacings of about 2 Å. "That night,

Bernal, full of excitement, wandered about the streets of Cambridge, thinking of the future and of how much it might be possible to know about the structure of proteins if the photographs he had just taken could be interpreted in every detail" (Hodgkin and Riley, 1968, 15). From that time on Bernal's goal was decided—the structure of proteins.

Bernal kept the interpretation to a minor key, as was to be expected. His plan was exploratory. What he was after was the basic structural features. Such a study of globular proteins gave him evidence of roughly spheroidal molecules. TMV gave him evidence of huge rod-shaped molecules with a sub-unit construction. But whereas Astbury's work suffered from too little diffraction data, Bernal's and Dorothy Hodgkin's suffered from too much! "The diffraction patterns contained thousands of reflexions that defied analysis by the techniques of the time" (Kendrew, 1970, 7). In retrospect Hodgkin felt they could have achieved more at that time, but they had imagined it would be more difficult to interpret the data than in fact it turned out to be (personal communication).

Bernal put his hopes in Patterson analysis. This gave the linear separation of the chief scattering atoms in the form of peaks. One could then test a variety of conceivable structures against this distribution. The choice could be further narrowed down by physico–chemical considerations from as wide as possible a field. But even then one had only very hypothetical structures. The proteins called for a concerted effort, a pooling of resources and "some form of central bureau for protein research which would facilitate exchange of information and material in this field, and assist in an ordered attack on the whole problem" (Bernal, 1939a, 38). Here, if ever, was a case where the individualistic, unorganized character of British research, so disliked by Bernal, should be replaced by the *planned* research, as found in modern socialist countries.

Bernal hoped in the future that it would be possible to apply Fourier methods, and in the case of chymotrypsin and haemoglobin he thought the existence of clear differences, both in spacing and in intensity, between wet and dry crystal diffraction patterns, would allow inter- and intramolecular features to be distinguished. In this way the diffraction effects due to the intermolecular pattern could be put on one side and a direct Fourier analysis of the intramolecular pattern achieved (Bernal, Fankuchen and Perutz, 1938, 524). Perutz made a very lengthy study of these shrinkage and swelling effects; they served to reveal a rough picture of the haemoglobin molecule, but they did not allow a direct Fourier analysis to be made.

We can form an impression of Bernal's approach in a letter which he wrote to Fankuchen, then in Manchester. After advising Fankuchen what rotation photographs to take in order to get a reliable indexing of the layer lines he went on:

> As to lattice dimensions, for the most part I do not bother to determine them more accurately than 1 per cent, and for that the Chart is good enough. When it is necessary, for molecular weight determination, to determine them more accurately, I use the distance between the highest order spots, that can be got on the equator, taking two exposures on either side, so as to get rid of centre error. With sterols, unfortunately, except for halogen derivatives and a few poly-hydroxyl compounds, the lattice is not good enough to give K doublets and permit the use of methods of high accuracy.
>
> (Bernal, 1935b)

Bernal's School

Although Bernal's Cambridge days must have been exciting they must also have been frustrating. First came the triumph of petrology at the expense of the "new crystallography" which remained in the hut until Rutherford graciously accepted it into the Cavendish laboratory in 1935. Then there was Rutherford's dislike of the subject being split off from physics, and his personal dislike of Bernal's politics (Wooster personal communication). There was a tinge of scorn in his banter with the X-ray crystallographers, as when he would put his face round the door and ask how the stamp collecting was going on! (Mrs. Fankuchen, 1970). To him those pretty diffraction pictures were just so many patterns. In 1935 Bernal was made Assistant Director of Research, in charge of the crystallographic laboratory, but no college fellowship came his way despite the efforts of his friends. C. P. Snow recalled his defeat at Christ's when an elderly don said: "No one with hair like that can be *sound*" (Snow, 1966, 24). Then Dorothy Crowfoot, Bernal's research assistant from 1932 to 1935 returned to Oxford, thus bringing to an end a most fruitful period of collaboration. Without her Bernal had no one experienced in the preparation of protein crystals.

In 1937 Rutherford suddenly died. He was succeeded by Sir Lawrence Bragg from Manchester. Blackett left Birkbeck College, London, to take the Manchester chair, and Bernal succeeded him. Now why did Bernal uproot himself and forsake the great Cavendish for the lowly kudos of a college devoted to "night school" education? It cannot have been the temptation of a higher salary—he never worried about money. Yet it was a natural move for him to make. Birkbeck stood for the sort of educational ideals which he prized. Like his friend Blackett, Bernal was politically far to the left. Continuity in the leftist tradition within the physics department was thus achieved.

About a year after Bernal's move to London, Fankuchen was able to join him. As Crowfoot and Riley were by this time in Oxford this left Perutz on his own in Cambridge, soon to be interned as an enemy alien and in 1940 shipped off to Canada. Of all those who worked with or under Bernal at that time Fankuchen was the one whose collaboration he most prized. Fankuchen came from Manchester to work with Bernal in 1936. He began work on the sterols, but quickly moved to TMV when Bawden and Pirie prepared it in the biochemistry department in 1937.

This X-ray diffraction study of TMV gives us some clues on the aims of both Bernal and Fankuchen in applying physics to biology. Fankuchen was definitely *not* interested in biological compounds for biological reasons. It was, recalled Bernal, "the physical interest rather than the chemical or biological that inspired his work . . . in X-ray investigations he was able to apply the physics in more and more refined ways to problems of interest in industry and biology" (Bernal, 1964, 917). The classic example was his study of intermolecular spacings in TMV gels under various conditions, using his very original X-ray monochromator. This was not a study of TMV so much as a study of the long-range forces which operate in gels.

In this approach Bernal seems to have shared, partly perhaps because the interpretation of intramolecular patterns proved so difficult, and partly because the gel structure revealed by the study of narrow angle data was so novel and interesting. One should remember that this work had been carried out before the advent of the electron microscope so the shape and packing of the rod-like particles had to be arrived at from the diffraction pattern. It was this possibility of bridging the gap between molecular structure and the cytological structure revealed to the optical microscopist which so intrigued Bernal. His was truly the *structural* approach in the field known as ultrastructure *but not the ultrastructure of solid elements but of the liquid crystalline state.* We might call it the biophysics of the nucleus. Bernal did not concern himself with the cellulose wall or the wool fibre but with the forces operating in liquids which might account for the movements of the chromosomes and the formation of the nuclear spindle (see Bernal 1940d and his Guthrie Lectures, 1951, 58–59). This was where his *dynamic ultrastructure* made contact with genetics.

Today it is not easy to appreciate just how striking was the discovery that TMV rods were some 2000 Å in length, and yet they were rigid structures displaying a remarkable degree of order, so much so that when these TMV rods were oriented in solution in a capillary tube and given a 400 hour exposure to X-rays, the resulting diffraction pattern was so like that of a single crystal that Bernal and Fankuchen went on trying to index the diagram and derive the unit cell when the data contained too many ambiguities. And yet Bernal was aware that he was dealing with material which was *almost* a single crystal, *but not quite*—that was what intrigued him, this biological world of the almost 3-dimensional crystal!

At first Bernal and Fankuchen thought the interpretation of the intramolecular pattern of TMV was going to be easy. It corresponded to a crystal with hexagonal lattice with reasonable dimensions, $a = 87$ Å, $c = 68$ Å. But as they improved their technique several spots showed up which demanded a larger cell. Still better photographs were taken:

> but the best obtained merely confirmed the impossibility of the simple interpretation. Orthorhombic and monoclinic pseudo-hexagonal cells were next tried, but they also failed to account satisfactorily for the observed reflections, while calling for many others that were not observed. It appeared that it was quite impossible to explain the pattern on the existing theory of X-ray diffraction from a crystal, for any cell large enough to give the observed spots was found to be larger than the size of the particle as inferred from all the intermolecular measurements.
>
> (Bernal and Fankuchen, 1941, 148)

What they lacked was the theory of helical diffraction. What they failed to see was that they were treating as spots what in fact were continuous bands due to the molecular transform of the TMV rods (see Chapter 18). When Crick pointed this out to Fankuchen in 1954 the latter refused to accept it. Consequently he would not accept the evidence for the α-helix! (Crick, 1968/72).

Having encountered these difficulties in arriving at the unit cell it occurred to Bernal and Fankuchen that they were dealing with *inter-molecular* spacings which merged into nearby intramolecular spacings. It was for the pursuit of this liquid–crystalline long-range order that Bernal so much wanted an electron microscope. In 1951 he wrote:

> Once the size of 100 Å or so is surpassed, new kinds of interactions, imperceptible against the thermal background at smaller sizes, become apparent. These are the long range forces assumed to account for many colloidal phenomena . . . Tobacco mosaic virus provided the first quantitative pictures of such forces, because it showed that the identical rod-shaped particles of the virus maintained themselves at equilibrium distances . . .
>
> The moment, therefore, that macro-molecules of this type are produced they must interact in the liquid and give rise to the physical unity which characterizes the individual organism.
>
> (Bernal, 1951, 42–43)

If this does represent the essence of Bernal's approach to biology then it is doubtful that in more favourable circumstances he would have achieved what a later generation of molecular biologists was to do. There is a hint of disappointment, of regret at not having concentrated on what later formed the centre of molecular biology, in the following passage:

> I have spent my time, when I wasn't doing things even more vague than molecular biology—things like the structure of liquids which have no proper structure at all—speculating on the meaning of those aspects of molecular biology which are run now by other people. I still have the delight in my memory of seeing some of the beginnings of these discoveries myself. But these days are over for me.
>
> (Bernal, 1966, 1)

He liked "to start something, drop an idea, get the first foot in—and then leave it for someone else to produce the final finished work . . . he suffered from a certain lack of the obsessiveness which most scientists possess and which makes them want to carry out a piece of creative work to the end" (Snow, 1966, 26).

Bernal's Research Institute

It was Bernal's fate always to work in old buildings, the crystallography hut, "the old Cavendish, the back stairs rooms and the cellars of the Royal Institution, and since the war two old houses, built in 1835, bombed and roughly patched up, behind the main building of Birkbeck College" (Goldsmith and Mackay, 1966, 14). But as far back as those early days in Cambridge he had planned a research institute for protein structure. Dorothy Crowfoot saw one such scheme when she visited Cambridge in the late 1930s, Bernal contributed to Needham's scheme, and in *The Search* C. P. Snow made the planning of an institute for protein structure the central plot of the book.

This institute was to be in London, and no doubt Bernal hoped his move to Birkbeck would bring the plan nearer to fruition. But it was not to be. When he left Cambridge in 1937 his work on the proteins was pushed aside by many other activities—his book *The Social Functon of Science*, his work for the Association of Scientific Workers. Before he had put the finishing touches to the papers on sterols and viruses, and before he had completed a lecture trip to the United States, Europe was plunged into war. He cut short his American trip, raced back to England and immediately put all his efforts into the task of defending Great Britain from the invasion and air attacks that seemed imminent.

By September 1939 Bernal was at the technical department of Civil Defence in Princes Risborough. Only that summer he had discussed with Edwin Cohn at Woods Hole how best Fankuchen could be helped to set up the X-ray study of protein in the States. He told Cohn:

> Events seem to have decided things and any co-operation in the protein field will have to count out most if not all Europe from now on. I am sorry because it seemed as if we might have contributed something to the solution of the main problems. It is particularly unfortunate with regard to X-ray work. I shall not be able to touch it for a long time, and I do not think anyone else here will either. My only hope is that it may be possible to get it well started in the States through Fankuchen, who I am sending over with all my materials to carry on if he can find some means of doing so.
>
> (Bernal, 1939b)

To Felix Haurowitz, the haemoglobin biochemist from Prague, he wrote describing his dash back to England. He urged Haurowitz to take Fankuchen on if possible. Some such support, he hoped, would materialize "because the work was in a promising stage and unless it takes root in America I am afraid it may be forgotten altogether in the stress of events" (Bernal, 1939c).

Then there was trouble over the immensely long virus paper. The Royal Society declined to publish it as it stood and Bernal had to get Fankuchen to place it in an American journal. "I am right out of science", wrote Bernal in despair, "so I cannot do any more myself" (Bernal, 1939d). Nor did Fankuchen quickly deal with uncompleted items, such as diagrams, get the paper into print, and set Bernal's mind at rest. Anxiously Bernal wrote again: "I hope you have been able to do something about the virus paper" (1940a). And in July: "How are things going? I have not heard from you for a very long time, and am anxious to know how you are getting on with the proteins . . . Have you been able to do anything about the virus paper?" (Bernal, 1940c).

The letters from which these passages have been taken are in the Fankuchen papers in the American Institute of Physics library in New York. From them we can understand how it came about that the virus paper was published so late, not in a British journal, but in an American journal founded by Jacques Loeb and run by the Rockefeller Institute, the *Journal of General Physiology*, with a strong tradition of physiological and biochemical papers, but no association with X-ray diffraction whatever. Bernal was definitely aggrieved by the delay on Fankuchen's part, and one can only guess what Bernal's reaction was to using the *Journal of General Physiology*.

From this correspondence, too, we can see how strong was Bernal's desire to continue the protein studies: "I feel so bad about being out of science and being able to do nothing about all the things that are happening in the protein field" (1940b). There is a sad ring to his account of the situation in 1945:

> I am back where I started, on the top floor of the Davy–Faraday [Royal Institution], with one X-ray tube which works about one day a fortnight between breakdowns, and vague prospects of getting three more and a few of my old cameras . . .
>
> Our proposed programme is to pick up more or less where we left off plus all the experience we can get from published work in the interval. Bill Pirie is going to supply us with viruses and I hope also to do some work on antigens and antibodies. The main idea is to link up the long range forces with the biological systems.
>
> (Bernal, 1945)

The last sentence confirms one's suspicion that Bernal was still primarily interested in what we have termed "dynamic ultrastructure". From this letter, too, we learn that his "new Institute is definitely embryonic. The organizers are there but there is very little to organize" (*Ibid.*). He battled on; the end of the war did not bring him academic peace, and when in the 1950s he was attempting to bring about a test ban treaty "he became involved in a sordid academic dispute and was challenged to show cause why his department and subject should exist at all" (Goldsmith and Mackay, 1966, 15).

The man who had opened up a new field in the 1930s only to be torn away by the war, could never again recapture the urge and patience he had once shown. Instead it was his hankering for something novel, something different, that got the upper hand. "What's new today" was an all too familiar question from Bernal, when he came into the lab.

In one sense Bernal never had a school, but in another sense he had one in the persons of Carlisle, Crowfoot, Fankuchen and Perutz. The reputation he established in the 1930s lived on and when Sven Furberg, Rosalind Franklin and Francis Crick turned to biology it was to Birkbeck that they went, in search of a place in Bernal's laboratory. In this sense, too, Bernal did succeed in achieving his ambition, for it was at Birkbeck that Furberg established the structure of the nucleoside, cytidine, so important for the structure of DNA, and Franklin established the structure of TMV.

In another sense Bernal failed, as we can see from his revealing review of *The Double Helix*.

There he wrote:

> I should say here that the distinction between the fully and the partially crystalline structures was fully recognized in practice between Astbury and myself. I took the crystalline substances and he the amorphous or messy ones. At first it seemed that I must have the best of it but it was to prove otherwise. My name does not appear, and rightly, in the double helix story. Actually the distinction is a vital one. The picture of a helical structure contains far fewer spots than does that of a regular three-dimensional crystalline structure and thus far less detailed information on atomic positions, but it is easier to interpret roughly and therefore gives a good clue to the whole. No nucleic acid structure has been worked out to atomic scale though the general structure is well known. It may be paradoxal that the more information-carrying methods should be deemed the less useful to examine a really complex molecule but this is so as a matter of analytical strategy rather than accuracy.
>
> A strategic mistake may be as bad as a factual error. So it turned out to be with me. Faithful to my gentleman's agreement with Astbury, I turned from the study of the amorphous nucleic acids to their crystalline components, the nucleosides.
>
> (Bernal, 1968, 324)

Entry of Perutz

Perutz had been trained as a chemist in Vienna (1932–36) under the alkaloid chemist, Späth. He found the organic chemist's approach unstimulating but was attracted by what he learnt of F. G. Hopkins' biochemical research in Cambridge. F. von Wessely referred to Hopkins' work in his course on physiological chemistry at Vienna. So Perutz asked Hermann Mark if he would try to find him a place in Hopkins' laboratory. Mark forgot Perutz' request when, in 1935, he was in England for a Faraday Society meeting. There he met Bernal who gave a paper on the carbon skeleton of the sterols. Bernal said he was looking for a research worker. The first single-crystal pictures of proteins had been taken in 1934 and he wanted to continue this work. On his return Mark apologized for his lapse of memory regarding Hopkins' laboratory but suggested Perutz go and work under Bernal. To Perutz' objection that he knew no X-ray crystallography, Mark in his characteristic manner replied: "You will learn it my dear boy!" Accordingly Perutz came to the Crystallography Laboratory in 1936.

When Perutz arrived in Cambridge that September he found the crystallography laboratory housed "in a few ill-lit and dirty rooms on the ground floor of a stark, dilapidated grey building. These dingy quarters were turned into a fairy castle by Bernal's brilliance and his boundless optimism about the powers of the X-ray methods" (Perutz, 1970, 152). But no protein crystals were available so at the suggestion of the new professor of mineralogy, Tilley, he worked on iron rhodonite from slag (Perutz, 1937). After ten months Perutz returned to Austria. That summer he visited his cousin, Felix Haurowitz, in Prague. Haurowitz showed him under the microscope the striking change in crystal form from the trigonal crystals of deoxyhaemoglobin to the monoclinic crystals of oxyhaemoglobin (Haurowitz and Perutz, personal communications), and he advised his cousin to go to Gilbert Adair in the physiology department when he got back to Cambridge. Sure enough Adair was able to supply him with beautiful 0.5 mm crystals of horse haemoglobin. At the same time like-sized crystals of chymotrypsin arrived from Northrop at the Rockefeller. Since the latter suffered from the problem of twinning, and the former did not, the decision was made in the autumn of 1937 to work on haemoglobin.

Now, as luck would have it, these crystals which Adair supplied were of methaemoglobin of the horse. This proved, wrote Perutz, "to have the simplest crystal structure of any protein of comparable molecular weight, with features so favourable that they make a crystallographer's heart leap with joy" (Perutz, 1949a, 135). Yet because of the absence of any direct method for obtaining atomic positions it seemed sheer madness to attempt so complex a molecular structure as that of haemoglobin; it was, wrote Perutz, "as promising as a journey to the moon" (Perutz, 1949a, 135).

Perutz has already described the story from the time of Bragg's arrival to fill the chair at the Cavendish left vacant by the death of Rutherford in 1937. With the arrival too of the metal structure experts Lipson and Bradley, Perutz felt forlorn among his haemoglobin crystals:

> I waited from day to day, hoping for Bragg to come round the Crystallography Laboratory to find out what was going on there. After about six weeks of this I plucked up courage and called on him in Rutherford's Victorian office in Free School Lane. When I showed him my X-ray pictures of haemoglobin his face lit up. He realized at once the challenge of extending X-ray analysis to the giant molecules of the living cell. Within less than three months he obtained a grant from the Rockefeller Foundation and appointed me his research assistant. Bragg's action saved my scientific career and enabled me to bring my parents to Britain.
>
> (Perutz, 1970, 152)

Bragg admitted (personal communication) that he was excited about haemoglobin not because it *was* haemoglobin but because of the challenge so complicated a structure offered to the X-ray crystallographer. "I was thrilled by them," he recalled, "and formed the ambition to get out as a final act in my X-ray analyst's life something as complicated as a protein" (Bragg, W. L., 1967b). Knowing that it would be better after the war to get permanent positions for the staff working on haemoglobin, he tried to get the

University to come to his rescue, but without success. Perutz recalled that, "Bragg felt that we must not rely indefinitely on the generosity of the Rockefeller Foundation which had supported me from 1939 onwards. So he put me up for an ICI Fellowship and found some temporary grant for Kendrew, but by 1947 these sources of support were approaching their end, and the University offered no help. At this desperate juncture D. Keilin, the late professor of biology, suggested an approach to the Medical Research Council. In traditional fashion, Bragg met Sir Edward Mellanby, the secretary of the MRC, for luncheon at the Athenaenum Club" (Perutz, 1970, 153). Bragg described the chances of getting out such a structure as haemoglobin as "infinitely small, but the importance of the results, if they did come out, was almost infinitely great, so perhaps it was worth the gamble. Mellanby played and gave us money from the MRC" (Bragg, W. L., 1967b).

Patterson Analysis

When Sir Lawrence Bragg described the history of the haemoglobin work at Cambridge he called it "How Protein Structures were not Worked Out" (1965, 1). He was referring to the strategy behind the earlier phases of the work, which laid great emphasis on the significance of Patterson diagrams. In 1935 A. L. Patterson had put forward a method for identifying important interatomic distances which did not require a knowledge of the phase of diffracted rays. It seemed to provide just what was needed in the study of complex organic molecules. It was the outcome of discussions and Patterson's long pondering over ways of analysing organic structures, first at the Royal Institution and later at the Kaiser Wilhelm-Institut für Faserstoffchemie.

The regularities in Patterson projections were seen as peaks of electron density. The distances between these peaks gave a direct measure of important interatomic distances in the structure, but because the "information" from the diffraction pattern does not include the phases of the diffracted rays these projections did *not* give the *actual distribution* of the density of scattering matter in the unit cell (Bernal, 1939e, 550). On the one hand symmetry features revealed by Patterson projections for insulin were missed because the then accepted molecular weight of insulin was twice the true value (Crowfoot and Riley, 1939), and on the other hand the projections for both insulin and haemoglobin seemed to support the hope that these proteins consisted of bundles of parallel rods, each rod being a polypeptide chain. This hope was a logical step from the parallel chain structures for silk and keratin, as described by Astbury. "Patterson synthesis", said Sir Lawrence Bragg, "was a guiding star which encouraged the investigations. As events turned out it was a false star" (Bragg, 1965, 3). Perutz had been well aware that the globin molecule might not consist of a system of layered chains, but instead it might be made up of "a complex interlocking system of coiled polypeptide chains where interatomic vectors occur with equal frequency in all possible directions", in which case the "Patterson synthesis would be unlikely to provide a clue to the structure" (Perutz, 1949b, 474). Had Perutz really thought this latter alternative was the more likely of the two, he would surely have given up the quest for the structure of haemoglobin. As it was, he arrived at the celebrated "pill-box" or "hat-box" model in which twenty polypeptide chains were stacked in four rows parallel to each other and separated by 10.5 Å. There had been consternation when Langmuir and Wrinch revived Frank's cyclol theory and asserted that Dorothy Crowfoot's Patterson-Harker sections of insulin supported a cyclol "cage" structure for the molecule (Langmuir and Wrinch, 1938; Wrinch, 1938). Bernal showed that this claim was based on the arbitrary selection of a certain number of vectors, and the agreement disappeared when all the vectors were taken into account (Bernal, 1939e). The Bernal—Perutz—Crowfoot picture of bundles of rod-like polypeptide chains which was favoured by Pauling and Astbury outlived the cyclol theory in crystallographic circles, but it too went the way of all such geometric pictures of protein molecules. It was attacked within the Cavendish laboratory by a newcomer to the field—a thirty-three year old postgraduate student by the name of Francis Crick.

CHAPTER SEVENTEEN

Pauling, Caltech and the α-Helix

> I have returned from a short vacation for which the only books I took were a half-a-dozen detective stories and your "Chemical Bond". I found yours the most exciting of the lot. I appreciate having a book dedicated to me which is such a very important contribution. I think your treatment comes nearer to my own views than that of any other author I know . . .
>
> (Lewis to Pauling, 1938)

Structural X-ray crystallography had flourished in England under the Braggs as nowhere else in the world. One of the main reasons for this in Sir Lawrence's opinion was that his father "had developed very accurate methods of measuring ionisation . . . It was he who built the famous X-ray spectrometer with which he got accurate quantitative measurements of X-ray diffraction" (Bragg, W. L., 1967b). Thus equipped, British scientists forged ahead with crystal analysis under the leadership of Sir William in London and Sir Lawrence in Manchester. "We got a lot of clever young men and so we established a corpus of X-ray crystallographers which was the strongest in the world, there is no doubt about it" (*Ibid.*). Despite this strength, the schools in Britain were several times anticipated by the work of a California scientist, one Linus Pauling. Born in Portland, he had spent his youth from the age of four in cowboy country where his father ran a drugstore in the town of Condon until his early death in 1910. In search of a career that would enable him to support his mother and give scope to his passion for chemistry, Pauling trained to be a chemical engineer at Oregon State College. From domestic duties and a spell of teaching he was able to supplement the family funds and continue his education so that by 1922 he had shown such promise that he was able to continue in postgraduate work, somewhat to the disappointment of his mother who was naturally so proud of her son's abilities as a road inspector and delighted that he should have been offered a permanent job in this capacity at a salary of $1000. Fortunately, Pauling went to Caltech and in the ensuing thirty years he built up a powerful school of structural crystallography well able to challenge British leadership. This school was of a special kind and we can best understand how it became so successful by looking back to the British schools and seeing what they lacked.

We have seen that the study of organic compounds flourished in London, and from this centre it spread to Leeds and Cambridge (Sir William had been at Leeds 1908–1915 but no school of structural crystallography survived him there). Whilst Astbury concentrated on the fibrous proteins—though not to the exclusion of the globular proteins—Bernal's emphasis was on the crystalline proteins and viruses.

Astbury and Bernal had been drawn to structural crystallography by Sir William, and like him they were keen to apply it, to put it to work; indeed Astbury was very much under pressure to show that it could yield results of value to the textile industry. Bernal, on the other hand, was virtually driven by his passionate desire to see science applied in the service of society. But of the two, it was Bernal who came closest to Pauling in his approach to the structure of proteins. In his study of water with R. H. Fowler, Bernal recognized the importance of "hydrogen-bridges" or bonds and he realized that the position of the proton might not be exactly mid-way between neighbouring oxygen atoms. Fowler and Bernal used quantum mechanical arguments for their structure, but unlike Pauling and Huggins this exercise did not lead them to the conclusion that the position of the proton *had* to be asymmetric with respect to the two neighbouring oxygen atoms. Again it was Bernal who assigned unit cell dimensions and space groups to amino acids and peptides as a preliminary exercise in the attack on the proteins. All the same, these British Schools seemed to lack an ingredient which their cousins in the United States possessed.

One suspects that this ingredient was the *confidence* that quantum mechanics could be used as a check and a guide in the selection of probable molecular structures, and in particular to accept or reject the bond distances and angles derived from X-ray diffraction pictures. This is not to say that British physicists played no part in the quantum revolution—that would be absurd. Neither does it imply that men like Bernal were unaware of what was going on in physics, nor that there existed no X-ray crystallographers in Britain whose work was chiefly directed to contemporary problems in physics. But here in Britain were flourishing schools of *protein* crystallography which had *not* been nurtured in an environment of the new physical chemistry. Bernal believed in the possibility of a *direct* attack on the crystalline proteins. He advocated the introduction of a heavy atom into the protein molecule, or studies of the effects of dehydration, to overcome the problem of phase determination. "The problem of the protein structure", he told an audience at the Royal Institution, "is now a definite and not unattainable goal . . ." (Bernal, 1939a, 556). In contrast to Bernal, Pauling emphasized the difficulties more than the possibilities. The great complexity of the proteins, he declared, "makes it unlikely that a complete structure determination for a protein will ever be made by X-ray methods alone" (Pauling and Niemann, 1939, 1860).

In America the situation was very different. No one was taking X-ray pictures of proteins save Ralph Wyckoff who took a few in the 1930s. One attempt from the 1920s is on record: Michael Heidelberger took his best haemoglobin crystals to Schenectady when he learnt that they had X-ray equipment there but he "came back disappointed because the technique was not up to it" (Hastings, 1970, 94). Even the techniques of structural crystallography themselves spread only slowly in the United States. During the 1914–18 War the General Electric Company's laboratories in Schenectady, the Massachusetts Institute of Technology (MIT) and the University of Cornell were the only centres, but as has been remarked, only one group—that at MIT—had a continuing existence in a university (Wyckoff, 1962, 432), and this only because Arthur Noyes moved from MIT, taking Burdick, Ellis and Dickinson with him to Throop College in Pasadena and there put James Ellis, Roscoe Dickinson and Linus Pauling on to structural crystallography. As a result of this action Noyes established a school in the subject which was different from the British schools because it was developed *intentionally* in the wider context of modern physical chemistry.

Throop College becomes the California Institute of Technology (Caltech)

In Chapter 16 we saw how an ancient university, outwardly governed democratically but really run by a small and powerful group of men, managed to avoid having a new institute of physico-chemical morphology or a separate department of crystallography thrust upon it. In science, fundamental particle physics, the new zoology, and the new geology called the tune. Throop College was as different as it was possible to be from Cambridge. The centre of the Caltech administration was an Executive Council consisting of five men: Millikan (Chairman), A. A. Noyes, Hale, and two local businessmen. By the late thirties Noyes was no longer a member but the Council had increased to nine with the addition of T. H. Morgan, Richard C. Tolman, Max Mason, William Munro and a third businessman. Pauling's view is that "it was this form of administration which gave Hale, Noyes and Millikan much freedom in putting their ideas about education and research into operation" (1965, 1). It was responsible for the great success of the Institute during the quarter century beginning 1920. Not until 1946 was the Executive Council replaced by a president. Caltech began in 1891 as a local school of arts called Throop Polytechnic Institute. On the Board of Trustees in 1908 was George Ellery Hale who foresaw the development in Pasadena of an institute of engineering and scientific research. By 1910 the trustees had separated out the elementary department, the normal school, and the academy, choosing instead to concentrate on the College of Technology which they moved to a fresh site on the southeast edge of town.

Between 1911 and 1921 the Institute grew and in 1919 attracted A. A. Noyes, already (1916) a research associate at Throop; he gave up his MIT professorship to accept the full time directorship of chemical research at Caltech. Millikan likewise was persuaded to change from part time to full time status in 1921. Meanwhile the name of the Institute was changed to Caltech.

This pattern of attracting an eminent scientist on a part-time basis, securing an endowment to build him research labs, and then persuading him to accept a full time appointment seems to be a wonderful idea, and for Caltech it seems to have worked like a dream. In the Throop days the budget had always been overspent; the finances of the Institute were rescued by contributions to Throop's Deficiency Fund from friends. By 1920 endowments had reached half a million dollars and in that year an additional $200,000 came in in the month of February alone. The Pasadena family of Gates gave funds for building the first two laboratories for chemistry in 1917 and 1927, and the Crellin family, also of Pasadena, gave the third building in 1937. Here we have to thank the wealthy people who came to California, were proud of their local Institute, and wanted to establish teaching and research on the western side.

Pauling, who had taken a degree in chemical engineering at Oregon Agricultural College because he didn't realize that one could take a degree in chemistry and get a job with it—is of the opinion that he had the greatest good luck in having gone to Pasadena in 1922. Among the institutions to which he had applied to do postgraduate work were Berkeley and Caltech, in Pasadena. "I do not think I could have found better conditions for a career in physical chemistry anywhere else in the world. Perhaps I would have got along just as well at Berkeley. I did not have to decide between Pasadena and Berkeley—I simply accepted the first offer" (Pauling, 1965, 2). Of course one could learn physical chemistry at many other universities—including Cambridge. What was special about Caltech's physical chemistry was that it was informed by the recent developments in quantum mechanics, which men like R. C. Tolman believed could be used to solve chemical problems.

Pauling's early years at Caltech covered a most exciting time in physics. The old quantum theory was running into difficulties and the new quantum theory was about to burst on the scene.

The Two Quantum Theories

The "old" theory associated chiefly with Bohr and Sommerfeld had to be patched up, *ad hoc* changes had to be made, so that the theory could do all that was required of it. There were the "forbidden drops", which Bohr could only explain in terms of the discarded atomic model of classical mechanics, the non-integral quantum of action which demanded the so-called "half-quantum numbers", the non-symmetrical electron orbits which Sommerfeld could only account for in terms of an "inner" and an "outer" quantum number for one and the same electron.

The new quantum theory arose from the challenge presented by these difficulties. Beginning with de Broglie's concept of "pilot waves" in 1925, the remaining years of the 1920s saw the brilliant establishment of the new quantum theory. Heisenberg developed the theory in the form later recognized as matrix mechanics while Schrödinger introduced his treatment in terms of the wave function ψ. Pauling experienced the "general feeling of excitement" generated by these new developments in quantum mechanics. Many puzzling questions were solved; others remained unanswered (Pauling, 1965, 9). Already in 1925 Pauling had known Heisenberg's work. Also he knew about "the problems of the fine structure of atomic spectra while . . . in Pasadena" (Pauling, 1973b). He had also been schooled in the relevance of modern physics to chemical problems by the seminars in physical chemistry at Caltech which R. C. Tolman organized. Moreover, Sommerfeld had lectured in Pasadena in the early 1920s.

In 1926, when Pauling went as a John Simon Guggenheim Fellow to Europe, he found himself at the centre of the quantum revolution. He attended Sommerfeld's brilliant lectures in Munich, and those of Schrödinger and Debye in Zürich. The aim of this visit had been to apply the recently discovered new quantum theory to the structure of the molecule and to the nature of the chemical bond (Pauling, 1970, 993). Whilst in Munich, Pauling wrote a paper for the *Proceedings of the Royal Society*, and another based on it for the American Chemical Society. The latter opened with a statement of the clarification which Heisenberg's work had brought, but his theory was abstruse. "It cannot", he wrote, "easily be applied to the relatively complicated problems of the structures and properties of many-electron atoms and of molecules" (1927, 765). Pauling therefore found it all the more gratifying "that Schrödinger's interpretation of his wave mechanics provides a simple and satisfactory atomic model, more closely related to the chemist's atom than to that of the old quantum theory" (*Ibid.*). With this wave mechanical treatment one could take a Lewis electronic configuration of a molecule, construct a corresponding wave function for it and calculate some of its properties, including of course its stability. This led to a fruitful interaction between theory and observation in which the quantum mechanical picture of the molecule was developed in a semi-empirical fashion.

The Theory of Resonance

In 1926 Heisenberg had introduced the concept of resonance in his treatment of the helium atom. A year later Heitler and London constructed a quantum mechanical picture of the covalent bond between hydrogen and oxygen in the water molecule. To do this they considered the molecule as resonating between two structures so that the Schrödinger wave function ψ represented "first one assignment, and then the other, of the two electrons, with opposite spins, to the two nuclei" (Pauling, 1970, 994).

Not long thereafter, Pauling showed that a similar resonance combination (or hybridization) between the $2s$ and $2p$ orbitals of the carbon atom could account satisfactorily for the equivalence and orientation of the four bonds in the carbon tetrahedron (Pauling, 1928).

During the 1930s many investigators were active in the task of preparing reliable measures of inter-atomic distances. It was becoming clear that these were *not* constant for the same element in different compounds. One could not simply add elemental atomic radii together to arrive at corresponding values for their compounds. Nor was it any longer wholly satisfactory to regard single, double, ionic and covalent bonds as distinct entities. Quantum mechanics had given a physical argument for the existence of such bonds, and at the same time it furnished by means of hybrid wave functions an argument for the existence of a continuous spectrum of bonds between the purely ionic and the purely covalent, and between single and double bonds.

In the past, organic chemistry had supplied examples of compounds in which the bonding could not be a straightforward one, and when X-ray crystallography showed that the interatomic distances were intermediate between those for single and double bonds quantum mechanics could be used to account for this fact. There resulted an interaction between theory and observation which was highly beneficial. If, for such bonds, one used a wave function which was a linear combination of the wave functions of several possible molecular structures—the canonical forms—then the properties of such bonds could be represented satisfactorily, and more important—one could make predictions about the structure of molecules from a knowledge of interatomic distances. Nor did the implications of this approach end there. In conformity with the second law of thermodynamics, one would expect that the stable configuration of a molecule would be that which represented the lowest free energy. If there was resonance between several forms, then the free energy of the latter had to be greater than that of the hybrid, and the difference was the stabilizing energy of the resonance hybrid. Pauling exploited fully this concept as a criterion for deciding between possible molecular structures. "The idea of resonance stabilization," he wrote, "was very important. It goes back to Heisenberg's paper on the helium atom. In my 1928 paper on the chemical bond, I made some mention of the resonance phenomenon, but it was not until about three years later that things became quite clear" (Pauling, 1973b). Interatomic distance gave a measure of the percentage single/double bond composition. Appropriate electronic structures were suggested. From a knowledge of bond energies, the total energy of their formation could be calculated and compared with the observed energy of formation of the compound. The difference could be "confidently interpreted as the resonance of the molecule among these electronic structures" (Pauling and Sherman, 1933, 606). The power of this approach lay in this ability to measure the resonance stabilization. This is what gave Pauling the conviction from the early thirties that the amide bond must resonate between the single and double forms, and later to assert its planarity.

Realizing the importance of interatomic distances Pauling became impatient with the tedious procedure of arriving at these distances by X-ray diffraction. This involved finding the arrangement of the molecules in the unit cell first, then extracting the required intramolecular distances. Was there a more direct method? In 1930 he visited Germany and was shown an electron diffraction apparatus in Mark's laboratory in Ludwigshafen. Mark did not plan to continue working with it and was happy that Pauling should develop its use in Pasadena. On his return, Pauling set Lawrence Brockway to build the apparatus.

As a result of Brockway's work, interatomic distances for organic compounds were published from 1933 onwards. In 1935, with the introduction of their radial distribution method, this work was accelerated and a survey of the partial double bonding in carbon compounds became possible. From a knowledge of the number and valency of nearest neighbours it was concluded that in benzene the C–C bond should have 50 per cent double-bond character, in graphite 33 per cent. When the carbon–carbon distance in these and in pure single- and pure double-bond compounds was plotted against the above single/double bond percentages a curve was obtained which showed scarcely any slope until the double-bond contribution had fallen below 50 per cent (see Fig. 17.1). A relatively small percentage of double-bond character, it seemed, had a large effect on the structure. This effect had really been exaggerated by the wrong value for the pure C=C bond at 3.36 Å instead of 3.33 Å. None the less, it was evident that restrictions imposed by pure double-bonds applied in large measure to partial double-bonds. Coplanar arrangements and angles of about 120° were therefore "a requirement for resonance in heterocyclic and hydrocarbon ring systems" (Pauling and Sherman, 1933, 617).

Resonance of the Amide Bond

In his concern to develop the concept of resonance to account for variations in interatomic distances in organic compounds Pauling was led to consider

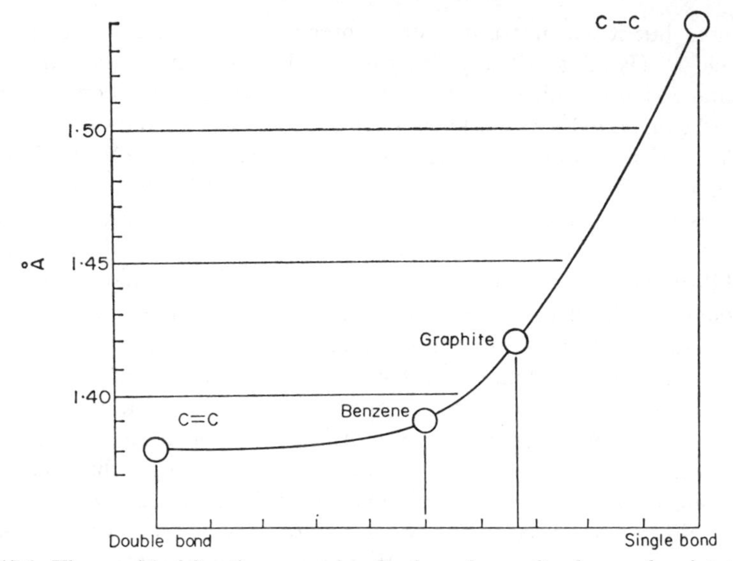

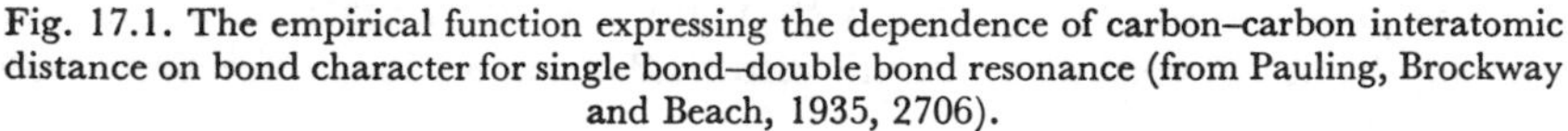
Fig. 17.1. The empirical function expressing the dependence of carbon–carbon interatomic distance on bond character for single bond–double bond resonance (from Pauling, Brockway and Beach, 1935, 2706).

the C–N distance in, and energy of formation of, urea, oxamide and oxamic acid. Parallel studies of the carboxylic acids gave him two lines of evidence for the existence of resonance in the carboxyl group thus:

and in the amide group:

(Pauling and Shermann, 1933; Pauling and Huggins, 1934).
The C–N distance in urea was 1.33 Å and its resonance energy 1.59 e.v. These data were taken to mean almost complete degeneracy, i.e., 30 per cent double bond character to both C–N distances because of resonance between the three forms:

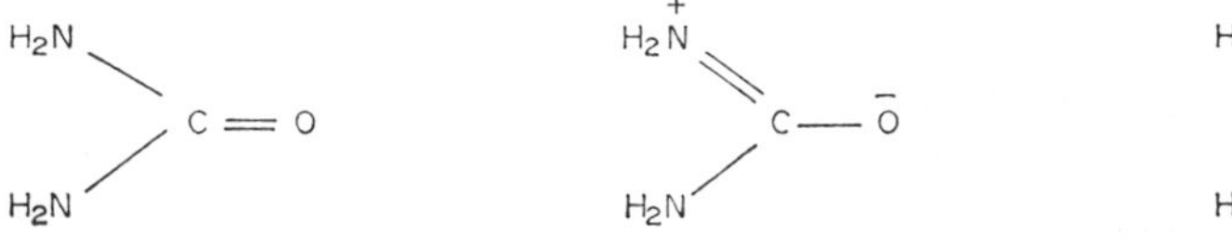

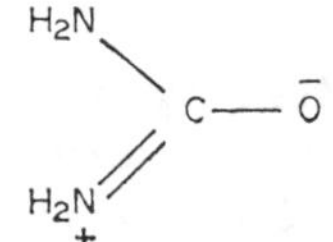

During the 1930s Pauling's interest, like that of many other chemists, was drawn to the proteins. In 1934 he began work on haemoglobin. In the summer of 1937 he turned to the question of how in α-keratin the polypeptide chain was folded. Like Astbury, he kept the folding of the chain in the same plane as the chain and he failed to form a convincing model. But when he recalled this early attempt many years later he was convinced that he had in 1937 assumed the planar amide group. "I am sure that my bond lengths and bond angles were good to a couple of hundredths of an Angstrom and a couple of degrees, and that I used only planar amide groups. But I did not have the simple idea of the helical repeat, the repeated operation of the rotatory translation" (Pauling, L., 1968).

Hydrogen Bonding

It is well known that in 1920 Latimer and Rodebush introduced the concept of a hydrogen atom being shared between two oxygen atoms in water, thus forming a weak bridge or bond between neighbouring molecules. Apparently Maurice Huggins had arrived a year before at an essentially similar idea in an unpublished thesis in Berkeley (Lewis, G. N., 1923, 109). Both Pauling and Huggins realized that in resonating structures an additional stabilization would result from hydrogen bond formation, and in 1937 both men constructed polypeptide chains with lateral hydrogen bonds between neighbouring chains. Indeed, before Pauling had built his models Huggins had stated: "In proteins, the chief forces (other than those in cystine and similar cross links) connecting the primary chains I believe to be due to hydrogen bridges . . . I firmly believe, NHO bridges are most important in joining the primary chains" (Huggins, 1937, 550). His structure (see Fig. 17.2) with five-valent carbon is, of course, an unconventional one.

The idea of hydrogen bonding as a structural feature of proteins was very clearly in Pauling's mind in 1936 when Mirsky and he published their famous paper on the denaturation of proteins. They attributed this process to the breakdown of a uniquely folded polypeptide chain by destruction of the hydrogen bonds.

Collaboration with Corey

Robert Corey was a research fellow at the Rockefeller under Ralph Wyckoff. There they took X-ray pictures of porcupine quill and other fibrous proteins, but in 1937 Wyckoff left the Institute and Corey was on his own with one year's terminal salary to come. He was given permission to spend the year where he liked and to take some equipment with him. He chose Caltech.

Pauling and Corey met in the Fall of 1937. The former had just constructed his models of folded polypeptide chains. Naturally they talked about proteins and decided that an attack on the amino acids and peptides was called

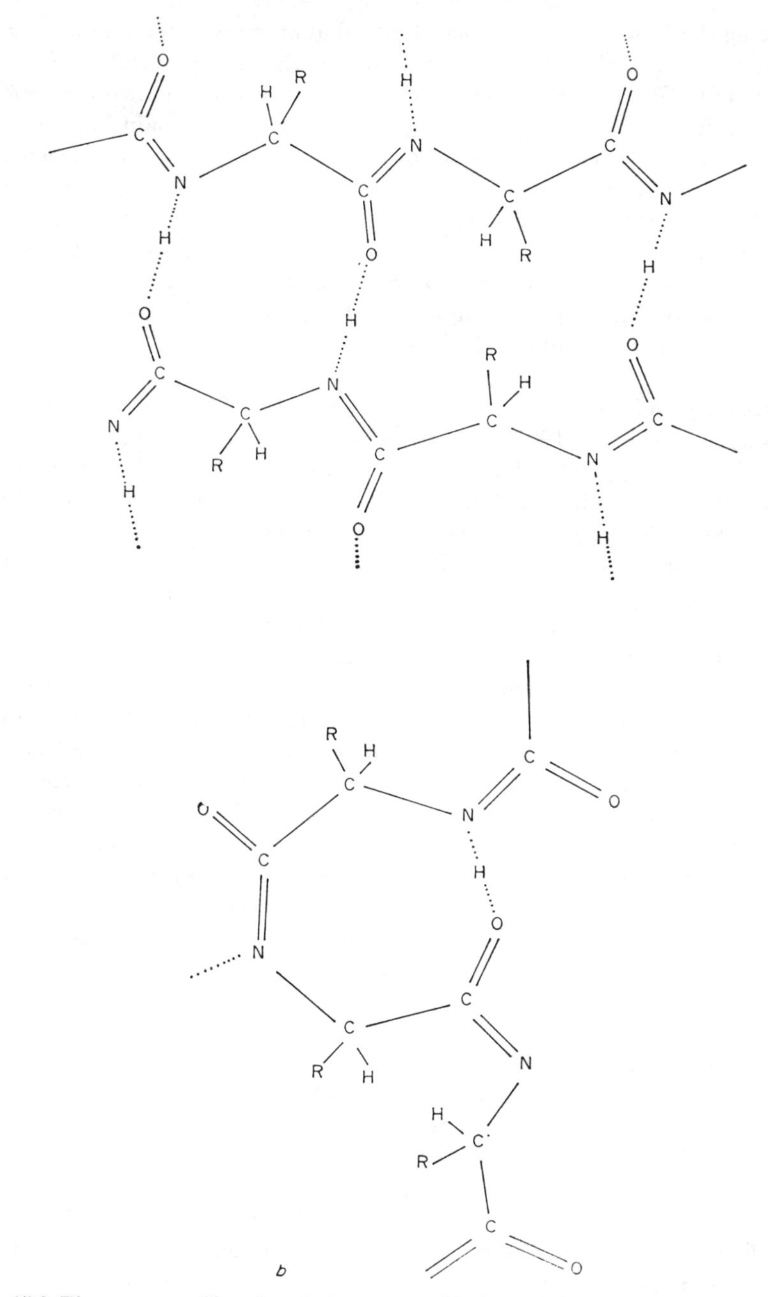

Fig. 17.2. The structure of beta keratin is represented in (a) and the structure of alpha keratin in (b). The dotted lines indicate hydrogen bonds (from Huggins, 1937, 449).

for. When Pauling looked through the literature he could find only two papers—in which the structures were both wrong—since Bernal's work of 1931. This was hardly surprising, for an attempt on the structure of even the simplest amino acid—glycine—involved five atoms other than hydrogen, and therefore fifteen parameters. That was no easy task. To be sure, Fourier analysis had been made less forbidding by the procedure of Patterson analysis, but in Pauling's opinion "the latter still isn't very valuable and it wasn't in those days" although his student David Harker had simplified it by his idea of Patterson–Harker sections (Pauling, L., 1968).

Among the crystals which Bernal had studied were the long needle-shaped monoclinic crystals of the cyclic dipeptide, diketopiperazine. Because of their negative birefringence Bernal concluded that these crystals must be

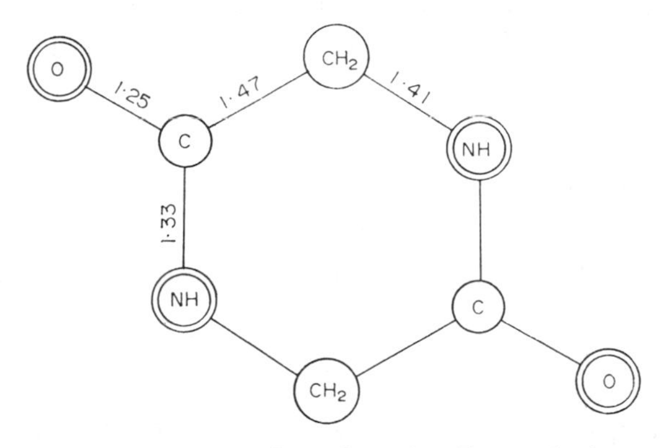

Fig. 17.3. Diagram of the molecule of diketopiperazine. Figures give interatomic distances in Å. Angles between all bonds are close to 120° (from Corey, 1948, 389).

"built from centro-symmetrical, almost flat, hexagonal molecules, linked together in ribbons by their residual electrical forces" (1931, 369). Corey found Bernal's words true. There was not much room for doubt as to the overall structure of the molecule, and because of the birefringence the molecules had to be parallel to the 101 plane. Thus assisted, Corey was able to give signs to the structure factors and carry out a two-dimensional Fourier analysis of the distribution of the scattering power in the unit cell projected on the 101 plane. Among the points established by this pioneer study were the coplanar arrangement of the molecules, with bond angles of 120°, and the C–N distances of 1.33 Å, just as in the resonating amide bonds of urea. Corey could also point out that if these C–N bonds were really pure single bonds then the ring should be puckered, as in other six-membered, single-bonded rings. "Thus the observed coplanarity provides strong evidence for resonance" (Corey, 1938, 1603). The bond distances in diketopiperazine are shown in Fig. 17.3, and the several Lewis electronic configurations [canonical forms] in Figure 17.4. Resonance between them accounted for the short C–N distance. If one accepted that diketopiperazine was equivalent to an open chain dipeptide *this would mean the planarity of the peptide C–N bond.*

In 1939 Albrecht and Corey published a structure for glycine in which there was a short amide bond suggestive of resonance. By the time war brought this programme to a close in 1941 Corey had put forward a structure for DL–alanine (Levy and Corey, 1941), and Edward Hughes and Walter Moore had begun their study of the dipeptide, glycylglycine (Hughes and Moore, 1942). The latter carried out a Patterson projection which they succeeded in interpreting. Meanwhile Corey had given a diagram of a fully extended polypeptide chain with dimensions and angles based on the assemblage of data up to 1939 (Corey, 1940). When a similar diagram was published ten years later (see Fig. 17.5) there was only one major change—the alteration of the abnormally short α C–N distance from 1.40 Å to 1.47 Å.

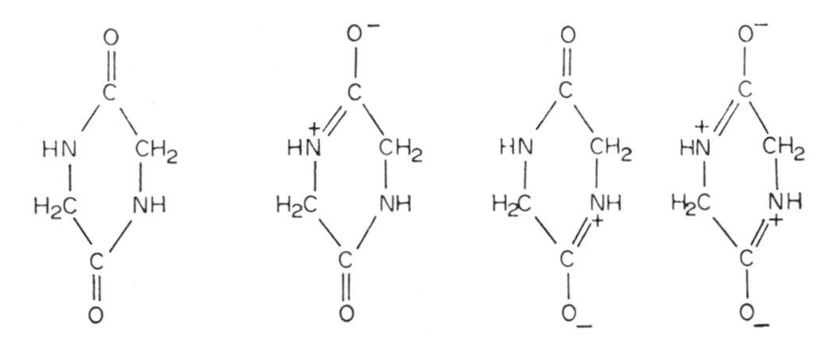

Fig. 17.4. The four Lewis electronic configurations for diketopiperazine.

Both configurations showed the planarity of the peptide C–N bond and of all the atoms pictured in Figure 17.6. Pauling was delighted with the close agreement between these observations and his expectations.

The α-Helix

Pauling's strategy in solving the structure of polypeptides was indirect. He did not, like Astbury, go to the extended chains in silk and keratin, and try to derive from their diffraction patterns a polypeptide chain structure. Instead he worked his way up from the amide bond. In one sense this was not by design—his work on resonance in amides antedated his interest in the structure of polypeptides. In 1932 he had pointed out that the resonance in amides involved an electronic configuration in which the nitrogen had no unshared pair of electrons. This accounted for the observation that amides did not show basic properties (Pauling, 1932, 296). The work on amino acids and peptides followed. The next stage was to build a polypeptide chain structure that would fit the data on the fibrous proteins. From the papers

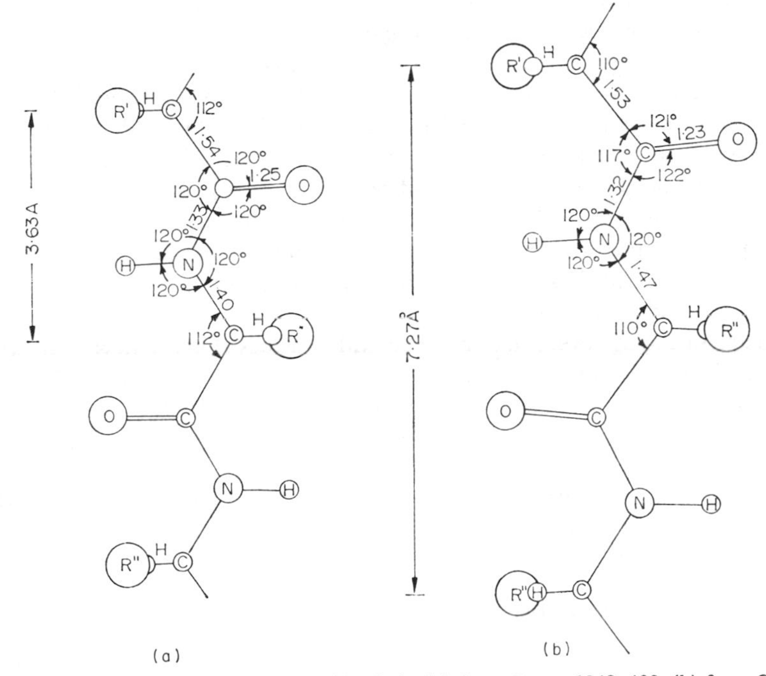

Fig. 17.5. Configuration of the polypeptide chain (a) from Corey, 1948, 400, (b) from Corey and Donohue, 1950, 2900.

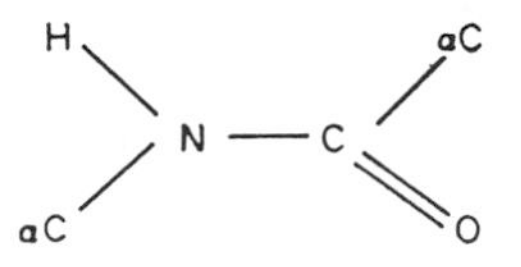

Fig. 17.6. Bonds in the polypeptide chain which are all in the same plane (from Pauling, Corey and Branson, 1951, 206).

published from Caltech between 1938 and 1950 it is quite clear that these studies were undertaken with the long term intention of solving the structure of proteins. This was also the reason for Pauling's grant applications to the Rockefeller Foundation in 1937 and to the National Foundation for Infantile Paralysis in 1946.

This programme was made even more long term by the entry of the United States into the Second World War. After Pauling's first attempt on the structure of α keratin eleven years passed; then in 1948 he returned to the problem when he came to Oxford as George Eastman Visiting Professor. Part of the course he gave there was devoted to biological specificity, and included anti-bodies, something on protein structure, but nothing on helical polypeptides. Then he fell ill and after whiling away the time reading detective stories he decided, for a change, to try to solve the structure of α keratin with only paper, pencil and ruler. His procedure can be summarized as follows: (1) Symmetry principles demand that the polypeptide chain when folded must assume a helical conformation. (2) All peptide residues are assumed to be equivalent. (3) The peptide C–N bond is planar, therefore there can be no rotation at either end of this bond. (4) Rotation is achieved by the dihedral angle NαCC at the αC atom. (5) The most stable, and therefore most propable, configuration is one in which hydrogen bonding takes place between residues along the chain. (6) Satisfactory hydrogen bonding is achieved by adjusting the pitch of the helix.

The results of this exercise were the α and γ helices. This achievement with such rudimentary equipment shows just how limited are the possibilities once the planarity of the peptide C–N bond is accepted. If the chain portion N–C–αC was denied rotation, then the number of possible conformations was very small.

Pauling simply drew a polypeptide chain across a sheet of paper, with the peptide bond in the plane of the paper and the αC atom rotated so as to bring all the carbonyl groups on to the same side of the chain (see Fig. 17.7).

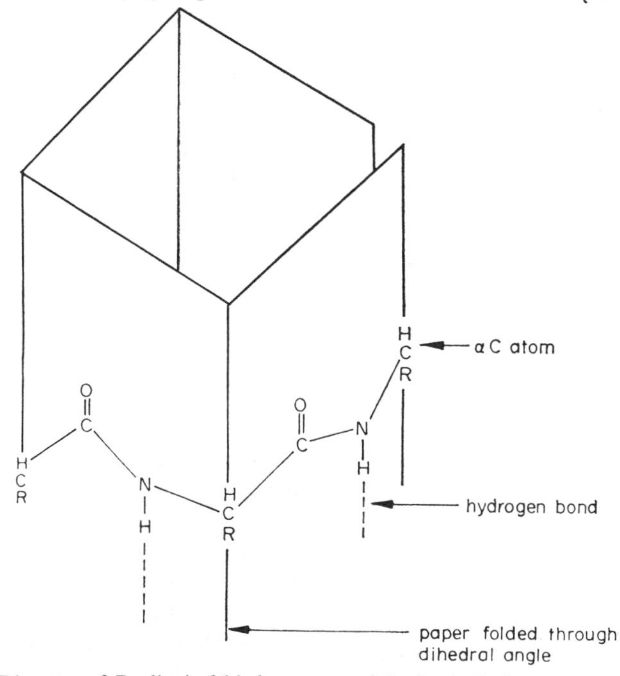

Fig. 17.7. Diagram of Pauling's folded paper model of a helical polypeptide chain (not drawn to scale).

Then he drew parallel lines through the αC atoms at an arbitrary angle to the chain and folded the paper along these lines through the dihedral angle (109°). This operation twisted the chain into a helix. It was then a matter of trial and error to find the orientation of the fold which brought carbonyl and amino groups into line for acceptable hydrogen bonding. In fact changing the orientation of this fold within the plane of the paper was equivalent to altering the pitch of the helix. The result was that as early as 1948 Pauling had arrived at a rough approximation to the 3.7 residue helix now known as the α helix. Each amino acid residue was hydrogen bonded to the residue three removed from it in one direction or the other. Because all the amino and carbonyl groups were directly above one another all hydrogen bonding was parallel to the helix axis.

About six weeks later Pauling was invited to give some lectures in Cambridge and he talked with Perutz who showed him the contour maps from the Patterson synthesis of haemoglobin in which there were prominent peaks 10 to 11 Å from the origin. These peaks excited Pauling for they were just about the right dimension for his α helix, but he was still sobered by the thought that the α helix did not repeat at 5.1 Å as the Astbury data on α-keratin demanded, but at 5.4 Å. Pauling recalled his visit to Perutz in Cambridge and the sight of those contour maps. "I thought to myself, that (10 to 11 Å) is just the right diameter for the α helix, you know, but I didn't say anything to him about it, because this 5.4 Å pitch per turn puzzled me, and I thought there is still a possibility that there is a real joker, you know, eluding me, that there is something wrong" (Pauling, L., 1968).

The following spring Pauling was back in Pasadena when Gene Carpenter and Jerry Donohue presented their fine paper on the structure of acetylglycine to the American Chemical Society meeting in San Francisco. They evaluated the peptide link at 1.323 Å with a margin of error not exceeding 0.030 Å. This and the carbonyl distance of 1.243 Å approached close to the theoretical values of 1.33 Å and 1.25 Å derived from the assumption of 60 per cent of canonical form I and 40 per cent of II.

Hughes and Moore also published the completed account of their study of β-glycylglycine at this time, but it will be remembered that their C–N distance at 1.29 Å was thought unreliable. Diketopiperazine and acetylglycine thus gave the only direct evidence in favour of the partial double bond character of the peptide bond. Acetylglycine also gave the first reliable evidence that the earlier data indicating an abnormally short αC–N distance were wrong. This bond was now 1.45 Å in length.

During 1949 Pauling's group were becoming increasingly confident. They had good reason to be. They had broken the back of the great labour presented by three-dimensional Fourier synthesis using all intensity data, so that for threonine and acetylglycine they had been able to refine the atomic positions suggested by Patterson analysis to an unprecedented degree of reliability. Donohue had gone back and refined the data on DL–alanine. In September Corey and Donohue addressed the American Chemical Society in Atlantic City on the dimensions of the polypeptide chain (see Fig. 17.5b). Pauling's α helix was in harmony with these dimensions, but Pauling still kept quiet. The 5.1 Å repeat still troubled him. The winter passed. It was 1950. The Atlantic City address by Corey and Donohue was published, including its appropriately cautious conclusion:

> Although the dimensions of the polypeptide chain in proteins probably depart little from those here indicated, more direct evidence based on precise determinations of atomic positions in crystals of higher linear peptides is greatly desired.
>
> (Corey and Donohue, 1950, 2900)

Over twelve years had gone by since Corey came to Pasadena and began work on diketopiperazine and glycine. Two years had passed since Pauling had the idea of the α helix. Then came the wonderful work on sickle-cell anaemia which was announced in April 1949 in Washington D.C., and Detroit (Pauling, Itano, Singer and Wells, 1949). This must surely have taken Pauling's attention off the helix at least for a few months. But by the following winter H. R. Branson and S. Weinbaum had been put to work on the exploration of all possible helical conformations of the polypeptide chain with planar peptide bond.

Instead of trying to make a model which would give the 5.1 Å repeat of α-keratin Branson simply applied the criteria that Pauling had used in Oxford in 1948, and by the use of accurate model building equipment they strove to achieve orientation of all hydrogen bonds parallel with the fibre axis. *The result was neither an integral helix nor the* 5.1 Å *repeat*, and yet in October 1950 Pauling and Corey sent a note to the American Chemical Society reporting that they had found the 3.7 residue and 5.1 residue helices. Why such confidence after all the caution? Well, there was the systematic exploration of planar peptide helices by Branson and Weinbaum. Then there were the data from N-acetylglycine available in March 1949. Also a series of papers had appeared in *Nature* on the synthetic polypeptides made by E. J. Ambrose and W. E. Hanby at Courtauld's laboratories in Maidenhead, Berkshire. First came the data on infra-red dichroism, indicating the parallel alignment of hydrogen bonds with the fibre (sheet) axis and their intramolecular character (Ambrose and Hanby, 1949; Bamford, Hanby and Happey, 1949a, 1949b). In July, F. Happey's data from the X-ray pictures were reported. The co-polymer of phenylalanine and methyl glutamic ester gave a polar arc at 5.26 Å, while the polymer of phenylalanine gave a similar arc at 5.28 Å (Bamford, Hanby and Happey, 1949a). *There was no indication of the Astbury* 5.1 Å *repeat!* The Courtauld scientists, however shrugged off the difference between 5.26, 5.28, and 5.1 Å, much to Astbury's annoyance. He denied that their pictures could result from a fold like that in α-keratin (Astbury, 1949), whereas Happey enthusiastically explored the analogy. When they succeeded in effecting an α–β transformation they became, to quote Happey, "intoxicated" with the correctness of the keratin model for these synthetic polypeptides, and for poly-γ-benzyl-L-glutamate they postulated a ribbon chain.

Whereas Bamford, Hanby and Happey *wanted* to find a near 5.1 Å repeat, Pauling and Corey did *not* want to. By July 1950, Happey had discussed his latest and best X-ray pictures with the group at Maidenhead. One picture in particular—of the α form of poly-γ-methyl-L-glutamate—showed quite clearly that the so-called meridional arc at 5.26 Å was in fact an off-meridional doublet, and they rightly put the true repeat distance at about $5\frac{1}{2}$ Å (see plate 13). Happey also noticed another feature of importance—the equatorial reflexions could be indexed on the basis of three or six-fold screw symmetry, such as would be possessed by a helical chain. This suggestion he did not pursue. Instead, he followed the conventional procedure which he had learnt under Astbury in 1930–32, that of attempting to index the diagram in order to deduce the unit cell, and from the latter to decide on the type of chain molecule. The result was the twofold ribbon chain, which he described to the Royal Society in November 1950 (published March 1951) as in Fig. 17.8.

A striking and significant feature of the picture illustrated in this Royal Society paper is the symmetrical arrangement of six arcs around the centre of the diagram (see plate 13). When Pauling saw the Happey pictures in March 1951 he at once spotted this feature, which he correctly described as "hexagonal or closely pseudohexagonal" (Pauling and Corey, 1951b, 243), a form of packing typical of helical rods when aligned parallel to the fibre—or as in this case, to the direction of shear. Perutz, who received an advance copy of this paper on synthetic polypeptides, and who did read it in 1950, found Happey's data confusing. He missed the pseudohexagonal packing. It just did not occur to him to check on the equatorial intensities. Because there was no 5.1 Å polar arc Perutz felt unhappy about the pictures. He was, he recalled, "confused by his slavish following of the 5.1 Å picture of α-keratin" (Personal communication).

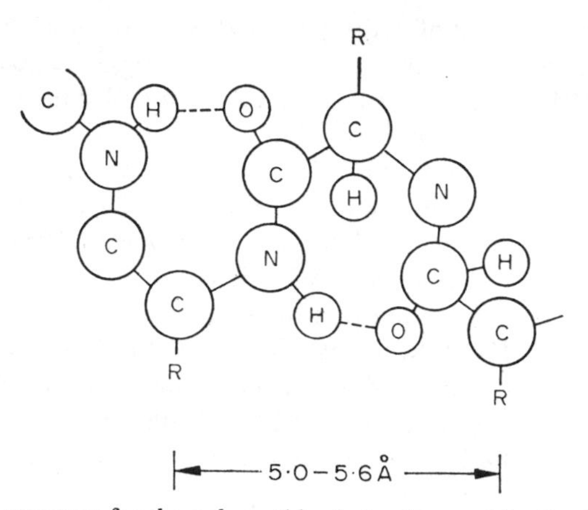

Fig. 17.8. The α_{II} structure for the polypeptide chain, discussed by Bamford, Hanby and Happey, 1951, 30.

Publication of the α-Helix and the γ-Helix

The note which Pauling and Corey published in 1950 was very brief and seems to have attracted little attention, but all the Cambridge crystallographers read it, even though this meant going to a library outside the Cavendish. Crick recalled that it was "rather cryptic, and we thought we would wait for the longer papers." After all, "Pauling was not infallible in what he did [as Watson seemed to think at that time]. He was often right, but he was often wrong. Therefore, the fact that he had announced a structure didn't mean that it was right" (Crick, 1968/72). But in 1951, the dramatic announcement in the form of one paper in the April issue of the *Proceedings of the National Academy of Sciences* (Pauling, Corey and Branson, 1951) and *seven* in the May issue (Pauling and Corey, 1951a–g) caused a stir, and in Cambridge, the home of protein crystallography, consternation.

The April paper announced the stereochemical requirements for a folded polypeptide chain:

> An amino acid residue (other than glycine) has no symmetry elements. The general operation of conversion of one residue of a single chain into a second residue equivalent to the first is accordingly a rotation about an axis accompanied by translation along the axis. Hence the only configurations for a chain compatible with our postulate of equivalence of the residues are helical configurations.
>
> (Pauling, Corey and Branson, 1951, 206)

By insisting on the planarity of the peptide bond they found only five angles of rotation per residue which would give models consistent with the stereochemical demands: 165°, 120°, 108°, 97.2° and 70.1°, and of these only the last two brought the C=O and NH groups into line for hydrogen bonding to take place parallel to the axis of the helix. These were the 3.7 residue helix (see Fig. 17.9) and the 5.1 residue helix, subsequently known as the

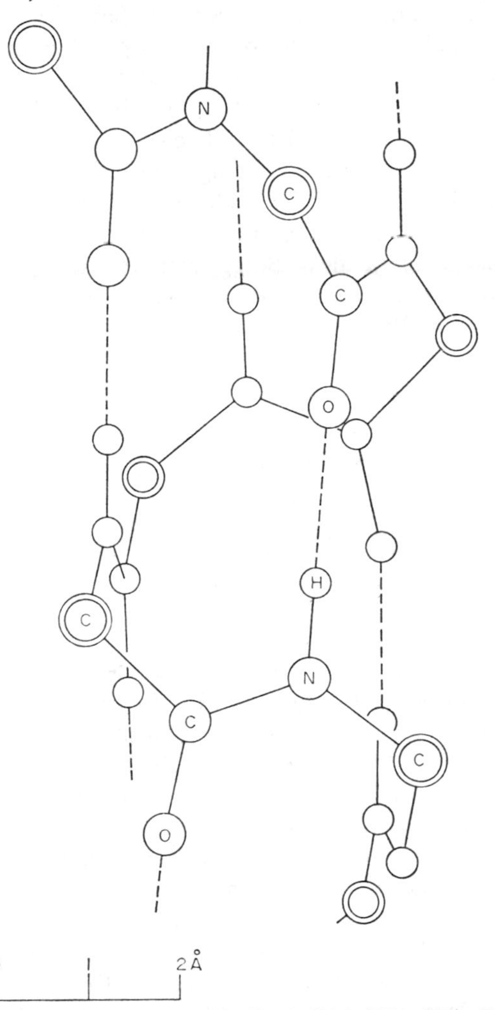

Fig. 17.9. A projection of part of the *α* helix (from Crick 1954, 209). Only the backbone atoms are shown. The *α* carbon atoms, to all of which the side chains are attached, are shown as double circles. Hydrogen bonds are dotted.

α and the *γ* helices, or better—the 3.7_{13} and the 5.1_{13}. The first of these, they thought, would be found in α-keratin, contracted myosin, and other fibrous proteins. It would be found to constitute "an important structural feature in haemoglobin, myoglobin and other globular proteins, as well as of synthetic polypeptides" (Pauling, Corey and Branson, 1951, 210). They expected the 5.1 residue helix would be found in the super-contracted form of keratin and myosin.

Pauling's former student and collaborator, Jerry Donohue, was later to show that the α- and γ-helices are not the only possible stable configurations for polypeptide chains. The 2.2_7, 3.0_{10} and 4.4_{16} helices, on theoretical grounds, did have a possibility of existing, but the α-helix remained the only one free from strain (Donohue, 1953). The γ-helix has never been found in nature, nor has it been produced among the synthetic polypeptides. But in 1951 Pauling was justifiably jubilant. There was no doubt that the Cavendish group had floundered, and their paper of 1950 on the configuration of polypeptide chains was a mess. Bragg later described it as "the most ill-planned and abortive in which I have ever been involved" (Bragg, 1965, 6). How wonderful that Pauling had been led to predict an axial repeat at

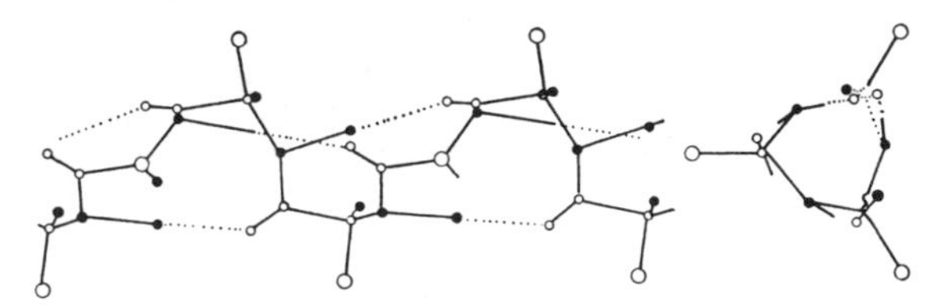

Fig. 17.10. The 3_{10} helix described by Huggins in 1943 (from Huggins, 1943). Repeat distance 5.1 Å.

5.4 Å, and that the Courtauld's scientists had inferred a unit cell for poly-γ-methyl-L-glutamate with a length of 5½ Å in the chain direction, and a density to fit with 3.7 residues in the repeat (Bamford, Hanby, and Happey, 1951, 33). Pauling's confidence and enthusiasm made him positively ebullient. The man with the slow speech, the delightful rural-sounding Oregon accent, knew when to keep his secrets and when to announce them. He knew how to command the attention of the scientific world, so that it listened to what he had to say, and those in the know, understood.

Pauling lectured on the α- and γ-helices first in Caltech and then in New York. At the end of the latter lecture someone in the audience near the back of the hall asked him whether the 3.7 residue helix was the same as one of those which he had described in 1943. The questioner had collaborated with Pauling in 1934. It was Maurice Huggins. Pauling replied that it was different. Huggins elaborated on some details and again said: "Isn't it similar?" Pauling then hedged a bit, but finished by asserting that it "differed in important respects" (Huggins, 1969a; Huggins, 1969b). Pauling had rightly stated that his two models differed from the 3_{10} helix which Huggins had described in 1943 (see Fig. 17.10), in which each residue was hydrogen bonded to a residue *two* to the left or right of it, instead of *three* to left or right. The *topology* (i.e., the connexions of the hydrogen bonds) was completely different. The fact that Pauling did not go into details suggests that he was not sure exactly what sort of a helix it was that Huggins had described in 1943 that was like the α helix. He had probably relied on the descriptions of Huggins' models given by Bragg, Kendrew and Perutz, and these authors had described them as integral. This would explain why Pauling, Corey and Branson wrote that "Bragg and his collaborators and Huggins discussed in detail only helical structures with an integral number of residues per turn . . ." (1951, 210). In fact Huggins had said of the model illustrated in Figure 17.10: "There are about three residues per turn of the spiral", and

> It may be noted that there is nothing about this structure which requires exactly three residues per turn of the helix. In fact, it would seem, from the models that have been made, that the bond distance and angle requirements are best satisfied by a slightly smaller number of residues per turn.
>
> (Huggins, 1943, 211)

Unfortunately, the model in question was drawn as if it were an integral helix. It returned to the same point in exactly three residues. *Furthermore Huggins did not assume a planar peptide link.* To be sure he knew about resonance of the amide bond, having discussed it with Pauling in their paper many years before (Pauling and Huggins, 1934, 225). But in 1943 he was not sure whether there was enough double bond character to the peptide link to prevent departure from planarity (Huggins, 1969b). In his 3_{10} helix this departure would appear to be about 40 per cent. Was Huggins really working along the same lines and on the same principles as Pauling? From a close examination of his papers it seems not. At no time did he discuss the resonance energy of the amide bond in connection with the peptide link, and when he drew the resonating structures concerned (Huggins, 1943, 196) he showed the hydrogen atoms in different positions. But resonance does not involve a change of position of the nuclei. Therefore he was confusing resonance with tautomerism.

Model Building

Despite these shortcomings, Huggins' 1943 paper had a profound influence on the Cavendish group. Crick recalled that it was "seminal", for here was spelt out more than before the stereochemical arguments basic to the practice of model building for long-chain molecules. In a paper on ionic crystals Pauling had introduced his "Rule of Parsimony" (Pauling, 1928b, 16). Huggins expressed this idea in simpler language, when he said: "atoms of the same kind crystallizing in the same environment, tend to be surrounded similarly. As a result of this all like atoms or ions are almost invariably surrounded in a similar fashion throughout a given crystal . . ." (Huggins,

1931, 1270). Then he mentioned the principle of close packing, which he enlarged upon in 1943.

> ... the most stable arrangement for an assemblage of molecules is one in which the component atoms and groups are packed together so that (a) the distances between the neighbours are close to the equilibrium distance, (b) each atom or group has as many close neighbours as possible, and (c) there are no large unoccupied regions. In other words, each structure tends to be as "close-packed" as possible, consistent with the "sizes" of its component atoms or groups.
>
> (Huggins, 1943, 198)

In the same paper he had emphasized the need for models of fibrous proteins to have a screw axis of symmetry (translation along the fibre axis with rotation around it).

> The unbalanced forces on opposite sides of a chain which has no screw axis—e.g., any of the earlier chain structures advocated for α-keratin by Astbury or the one that he has most recently proposed for collagen—would tend to bend it continuously in the same direction ...
>
> A principle which is logically reasonable and which has been amply verified by structure analyses of a great many substances is that like atoms or atomic groups tend to be surrounded in a like manner. Because of the variety of R groups present in any given protein, some differences between the environments of corresponding groups must be expected, but these differences should be minor ones. In general, a structural pattern for a protein in which like groups are all surrounded in a like manner, except for differences between the R groups, is more probable than one in which this is not the case.
>
> (Huggins, 1943, 197)

No one can say that the principles leading to helical conformations were not spelt out by Huggins with clarity. Nor can it be said that he failed to emphasize the importance of hydrogen bonds both in 1937 and in 1943. Had he accepted the planarity of the peptide C–N bond he must surely have discovered the helix, for as Bragg was later to remark: "With this condition the helix immediately follows", for the chain can then turn a corner only at the α C atom (Bragg, 1965, 6).

Pauling differed in his approach from Huggins not only in his conviction of the correctness of the conclusions derived from resonance theory, but in his personal involvement in the programme of structural studies of amino acids and peptides at Caltech, in which rigorous methods were used. Patterson analysis was kept in its rightful place as a preliminary guide to the most acceptable structure. This was then built using accurate equipment, and the diffraction pattern to be expected from this model was calculated by the method of Fourier synthesis. This theoretical pattern was compared with the observed pattern. If necessary the model was refined and the process repeated. Because Astbury never attempted such accurate determinations his work was left far behind. Because Huggins was not involved in such work he lacked Pauling's confidence at its results.

In their model building both Huggins and Pauling recognized that helices might be non-integral, but there is no doubt that current thinking in crystallographic circles was set in a groove and required more than an odd sentence from Huggins to throw it out. To many a worker it was not a question of non-integral helices being considered and rejected, but not even being considered. When one thought of a chain repeating a pattern and thus generating symmetry, it was obvious to think in whole numbers. Were there not only five symmetry types—1, 2, 3, 4, and 6-fold—all of them integral? When Tony North learnt his trade in the early fifties, this was the way he was taught to think (personal communication). Just how much of a surprise to many workers was Pauling's announcement of 3.7 and 5.1 residues in the unit cell in the fibre direction can be guaged from Bernal's remark that biomolecular studies broke and shattered "formal crystallography" completely.

> We clung to the rules of crystallography, constancy of angles and so forth, the limitation of symmetry rotations to two-, three-, four-, and six-fold, which gave us the 230 space groups, as long as we could. Bragg hung on to them, and I'm not sure whether Perutz didn't too, up to a point, and it needed Pauling to break them with his irrational helix.
>
> (Bernal, 1966, 3)

This was Bernal's impression. But it was not true of Bragg. Why, then, did the Cavendish group not discover the α-helix? Strange to relate, they *did* build a helix with planar* peptide link—their 4_{13} model—and they published it in 1950 (see Fig. 17.11). It is almost identical with Pauling's and Corey's α-helix! Instead of repeating in 3.7 residues it repeats in 4—thus having four-fold symmetry. Having come so near, they proceeded to reject it on the grounds that it repeated in $5\frac{1}{2}$ Å instead of the 5.1 required by α-keratin, and the similar repeat distance derived from Perutz' Patterson projections for haemoglobin and Kendrew's for myoglobin. From these projections, too, they had deduced polypeptide rods which, on the basis of density of the crystals, had three, not four, residues in the repeat. In desperation they concluded their Royal Society paper by supporting the ribbon-type twofold chain described by Astbury and Bell in 1941!

Consternation in Cambridge

When Perutz' Patterson maps of haemoglobin suggested the presence of parallel polypeptide chains in folded conformation (1949) Sir Lawrence Bragg became "deeply interested" and he "speculated on the form of the folded . . . chain" (Bragg, 1965, 5). He favoured a Huggins-type helix "because it placed each amino acid residue in the same kind of position in the chain" (*Ibid.*). He and his colleagues realized that the "chains may possess a twofold, threefold, fourfold or higher screw axis of symmetry" (Bragg, Kendrew and Perutz, 1950, 330). The symmetry rules *could* be relaxed. But how far? To non-integral helices? Yes, providing the diffraction data allowed, but it did not! Astbury's α-keratin picture, with its 5.1 Å repeat on

* Almost but not quite planar.

a layer line, stood in the way. Accordingly, the Cavendish group rejected non-integral helices. On the other hand, they considered models with planar and pyramidal peptide C–N bonds.

Bragg recalled how they had invited their "chemical colleagues to look at what they felt were our most promising structures and [to] tell us of any criterion which would make one more probable than the rest, but we completely missed the real clue" (Bragg, 1965, 6). Perutz confirmed that Lord

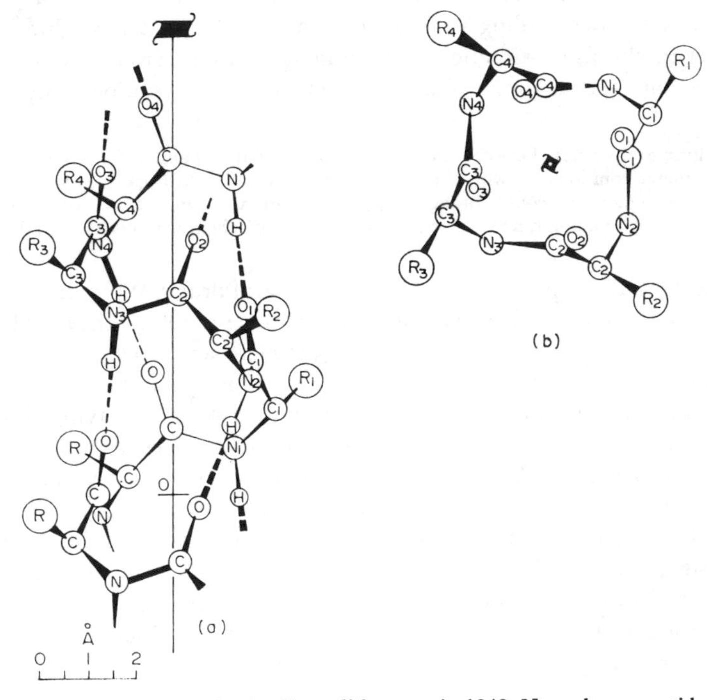

Fig. 17.11 The 4_{13} helix built by the Cavendish group in 1949. Note planar peptide link. (a) Viewed from the side (b) in projection down the helix axis (from Bragg, Kendrew and Perutz 1950, 336).

Todd was invited over, "but we did not ask him about the peptide bond, and he evidently did not note the fact that in the model before him they were not planar. Of course he knew they had to be planar" (Personal communication). Lord Todd also believed that he would have told them about the planar bond if only they had asked him. What are we to make of this? Did they show him their 4_{13} helix with planar peptide C–N bonds? I doubt it. Was Todd so aware of the planarity of these bonds as Perutz has suggested? Again, it is doubtful. Did they show him their models? Todd states that he "was never invited" to look at them.

Crick was himself a witness to a discussion on the planar or pyramidal character of the peptide link, and is firmly convinced that the participant who argued for the pyramidal version was Charles Coulson, then professor of theoretical chemistry at King's College London. Coulson had argued from the case of ammonia in which the third hydrogen atom can flip from one orientation to another. Crick did not know enough structural chemistry at that time to contribute to this discussion and the Cambridge physical chemist, Manley Price, who taught the planarity of the peptide bond, was not asked for his opinion. Cohn and Edsall gave the planar configuration in their great book *Proteins Amino Acids and Peptides* (1943), but this was no more than a restatement of Corey's opinion based on diketopiperazine, which is not a peptide (Cohn and Edsall, 1943, 322).

Perutz believes that the situation might well have been different had his request to visit centres in the United States been granted by Sir Edward Mellanby, Secretary of the Medical Research Council. In 1948 he had written asking for leave of absence for the first half of the year 1949 so that he could pursue the suggestion of Weaver and Pomerat that he apply for a Rockefeller Travelling Fellowship. He received the following reply:

> I do not think the Council would look favourably on your suggestion that you should go to the United States for six months, with a Rockefeller Travelling Fellowship. They take the view that a Unit of Research having been established, the Director should not only remain at his post, but that he should be of sufficient standing to attract Americans to his laboratory, rather than that he should go and seek their help. The next stage, which they welcome, is an invitation to go to America to occupy some well-known lectureship, in recognition of the high standard of research.
>
> These are the stages that I think you should aim at, rather than ask for a Rockefeller Foundation Travelling Fellowship. I do not, therefore, propose to forward your application . . . and I return it herewith . . .
>
> (Mellanby, 1948)

Hence it was not until March 1950, when Perutz visited Pennsylvania State College and other places, that he heard from E. W. Hughes about the recent work at Caltech on acetylglycine and the improved data on amino acids. Not until 1953 did Perutz visit Caltech itself. Meanwhile the model building exercise in Cambridge had been completed. And on his return from the United States in 1949 Perutz' concern was not with the chain fold in polypeptides but with sickle cell anaemia, about which he had learnt during his American trip. At once he started experiments to try to demonstrate crystallographic differences between the normal and sickle-cell haemoglobin crystals. This was only partially successful and was followed by the programme on isomorphous replacement.

Clearly there were many problems being attacked at that time, and only with hindsight does one find it curious that what we now see as so significant was not appreciated as such at the time. Certainly Bragg took the question of the peptide bond seriously, and about 1947 or '48 he initiated work on the dipeptide, cysteyl–glycine–sodium iodide by William Cochran's student H. B. Dyer, who was in Cambridge on a Croll scholarship from South Africa. Sir Lawrence took an active interest in this work, as can be seen from his correspondence in 1949. In January he told H. S. Lipson that Cochran had just got out quite a pretty structure. "It is the structure of a dipeptide of glycine and cystine, combined with sodium iodide. By an extraordinary lucky fluke the iodines lie on the twofold axes, so the Patterson projection along this axis immediately tells one the whole structure of the molecule" (Bragg, 1949a). In February he wrote to Sir Charles Harington, who had supplied the crystals for Dyer's work:

> We have been thinking hard about the protein chain in the light of Perutz' latest results on haemoglobin. If we can trust the information they give that the fundamental thing about the protein chain is a repeat of three amino acids in every 5 Å of length, the possibilities narrow down to very few, only two or three in number. An essential piece of information in building structures for the chain is a knowledge of whether the three bonds round the nitrogen atom are coplanar or inclined at tetrahedral angles. We hope to get this from the dipeptide, the only case as yet where a bit of the actual chain has been analysed. It ought, however, to be checked on other compounds.
>
> (Bragg, 1949b)

In May Sir Lawrence wrote to Harington again asking for more peptides:

> Dyer is now completing the analysis of the cysteyl–glycine, and is looking for fresh worlds to conquer. We want to do some of the simpler peptides so as to get a clearer idea of the orientation and length of the peptide chain, which we must incorporate in our models (of polypeptides). For instance it is not yet certain whether the three bonds from the nitrogen atom in the chain are coplanar and 120° tó each other, or are directed to the three corners of a tetrahedron. The solution of a few peptides might settle this point. Unfortunately in cysteyl–glycine the iodine masks the N atom in rather an awkward way in the vital projection.
>
> (Bragg, 1949c)

He also told Heilbron how eager he was to "get some more material in the shape of crystalline di, tri, tetra and penta-peptides, which I think we could analyse rapidly. We are hoping, of course, that when we get examples of the long peptide chains they may hint as to how the chains are folded in the proteins themselves" (Bragg, 1949d).

Instead of a full three-dimensional Fourier synthesis Dyer used a limited form of three-dimensional Patterson synthesis. He confirmed the short C–N distance for peptides which in his structure came out at 1.32 Å, but the αCNC angle was 139°—an increase of over 16 per cent on the Caltech figure of 120°. "This cannot be entirely ascribed to experimental error," wrote Dyer (1951, 46). His abnormally large value for this angle was referred to by Bragg, Kendrew and Perutz in their Royal Society paper (1950, 328) in which they assumed for their polypeptide models a peptide N angle of either 109° 28′ (tetrahedral) or 120° (planar). Justifying this decision in a footnote they wrote:

> No complete structure analysis of any compound containing nitrogen in proteins of the α-keratin type is yet available. We have therefore thought it better to leave this question open for the present, and to examine, in each type of structure, the effect of the two configurations given.
>
> (Bragg *et al.*, 1950, 329)

Presumably this conclusion was not questioned when in February 1950 Kendrew gave a talk on polypeptide chains to the informal "Hardy Club" which R. D. Keynes, Michael Swann and J. M. Michison had founded in 1949. Besides the three founders, the following came to hear Kendrew: K. Bailey, H. Barlow, F. H. C. Crick, A. L. Hodgkin, A. F. Huxley, P. R. Lewis, M. F. Perutz, L. Picken, J. W. S. Pringle, M. G. M. Pryor, J. R. Robinson and Lord Rothschild. One notes no organic or physical chemist in this group. The biologists were surely not qualified to question Kendrew's conclusions. One assumes that Crick and the other crystallographers present accepted Perutz' view (personal communication) that Dyer's data made the Caltech evidence on the planarity of the peptide bond seem questionable. Indeed, Crick read Dyer's thesis but did not question his conclusions (Crick, 1968/72).

Perutz Observes the Residue Repeat

It was about the end of May 1951 that Perutz received the eight papers on the α- and γ-helix. It was a Saturday, and by lunchtime he had read through them and realized that if the α helix was the structure of the polypeptides in fibrous proteins and haemoglobin, there should be a wide-angle reflection on the meridian corresponding to the very small axial distance between neighbouring amino acids in the helix of 1.49 Å. That afternoon he went to the laboratory, found a horse hair and a porcupine quill and took oscillation photographs on a cylindrical film of 3 cm radius "instead of the flat plates normally used". To his delight the photographs which he developed that afternoon showed the predicted reflexion. His Saturday labour had been well worthwhile. Next week he met A. Elliott in London and told him of his discovery. Elliott sent him a sample of the synthetic polypeptide, poly-γ-benzyl-L-glutamate from which he again obtained the new reflexion. His haemoglobin data went down to 2 Å resolution and could not therefore show the 1.5 Å reflexion, but on further studies along the x axis he detected a faint bulge "with a distinct maximum of intensity at 1.5 Å". This finding, taken with other evidence stated by Pauling, Corey and Branson, though not proving the structure "leaves little doubt about their structure being right" concluded Perutz. Here then, said Perutz, is a discovery, showing:

> that even relatively disordered substances like hair may contain an atomic pattern of such high intrinsic regularity that it gives rise to X-ray diffraction effects at spacings where they had never before been suspected.
>
> (Perutz, 1951, 1054)

The unflappable Perutz had lost no time. He did not spend the week-end wallowing in regrets that they had missed the boat over the α-helix, but immediately tested the predictive power of the structure. To Dorothy Hodgkin (personal communication) Perutz' demonstration that there are stretches of α-helix in haemoglobin, and therefore in a functional protein, meant more than Pauling's demonstration that one could build plausible models of a helical polypeptide—either the α- or the γ-helix, one of which proved right and the other wrong. This view reflects her own earlier recognition that chain configurations in proteins must be helical. In the 1940s, led on by the trigonal symmetry of insulin crystals, they had built a helical chain with a threefold screw axis of symmetry. This they soon found had been done by Huggins in 1943 and H. S. Taylor in 1941. They went on to place short lengths of this model "parallel with the threefold axis in positions suggested by Bernal's observations on the insulin Patterson projections . . ." (Hodgkin and Riley, 1968, 26). But a visitor to the laboratory soon "destroyed this model practically at birth." Now at last, in 1951, it seemed the position and character of these chains in proteins was being revealed.

The Impact of the α-Helix

We have cited a passage from an address by Bernal which captures very accurately the impact of Pauling's *non-integral* helices on the profession of structural crystallography. From 1951 onwards the way of thinking of the structural crystallographer was broadened. The rules of the game were reinterpreted.

Although Pauling made helical models fashionable it can hardly be said that he introduced them, and he would certainly not claim this. Helical models were already advocated for selenium, tellurium and many other organic polymers (see Chapter 5). Pauling introduced a helical model for fibrous sulphur in 1949, Huggins advocated helical conformations for the fibrous proteins in 1943 and he influenced the group at Cambridge. As Crick recalled, the helical conformation:

> was very much in the air just before the α-helix . . . Pauling obviously thought that way; Bragg was convinced things were helical. They were all building helical models. I would say, you would be eccentric, looking back, if you didn't think DNA was helical.
>
> (Crick, 1968/72)

But there is no doubt that Pauling's and Corey's work made sure the helix became fashionable. Instead of being one of many conformations met with in chemistry, it became the most common structure for the student of biologically important molecules. Such a conformation of long chain molecules was now to be expected in natural and synthetic fibres, but since one could rarely achieve single crystal orientation in these fibres there was a growing need for the development of a predictive theory for helical diffraction. The result was that at the end of 1951 the Fourier transform of a helix was worked out. The interpretation of fibre diagrams then moved on to a fresh level, and it was in this context that the structure of DNA was established.

CHAPTER EIGHTEEN

Watson and Crick

> I'm Watson, I'm Crick,
> Let us show you our trick,
> We've found where the seed of life sprang from.
> We believe we're a stew
> Of molecular goo
> With a period of thirty-four Ångstroms.
>
> (E. S. Anderson *et al.*)

When Watson and Crick met in 1951 they presented a striking contrast. Crick was a confident, ebullient, articulate Englishman from the middle classes, with a loud laugh, an insatiable curiosity, who could talk for hours on end, and was decidedly extroverted. Watson was diffident in manner, his words were brief, his curiosity was strictly confined to scientific subjects and ornithology. He was introverted, he appeared as a "loner". Whereas Crick at 35 had still not completed his Ph.D. thesis Watson had completed his in 1950 at the age of 22.

Long ago Watson had been discovered by Louis Cowan, producer of the Chicago Quiz Kid show. The University of Chicago took him for a Bachelor's course in science when he was only 15 years of age (as part of an experimental policy of early intake from any part of the country). Chicago gave him the kind of courses in biological sciences that reflected its strength in embryology. But Watson was not "switched on" by them. His boyhood zest for ornithology remained dominant. Paul Weiss, who taught him embryology and invertebrate zoology, recalled how remarkable Watson had been in those three years: "He was [or appeared to be] completely indifferent to anything that went on in the class; he never took any notes and yet at the end of the course he came top of the class" (Personal communication).

Apparently Watson learnt little genetics at Chicago, though the population genetics of Sewell Wright "intrigued him". But from the moment that he read Schrödinger's book *What is Life?* he "became polarized toward finding out the secret of the gene" (Watson, 1966, 239). One may doubt how strong this polarization was at the time—it did not prevent him from enrolling in a summer school for advanced ornithology and systematic botany at the University of Michigan in 1946, and when he applied to do graduate work at Indiana University it was ornithology that he put as his major interest. The dean of sciences and chairman of the zoology department—Fernandus Payne—recalled that:

> After James graduated at the University of Chicago he wrote me about the possibility of a fellowship and stated that his interests were in ornithology. I replied and said that the department of zoology was not prepared to offer graduate work in ornithology and advised him to consider Cornell. I further stated that we were qualified to offer graduate work leading to the Ph.D. degree in genetics and experimental embryology . . . Following this correspondence James and his father came to the university for a personal visit . . .
>
> (Payne, 1973)

Payne's show of straight-talk seems to have impressed Watson. He came to Bloomington for an interview, and was awarded one of the University's dozen-odd graduate fellowships. Thus, the quiz kid from Chicago, the bird-watcher, could be seen in Bloomington, a strange figure, always clad casually, usually wearing tennis shoes, tall, thin and awkward-looking. Watson did not have a fund of small talk and he lacked an affable friendship with the other graduate students in his year. With David Nanny (now at Illinois) he did form a close friendship based on mutual respect, but it was characteristic of Watson to seek out older more experienced men and talk with them. Some of his fellow students thought of him as "way out". He would walk past them on the corridor with that far-away look in his eyes. To those with whom he did not wish to talk he could be reserved, almost disdainful. This young aspiring scientist had come up quickly to graduate at 19. He was used to the companionship of bright lads at the University of Chicago and had no time for numskulls.

Tracy Sonneborn allowed Watson to join the Friday evening seminars at his house in Bloomington where his graduate students discussed their work. Not all of them liked the way Watson would turn discussions in a direction which was of interest to him, and none liked his habit of opening a book to read when the speaker proved dull or unintelligent.

Watson thrived on scientific discussions. He was a man of single-minded purpose, and had no inhibitions about going up to the great men of science to engage them in discussion, although he was but 19 years of age. After a seminar he could be seen in conversation with Muller, Sonneborn or Luria. At scientific meetings it was men of their calibre and experience that he sought out in his thirst for the mysteries of the gene.

The Nature of the Gene

What was the source of this curiosity of Watson's? We have mentioned his reading of Schrödinger's book *What is Life?* and Payne's insistence that he must not devote himself to ornithology at Indiana. We may also mention his reading of Sinclair Lewis' wonderful satire on medical research: *Martin Arrowsmith.* Martin's aspiration to follow in the steps of the dedicated research scientist, Max Gottlieb, and his discovery of phage, is said to have given Watson the urge to discover something great too, and to find it in the science of microbiology. But as Payne no doubt impressed upon Watson, Indiana had one of the greatest living geneticists of the age—Herman J. Muller. What better than to study at the feet of the great man himself.

So Watson began graduate work in the zoology department, with the understanding that he would probably carry out research on *Drosophila* genetics under Muller. Soon Watson judged that, despite the presence of Muller, *Drosophila's* "better days were over and that many of the best younger geneticists, among them Sonneborn and Luria, worked with micro-organisms" (Watson, 1966, 239). Watson completed Muller's course on "Mutation and the Gene" and got an A for it, but it was to Salvador Luria and his work that Watson was attracted when he attended Luria's lectures on viruses. Accordingly he spent much of his time in the bacteriology laboratories on the attic floor of the language arts building.

Luria had recently discovered the production of active phage particles when *many inactivated* phages were allowed to infect bacteria. He called this multiplicity reactivation and he explained it in terms of the existence of a "gene pool", formed by the independent replication of discrete genetic units, from which active phage particles were assembled (Luria, 1947; 1950). The infecting phage was assumed to break up into such units once it entered the host cell. The inactivity of ultraviolet irradiated phage was attributed to damage to one or several of these units. Damaged units of one type, Luria believed, could be replaced from the gene-pool by undamaged units of another type, and active phage particles successfully assembled. By quantitative techniques Luria and Dulbecco were able to suggest figures for the number of such sub-units in the various phages (Luria and Dulbecco, 1949, 113).

Watson's Thesis

Luria therefore felt encouraged to pursue this promising approach further—deriving a model for phage replication from the genetic features of irradiated phage. So far reactivation of phage was restricted to particles treated with ultraviolet light. Luria failed to detect it for particles treated with hard X-rays. This difference looked interesting. Further examination might reveal more features of the replication process. Or in the words of Watson's thesis:

> It is probable that inactivating agents other than ultraviolet light can cause partial damage [to phage]. More interesting is the possibility that these agents will cause different types of damage, which will block virus reproduction at different stages. We might therefore be able to reconstruct the successive steps in host virus interaction by studying at what stages in synthesis the multiplication of the inactive phage is blocked. To test these possibilities we have begun to study bacteriophage inactivated by X-rays.
>
> (Watson, 1950a, 1)

Unfortunately the programme did not live up to these expectations, and its chief value to Watson lay in the conviction it gave to him and to Luria that the approach through radiation genetics was not promising. The scheme of replicating sub-units seemed to Luria in his essay on virus reproduction "to be as far as we can go at present in analysing phage production from evidence supplied by the end products" (Luria, 1950, 509).

Graduate Course Work at Bloomington

One might conclude from *The Double Helix* that Watson's Indiana studies led to no knowledge of chemistry and physics. This would be wrong. For his thesis he had studied and understood the literature on the target theory of radiation genetics. From Chicago he had credits in inorganic, qualitative and organic chemistry, and from Indiana credits in advanced organic laboratory preparations (despite the ether incident described in *The Double Helix*), and in the lecture course on proteins and nucleic acids. Although he only scraped home with a C in scientific German, he used this expertise to good effect in his reading of scientific literature both in Bloomington and in Cambridge.

An obvious question is whether or not the course on proteins and nucleic acids gave Watson his conviction of the importance of nucleic acids in heredity, but as the course professor, Felix Haurowitz, later remarked (personal communication) the effect, if any, on Watson would have been the negative one of stimulating him to show that Haurowitz was wrong. These lectures were written up as given, and delivered to Academic Press in the summer of 1949 after which only minor additions were made. Although there is a reference to Gulland's statistical tetranucleotide and Chargaff's 1948 results on base composition, Haurowitz declared: "The present state of knowledge does not permit us to say whether or not nucleic acids are species specific" (Haurowitz, 1950, 223). If they were not, then the nucleic acids could have but a "secondary importance". Haurowitz was convinced that the specification of such complex molecules as the proteins could only be achieved by equally complex molecules—by proteins. Not until Crick and his co-workers demonstrated the phased unidirectional reading of a triplet or multi-triplet nucleotide code in 1961 did Haurowitz accept that DNA is the hereditary substance. His scheme for protein synthesis is shown in Figure 18.1.

There was, in fact, very little about nucleic acids in these lectures. A student who wrote about nucleoproteins for his terminal essay gave no hint of the importance of DNA in heredity, and Watson had no occasion to make a special study of the literature on nucleic acids because his terminal essay was on lysozyme (Watson, 1949a).

Whilst in Bloomington, Watson wrote a 25 page essay for Sonneborn entitled: "The Genetics of Chlamydomonas with Special Regard to Sexuality". Here he analysed the work of F. Moewus and R. Kuhn, concentrating

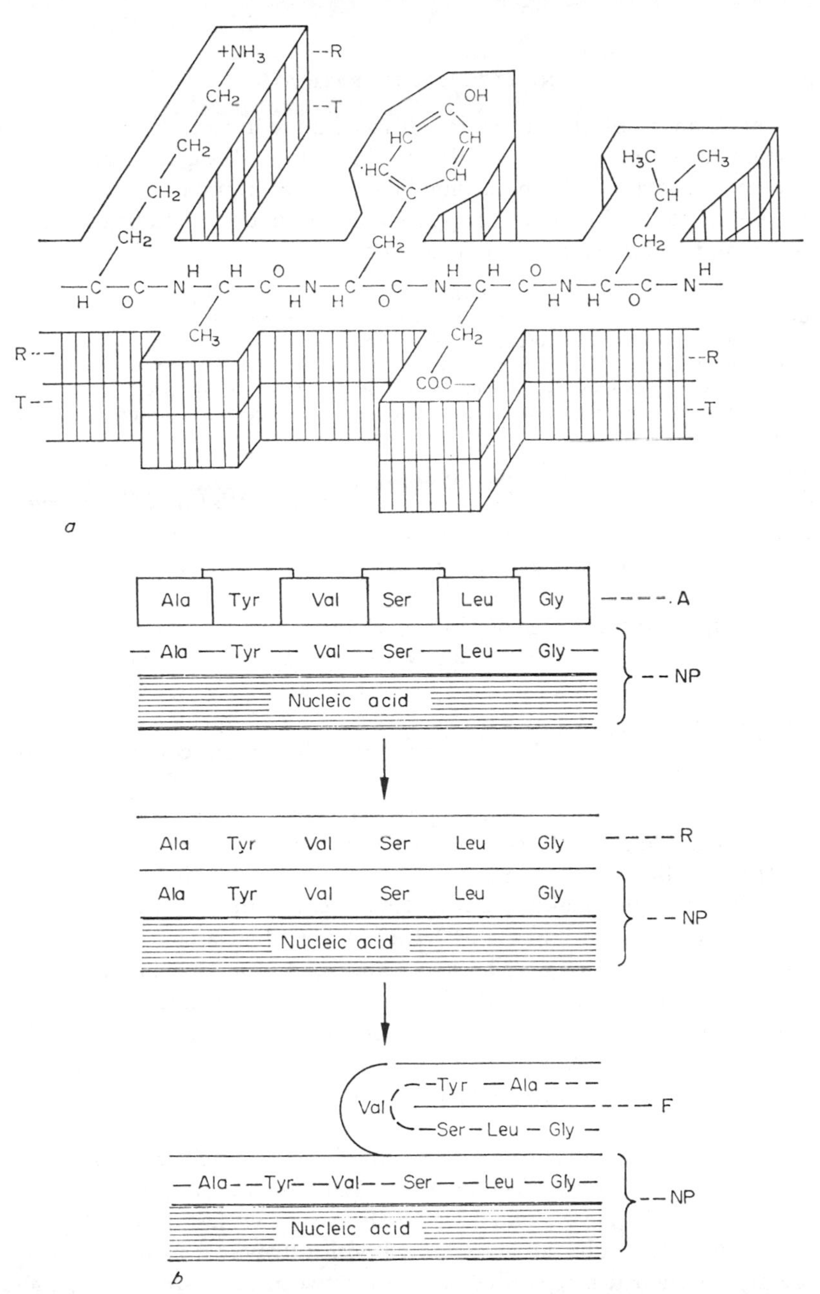

Fig. 18.1 (a) Duplication of a protein according to Haurowitz, showing like-with-like attraction between amino acids of the template polypeptide chain (T) and free amino acids which have become linked together by non-specific enzymes to form a replica chain (R) (from Haurowitz, 1950, 345). (b) Duplication of protein template by like-with-like attraction, the nucleic acid acting as a stretcher holding the template chain in the extended configuration. A = adsorbed amino acids, NP = nucleoprotein template, R = replica chain, Ala = alanine, Tyr = tyrosine, Val = valine, Ser = serine, Leu = leucine, Gly = glycine (from (Haurowitz and Crampton, 1952, 48).

his attention on the biochemical side. This led him to introduce a new scheme for the genic control of the sexuality mechanism in *Chlamydomonas*. He admitted that it lacked "the basic simplicity of the original Kuhn–Moewus idea", but he claimed for it fewer apparent contradictions, and he concluded his essay on the following enthusiastic note:

> In spite of these changes the Moewus–Kuhn work on sexuality remains a first rate undertaking of most gigantic magnitude. However, with regard to the genetic analysis [by Moewus] one often has the partial feeling that some of the statements reported as facts were merely wishful thinking. It is hard to imagine how all of [the] work reported was ever done. Certainly some experiments were either never done or reported falsely. Again, one is forced to repeat the usual statement that work of this great importance with regard to gene action should be repeated independently.
>
> (Watson, 1949?b)

This would appear to be the first clear expression by Watson of his recognition of the fundamental importance attaching to the chemistry of gene action. Subsequently Moewus himself had to admit that his work was unrepeatable.

Watson spent the summer of 1948 in Cold Spring Harbor on his phage experiments. There he came to know Ernst Mayr and his family. They enjoyed this prankish, ever-active boy and they respected his serious devotion to science. Now according to Mayr everyone at Cold Spring Harbor recognized the importance of Avery's work, and they were convinced of the involvement of DNA in the determination of genetic specificities (personal communication). Add to this Luria's personal acquaintance with Avery from the visits he made to the Rockefeller when he was working in New York in the 1940s. "I had great admiration for him", he recalled, "and I also enjoyed his as a personality" (Luria, 1968). Everyone there knew Avery's work "was exceptionally important" (*Ibid.*). Add to this the fact that Delbrück and Avery's brother Roy, were both at Vanderbilt University, and that Roy Avery showed Delbrück his brother's famous letter about the identity of the transforming principle (Delbrück, personal communication), and the fact that Delbrück visited O. T. Avery at the Rockefeller. Is it then surprising that Watson should have become convinced of the importance of DNA in heredity?

How are we to equate this with the absence of any telling references to DNA in Watson's work at Bloomington? Why are the only comments on DNA in his thesis non-commital? (On the very low frequency of photo-reactivation of X-irradiated as compared with ultraviolet irradiated phage he simply mentioned the possible *connexion* between photo-reactivation and the specific absorption of ultraviolet light by the nucleic acid fraction of the phage. This connexion "cannot yet be decided" (Watson, 1950a, 39). And he found Seymour Cohen's discovery of the absence of DNA synthesis in cells infected with ultraviolet irradiated phage "of interest" (*Ibid.*, 42). How could Luria remark that "DNA may be involved mainly in the final steps of the 'baking' of active particles" (Luria, 1950, 509) if he really believed all along that it was the phage's hereditary substance?

The answer is surely that for Luria and Watson, as for other members of the Phage Group, the question of *the precise role* of DNA was regarded an *open* one. It was up to the chemists to settle it, not the biologists. The latter were busy exploring the mechanism of gene duplication from their side. Luria and Watson were hoping that the more they learnt about bacteriophage the closer they would get to the gene. Luria felt that "radiation work had been so suggestive of the properties of the gene; therefore I felt that interpreting radiation effects at the level of something as simple as a phage would help interpret radiation effects on the gene" (Luria, 1968). This was the thrust behind most of his work until 1950. Luria also admitted that many of them in the Phage Group "were so biased by the idea that the information would somehow have to be in protein that we probably kept our minds too open." This open attitude was continued longer than was justified because "as long as you could work on bacteriophage and were not doing much chemistry . . . you could say: Which one (nucleic acid or protein) will be the critical one we do not know . . ." (*Ibid.*). By the year 1950 three events had played their part in changing Luria's attitude: the discovery that the contents of phage particles can be liberated by osmotic shock leaving phage "ghosts" (Anderson, 1949), the discovery of photoreactivation in phage (Dulbecco, 1949), and the indecisive results of Watson's research. The time had clearly come for a change of strategy.

Separate Functions for Membrane and Contents of Phage

Long before Watson came to work with Luria it had become clear that the concept of phage as some "primordial gene-stuff" (Hershey, 1966, 100) composed of nucleoprotein and functioning as an autocatalytic enzyme (see Chapter 9) had been discarded. The early electron micrographs of sperm-like phage particles obtained by Ruska (1941) and by Luria and Anderson (1942) showed clearly that the phage had an organization. On the other hand the metabolic inertness of the free phage particle did not encourage the members of the Phage Group to think of it as an intracellular particle. The result was Luria's suggestion that it broke down into smaller units within the host cell. The natural step to take, surely, was to study the chemistry of this process? Here, those who were pursuing Delbrück's line kept to genetics and continued to put their faith on such phenomena as multiplicity reactivation. The impression that Delbrück gave at the time was not so much that he disliked enzymes and intermediary metabolism but rather that he was apathetic about them. Luria, on the other hand, made no bones about expressing his lack of optimism for metabolic studies. Yet biochemists were not excluded from the Group. True, they found the task of putting over their work to the others hard going. They had to explain their work in non-technical terms. One of them—Seymour Cohen—often had the feeling that the biochemical work was not understood. As we shall see, the attitude of the Group to chemistry did change by 1951, but then it seemed that metabolic studies were not going to yield the clues to phage replication. The first biochemist in the Group was Seymour Cohen, who had been trained in the biochemistry department of Columbia University, where isotopes were used. As Chargaff's first doctoral student he worked on the chemistry of thromboplastin and of the Rickettsiae and was involved in extracting nucleic acid constituents. Then he had gone to Princeton to work with Stanley on plant virus nucleic acids. War work brought him back to Columbia to work on the Rickettsiae and when this work was transferred to Philadelphia, Cohen went there to continue it.

During this time under Stanley, Cohen became interested in the mechanism of viral infection and on many occasions he asked Stanley "Why there was no work going on in the multiplication of TMV." Stanley's reply was that they had enough to do concentrating on the virus itself, believing as they did that "by analysing it you would obtain clues about how it operated" (Cohen, 1973a). Cohen could see that he was not going to get what he wanted "just from looking at the virus, especially since at that time we thought it was the protein we had to worry about and not the nucleic acid" (*Ibid.*).

After the war Cohen tried to use tissue culture cells for studies of viral infection, but failed. Then he began to work with Thomas Anderson who was using the electron microscope to study influenza virus and Rickettsia. Anderson taught him the techniques of phage work and Cohen began an orienting piece of research.

> I was at that point [1945] merely trying to outline the general parameters of the phage-infected cell—what was the same as the uninfected cell and what was different. I began with respiration and showed that that had not changed at all. I was looking for some "handle" and doing those things which came easiest . . . I had a rather stiff experience with nucleic acids from Stanley and a good portion of my work with Rickettsiae related to them. We had fairly sensitive colorimetric techniques for doing nucleic acids and the protein methods were relatively insensitive.
>
> (Cohen, 1973a)

In December 1945 Cohen visited Vanderbilt University and described to Delbrück, Luria and other phage workers the results of his colorimetric study of total nitrogen incorporation and DNA synthesis in the infected cell. He had discovered the seven minute lag after infection, followed by a rapid increase in the rate of DNA synthesis. His audience did not know what to make of this seven minute lag and whether it was true.

> They were wondering about this character who was producing a new kind of data, and perhaps about how valid this data might be. Was I really firm on this and who was this

guy coming into the group. None of them had done nucleic acid estimations—few people had. There certainly was not any discussion of it in terms of Avery, far from it!

(Cohen, 1973a)

The following summer Cohen attended the second phage course at Cold Spring Harbor and in 1947 he contributed to the Cold Spring Harbor Symposium. In 1948 he published the first study of ^{32}P incorporation into phage, when he showed that most of the isotope in the phage progeny derived from material assimilated from the medium by the host cell *after* infection (Cohen, 1948). Lloyd Kozloff, too, had been working on the metabolism of the phage-infected bacterial cell, at the suggestion of Earl Evans, professor of biochemistry at Chicago. He had attended Delbrück's phage course in 1946, and, like Cohen, had also been exposed to isotopic studies at Columbia. Kozloff was soon joined by Frank Putnam, and in the environment of biochemistry at Chicago: this group set to work to trace the sources of phosphorus. Two surprising discoveries resulted: some 20 to 30 per cent of *host* bacterial DNA was incorporated (after degradation) into phage DNA, and only between 20 and 40 per cent of the phosphorus in the infecting phage was incorporated into progeny phage.

Non-specific Transfer

The discovery that less than half the phage phosphorus could be found in the progeny phage was made in 1949. Watson was working in Cold Spring Harbor in the summer of 1950 when Kozloff and Putnam reported this result. He recalled that he was "most affected" by their talk. It was certainly disappointing if, as Watson claims, Luria and he felt that DNA "smelt" like the essential genetic material (*Helix*, 23) at the same time as "they were not at all sure that only the phage DNA carried genetic specificity (Watson, 1966, 240). Perhaps, it was argued, phage DNA consists of genetic and non-genetic parts. Cohen then suggested testing this "bipartite" theory by carrying out a second generation experiment. If correct, transfer of ^{32}P from first to second generation should be not 30 per cent but 100 per cent. This was the experiment which Watson carried out with Ole Maaløe in Copenhagen in the winter of 1950/51. The transfer was not 100 per cent but again 30 per cent.

In their paper describing this conclusion Watson and Maaløe concluded that the bipartite theory was ruled out, but: "A different answer might be obtained with a label like sulphur, that would label specifically the protein moiety of the phage" (Maaløe and Watson, 1951, 508). Luria was having his worst premonitions about the value of biochemical studies confirmed. Watson's major contribution in this tracer work, wrote Luria to Paul Weiss, "may well prove to be the evidence he provided for the inadequacy of the tracer technique for the analysis of the chemical basis of genetic continuity in virus reproduction" (Luria, 1951a). About five months later (March 1952) Luria confided in Hershey that he had assumed "since last summer that ^{35}S was also transferred by 30–50 per cent, and this led me to the obvious idea that all transfer is non-specific (at least in the sense that the genetic material itself may turn over rapidly) and therefore that one could tentatively forget transfer experiments as evidence for genetic continuity" (Luria, 1952a).

Maaløe and Watson did not stop here but went on to test the transfer of ^{14}C labelled adenine and glycine. They found a maximum transfer for adenine of 50 per cent whilst in the case of glycine only 10–15 per cent appeared to be transferred (Watson, 1952b, 114). Describing this work to the National Foundation for Infantile Paralysis in August, 1951, Watson, wrote:

> We have been concerned with nucleic acid metabolism, in particular, the structural stability of nucleic acids during the reproduction of genetic units. That is, does reproduction occur in such a way that the original unit (gene ?) serves as a "template" for the newly formed particles and is then lost upon the completion of replication, or is replication accomplished by a process which necessarily involves the incorporation of the original particle into the progeny particles . . .
>
> Our results at present . . . are quite surprising. While approximately 50 per cent of the nucleic acid is transferred to the progeny, only 10–20 per cent of the parental protein appears in the progeny. Moreover, the isotopic transfer may not be specific since we find that agents such as ultraviolet light and X-rays which destroy the ability of the virus to transfer genetic specificity do not strongly effect isotopic transfer.
>
> (Watson, 1951a)

Watson as a Merck Fellow

In the autumn of 1949 the question of Watson's future plans was discussed. In the Luria–Delbrück circle "the constant reference to their early lives" left Watson with "the unmistakable feeling that Europe's slower paced traditions were more conducive to the production of first-rate ideas." It was natural therefore that Luria should think of a European scientist who had attended the Phage Course—Herman Kalckar of the State Serum Institute in Copenhagen. At this time (1949) Luria was still hopeful that the study of the metabolism of phage-infected bacteria might provide the clues to the nature of phage reproduction. Later he was to recall to Paul Weiss that he had felt Watson

> ought to receive not only a first class training, but a highly diversified one, to enable him to evaluate, choose and use a variety of approaches to the central problems of biology. In suggesting that he applies for a fellowship to spend one or two years in Copenhagen, we thought, on the one hand, that he would learn the modern techniques of enzymatic and isotopic studies on nucleoprotein synthesis; on the other hand, that he would develop his knowledge of advanced physics . . . to the extent of being conversant with the physicist's approach to molecular structure.
>
> (Luria, 1951a)

It is noteworthy that whereas Luria described the subject of Watson's work in Copenhagen as "enzymatic and isotopic studies on *nucleoprotein synthesis*",

Watson spoke of "*nucleic acid* metabolism" (Watson, 1951a). Significant also is the different wording in the introduction to the manuscript and published versions of the Maaløe and Watson paper on phosphorus transfer. Whereas the authors wrote confidently that by using ^{32}P "the genetic material can be labelled", in the version which appeared in the *Proceedings of the National Academy of Sciences* (after Delbrück had rewritten the introduction) it was possible to "label the virus *particle*" (Maaløe and Watson, 1951, 507).

In a further paper with Maaløe, Watson described how they had succeeded in raising the 30 per cent transfer of ^{32}P and ^{14}C-labelled adenine to a maximum of 50 per cent, the majority of which went to early formed phage. This result, they admitted, did not allow them to decide between transfer "via extensively degraded parental material or via large, genetically specific units" (Watson and Maaløe, 1953, 441). The state of these isotopic studies has been well described by Kozloff: "It was realized that there might be a direct connexion between the transmission of genetic information and the transmission of parental material. But all these early experiments failed to show any such connexion" (Kozloff, 1966, 111). It was unfortunate that the choice of the T-even phages was a good one for the identification of their chemical constituents, but a bad one for tracing the transmission of these constituents. These phages have far more nucleic acid than is used in their replication. The resulting redundancy obscures the connexion between genetic and chemical transfer.

We might ask what impressions those around Watson gained as to his conviction that DNA was the genetic material. Maaløe was non-committal, but he did recall how often Watson went to the library to read genetical literature. Watson even confided to Maaløe that the mechanism of crossing-over would not be understood until we knew the structure of DNA (Maaløe, 1974). Crick's opinion of this piece of evidence was not encouraging—a typical Watson remark, a "profound" statement thought of five minutes before (personal communication). Does it not reflect an association in Watson's mind between DNA and the behaviour of the genetic material? Surely it does, but this association was a hope, and the question of its validity was very much an open one during Watson's time in Copenhagen and for his first seven months in Cambridge. Not until Hershey-Chase did this situation change.

Watson's Move to Cambridge

We have no evidence that in the winter of 1950/51 Watson was aspiring to switch from microbial metabolism to structural chemistry. His experiments with Maaløe showed him that he "had not done anything which was going to tell us what a gene was or how it reproduced" (*Helix*, 28). In the back of his mind he feared that the gene might possess such a "fantastically irregular" structure that a straightforward attack on its structure would fail to solve it. What completely changed his attitude was the contribution of Maurice Wilkins to the meeting which Watson attended in Naples in the spring of 1951. Wilkins showed an X-ray picture of the crystalline form of DNA and expressed the hope that the solution of its structure would contribute to our understanding of how genes work.

Suddenly Watson became excited about chemistry. Now he knew, he recalled, that "genes could crystallize; hence they must have a regular structure that could be solved in a straightforward fashion" (*Helix*, 33). At once he began to think of working with Wilkins, but failed to interest Wilkins when he approached him and started talking to him about the gene. (He never asked Wilkins if he could come to work at King's College, London.) What happened next we learn from Luria's letter to Weiss, then chairman of the Merck Fellowship Board:

> The first part of the programme [Watson's isotopic studies] was successful, and Watson learned techniques and applied them very profitably to the isotopic analysis of virus synthesis. At the same time, Watson felt throughout the year that in this biochemical work he was not getting, for a variety of reasons beyond his control, the inspiring guidance he had expected. He did most of his successful work during the past year in Copenhagen with Dr O. Maaløe of the State Serum Institute, who is now spending one year in the United States. He felt that even if he remained in Copenhagen he would have to shift his programme more or less completely, and I suggested to him that Copenhagen might not be the best place to stay and pointed out the value of an exploratory visit to England and Sweden during the summer.
>
> At this point I happened to meet Dr Kendrew of the Cavendish Laboratories at Ann Arbor and heard of the remarkable progress recently made there on protein structure analysis and of Perutz's interest in developing more and more the structural analysis of molecules of biological importance. Kendrew seemed keen on the idea of a man with Watson's interests spending some time with them, and we considered the possible role of Dr Roy Markham of the Molteno Institute, a specialist in virus nucleoprotein, in such a plan. Watson, upon my suggestion, visited Cambridge, was encouraged to go there, was greatly stimulated by the group in question, and planned accordingly. His plans, which I suspect he failed to make adequately clear in writing to the Committee, involve both biochemical work on viruses, along lines somewhat different from those followed at Copenhagen, and the learning of the theory and techniques of X-ray diffraction analysis and their application to the virus problem.
>
> I personally feel that this approach will prove more fruitful than the isotopic one in the study of the basic structures involved in biological replication, and have myself considerably shifted the emphasis in my laboratory toward structural analysis of virus material. . . .
>
> In conclusion, I feel Watson's present plan, far from being a drift into the wilderness, is a considerate search for the type of preparation that may improve his usefulness to biology. A Merck Fellowship, with the opportunities it provides for cross-the-border line training without emphasis on research accomplishment under tenure, is an ideal chance for this plan. Watson is exceptionally young (only 23, I think) and has plenty of time to "do things". The broader and sounder his base, the farther he will go, and I think we will all be proud of him.
>
> (Luria, 1951a)

This letter, it should be noted, offers no evidence that Watson wished to move to Cambridge in order to discover the structure of DNA. What it does suggest is that Luria and Watson believed the most fruitful line to pursue in search of the gene was the molecular structure of nucleoproteins and nucleic acids as extracted from the cell and from bacterial and plant viruses. Recently, Paul Weiss consulted the correspondence over Watson's fellowship which is preserved in Washington, and he wrote:

> On the factual side, there is no question that his main interest and activity in his first fellowship year was focused primarily on the question of stability or instability of *nucleic acids* in the reproduction of bacteriophage in infected cells. I am satisfied that from the very start he has been searching for some clues about the function of nucleic acid in the replication of macromolecular structures, such as viruses and genes. In this sense, his application for an extension of his fellowship for a second year was a logical extension of his original *biochemical* programme. As Luria expressed to me in his letter of October 20, 1951, he and Max Delbrück were the ones who prodded him to turn more in the direction of the *physical* side of molecular structure . . . It is in this connection also that *nucleoprotein structure*, rather than merely *nucleic acid synthesis*, has come in the foreground . . .
>
> (Weiss, 1973)

Perhaps it was naïve of Watson to expect the Merck Fellowship Board to accept his change of plan. Watson has told us how "spendidly co-operative" Kalckar was in writing enthusiastically to endorse his plan to study molecular structure in Cambridge (*Helix*, 43), but permission was refused. The Chairman of the board, Paul Weiss, later recalled that they had only ten Merck Fellowships to offer per year. Extension for a second year was on the understanding that the work would be a continuation of the first year's programme. At that time a post-doctoral fellowship was a novelty. The Merck Fellowships were intended for those scientists who wished to leave the discipline in which they had worked as postgraduates in order to enter a new discipline. Normally they would have continued research for at least three years after their doctorate, before being considered by the Board. And here was a 23-year-old biologist wanting to use his Fellowship extension in a field for which he had received no preparation whatsoever. Oh no! His case had to be considered on a level with fresh applicants, which included those better equipped than he to profit from exposure to structural crystallography. If Watson would go to Caspersson's Institute of Cytochemistry in Stockholm, on the other hand, the grant could run for a full second year. But Watson remained in Cambridge—and made no attempt to hide the fact. When his mother heard of the troubles her son was having she telephoned Weiss "in Chicago long-distance from Indianapolis and implored me 'please stand your ground, for that boy for once needs to learn a lesson' " (Weiss, 1973). In the event a compromise was arrived at. Weiss submitted Watson's case "to the whole membership of the Board with a recommendation that, rather than rescinding the fellowship, they merely reduce its duration by a certain fraction" (*Ibid.*). It ended in May instead of in September, the grant for that second year being reduced from $3000 to $2000.

Watson thus escaped from cytochemistry, microbial metabolism and radiation biology. In Cambridge he met Francis Crick who had escaped from experimental cytology. Crick said of Watson in 1951 that

> . . . he was the first person I had met who thought the same way about biology as I did. Perutz and Kendrew were really interested in crystallography and proteins. They were not really interested in genetics as such. Whereas I decided that genetics was the really essential part, what the genes were and what they did. And Watson was the first person I had met with exactly the same ideas as I had, but I cannot remember in detail what they were; I just remember this general impression. The interest was that his background was in phage work which I had only read about and did not know first hand and my background was in crystallography which he had only read about and did not know first hand. So our ideas of a general nature were already formed when we met, and we merely, as it were, went on to discuss the detail—what were genes made of and so on.
>
> (Crick, 1968/72)

Francis Crick

A biographical essay on Dr Francis Crick has already appeared in *Daedalus* and in a revised form in *The Twentieth Century Sciences*, edited by Gerald Holton (Olby, 1970; 1972). All we need note here is that Crick was educated at Mill Hill School and University College London where he obtained a second class degree in Physics (1937) and began research on the viscosity of water. This work was terminated by war service in the Admiralty Research Laboratory at Teddington, later at Havant, near Portsmouth. When peace came, Crick stayed on as an Admiralty scientist, but he hoped eventually to go either into fundamental particle physics or into biology. One event which stimulated his interest in the latter was Schrödinger's book *What is Life?* Another was the "religious" one of wanting "to try to show that areas apparently too mysterious to be explained by physics and chemistry, could in fact be so explained" (Crick, 1969). A further aid was Sir Harrie Massey who put him in touch with A. V. Hill, the biophysicist, and through him Sir Edward Mellanby, Secretary of the MRC.

Experimental Cytology

If Bernal had been in London at this time (1947) Crick would surely have entered at once the field of protein X-ray crystallography. But Bernal was abroad, and more important, Mellanby was opposed to his going straight into this field and A. V. Hill thought a wiser plan than London was Cambridge, where the experimental cytology of Dame Honor Fell's group at the Strangeways Laboratory would serve admirably to introduce Crick to biology and the biophysical techniques he would need in his chosen field: "the division between the living and the non-living, as typified by, say, proteins, viruses, bacteria and the structure of chromosomes" (Crick, 1947).

At Strangeways Crick performed "twist", "drag" and "prod" tests using a magnetic field on particles ingested by cells in tissue culture. The inconsequential results, plus a theoretical discussion by Crick, appeared in

Experimental Cell Research (Crick and Hughes, 1950). Crick's and Hughes' comment on the classic experiments of the botanist, Heilbronn's insertion of iron fragments into slime moulds was that "whatever else was being measured, it was certainly not the viscosity" (*Ibid.*, 73). Robert Chambers' observation of differences in the consistency of different protoplasms was "not even approximately quantitative" (*Ibid.*). They went on:

> We do not see why the presumably co-ordinated chemical actions of the cytoplasm *must* imply an organized frame-work in the cytoplasm, . . . The "structure" we have established may well be a rather non-specific, transient affair for a normal cell . . .
>
> If we were compelled to suggest a model we would propose Mother's Work Basket—a jumble of beads and buttons of all shapes and sizes, with pins and threads for good measure, all jostling about and held together by "colloidal forces".
>
> (Crick and Hughes, 1950, 50)

In time Crick grew to dislike this model; in fact they must have been pulling, stretching and tearing the endoplasmic reticulum.

The Cavendish Laboratory

Early in 1949 Crick was seeking to move to the Cavendish to join Perutz' group. " . . . he has always been keenly interested in the problem of protein structure," wrote Perutz to Mellanby, "and would have liked to join our unit from the start, but was advised to gain some experience with living materials before making a final decision about his future line of research" (Perutz, 1949c).

In June 1949 Crick joined the Cavendish staff. As we know from Watsons' account it was not long before Crick began to cause trouble in his well-intended attempts to introduce vigour into the interpretation of the X-ray data on haemoglobin. He had taught himself the principles of structural crystallography, but from a different starting point than that of Perutz, Bragg and Kendrew. Consequently he saw their work in a different light.

The Fourier Transform of a Helix

The result of Crick's approach to X-ray diffraction was not only to cast doubt on the reliability of the evidence for Perutz's hat-box model (see Chapter 17) but also to give him the confidence to attempt the derivation of the Fourier transform of a helix.

Bragg had been urging Cochran to work out the Fourier transform of a helix since Pauling's α-helix had been published, but nothing had come of his request. Then V. Vand sent a paper to the Cavendish at the end of October 1951 in which he had derived a partially correct transform. This was shown to Cochran and Crick by Bragg. It was all that was needed to spur them on to find the correct solution. Cochran was not greatly excited by his success, although he had by this time taken pictures of synthetic polypeptides. His main concern was to improve the techniques of structural crystallography.

By 1950 he saw his research aims as:

> . . . to make X-ray measurements with sufficient accuracy to give information about valency bonds and hydrogen bonds, and to try to develop systematic methods of solving crystal structures. In July, 1951 I arrived back from six months in U.S.A., via a conference in Stockholm. I had heard about Pauling's work on the α-helix but I do not remember now whether I had then read his papers. Knowing from experience how difficult it was to determine the crystal structure of even a purine I regarded his work as speculative and was convinced that it was impossible to determine protein structures by X-ray methods. Consequently I was resistant to Bragg's oblique attempts to draw me into his work. He then suggested that I take some photographs of a specimen of poly-methyl-glutamate which he had obtained from the Courtauld's group and I did so without much enthusiasm—this was in September. I don't believe I knew of Wilkin's work at this time and if I had I would have dismissed it as a lost cause. . . . When I was shown Vand's paper I therefore treated the problem as a fairly academic one. Vand had correctly worked out the Fourier transforms of a continuous helix but had gone wrong in taking the step to a discontinuous helix. With his result and my own experience of the properties of Fourier transforms it was easy to get the right answer. Crick got it too by a tougher route; Watson's account is substantially correct at *this* point. Somewhat later it struck me that our calculations fitted the photographs which I had taken.
>
> That was the full extent of my participation. Our note was on the structure of poly-, methyl-glutamate, and despite this success I do not believe it ever occurred to me that the theory was relevant to DNA, or that I knew at that time that there was experimental data for DNA just as good as the data for poly-methyl-glutamate. I went back to my own work, and I did not regard our note to *Nature* as being of much importance—a diary entry dated 17th Dec. includes " . . . this is the most uninspiring term I've ever had in Cambridge as far as research is concerned"! I remember Francis very excitedly showing me a model of the DNA structure, my recollection is that this was fairly soon after our joint note was published. If so, I now realize from Watson's account that this was their incorrect version but in any case I was not impressed, since Francis was always throwing out ideas and as far as I knew it was entirely unsupported by experimental evidence.
>
> (Cochran, 1968)

Crick was involved in this work because he often talked with Cochran in the dark room on the top floor of the Old Cavendish (before he moved into the MRC hut nearby). They talked about Vand's paper and Cochran said: "Well the obvious way to do it is to work it out for a helix of just one atom. When you have done that you can do it for a whole group of atoms" (cited by Crick, 1968/72). Now because Crick planned to attend a wine tasting that evening (Oct. 31) and was not feeling very grand he went home early from the laboratory, got bored, and set to work to derive the Fourier transform. Unbeknown to him, Cochran also set to work, but whereas Cochran used a table of Bessel Functions and quickly saw the significance of the decreasing amplitudes of the waves as he went from zero order functions to higher order functions (see Fig. 18.2), Crick struggled in a clumsy fashion without a table and did not see the significance of his results. Next morning Crick learnt that Cochran "had done a much more elegant derivation, by deriving the transform for a continuous helix then multiplying it by a series of planes. I had done this enormously clumsy thing and our answers were both the same except that each of us had made a different slip which we soon corrected" (Crick, 1968/72).

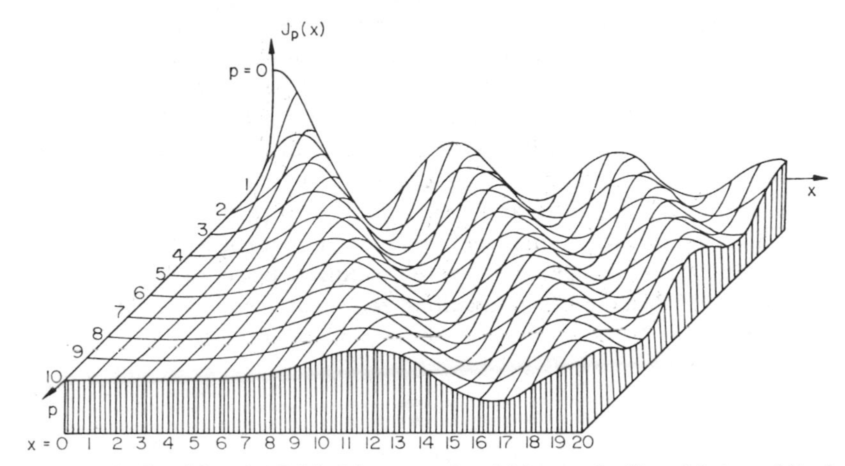

Fig. 18.2. The Bessel function $J_p(x)$ of the two real variables x and p (from Jahnke and Emde, 1933, 192).

Why was this derivation so significant? Three reasons stand out. First it demonstrated the fact, not sufficiently appreciated, that in fibre diagrams, there is no three-dimensional lattice, therefore the fibre diagram cannot be treated like a single crystal where all the diffraction maxima are spots which can be indexed thus leading to the derivation of the unit cell. Crick explained the situation thus:

> Now, what happens if you have an array of vertical helices which are equidistantly spaced and parallel but not particularly in phase up and down? The answer is that if you look in projection, downwards, it doesn't matter that the helices are "up and down", everything is regular. Now the diffraction pattern that corresponds to that is on the equator. In that case you get spots on the equator, but on the layer lines you won't get spots, because there is no consistent phase relation at all the other angles. Instead you get "bands".
>
> (Crick, 1968/72)

These bands do not represent diffraction due to a lattice but the periodic maxima in the continuous structure factors due to the helical *molecules* themselves—the molecular transform. The importance of this fact can be seen by turning to the account of Bernal's and Frankuchen's interpretation of their diffraction patterns for TMV on p. 260.

The second reason for the importance attaching to the Fourier transform of a helix is that it allows predictions to be made, and that these predictions lend support to the α-helix for polypeptide chains and to the Watson–Crick model for DNA. It became possible to arrive at the same structure "by two distinct roots" (Crick, 1954, 214), model building and the general features of the X-ray diagram. Thus Bessel functions of higher order (see Fig. 18.3) "remain very small until a certain value of $2\pi Rr$ is reached, and . . . this point recedes from the origin as the order increases" (Cochran and Crick, 1952, 234). Now for a helical molecule the maximum radius sets an upper limit to the value of $2\pi Rr$. One would expect, therefore, that no set of atoms could make "an appreciable contribution to the amplitude of a reflexion occurring on a layer line with which only high-order Bessel functions are associated, because $2\pi Rr$ comes within the very low part of the curve in the graph" (*Ibid.*). Cochran and Crick went on to predict that meridional reflexions could occur only on layer lines involving zero order Bessel functions (in DNA it is the tenth layer line, where the diffraction repeats), and that the "bands" would, like the Bessel functions, move out from the meridian as the layer line number goes up. Putting these predictions together we have the

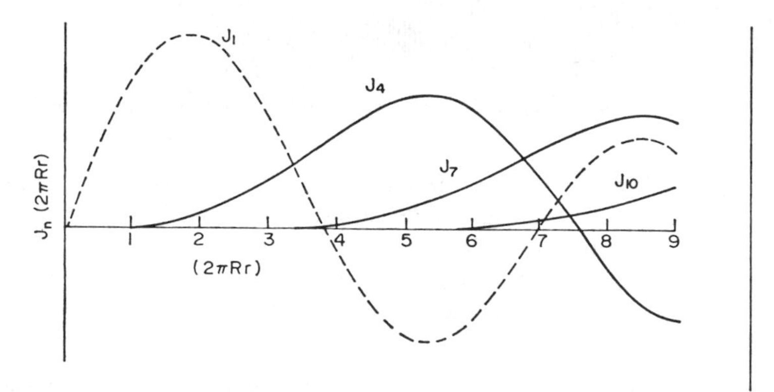

Fig. 18.3. The march of higher order Bessel functions (with J_1 added dashed) (from Cochran, Crick, 1952).

meridional absences and cross-ways pattern now so familiar for helical structures. This much, it seems, had been discovered by Wilkins' colleague, A. R. Stokes, who had plotted out the Bessel functions for a helical structure back in the summer of 1951 (see Fig. 18.4). When he received an advance copy of Cochran's and Crick's *Nature* paper he told Crick that the "conclusions you give appear to be mainly the same as those I have arrived at", but he did not see the need for mentioning this fact in *Nature*, "especially as I only got some of them very recently" (Stokes, 1951).

The third reason for the importance of the Fourier transform of a helix lay in the tool it provided for rapid calculation of the diffraction pattern that a given helical model should yield. Hitherto this task had been too arduous to be attempted. Pauling had confined his predictions from the α-helix (1951) and later from his structure for DNA (1953) to the form factor on the equator (see Fig. 18.5). Now the task which might take months could be achieved in hours.

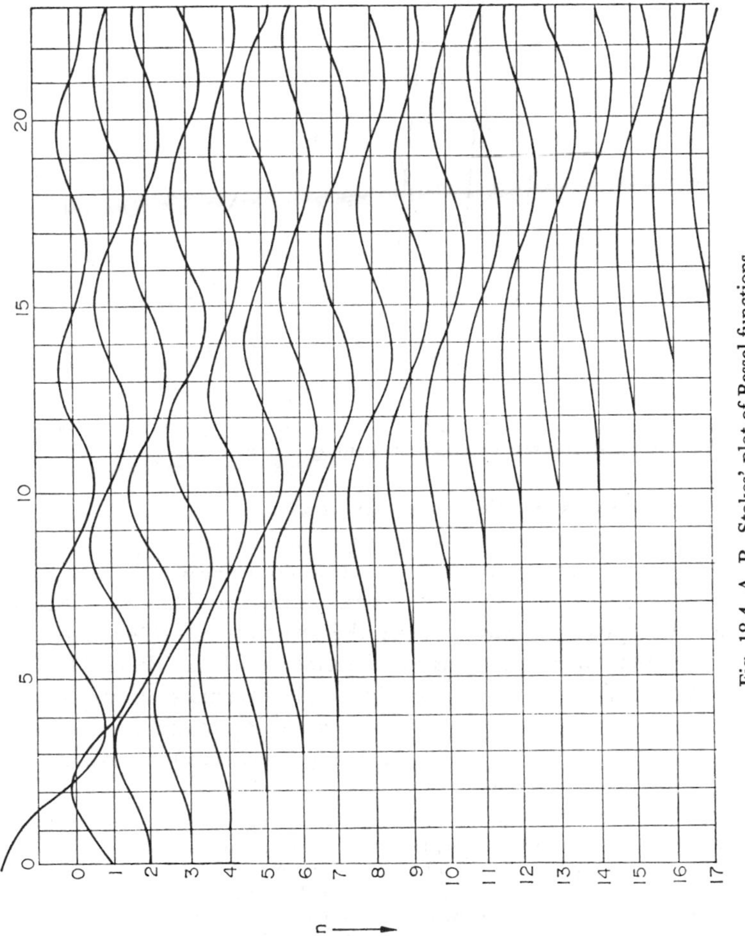

Fig. 18.4. A. R. Stokes' plot of Bessel functions.

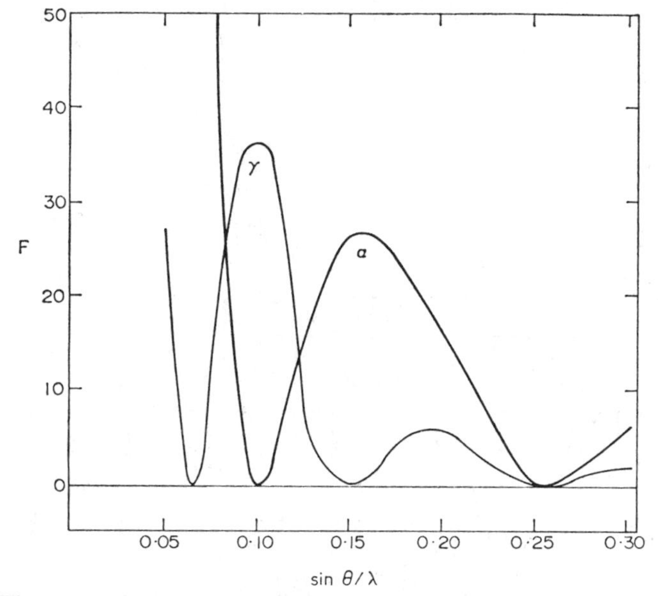

Fig. 18.5. The square of the form factor for the 5.1 residue helix (γ) and the 3.7 residue helix (α) for equatorial reflections (from Pauling, Corey and Branson, 1951).

Collaboration between Watson and Crick

It is true that Crick possessed the expertise in crystallography and Watson the expertise in phage genetics but their collaboration was no mere addition of these two ingredients. Rather, its character derived from the nature of the problem. The problem was to *guess* the structure of DNA and then to demonstrate that their guess was in conformity with the data. The "guess" involved a crystallographic argument by a series of steps leading to a theorum —the model. If any one of these steps was wrong the model would be wrong. The advantage of having two men working together on such a task is obvious. "If, for example," said Crick,

> . . . I had some idea, which as it turned out . . . was going off at a tangent, Watson would tell me in no uncertain terms [that] this was nonsense, and *vice versa* . . . It is one of the requirements for collaboration of this sort that you must be perfectly candid, one might almost say rude, to the person you are working with. It is useless working with somebody who is either much too junior than yourself, or much too senior, because then politeness creeps in and this is the end of all good collaboration in science.
>
> (Crick, 1962, 13)

Under Crick's excellent tutorship Watson learnt a lot of crystallography and as we shall see in Chapter 21 he made his own contributions to the crystallographic arguments. Also he continued to display that independence of judgement which had stood him in good stead hitherto. Not even Crick's forceful arguments could shift Watson when he had a mind to hold to his convictions. But what can we say of Watson's contribution to Crick's knowledge of genetics?

The Hershey–Chase Experiment

Crick felt that the pursuit of the structure of DNA was worthwhile simply because nucleic acid formed a part of the nucleoprotein complex widely regarded as the hereditary material. Indeed, the fact that DNA was a compound of unknown structure was enough to justify working on it. For Watson more biologically-compelling reasons were required. We have discussed the evidence supporting the view that in 1951 he definitely *wanted* to show that DNA was the hereditary substance. We may now note the fact that in 1951 Luria had virtually rejected this conclusion and spoke of a "DNA cytoplasm" which was added to the genetic protein units of phage during the "final baking" of phage particles (Luria, 1952). If Watson did not follow Luria, at least he communicated Luria's findings to Crick. The latter can recall the change in Watson's attitude to DNA brought about by the celebrated experiments of Al Hershey and Martha Chase at Cold Spring Harbor in 1951/2.

These experiments represented a logical development of Thom Anderson's discovery of osmotic schock (Anderson, 1949), the inactivation of phage after suspension in high concentrations of salt solution followed by rapid dilution. The resulting inactive phage was visible in electron micrographs as tadpole-shaped phage "ghosts". Two years later Roger Herriott studied the properties of these ghosts. He found they could absorb to bacteria, inhibit their reproduction, and lyse them; chemical and enzymatic analysis showed the virtual absence of phosphorus (P/N = 0.01) and suggested their protein nature. Herriott concluded that phage nucleic acid was not essential for the functions of the ghosts. The site of the latter functions was hence limited to "the protein portion or the small amount of lipid reported by Taylor. This then is a beginning in the physical separation of the various biological functions of a virus and the correlation of these functions with certain morphological and chemical properties" (Herriott, 1951a, 754). That year Hershey began to follow up Herriott's discovery, and in the correspondence between them we find the following

> I've been thinking—and perhaps you have, too—that the virus may act like a little hypodermic needle full of transforming principle; that the virus as such never enters the cell; that only the tail contacts the host and perhaps enzymatically cuts a small hole through the outer membrane and then the nucleic acid of the virus head flows into the cell.
>
> If this is so, then two experiments suggest themselves (a) one should be able to get virus formed by the nucleic acid alone if one only knew how to get into the cell—and that, of course, is the $64 problem, and although I have some ideas on how to approach it, I'm not very proud of them. The other thing is that if the above notion is correct, then one should find the ghost in the cell debris after lysis. The latter shouldn't be hard to determine. I got ^{35}S to do it and heard recently that you have done some experiments very

much like this. If you are working along this line, I'll work on something else for there are plenty of things to be done. If you are on a different trail entirely, I'd like to answer the problem or idea posed above.

(Herriott, 1951b)

About this time Hershey had confided to Delbrück that: "Herriott claims there is no basic protein released by osmotic shock, but of course this does not signify" (Hershey, 1951a). When he received Herriott's letter describing the hypodermic needle action of the phage he wrote back explaining to Herriott how he had been led to a like conclusion, but not from starting out with the assumption that it was the DNA which entered.

I have arrived at your notion the hard way, namely by doing the ^{35}S experiments you planned . . . There is little or no ^{35}S *in* the phage progeny, because we are able to separate it from them sometimes. I believe we now have a reliable method for doing this . . .

Your idea about the nucleic acid is an intriguing one. I haven't any notion how to prove it. At present we can say that the intracellular prophage is either very small or very fragile . . .

I might add my opinion, for what it is worth, that the intracellular phage cannot be soluble nucleic acid, but is more likely to be a small, highly organized nucleus. The surprising thing is that it probably contains little protein.

(Hershey, 1951b)

The method for separating the ^{35}S fraction of the phage from the infecting material invented by Hershey was agitation in a Waring blender. They explored a variety of experimental situations by which they sought to separate phage components and identify their functions and chemical constitution. The famous paper reporting this work they entitled: "Independent Functions of Viral Protein and Nucleic Acid in Growth of Bacteriophage." They concluded that sulphur-containing protein was confined to a protective coat responsible for "adsorption to bacteria, and functions as an instrument for the injection of the phage DNA into the cell. This protein", they said, "probably has no function in the growth of intracellular phage. The DNA has some function" (Hershey and Chase, 1952, 56). Their evidence was not all that convincing, certainly their chemical data was inferior to that of Avery, MacLeod and McCarty; they could say nothing about sulphur-free protein. The percentage of the sulphur label in the protein ghosts was 90 per cent not 99 per cent, and only 85 per cent of the phosphorus label was demonstrable in whole phage. Why then, were their results so well received? The answer would in part be the simple one that here was the first step leading to the production of vegetative phage exposed to view—injection of DNA into the bacterial cell leaving behind the protein ghosts. What little protein appeared to go in was insignificant by contrast with the DNA and with the protein left outside. "Their experiment", wrote Watson, "was thus a powerful new proof that DNA is the primary genetic material" (*Helix*, 119).

Reception of the Hershey–Chase Experiments

Hershey must have sent accounts of the experiments to his friends in February or March. Maaløe replied describing it as a "very beautiful piece of work, and it gives all of us a lot to think about" (Maaløe, 1952). For Delbrück it was "the best paper you ever wrote, as to substance, I mean. For once I am really envious ... Weidel's paper had some very suggestive evidence that bacterial membrane can do something to the phages, namely cause a phage substance to be released. He swore it was not DNA, but I think it must have been DNA..." (Delbrück, 1952a). Only Luria seemed unwillingly to concede Hershey's conclusion. He had only recently submitted a paper to the Society for General Microbiology in England in which he still favoured the picture of vegetative phage as a DNA-poor particle to which a DNA-rich cytoplasm was later added. At the time when he wrote it Luria was under the impression that between 30 and 50 per cent of the ^{35}S in labelled phage was transferred to the progeny. On March 3 he wrote to Hershey:

> Your note on ^{35}S not transfer came in time to delete reference, but too late to change my whole attitude towards protein and DNA. Then your report made everything perplexing.
>
> I must say your story sounds convincing and plausible, and I am particularly impressed by the evidence on heated bacteria, as to the early separation of DNA from protein. There are, however, several facts that make me keep an open mind as to whether DNA or protein rules reproduction ...
>
> As you see in my MS, my idea of a DNA-cytoplasm was presented with many hesitations and qualifications, although the circumstantial evidence for it was gathered ... When I had your cryptic note, I was slow in guessing the type of evidence you had about ghosts, etc. I still think there is a pretty good chance that a good deal of protein enters and acts genetically, but that transfer of ^{35}S may not be present because of any one of a variety of reasons ...
>
> (Luria, 1952a)

One of the facts which caused Luria to keep an open mind was the absence of DNA from the supposed vegetative phage or "donuts". "One would have to imagine that the DNA escapes from them in preamature or proflavine lysis", he objected—precisely what must have happened!

Watson's reaction to Hershey's news was very different from Luria's. The latter had been barred for political reasons from attending the Oxford meeting of the Society for General Microbiology in April, 1952. Luria had been able to add a short passage on the work of Herriott and Hershey and Chase (Luria, 1952b, 110) but it was Watson, who in a long address at the start of the discussion, gave the Hershey–Chase experiment full justice. "It is tempting", he declared, "to conclude that the virus protein functions largely as a protective coat for the DNA and that the perpetuation of genetic specificity is largely or entirely a function of the DNA" (Watson, 1952b, 114). When countered by an interpretation of this work in terms of nucleic acid–protein interaction Watson retorted that he thought the relationship was more likely "like that of a hat inside a hatbox" (*Ibid.*, 116). To the majority of the audience the sight of this gangling young American waving an air mail letter and going on about DNA did not inspire confidence. This reaction was reinforced by Sir Frederick Bawden's suave and amusing oratory.

Nor can it be said that the Phage Group as a whole jumped on the DNA trail in 1952. At the phage meeting at Royaumont, near Paris, Hershey expressed himself with typical caution: "Parental DNA components are, and parental membrane components are not, materially conserved during reproduction. Whether this result has any fundamental significance is not yet clear" (Hershey, 1953a, 110–111). He did not sound over-enthusiastic, indeed he almost apologized for the lack of clear cut discoveries:

> The study of viral infections of bacteria is continually yielding results that could not have been predicted from results that went before. At best this means that progress is being made. At worst it means that our field of endeavour is in a healthy condition, amenable to progress.
>
> (*Ibid.*)

In Cohen's opinion work on the T-even phages had reached a temporary plateau by the end of 1951 (Cohen, 1968, 17) and even by the time of Royaumont phage researchers *did not know*, contrary to what Guntha Stent has asserted, "that the phage DNA is the sole carrier of the hereditary continuity of the virus and that the details uncovered hitherto were to be understood in terms of the structure and function of DNA" (Stent, 1966, 6). What changed the situation was the discovery of 5-hydroxymethyl cytosine (Wyatt and Cohen, 1952; Hershey, 1953) which made separation of host from viral DNA synthesis possible and suggested the existence of specific viral enzymes not found in the host. The power to produce viral enzymes could then be used to identify genetic markers. On the other hand, these developments did not suddenly start in 1952. Something else was needed to clear up the confused field of phage biology and this was provided by the Watson-Crick model for DNA.

SECTION V

HUNTING FOR THE HELIX

WILKINS, GOSLING, FURBERG, FRANKLIN, PAULING, WATSON AND CRICK

CHAPTER NINETEEN

DNA as a Single- or Multiple-Strand Helix

At the end of the Second World War the Society of Experimental Biology decided to hold their first symposium. It was to be on respiration, but because David Keilin had a "dustup" with other members over the organization of the meeting James Danielli was able to step in and suggest a different topic—nucleic acids. Such "dustups" were not infrequent. Many a time Danielli would come to Keilin's office "and sit down while Keilin went into a tirade about something that was annoying him; he would suddenly realize that I was there sitting smiling at him. Then he would be quite charming" (Danielli, 1969).

So we have to thank Keilin for his show of temper that in July, 1946 a symposium took place in Cambridge on the nucleic acids at which Gulland presented his evidence for hydrogen bonding between bases and Astbury surveyed the X-ray crystallographic work done at Leeds. He could add little that was fresh since the pre-war days, but he did publish the pattern produced by a very well oriented fibre of the sodium salt of DNA hitherto reproduced only in Bell's thesis. We now know that this was a mixture of two different molecular forms. Besides the equatorial spacing of 16.2 Å it gave a spacing on the first layer line at about 27 Å. Now all Bell's pictures showed a prominent meridional reflexion at 3.34 Å, and since $27/3.34 \simeq 8$ Astbury concluded that "*the arrangement in space* is repeated along the fibre axis every eight (or sixteen) nucleotides" (1947, 67). What arrangement? Astbury was cautious—it was a sequence of bases determined "by either chemical or geometrical considerations, or a combination of both." But then he could not restrain his desire to jump to a conclusion, to link up chemical and crystallographic observations.

> It hardly seems likely, though, that the fact that the intramolecular pattern is found to be based on a multiple of four nucleotides is unrelated to the conclusion that has been drawn from chemical data that the molecule is composed of four different kinds of nucleotides in equal proportions.
>
> (Astbury, 1947, 67)

To him it was improbable that a fibre which gave such fine diffraction patterns could consist of polynucleotide chains with the bases distributed at random. "Rather must they follow one another in some definite order—at least, in the more crystalline regions of the structure . . ." (*Ibid.*). Thus did Astbury reverse his earlier decision against the tetranucleotide theory.

Mansel Davies' Model Building

Where Astbury's Cambridge paper showed an advance on his Cold Spring Harbor paper of eight years before was in the support given to his pile-of-pennies model which Mansel Davies provided from model building. He joined the textile physics laboratory in 1942 to carry out research for British Celanese Ltd. In 1945 this programme came to an end and Astbury obtained from the Rockefeller Foundation a grant to set up infra-red equipment for the examination of proteins. While waiting for this equipment to arrive Astbury suggested to Davies a continuation of Bell's X-ray studies of nucleic acids which the Rockefeller (?) had supplied. According to Davies the resulting pictures were better than Bell's, but as none has been published or preserved it is not possible to confirm this opinion. Davies had already built models of cellulose for Astbury in 1944 "and we had shown that in alginic acid as well as in cellulose the 'arm-chair' form of the furanose ring was consistent with the X-ray picture" (Astbury, 1945).

Here at last was the collaboration needed in Leeds to push the DNA work to a deeper level. But something went wrong. Davies studied Meyer's and Mark's *Der Aufbau der hochpolymeren organischen Naturstöffe* (1930) and H. A. Stuart's *Molekülstruktur* to decide on bond angles and lengths. A day or two later the wooden balls used in those days were drilled in the workshop. For two reasons they could not be too precise about bond angles. Davies later claimed that the data at that time permitted a variation in bond angle of up to 5° in lesser known types, and this was supported by infra-red studies which confirmed the ease of distortion of these angles. Second, "the workshop could set for a tetrahedral or 120° angle within plus or minus 2°, but other values were 'special settings' and not to be required too often" (Davies, 1967).

The task Astbury had set Davies was to build a model that incorporated the 3.3_4 Å, or as he now called it "about 3.4 Å", repeat along the backbone. Davies relied upon Lythgoe's and Todd's establishment of the β-linkage for the glycosidic bond, Gulland's evidence from spectroscopy for the purine linkage at N_7 (Gulland, Holliday and Macrae, 1934), pyrimidines at N_3 (Levene and Tipson, 1932), and for the phosphodiester link between adjacent sugars the absence of a hydroxyl group on C_2 in the desoxyribose of DNA left only C_5 and C_3, C_4 being involved in ring formation. Although this conclusion was not established until 1952 by Todd, Astbury and Davies had guessed rightly in 1946 (see Fig. 19.1). It was clear to Davies that the glycosidic link was not in the plane of the sugar ring but at the tetrahedral angle, so the earlier assumption of Astbury and Bell that sugar and base were co-planar just had to be wrong. Astbury accepted this important revision of his model, but later forgot all about it, thus justly earning the continued association of his name with the co-planar model!

So far so good. But when Davies came to join up the sugar rings by the phospho–diester linkages he ran into difficulty, as indeed he should have. To get the bases over one another separated by only 3.4 Å brought the oxygen and OH group bound to the phosphorus atom too close to other atoms in the

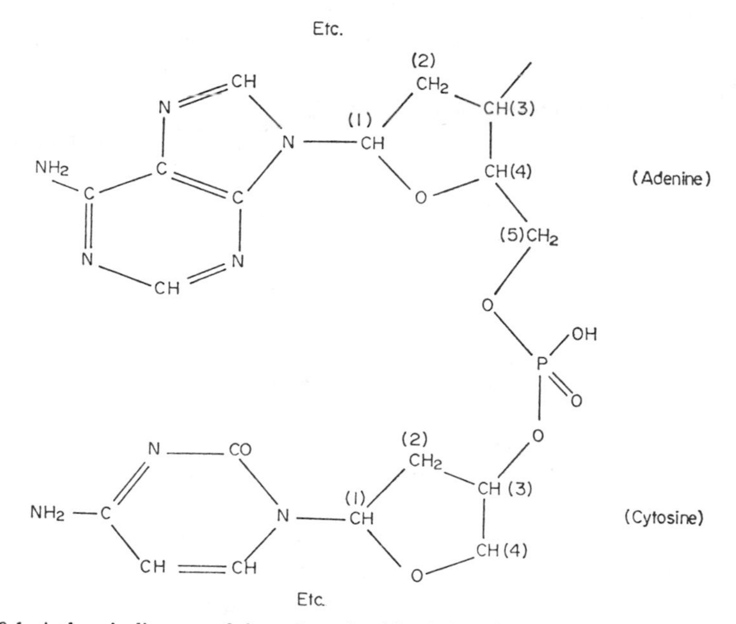

Fig. 19.1. Astbury's diagram of the polynucleotide chain in DNA showing the correct base-sugar-phosphate linkages. This diagram was not intended to show the conformation of the chain (from Astbury, 1947, 69).

chain. Perhaps, argued Astbury, the strong van der Waals attractions between the flat rings of the bases were sufficient "to distort the oxygen van der Waals distance somewhat . . . making the polynucleotide column more labile than would otherwise be the case" (Astbury, 1947, 70). But there must have been something else wrong with Davies bond angles, for with present day equipment it just is not possible to bend the awkward backbone of DNA into a configuration that brings the bases over one another at 3.4 Å without adopting a helical conformation.

Astbury did consider a helical configuration for DNA, but only one in which the bases were on the outside, and the helix contained but a single chain. Such a structure would not have brought the bases to lie over one another, and only by interpenetration of neighbouring helices would the high density of 1.63 g cc be achieved.

> . . . nucleotides of neighbouring molecules must be closely interleaved in a surprisingly regular fashion. This seems highly unlikely, for the stabilizing forces of the structure will be chiefly the strong total attractions between the large flat areas, and the simplest and most effective way of reducing the potential energy to a minimum is for such areas to lie, if possible, directly on top of one another, as happens from other structures built from flat units.
>
> (Astbury, 1947, 68)

Davies recalled how repeatedly he had asked Astbury to get someone in to continue the nucleic acid work, but no further hands were available. There had been other attempts in the 1940s, but these were abortive. R. D. Preston's brother, Clifford, attempted work on live sperm.

> In 1940 he [Astbury] asked me to try to take an X-ray picture of oriented sperm. I remember the occasion vividly. He said he had little precise idea how to do it—I would need to fend for myself completely; . . . The idea was to attempt to orient the sperm by streamlined flow along a thin-walled glass tube (lithium glass, I think) using a tiny peristaltic pump which I made. At an early stage of this [work] Kenneth Bailey at Cambridge succeeded in producing long thin needles of egg albumen and Astbury decided to switch to this and leave the sperm for a bit.
>
> (Preston, C., 1968)

This failed partly because the tails of bull sperm are too short to give effective orientation but with a film of dried sperm A. M. Melland working in collaboration with John Buck at the Carnegie Institution and subsequently in the zoology department, Cambridge, UK, succeeded in producing X-ray pictures from a block consisting of 1230 oriented giant salivary gland chromosomes. Unfortunately, as Astbury reported at the time, it was impossible to decide the source of the two faint rings produced (Buck and Melland, 1942, 183). This joint project had been supported by a grant from the Rockefeller Foundation to C. W. Metz at Rochester University and to Astbury in Leeds.

In the next chapter we shall refer to the fine pictures taken by E. Beighton in 1951, but it is evident that until that time virtually nothing of any weight had been done in Leeds. According to Davies this state of affairs was partly due to the refusal of the Medical Research Council to support Astbury's "major application for systematic biomolecular studies which the laboratory had proposed as a coherent programme . . ." (Davies, 1967).

MRC support for Biophysical Studies

It is true that at the end of the Second World War there was a movement amongst scientists to persuade the Government to support major projects in fundamental research. Whereas John Randall, later Sir John, and Sir Lawrence Bragg were successful in their applications to the MRC, Astbury was not. Just why Astbury failed to get the support he sought was explained by Sir Harold Himsworth.

> In 1945 Mellanby and Astbury happed to sit next to each other at "The Royal Society's dinner". I presume that this was the annual dinner, because the date of the letter is December 8th and in it Astbury refers to his having been elected a member of the Royal Society Council that afternoon. Presumably in the lean year of 1945 only Fellows dine together: there was not enough food for guests.
>
> It is evident that they had talked shop and that Mellanby had been very attracted by some of the ideas that Astbury put forward. The upshot was that he encouraged him to think of support from the M.R.C. Astbury explained, in correspondence, that he had had lots of little grants from numerous sources and what he really wanted was long-term assured support so that he could get down to work.
>
> Today the upshot of such developing contacts would have been a request from Mellanby to him for a memorandum setting out his scientific proposals and requirements. But there is no reference to this ever having been asked for, or for Astbury having volunteered one. All the correspondence was on ways and means.
>
> The upshot of the above conversations was that Mellanby got the Council to ask Astbury to come to one of their meetings and speak to them. This he seems to have done in January, 1946. It is evident that Astbury did not convince the Council, and although the record is uninformative as to why this was, the subsequent correspondence with Mellanby is more revealing.
>
> From the correspondence it seems that Astbury managed to convey the impression to the Council that his real interest was in structures like cellulose and artificial fibres. My inference from the phraseology is that he must have taken something like the line that we must learn to walk before we could run; and biologically important macromolecules were very difficult. Anyway, it is quite clear that he insisted hard on the basis of his whole scheme being investigation of synthetic plastics; and then later attention might be given to more complicated materials such as biological products.
>
> I think this probably explains the point that puzzled you: why Council did not support Astbury and yet a few months later supported Randall. It is evident from other correspondence at that time that Council were looking for opportunities to support work on macromolecules of biological importance. They seem to have been put off by Astbury's insistence on working on synthetic plastics, and it is quite likely that they jumped at Randall's programme which was going straight for biologicals.
>
> Lastly, throughout the whole of the file there is not a single reference to nucleic acids. But I did find one thing: a reprint of Astbury's attached to the file. This was a reprint of an article that he had contributed to the book *Essays on Growth and Form* presented to D'Arcy Wentworth Thompson.
>
> (Himsworth, 1968)

The approach through synthetic fibres was no doubt a very sensible one, but clearly it did not appeal to the MRC and it was not Astbury's intention, were he supported by the MRC, to go full steam ahead on DNA. Rumour has it that when Astbury attended a meeting of the Council he became very curt, and when asked who would be the biologists in his unit, he replied that *he* was his own biologist. It may well have been a repeat performance of his interview in Cambridge many years before!

Randall's Application

The movement among scientists which had as one of its fruits the establishment of Randall's MRC unit in 1947 can be traced back to 1943 when Sir

Ralph Fowler and P. M. S. Blackett wrote to the Council of the Royal Society expressing their conviction that fundamental physics might suffer in comparison with applied physics if its development were left solely to the local initiative of the universities. They urged Council to give some guidance "if the case for increased resources is to be adequately put to the relevant Government authorities" (Royal Society, 1940–1945, 334). The result was the formation of a committee in November 1943 to advise on the post-war needs of fundamental physics. By the following spring the number of such committees had increased to six. Now biology, biochemistry, chemistry, geology, geography, geophysics and physics were included. By the close of 1945 the Royal Society had approached the Government for funds and the Treasury had agreed to consider applications. These were intended to be for much larger sums than were supplied through the parliamentary grants in aid of scientific investigations (established in 1850).

Randall's search for funds had begun independently of Council's action. In the early 40s he had cherished the idea of a biophysics research programme. He found a lecture which Bernal gave in Birmingham in 1941 on the structure of proteins very stimulating. In conversation afterwards Bernal advised Randall to visit C. D. Darlington, then at the John Innes Horticultural Institute. Shortly after this occasion Randall received some fine pictures of sperm heads under the ultraviolet microscope, which showed the high absorption by the chromatin packed heads. Thereafter he looked forward to the opportunity of starting up a programme of studying the cell by all possible optical techniques—ultraviolet, infra-red, interference microscopy—together with some X-ray crystallography. At the University of St. Andrews, where he was appointed professor of natural philosophy in 1944, he was able to make a start on a small scale, but reliance on Admiralty grants imposed war-oriented research for most of the time.

Maurice H. F. Wilkins, who had been working on the atomic bomb Manhattan Project, joined Randall in 1945. He recalled this move in his Nobel Lecture as follows:

> During the war I took part in making the atomic bomb. When the war was ending, I, like many others, cast around for a new field of research. Partly on account of the bomb I had lost some interest in physics. I was therefore very interested when I read Schrödinger's book *What is Life?* and was struck by the concept of a highly complex molecular structure which controlled living processes. Research on such subjects seemed more ambitious than solid state physics. At that time many leading physicists such as Massey, Oliphant, and Randall (and later I learnt that Bohr shared their view) believed that physics would contribute significantly to biology; their advice encouraged me to move into biology.
>
> (Wilkins, 1963, 127)

At the request of A. V. Hill, the biophysicist, in the winter of 1945/46 Randall had submitted a scheme for research in biophysical techniques applied to the cell. Hill replied enthusiastically advising him to double his estimates (Randall, personal communication), thus bringing them to £22000. During 1946 Randall's scheme was studied by the Royal Society Committee which had been set up by Hill and was now chaired by Sir Edward Salisbury. Sir Edward was also a member of the University Grants Committee and in this capacity he visited the University of St. Andrews where Randall was professor of physics. It required little effort to see that a biophysics programme would not flourish in this fine old university (founded in 1410) remote as it is from other centres of learning, save Edinburgh, and boasting in biology the legendary figure of Sir D'Arcy Wentworth Thompson, appointed in 1897, and like Johnny Walker still going strong at the age of 86, a vigorous dancer at Saturday night "hops", but hardly likely to make a fruitful contribution to current work on cell structure. Salisbury therefore took Randall on one side and advised him that if he wanted money for a biophysics unit he would have to move to London.

Fortunately the Wheatstone Chair of Physics at King's College London fell vacant at the end of 1945 on the move of C. D. Ellis to the Coal Boards This department was justly famous, both as the first teaching institute of its kind in England and for the distinction of its professors, which had included Wheatstone, Maxwell, Wilson and the three Nobel laureates, Barkla, Owen Richardson and Sir Edward Appleton; but in 1946 it was in a sorry state. From 1934–43 it had been evacuated to Bristol; meanwhile its buildings in the Strand had served as headquarters for a detachment of fire engines, its engineering laboratories for shell production and trainee courses in the use of machine tools. In 1940 the armament firm of Armour & Co., whose factory had been bombed, were given facilities and space in the King's laboratories until 1944. A crater 27 ft deep and 58 ft long marked the site of a direct hit in October 1940 on the quadrangle. To make matters worse the Wheatstone professor, Charles Ellis, had been granted leave of absence from 1943–46 to act as a scientific advisor to the Army Council. In his absence the former holder of the chair, Sir Edward Appleton, had directed the department until 1944.

King's College then presented a considerable challenge, which Randall accepted. He took up his duties there in the autumn of 1946, by which time his application to the Royal Society had been approved and submitted to the Treasury. There, Sir Alan Barlow advised passing the scheme on to the Medical Research Council as the more appropriate body. In due course its secretary, Edward Mellanby, interviewed Randall, approved of him, and the grant was given in March 1947.

In the succeeding two decades the staff of the biophysics laboratory has grown from a modest 40 to over 120. New accommodation has twice been provided for it, first in the new basement laboratories constructed under the old Quad (with physics and engineering) and second in a building of its own in Drury Lane. This growth record was achieved by the single-minded determination and authority of Randall, who would not bow to opposition or brook delaying tactics. The day after his appointment to the Wheatstone chair he telephoned the Secretary of King's from St. Andrews to state his requirements. To the Secretary, who under war conditions had practised stringent economy and in the disturbed state of the College had taken an active part in decision making, this was needless expense—a letter would have sufficed—and an unwelcome touch of the whip lash. To a colleague he complained, "I have just heard from that man Randall; I fear we are going to have trouble from him."

That opposition to Randall's Biophysics Unit did develop in King's is hardly surprising, for here was an outsider coming into a war-ravaged College and getting outside funds of a size hitherto undreamt of in King's. Randall was well aware of opposition to his plans both "within and without the College, but", he recalled, "I had no direct conflict with any head of department. . ." (Randall, 1974). A part of the task of establishing the programme fell on Danielli. He encountered opposition at the level of the Committee of Senior Fellows and at that of the Professorial Board. In their attitude at College meetings Donald Hey (chemistry) and John Semple (mathematics) showed their disinclination to come to Randall's aid if he ran into difficulties (Danielli, 1969). Despite this lack of support Randall succeeded, with the result that from 1947 onwards his department "was the only physics department in the country to have a major research interest in biophysics" (Randall, 1974).

Randall's Programme

Evidently it was the ideas which Bernal and Darlington had on cell division and chromosome mechanics which caught Randall's imagination. Accordingly, the programme "centred on investigation of the physical factors affecting mitosis and cell division, by direct and indirect methods" (MRC, 1972). Caspersson's work on ultraviolet microscopy had to be checked. A search for X-ray patterns from oriented sperm heads was to be made. Cytoplasmic streaming, the elasticity of chromosomes, long-range forces in gels, the intermolecular spacings of protein and virus particles in gels, the effects of ultrasonics on mutations were all to be studied. In short, Randall proposed to take up all lines that had been pursued in the inter-war years on the biophysics of the cell and thus to build on the foundations layed by Caspersson, Bernal, Darlington and Honor Fell.

Randall's aim was to distill from the many approaches to cell studies an inter-disciplinary attack on the secrets of the chromosomes and their environment. "Our experiment in biophysics," he said, "is perhaps more directly described as an experiment in cell research" (1951, 2). Much of the early work went into the examination of optical principles and the design of apparatus. Wilkins, William Seeds and K. P. Norris developed spherical mirror reflecting objectives which had the achromatic qualities necessary for micro-spectrometric work (Seeds and Wilkins, 1949). These were used in ultraviolet dichroism studies of TMV, nucleic acid and nucleoprotein (Seeds and Wilkins, 1950), and in infra-red studies of nucleic acid (Fraser, 1950). Meanwhile H. G. Davies and P. M. B. Walker constructed a refracting ultraviolet microscope (1950) with which they established the interphase timing of DNA duplication (Walker and Yates, 1952) independently of Swift at Chicago. Walker's new densitometer was later marketed by Loebl as the Joyce Loebl microdensitometer. E. M. Deeley then developed an integrating microdensitometer for his King's Ph.D. thesis, and Barr and Stroud later marketed this design.

By the summer of 1950 it appears that the unit had progressed from the exploratory stage to a more clearly conceived programme. No more was heard of ultrasonics, and the Heath–Robinson apparatus which Wilkins had used in their studies was consigned to oblivion, Majorie McEwan's and D. L. Mould's studies of long-range forces in gels of clay and mineral particles was recognized as a far cry from the forces operating in the cell (Randall, 1951, 3). McEwan's attempt to show that the metaphase plate was composed of non-newtonian liquids was dropped. Instead the Unit concentrated on DNA, TMV and nucleoprotein.* Wilkins' own shift of interests undoubtedly played the major part in this change. In his Nobel Lecture he recalled the fascination he felt when looking at chromosomes in cells "but I began to feel that as a physicist I might contribute more to biology by studying macromolecules isolated from cells. I was encouraged in this by Gerald Oster who came from Stanley's virus laboratory and interested me in particles of tobacco mosaic virus" (1963, 127).

The First X-ray Patterns from DNA

On the 12th of May, Wilkins attended a one-day meeting of the Faraday Society at which Rudolf Signer described his work in Bern on the preparation and physical properties of the sodium salt of DNA. From his collaboration with Caspersson in the 30s Signer had been convinced that DNA was a long chain molecule which was easily fragmented during extraction. At the end of the war he set about devising better extractive procedures, and by 1949 his student H. Schwander, using a modified version of the method of Mirsky and Pollister produced remarkably undegraded calf thymus DNA.

Signer brought to this Faraday meeting a bottle of the best DNA he could find, and distributed it. One of the fortunate recipients was Wilkins. Another was Paul Doty. His molecular weight determination by light scattering

*Also muscle and collagen (Wilkins, 1986).

gave a value of 7 million (Reichmann *et al.*, 1953). By the same method Peterlin arrived at 6.7×10^6 (1953). No wonder it gave the best X-ray pictures! Wilkins wanted the Signer DNA to use in his programme of optical studies with the reflecting microscope. He was not at this time involved in X-ray diffraction studies, nor had he any training in that technique. But in the physics department there was A. R. Stokes who had done X-ray crystallographic research on carbon compounds in Cambridge. Randall had assigned a graduate physicist—R. G. Gosling—from University College London to Stokes, who was to supervise his X-ray study of ram sperm heads,

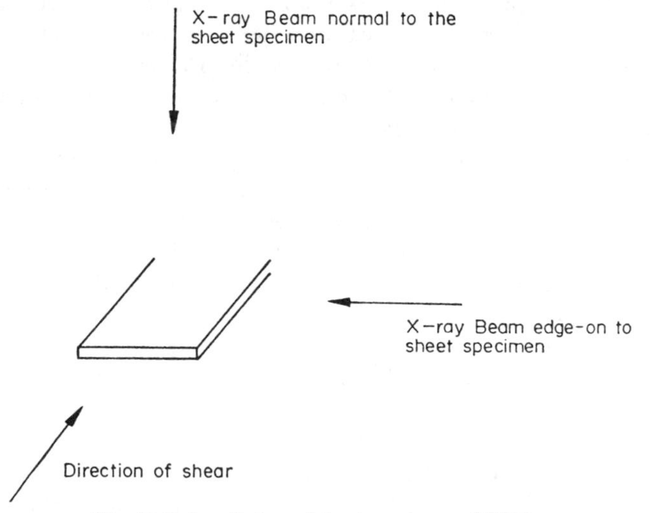

Fig. 19.2. Irradiation of sheet specimen of DNA.

thus complementing Randall's electron microscope studies of sperm. "It was," wrote Randall, "because Gosling wanted to get clear specimen patterns of DNA for comparison pictures that he approached Wilkins for some of his Signer material." (Randall, 1974.) Gosling tried to get diffraction patterns from sheets of the dried gel of Signer's DNA. Gosling mounted sheets, which had been sheared between glass plates perpendicular to the X-ray beam (Fig. 19.2). At Randall's suggestion the relative humidity was kept at 90 per cent by bubbling hydrogen through water and then passing it into the camera. Poor diffraction patterns resulted. A specimen of herring DNA prepared at King's by Mary Fraser a year later gave Gosling a similar pattern.

Crystalline Patterns from DNA Fibres

In the course of his optical studies of the Signer material Wilkins saw:

> in the polarizing microscope extremely uniform fibres giving clear extinction between crossed nicols. I found the fibres had been produced unwittingly while I was manipulating

DNA gels. Each time that I touched the gel with a glass rod and removed the rod, a thin and almost invisible fibre of DNA was drawn out like a filament of spider's web. The perfection and uniformity of the fibres suggested that the molecules in them were regularly arranged.

(Wilkins, 1963, 128)

The negative birefringence of these fibres was remarkably strong, thus indicating good orientation of the DNA molecules. So off went Wilkins to Gosling to get an X-ray diffraction pattern from them. Unfortunately the equipment available to Gosling was ill-suited for fibre work. The specimen to film distance was too large and the X-ray tube too weak to yield a pattern from a finely drawn fibre with a reasonable exposure time. So Gosling cemented together a bundle of about 35 fibres and gave them a four hour exposure, using cobalt instead of copper radiation to increase the resolution. The result was amazing. Gosling had produced a pattern with more diffraction maxima than had ever been seen before from DNA.

The next step was obviously to see how changes in humidity affected the pattern. At 15 per cent R.H. the pattern was still of the same type, but the three-dimensional order had clearly decreased, for the diffraction maxima had spread considerably. When dried over P_2O_5 and heated to 80°C before exposure, the same bundle of fibres showed no evidence of three-dimensional order whatever. Great must Gosling's and Wilkins' surprise have been when these same fibres gave the finest diffraction pattern yet, after merely returning them to 90 per cent R.H.*

The Cavendish Meeting

This last picture was undoubtedly an advance on the work of Astbury and Bell. It was the first clearly crystalline pattern with no phase other than crystalline DNA present. Wilkins commented later that:

It was therefore unavoidable that DNA was really crystalline. It should be recognized that Astbury's patterns did not establish this unambiguously, though they gave a very good indication. There are several instances of crystalline spots superimposed on a diffuse pattern from material of biological origin, being due to crystalline impurity which sometimes need not be present in more than a few per cent.

(Wilkins, 1968)

It was now June. All seemed set for a real advance, until Sunday the 4th, when the Siemens X-ray tube on indefinite loan from the Admirality broke down (Stokes, 1967a). Meanwhile Wilkins had discovered the reversible stretching of the Signer DNA fibres to double their original length, with a change from negative to positive birefringence and a fall in the dichroic ratio (see Fig. 19.3). In a humid atmosphere the fibres regained their original length and optical properties (Seeds and Wilkins, 1950, 422). Excitedly he wrote to Markham in Cambridge telling him about the X-ray pictures "much better than Astbury's, and almost like single crystals, with

*These early "Gosling" patterns were all the result of his collaboration with Wilkins (Wilkins, 1986).

about 100 spots. These fibres stretch reversibly, forming necks during stretching, the necks separating apparently two phases. I believe the molecule is an extensible one comparable with keratin" (Wilkins, 1950a). From this letter it is evident that Wilkins had been poring over the X-ray pictures in Astbury's 1947 paper. These had given Wilkins the hunch that highly polymerized RNA, as in TMV, "may be similar to our stretched DNA . . ." (*Ibid.*). So he wanted Markham to supply him with TMV nucleic acid.

There is no escaping the approach here; it rings of Astbury's line on α-keratin and the analogy between stretched or β-keratin and silk. I conclude

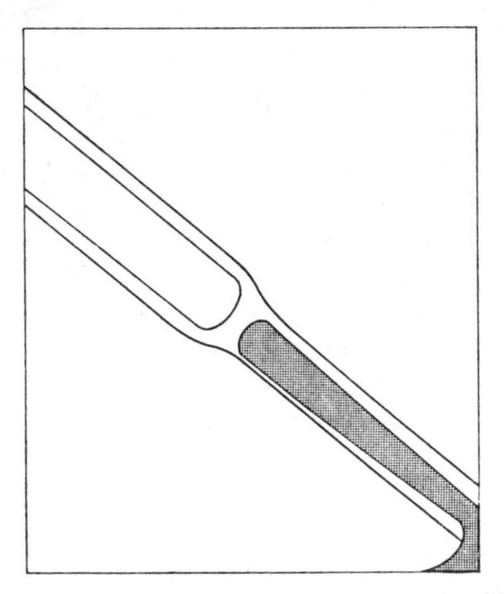

Fig. 19.3. Fibre stretched beyond the yield point to give a 'neck'. When viewed under crossed Nicols the narrow 'neck' is positively birefringent as in the right lower part of the diagram (from Wilkins, Gosling and Seeds, 1951, 760).

that Wilkins was picturing DNA as a folded long-chain molecule like α-keratin which could be uncoiled by stretching, and there seems no escape from the conclusion that this was the state of his thoughts when he contributed a talk on nucleoproteins to Perutz' unpublished conference at the Cavendish in July, 1950. As the notes which Wilkins wrote on the back of his copy of the programme show, he described his ultraviolet, dichroism and birefringence studies of TMV and DNA. Then he reported his X-ray studies with Gosling of DNA showing a "disordered" pattern (with 3.4 Å reflection) and a crystalline pattern (with reflections at 3.3, 3.53 and 4.2 Å). Finally, he mentioned the changes in cross section and in length of DNA fibres on drying, and the effects of stretching. This led him to his "Hypothesis", which we may infer was that of an intramolecular change like that in the $\alpha \rightleftharpoons \beta$ transformation of keratin.

These early pictures can be identified by the large white disc at their centre caused by use of an unnecessarily large lead shield. Beside the Signer DNA, they include calf thymus and herring sperm DNA prepared by Mary Fraser at King's, also human DNA and oriented *Sepia* sperm. Wilkins did not regard these two types of X-ray pattern as representing different molecular structures. Instead he thought of what Gosling had called the "sheet" patterns as merely "disordered" DNA which when ordered gave the "crystalline" pattern.

Infra-Red Spectroscopy

Shortly after the Cavendish meeting the MRC student, Bruce Fraser, included nucleic acid in his infra-red studies (in Wilkins, 1950b). The resulting spectra confirmed the generally held view that the bases in DNA are perpendicular to the axis of the molecular chain and of the fibre. That autumn Wilkins, Seeds and Fraser described their work at a Faraday Society meeting in Cambridge (September 25–28). Nine days later Randall, Wilkins, Gosling and Stokes met C. W. Bunn, the expert on the X-ray diffraction of synthetic fibres at ICI's research laboratories. Wilkins "had some quite good photos with him", wrote Stokes in his diary, "but we didn't get far on interpretation" (Stokes, 1967a). Apparently Bunn had suggested tipping the fibre during the course of exposure to get a spread in the diffraction maxima which might help to index them, and also suggested trying to make specimens with double orientation. After this meeting Wilkins wrote to Markham: "Bunn was quite definite in supporting the stretching molecule idea. Astbury was quite wrong in comparing nucleic acid necking with that of nylon" (Wilkins, 1950d). Evidently Wilkins must also have talked to Astbury about necking and Astbury had attributed it to the shearing of the long chain molecules over one another as in the case of a nylon fibre in which the chains were already fully extended. The reason why Wilkins was excited about necking in DNA was that he believed it to be an expression of an *intramolecular* change, and Bunn agreed with him. Just as Katz had won a glimpse of the intramolecular structure of rubber and Astbury of α-keratin by stretching the fibres, so Wilkins hoped to gain clues on the structure of the DNA molecule from these stretching experiments. In February 1951, Gosling and he sent a note to *Nature* announcing their confirmation of Astbury's conclusion that the bases in DNA have their planes at right angles to the fibre axis, separated by a distance of 3.4 Å, and they added their observation of "an unusual stretching phenomenon with fibres which suggests that these long molecules may be extended into a second form" (Wilkins, Gosling and Seeds, 1951, 759). They reckoned the "slipping of molecules over one another" as an unlikely cause of this reversible extension. "The optical observations show that the purine and pyrimidine rings have rotated during this process and lie on the average at about 45° to the length of the fibre" (*Ibid.*)

Despite these results the programme as a whole was still at the exploratory level; ". . . a good deal of the first three years or so had been taken up in instrument building and [in] sorting out [those] problems we were best able to tackle" (Randall, 1974). When, therefore, Randall gave his lecture, 'An Experiment in Biophysics', to the Royal Society, he asked that the efforts of his unit "will not be judged too severely at this early stage, as compared with the impressive achievements of much longer-established institutions . . ." (Randall, 1951, 2). This address provides the historian with one key piece of information—the sort of DNA model which was favoured at King's in the early part of 1951. Gosling and Wilkins were already aware of the work of the Norwegian crystallographer at Birkbeck College, Sven Furberg, in which the orientation of sugar at right angles to base was established in the nucleoside, cytidine, and the conclusion drawn that in DNA this orientation was preserved, thus making the plane of the sugar rings parallel to the axis of the molecule (Furberg, 1950a, 760). The King's group must then have studied Furberg's London Ph.D. thesis and compared Fraser's results from the infrared spectroscopy of DNA with the two models proposed by Furberg, one of which was a single helix enclosing the bases, the other a ribbon chain with the bases on the outside. Now Furberg used skeletal model building equipment to construct his models (Furberg, 1968), but the model illustrated in the *Proceedings of the Royal Society* is a space-filling model. So we can be confident that this model was built at King's in the winter of 1950/51. It was a slightly modified version of Furberg's ribbon chain with the bases on the outside. So there seems little doubt that at this time Wilkins was not thinking of a wide helix which could be pulled out on stretching but of what Furberg called at the time a "zig-zag" chain (Furberg, 1949b, 93), in which, as he later put it, "the ribose rings and phosphate groups form a flattish central column..." (Furberg, 1952, 638).

Furberg's Contribution

In 1947 Sven Furberg came to work at Birkbeck College at the suggestions of his teachers—Hassel and Finbak—who admired Bernal's work in structural chemistry. Through the Nuffield Foundation, Bernal had at last got support for setting up a laboratory for biomolecular research. Armed with a British Council scholarship Furberg came to London and met Bernal's colleague C. H. Carlisle, whose interest had been recently drawn to the nucleic acids by reading Gulland's work. No doubt it was Carlisle who impressed upon Furberg the importance of the nucleic acids as "midwife" molecules, so that Furberg's thesis contains the following quotation from Darlington's contribution to the symposium of the Society for Experimental Biology:

> We are now witnessing, after the slow fermentation of 50 years, a concentration of technical power aimed at the essential determinants of heredity, development and disease. This concentration is made possible by the common function of nucleic acid as the molecular midwife of all reproductive particles. Indeed it is the nucleic acids, which, in spite of their chemical obscurity, are giving to biology a unity which has so far been lacking, a chemical unity.
>
> (Darlington, 1947, 266; cited in Furberg, 1949b, 3)

G. Pitt had been working on the structure of hydroxypyrimidines at Birkbeck College. Now Carlisle wanted to carry the work a stage further in the direction of the nucleic acids. It was obvious to him that if Pitt's structure for the pyrimidine was linked to a ribose ring the two units could not be coplanar. So he put Furberg on to the structure of cytidine supplied in single crystal form by D. O. Jordan in Nottingham. Considering the techniques of those days, Furberg's solution of this molecule's 3-dimensional structure based on 2-dimensional Fourier projections was "a very remarkable piece of work. It was in fact such a difficult problem that if he had asked advice he would have been told that it was not possible to solve it" (Crick, 1973). He established the orientation of sugar to base as nearly perpendicular (see Fig. 19.4), and he went on to build two models of DNA in which his conformation of the nucleoside was incorporated. At the same time he had suggested a conformation for the phosphate group in cytidylic acid in which the P-O_3 link to the sugar ring was perpendicular to the plane of the base. On the basis of Astbury's data: a 3.4 Å separation of the bases and a repeat of some sort at 27 Å, Furberg built a helical model with eight nucleotides in the pitch of 27 Å, each residue being separated from its neighbours by a translation along the axis of 3.4 Å and a rotation around it of 45° (see Fig. 19.5a). The orientation of the phosphate groups gave him a separation of the phosphorus atoms of 5 Å. This was the first helical conformation with the bases on the inside of the helix, and the first model to account for the 27 Å repeat in terms of helical pitch. Furberg preferred this model to his second model (Fig. 19.5b) because the bases were brought over one another in the manner suggested by Astbury, thus allowing van der Waals forces between the rings of the bases to give stability to the structure.

In Furberg's second structure the residues occupied identical crystallographic positions every 6.8 Å. The 27 Å was not accounted for by a structural feature of the model. But Furberg, following Astbury, wrote that both models may "be made to have repeat units of this magnitude by appropriate choice of the sequence of the nucleotides" (1952, 639). This statement comes in a paper which Furberg submitted in March, 1952, when he also said that the evidence available on the internal structure of DNA did "not admit a choice to be made at this stage" between his two models or some of their intermediates (*Ibid.*). Evidently Zamenhof's and Chargaff's work on DNA cores was unknown to Furberg or else he would surely have rejected his

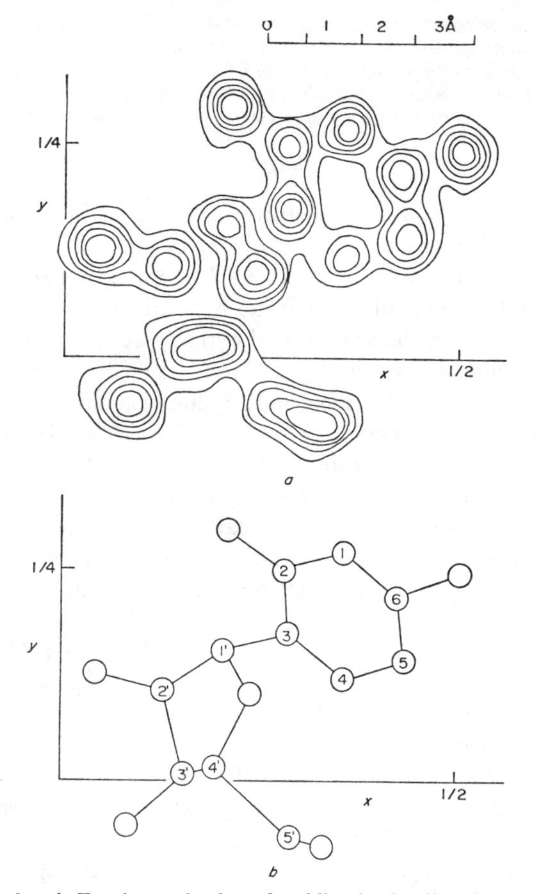

Fig. 19.4. (a) Furberg's Fourier projection of cytidine in the direction of the c-axis (b) his model of cytidine corresponding to the above projection. Note that the distance between C_3, and C_4, was put at 1·4 Å (from Furberg, 1949a, 22).

second model and accounted for the 27 Å repeat *only* by the pitch of a helix.

The Naples Meeting

We can gather some clues as to the state of Wilkins' work on DNA and nucleoproteins in early 1951 from the paper he gave during a four day meeting on "Submicroscopical Structure of Protoplasm" (May 22–25, 1951) at the Naples Zoological Station. His paper opened with the following statement of aims:

> The properties of crystals reflect the properties of the molecules of which they are composed. Hence, when living matter is to be found in the crystalline state, the possibility is

increased of molecular interpretation of biological structure and processes. In particular, the study of crystalline nucleoproteins in living cells may help one to approach more closely the problem of gene structure.

(Wilkins, 1951a, 105)

In the audience was the young post-doctoral American, J. D. Watson. No wonder he pricked up his ears! Wilkins described how the group at King's were concentrating on two aspects of the nucleic acids: their structure and their molecular orientation in sperm heads. His message was that extracted DNA and cellular DNA were the same, since their properties were so similar. No X-ray patterns from *Sepia* sperm heads had yet been taken, but optical studies revealed striking parallels in the arrangement of the nucleic acid molecules in the sperm head and in the fibre, the orientation of the bases and

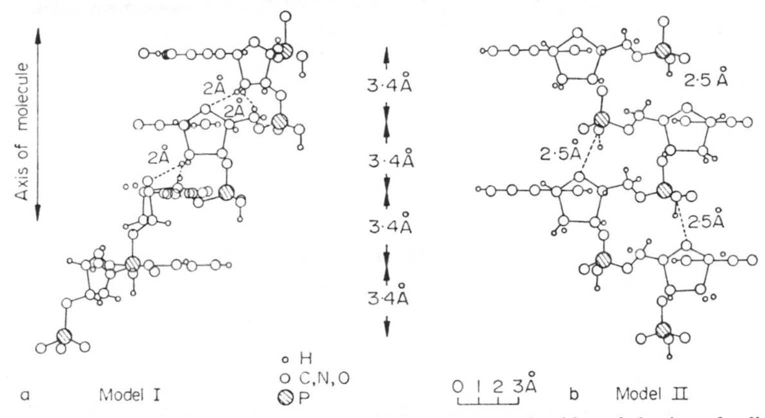

Fig. 19.5. Two models of thymonucleic acid based on nucleotides of the 'standard' configuration. The planes of the purine and pyrimidine rings are perpendicular to the plane of the paper (from Furberg, 1952, 637, also in his thesis, 1949b, 92).

the presence of an extensible molecular chain structure. Again it is clear that Wilkins was attempting to do with the nucleoproteins what Astbury had done with wool. The key to the structure was to come from studies of stretching by which an extended configuration of the molecules was achieved. Evidently Wilkins hoped to connect the normal and stretched state with two types of diffraction pattern. He showed the crystalline pattern and perhaps referred to the possibility that a stretched pattern might be found, for at the end of the discussions which followed Wilkins' paper Astbury said: "I am especially interested in the phenomena he [Wilkins] has observed on stretching thymo-nucleate fibres, since we ourselves, at Leeds, had obtained two different X-ray diagrams even before the war . . ." (Astbury, 1951, 113). This remark implies that Wilkins had talked of *two* X-ray patterns. What is also clear from this meeting is the fact that Wilkins did not discuss helical molecules. He did discuss the *helical packing* of molecules. Both in the TMV crystals which he studied *in vivo*, and in hydrated DNA fibres (after extension), he noted banding when viewing them between crossed nicols. This pattern indicated a regular change in the optic axis along the crystal and the fibre, suggestive of either a zig-zag or a helical packing of the long-chain molecules.

For Watson the real excitement came when the X-ray picture of DNA—of the crystalline or A form—"was flicked on the screen near the end of his talk" (*Helix*, 32). Although Wilkins' "dry English form" obscured any enthusiasm, his statement that "the picture showed much more detail than previous pictures and could, in fact, be considered as arising from a crystalline substance" caused Watson to make a *volte face* and get

> excited about chemistry. Before Maurice's talk I had worried about the possibility that the gene might be fantastically irregular. Now, however, I knew that genes could crystallize; hence they must have a regular structure that could be solved in a straightforward fashion.
>
> (*Helix*, 33)

Wilkins' Helical Model

Whereas we have no evidence that Wilkins' seriously considered Furberg's helical model for DNA up to May, 1951, both Wilkins and Stokes recalled discussion in the summer of that year when a single-chain helix was favoured. This development could not have arisen from any new and striking X-ray pictures, for since the Siemens tube had burnt out in the previous July no pictures appear to have been taken; moreover the arrival of a professional X-ray crystallographer, Rosalind Franklin, did not change this state of affairs quickly. In February, when Franklin had been at King's for a month Wilkins wrote to Markham:

> We are now beginning some more real effort on nucleic acid and the infra-red people have some results on position of rings for publication . . . We now have Miss Franklin for X-ray work and hope to get something really done, as almost no progress was made since the summer.
>
> (Wilkins, 1951b)

By this time the Ehrenberg–Spear tube had been put into working order. A few prototypes of these tubes had been made at Birkbeck College where the design had been developed. King's was fortunate to be given one of these in May, 1950 before commercial production (Stokes, 1967a). At first Franklin had been busy writing up her big paper on carbons for the Royal Society. She finished revising it in May (Franklin, 1951a). Meanwhile Wilkins was at work in Naples with B. Battaglia on the macromolecule–like behaviour of *Sepia* sperm in sheets and fibres (Wilkins 1951c, 214). That summer there were several important meetings to attend. Franklin went to the crystallographers' international congress in Stockholm at the end of June, Wilkins to the annual meeting of the Faraday Society at the British Iron and Steel

Federation's estate, Ashorne Hill, near Leamington Spa (July 18–20), and to the Gordon Conference in the United States (August 27–31). So it is not surprising that no X-ray pictures were taken that year until September.

There had, however, been an important development in Wilkins' thoughts on DNA which he expressed at the second Protein Conference held by Perutz in mid-July, 1951. Wilkins now saw as significant the meridional absences and indications of a "cross-ways" in the DNA patterns, and had intuitively deduced from them the helical character of the DNA molecule.* The slope of the "cross-ways" gave the gradient of the helix (see Fig. 19.6).

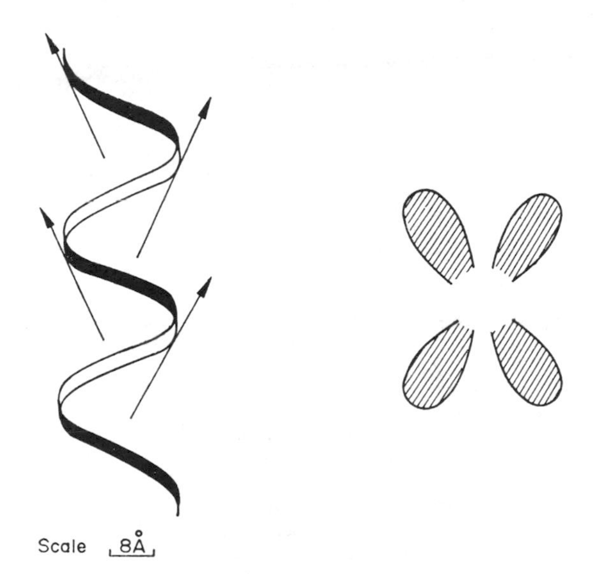

Fig. 19.6. A diagram of the sort shown by Wilkins in Cambridge in 1951.

The man who had been thinking of the helical packing of DNA molecules in Naples, and who saw that there might be "some connexion between the helical structure of TMV crystals, spiralized chromosomes and the helical shape of many sperm heads" (Wilkins, 1951a, 109), now began to consider the possibility that the DNA molecule, like the polypeptide chain of the α-helix, might itself be helical. It was the meridional absences which had so impressed him, and he asked his colleague Stokes what molecular structures would account for them. Stokes had suggested a helix. By the time Wilkins had returned from the Cambridge and Leamington meetings in late July, 1951, he felt the need to go beyond this intuitive interpretation and he asked Stokes to work out the theory of helical diffraction.

Stokes could not remember how long this took,

*"Stokes was the first to point out the evidence that DNA might be helical. . . . It was Stokes not I who was impressed first" (Wilkins, 1986).

but I think a day or two is a safe estimate—once you know just what you want to work out, the actual calculation does not take long and, mathematically, there is nothing very original in it. I feel sure that if the "anti-helix cloud" had not descended when it did, we might have made our ideas known in a short time, but unfortunately we were discouraged for a time.

(Stokes, 1967b)

Stokes had written out his derivation of helical diffraction on one sheet and on another he had plotted the Bessel functions on the first seventeen layer lines. Wilkins pinned these on the notice board at King's. This graphical representation of the Bessel functions showed how the diffraction maxima moved out from the origin at an angle with the meridian equal to that of the slope of the helix, leaving an absence of intensity along the meridian itself. When all four quadrants of the fibre diagram were considered this gave the now familiar "cross-ways" which could be detected in the 1950 "sheet' patterns. In the 1950 "crystalline" pattern the meridional absences were unmistakable, but there was no obvious "cross-ways".

It must have been while Wilkins was in the United States, late in the summer, that he wrote a letter, a part of which has been preserved in Franklin's papers. It tells us much about the ideas he had towards the end of 1951 and reads as follows:

Stokes has supplied a few good things on helices and I have done more swelling and shrinking of fibres. The cross section increases by a factor between 3.7 and 4 on swelling to 100 per cent humidity. Length 40–30 per cent increase.

I think these simple *volume* experiments probably good enough to tell the number of chains per unit cell. We assume the density of the dry fibre is 1.6. Even if it has holes in it the chains can't very well be 2 per unit cell, but the number must be reduced towards one. There are many assumptions in this but I think the evidence points strongly towards 1 rather than 2.

Riley's published (Faraday) figures are in *volumes*.

At the moment it does not seem very important to know the density of the water, only the fraction of unit cell *volume* occupied by it. Of course percentage crystallinity determination is basic in all this work and doubtless your experiments with monochromatic radiation are the only way to clear that up. No density or volume experiment has much meaning unless one assumes a figure for percentage crystallinity.

The Patterson seems to have got the wrong symmetry due to some slip and as a result I have left it behind.

The structure might be a 40° pitch single helix, one per unit cell, the layer lines being given by the pitch of the helix and the nucleotides uniformly spaced along the helix. Stokes can explain the approximate 006 007 008 spots being three and not one in terms of a single helix. However that is all conjecture.

Huxley camera (hope you will check this) in Norris's hands, he volunteered. Raymax ordered and Broad's setup in hand too.

DSIR promising for X-ray technician as this is a *new development* outside the original MRC plan.

Hope you have a good holiday

M.W.

(Wilkins, 1951d)

From a discussion of this letter with Wilkins I understand that before he left for the United States of America in 1951 he had asked Franklin to attempt a

Patterson analysis of the crystalline pattern. The extraordinary feature of this letter—that Wilkins was still favouring a single-chain structure—was explained by him as due to a large overestimation of the water content of the crystalline DNA. (His value of 70 per cent was not obtained from the crystalline specimen which had been X-rayed but from material used in the swelling and shrinking experiments.) We also learn from this letter that Wilkins knew Riley's and Oster's work on DNA gels. The reference to "(Faraday) figures" was presumably to their contribution to the Faraday discussion at Leamington. This letter also shows that Wilkins was basing his conception of the molecule on the details provided by the 1950 pictures. Hence the pitch of the helix was about 45°. If this represents his conception of the DNA molecule in July, 1951, then he must have been interested in the contribution of Riley and Oster to the annual meeting of the Faraday Society.

The Leamington Spa Meeting

The Riley and Oster paper was on X-ray and light scattering by colloidal and macromolecular systems (1951a). The section dealing with DNA was a resumé of the authors' much longer paper in *Biochimica et Biophysica Acta* which appeared during the winter of 1951/52. Like Wilkins, Riley and Oster had been the lucky recipients of Signer DNA, but unlike Wilkins they were working in an institution well supplied with X-ray equipment—the Royal Institution, once the home of Sir William Bragg. They felt that "knowledge of molecular interaction in environments similar to those which exist in the living cell may be more important to an understanding of gross cytological phenomena than even a complete picture of the molecule taken in isolation" (Riley and Oster, 1951b, 526). So their aim was not "to achieve a detailed analysis of the nucleic acid molecule itself, but to study molecular arrangement in aqueous systems at relatively high concentrations" (*Ibid.*). Since they did not orientate their specimens the X-ray pictures they obtained were like powder and liquid patterns. Their procedure was to start with the dry DNA and to add water, a little at a time until they reached a wet gel. Before the latter state was achieved they noted two types of X-ray pattern—one they called crystalline, the other moist crystalline (in their Faraday paper, "liquid crystalline"). As they were looking for changes in intermolecular packing they concentrated their attention on narrow angle spacings. The many wide angle spacings which they reported for the crystalline pattern were not paralleled by any observations on the wet-crystalline pattern, and they conceived of the change from the one to the other as a continuous one. To account for the alterations in narrow angle spacings on hydration they advanced two explanations. Either there was an intra-molecular unwinding of a tight helical molecule into a looser broader helix, or there was a system of micellar rods, each consisting of a hexagonal array of seven rod-shaped molecules, which could so move apart as to create holes into which the water went. The molecule of Signer DNA itself they pictured as a rod 16 Å in diameter and 8000 Å long (Riley and Oster, 1951a, 116). The diameter was arrived at from a consideration of the micellar systems, but they noted an X-ray diffraction spacing of 16 Å, close to the equatorial 16.5 Å spacing in the Astbury and Bell pictures (Riley and Oster, 1951b, 545).

Wilkins' Return to Kings

In the United States Wilkins had visited Chargaff who gave him DNA from *E. coli* wheat germ, and pig. Thus enriched, he looked forward to rejoining the King's group. With Franklin at last well equipped and more DNA material to hand, surely they would be able to forge ahead and at last get down to some fruitful and collaborative experimentation.

When Wilkins arrived back he soon found his hopes shattered! To be sure, Franklin had been busy. She and Gosling had taken a lot of X-ray pictures that September, and in their efforts to get more crystalline patterns they had increased the relative humidity to over 90 per cent. But instead of getting more detail on the diffraction pattern there was less, for it was an altogether different array of reflexions! No longer was it crystalline, but paracrystalline, or as they called it "wet". About a year later this form was called "structure B" in contrast to the crystalline form which became "structure A". The wet form was similar to the ill-defined "sheet" diagrams obtained by Gosling and Wilkins in 1950, but now the cross-ways could be seen to consist of a series of smeared spots. There was a complete absence of intensity along the meridian until 3.4 Å when an arc of intensity dominated the whole diagram.

It was hardly surprising that Wilkins became very excited at the sight of this picture. Here was the most striking evidence, albeit on an underexposed plate, that DNA was a helical model—the meridional absences, the repeat at 3.4 Å. This was the best "double-diamond" pattern yet! To his amazement his enthusiasm was not shared by Franklin. "How dare you interpret my data for me", she snapped. Here was Wilkins, back from yet another of his many trips, not content to send her silly advice about single-chain helices and volume measurements he now wanted to push in on her programme of work just when it had begun to look promising. Had he not already shown signs of his amateurism—the overlarge lead shield, the control of humidity by passing hydrogen through water instead of through the appropriate salt solution, his use of single-crystal equipment for fibre work? And as for his attempts at interpretation they smacked of the intuitive approach of Astbury, and where had Astbury got, he had not the solution of a single structure to his credit. Such may well have been her feelings at the time.

Rosalind Franklin's approach

When Franklin came to King's in January, 1951 she brought with her an expertise developed over the preceding eight years in her studies of coals and chars, first at the laboratory of the British Coal Utilization Research Association and second at the Laboratoire Central des Services Chimiques de l'État in Paris. In England she had developed the hypothesis of molecular sieves to account for the real and apparent densities of the coals she studied, and in France she had used X-ray crystallographic techniques to arrive at a more precise quantitative model of the micellar packing. She was therefore experienced in density determination and in X-ray diffraction by amorphous substances at the low angles which reveal micellar packing. She had had no previous contact with biology and substances of biological importance before she came to King's, and it seems that any suggestions along the lines that chromosomes and spermatozoa presented helical morphology, therefore DNA is likely to be helical, would have found no sympathy with her. What then brought her to work on DNA?

Franklin had been radiantly happy at the Paris laboratory. There she was treated as a colleague and equal despite the fact that she was a woman. She was straight, tough in argument, she did not waste words or tolerate fools and amateurs. Though she loved hill walking and climbing and stood for the Left in politics, there seems to have been little to distract her from her serious and ardent pursuit of science. In Paris her colleagues stood up to her. Her bark, when it came, never froze them, instead it stimulated them, and there resulted plenty of high-spirited exchanges. But her family wanted her back in England. She was nearly thirty by now and began to feel that she ought to get back into the British scientific community before it became too difficult to make the return. Early in 1949 she had visited Charles Coulson, then at King's. A year later she was scanning the advertisements in *Nature* and wrote to Coulson for advice about the I.C.I. Fellowships available at a number of universities in the U.K. Coulson explained to her that one had to find an appropriate department and get the consent of the departmental head before applying. He thought Birkbeck or King's the most suitable places to go to in London. "If you are interested in possible biological applications of the technique that you now know so well there could be quite a lot to be said in favour of King's" (Coulson, 1951). This was clearly the first time that a biological topic came into the correspondence and Franklin accepted Coulson's offer to take the matter up with Randall, though she warned him:

> I am, of course, most ignorant about all things biological, but I imagine most X-ray people start that way. I am certainly interested in the biological X-ray work. I don't know anything about what the work is at King's, but since at Birkbeck the emphasis is on single-crystal work I imagine that at King's it will be on other aspects. This would please

me more, not because the results of single-crystal work are less interesting, but because the actual technique doesn't appeal to me greatly.

(Franklin, 1950)

It must have been about the end of April that Franklin came to King's to see Randall. At that time there was no work proceeding on DNA as such and Signer had not yet brought his material to London. Evidently it was decided that a study of proteins in solution would provide the best topic for the X-ray diffraction technique she had been using in Paris. This agreed, Franklin was interviewed by the Fellowship committee in June and was awarded a Turner–Newell Fellowship for three years starting that autumn. In order to finish off work in Paris she arranged to delay the start of the fellowship until the New Year, and in November, 1950 she wrote to Randall asking about the work in progress and how it related to her project. Randall's reply is most revealing. Since Franklin's visit in the spring the results obtained from the Signer DNA were known and were felt so promising that, as Randall put it, "the slant on the research has changed somewhat . . ." Franklin's letter must have prompted Randall to hold a discussion with his colleagues for he went on:

After very careful consideration and discussion with the senior people concerned, it now seems that it would be a good deal more important for you to investigate the structure of certain biological fibres in which we are interested, both by low and high angle diffraction, rather than to continue with the original project of work on solutions as the major one.

Dr. Stokes, as I have long inferred, really wishes to concern himself almost entirely with theoretical problems in the future and these will not necessarily be confined to X-ray optics. It will probably inolve microscopy in general. This means that as far as the experimental X-ray effort is concerned there will be at the moment only yourself and Gosling, together with the temporary assistance of a graduate from Syracuse, Mrs. Heller. Gosling, working in conjunction with Wilkins, has already found that fibres of desoxyribose nucleic acid derived from material provided by Professor Signer of Bern gives remarkably good fibre diagrams. The fibres are strongly negatively birefringent and become positive on stretching, and are reversible in a moist atmosphere. As you no doubt know, nucleic acid is an extremely important constituent of cells and it seems to us that it would be very valuable if this could be followed up in detail. If you are agreeable to this change of plan it would seem that there is no necessity immediately to design a camera for work on solutions. The camera will, however, be extremely valuable in searching for large spacings from such fibres.

I hope you will understand that I am not in this way suggesting that we should give up all thought of work on solutions, but we do feel that the work on fibres would be more immediately profitable and, perhaps, fundamental.

(Randall, 1950)

In view of this letter it is very understandable that Franklin thought she and Gosling would be the only ones working on DNA fibres. The brief was clear. Interesting but preliminary results had been obtained. She was to take over the work and intensify the X-ray diffraction analysis. When Randall held the discussion in November, 1950 with his senior colleagues, Wilkins may no doubt have agreed that Franklin should take up the work he and Gosling had done some five months before.* It was not clear to him at that time that the

*But Wilkins did not see Randall's letter to Franklin.

key to the structure of the gene lay in the structure of *extracted* DNA. He was very keen to continue his study of oriented sperm, living sperm, TMV nucleoprotein and TMV RNA. It should be remembered that Watson and Crick occasionally worried that the DNA structure might prove biologically boring (*Helix*, 188), and Riley and Oster deliberately turned their attention away from the intramolecular structure because they expected biologically significant aspects of DNA lay elsewhere. But Wilkins never imagined that Franklin would be unwilling to collaborate with him and exchange information. When she came to King's, Wilkins was away. Several weeks elapsed "before he began to demonstrably take an active part in the interpretation of the X-ray data. It was perhaps natural under these conditions that she felt, rightly or wrongly, that the problem had been assigned to her". (Randall, 1974).

We can see now that Wilkins and Franklin were almost bound to start off on the wrong foot. All was well while Wilkins kept to his optical studies or attended conferences. But when he returned from Chargaff's laboratory and expected to join in the X-ray work Franklin was outraged, and matters between them were only temporarily smoothed over by Wilkins agreeing to leave the Signer DNA to Franklin and Gosling whilst he used the Chargaff DNA. Little did he know then that he would fail to get the "crystalline" pattern (A structure) with the Chargaff material, or that this material would prove more difficult to orientate than Signer DNA for the production of crisp "wet" patterns.

When Crick invited Wilkins for the week-end after his return from the United States he gladly came. Watson was present when they talked about DNA which they all agreed was a helical molecule. According to Watson, Wilkins referred to the meridional absences in the "wet" pattern which Stokes "had told him was compatible with a helix. Given this conclusion, Maurice suspected that three polynucleotide chains were used to construct the helix" (*Helix*, 56). This being so, we conclude that by the autumn of 1951 Wilkins had rejected his earlier belief in a single chain helix and now accepted a triple helix. If he did tell Watson and Crick about the "wet" pattern that Franklin and Gosling had produced during September it is doubtful that he gave them the impression that here was a striking new discovery. This early B pattern was underexposed and nothing like so arresting as the example produced the following summer. Wilkins later recalled that Franklin "resolved layer lines in the cross-ways pattern during, I think, September, 1951 while I was away in the U.S. collecting DNA from Chargaff. The layer lines and the distribution of intensity agreed strikingly with Stokes' calculated diffraction from a helix" (Wilkins, 1968). But Wilkins' attempt to enter into the work had been resisted, and as a result he no longer knew what they were doing, although they all worked in the same basement building in the College. Fortunately there was to be a colloquium in Randall's department when Franklin would give a report.

The 1951 Colloquium

The entry in Stokes' diary for November the 21st reads: "Colloquium on Nucleic Acid Structure, at which Wilkins, R. Franklin and I made contributions (mine being on general helical theory)" (Stokes, 1967a). One assumes that Wilkins talked about his study of the extensibility of the nucleic acid fibre, its optical properties, and his work on orientated films of sperm in Naples the previous spring. Stokes would have described the general properties of the Fourier transform of a helix, and how the theory could be used to interpret diffraction patterns for DNA.

Let us assume that Franklin did use the notes for this colloquium which her collaborator, Aaron Klug, has preserved together with all her other research notes. Then a picture of her contribution emerges which differs strikingly from Watson's recollections (Klug, 1968). She reported three "more or less well-defined states", (1) wet (2) crystalline (3) dry, and showed slides of their diffraction patterns. On the first she drew attention to the 3.4 Å arc on the meridian and the two "oblique smears at about 40° to it. On the equator there was a "*sharp* intense spot" and she concluded that only the equator showed "high order". Since no further details were given of the "wet" pattern we may be confident that Franklin did not discuss the number of layer lines visible, and made no assignment for the 3.4 Å arc. On the dry form she noted the gradual disappearance of the equatorial intensities, leaving only the 3.4 Å meridional arc and two side arcs. When she came to the "crystalline" pattern she gave no details other than the 27 Å spot. "The amount seen", she noted, is "at present limited by experimental difficulties". The Phillips microcamera revealed more spots, as did better specimens, but good fibres were small, and small fibres demanded long exposure times. Her notebook for 1951 shows how many interruptions the work suffered from. A long run was difficult under such conditions, the underexposed "wet" pattern had been a six-hour exposure of a bundle of "several fine fibres" (Gosling, 1954, 63, Plate 2).

The Crystalline to Wet Transformation

Wilkins and Gosling had not observed this conversion of their crystalline pattern, into a quite distinct "wet" pattern. This failure may have been due to the salt content of Franklin's fibres which may have been raised by the use of salt solutions for control of the humidity. And Franklin and Gosling reported difficulties in getting some fibres to go through the change from "wet" back to "crystalline".

At the Colloquium Franklin concentrated on the questions: how much water does the DNA absorb from the dry condition over P_2O_5 to the crystalline state at a relative humidity between 70 and 80 per cent? Her answer was that unpulled fibres contained 42 per cent by weight of water. Her measurements of the further water uptake before the "wet" form was produced were not yet completed, but she noted a "large uptake at relative humidities above 80 per cent" (Franklin, 1951b). Her 1951 estimate for the unit cell was only preliminary, but she could index it as "monoclinic, face-centred. This is nearly but not quite hexagonal in projection" (*Ibid.*). Now she could assemble the data required to determine the number of nucleotides in the unit cell. Her notes on this operation are as follows:

> Using Astbury's value, 1.63, for density of dry DNA, measured value 42 per cent for H_2O content of crystalline structure and 330 for mean molecular weight nucleotide gives 46 nucleotides per unit cell (face centred), i.e., 23 nucleotides per primitive cell, equivalent probably to 24 per primitive cell.
>
> (Franklin, 1951b)

Franklin's Interpretation

In the light of subsequent events, Franklin's interpretation of her results at this stage deserves a detailed description. The following text is an edited version of her notes.

General Hypotheses
Chain Groups: The structure is very nearly hexagonal in section perpendicular to the fibre axis. The molecular chains presumably run parallel to the fibre axis. This suggests that the structure consists of only slightly distorted cylindrical units in nearly hexagonal close packing.
Evidence for Spiral Structure: (1) A straight chain, untwisted, is highly improbable because of unbalanced forces. (2) The absence of reflections on the meridian in the crystalline form suggests a spiral structure, in which the electron density projected onto fibre axis is nearly uniform. (3) The presence of a strong 27 Å period. This is much too marked to result merely from different nucleotides, and must mean that nucleotides in equivalent positions occur only at intervals of 27 Å. This suggests 27 Å as the length of a turn of the spiral. The near-hexagonal packing suggests that there is only one helix (containing possibly >1 chain) per lattice point. Density measurement (24 residues/27 Å) suggest >1 chain.
Change Crystalline ⇆ Wet: In the wet diagram only the equator shows sharp reflections. This suggests the presence of cylindrical units randomly displaced parallel to the fibre axis. The diagram (other than on the equator) then represents the form factor of a single lattice-point unit.

In the change from one form to the other the equatorial spot shifts only by about 10 per cent. This suggests that the group of chains associated with a single lattice point remains intact; i.e., it confirms that the (2, 3, or 4?) chains are bonded together more strongly than are the chains in different groups, and are not separated by the action of water. It is in this change that a large length change occurs. So the isolated helix has not got the *same* structure as in the crystalline form. The latter involves some strain of the helix. [Cf. Pauling.]
Crystalline Structure: Here the inter-unit bonds are as important as the intra-unit ones and must be considered. They might be: (1) Base to base (NH—CO bonds?) (2) Base to phosphate (3) Phosphate to phosphate. They are the bonds which are destroyed by water

at relative humidities > 80 per cent. Therefore (1) is entirely ruled out, (2) is doubtful and (3) Phosphate to phosphate bonds, is highly probable. The presence of

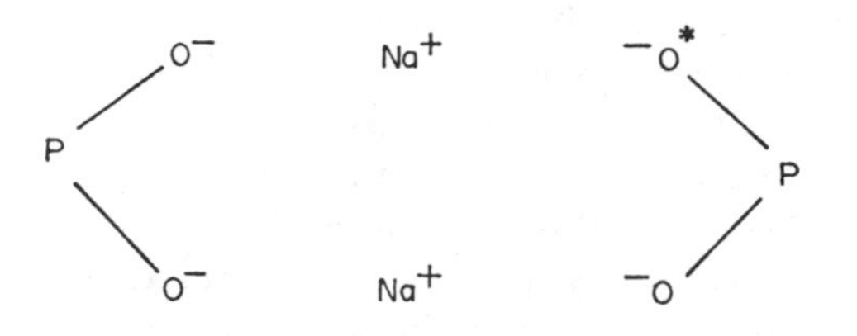

(and variants) as inter-unit links would account well for the sensitivity of the structure to the water content and for the large amount of water in the crystalline structure (probably held in the neighbourhood of the phosphate groups). The water content of this form suggests 8 molecules of water per nucleotide.

The Change Crystalline ⇆ Dry: Thorough drying stabilizes the crystalline structure subsequently formed; i.e., although, on drying, crystalline order disappears, the inter-unit bonds responsible for it are preserved and strengthened in the dry state. On drying, hkl and hko reflections gradually fade out, decreasing to zero in intensity while getting only slightly more diffuse. This suggests that the three-dimensional skeleton remains but becomes strained and buckled owing to the absence of water, this loss presumably making holes in the inter-unit spaces. This conclusion is confirmed by the low angle scattering in a very dry photo.

Conclusion: Either the structure is a big helix or a smaller helix consisting of several chains. The phosphates are on the outside so that phosphate–phosphate inter-helical bonds are disrupted by water. In this external position the phosphates would be available to proteins.

Difficulties: Although the figure of 24 nucleotides per unit cell has been deduced one does not know how homogeneous the structure is. Since there are no amorphous rings on the diffraction pattern one cannot measure the degree of disorder.

(Franklin, 1951b, ed. Olby)

Watson's Visit

How was it possible for Watson to attend this colloquium and yet in his recollection of it to say not a word about Franklin's advocacy of helices? Can we get around this difficulty by assuming that Franklin's notes were not used for the colloquium? This hardly seems likely, bearing in mind that they are headed "Colloquium Nov. 1951". Wilkins' commented: "My own memory is that Franklin said nothing about helices, but I cannot be certain. I think it would have been out of character to present in a talk the speculations in the notes" (Wilkins, 1972). But these speculations about helices were very limited, and they were supported by her realization that she had to do with near-hexagonal packing, indicative of cylindrical molecules. Surely she had every reason to refer to this feature and its significance in her talk?

Let us therefore start by noting that Watson said nothing about Franklin's anti-helical views at this stage. He already knew from Wilkins and from Astbury and Bell that there was a chain repeat at about 27 to 28 Å, that the

* In fact there is only one ionizable oxygen per phosphorus atom at neutral pH.

bases had to be 3.4 Å apart, and Wilkins had told Watson and Crick that he favoured a 3 strand model. So when Franklin repeated these pieces of information they struck no surprise in Watson.

At the beginning of November, Crick and Cochran had worked out the Fourier transform of a helix. Watson was infected with Crick's enthusiasm for the potential this theory had for the solution of fibre diagrams, and with Crick's assessment of the importance that should be attached to the approach that Pauling used in his successful attack on the structure of polypeptides. Watson came to London expecting to hear Franklin talk about model building, the details of the layer line intensities, the likelihood of a simple solution. Instead he was treated to a sermon in caution, a lecture on the technical difficulties, and a long and boring enumeration of data about the water content of the specimen over a range of humidities. What had this to do with the model? Who cared about phosphate–phosphate interactions *between* molecules. It was the structure of the molecule itself that Watson was after! By his own admission much of what Franklin said passed over his head. His attention began to wander and in conformity with his normal practice he made no notes.

We have therefore to distinguish between Franklin's provisional acceptance of a helical structure for the A and B forms of DNA in the winter of 1951/2 and her evaluation of the weight to be given to the evidence at that time. According to Watson scarcely a week passed after the colloquium before Franklin was declaring "there was not a shred of evidence that DNA was helical" (*Helix*, 94). Such an attitude was not inconsistent with her continued provisional acceptance of a helical conformation for the DNA molecule, and in February, 1952 she submitted her first report to the Fellowship board. There she again accepted helices for both forms.

The Advance on Astbury and Bell

By the close of 1951 there had been a significant step forward from the position described by Astbury in 1946. This had been made possible by the availability of the Signer Schwander DNA and the new X-ray equipment consisting of Ehrenberg–Spear tube and Phillips microcamera. Three important observations were made at Kings: well oriented fibres could be made, single phase crystalline patterns could be produced, there were two structural forms of the molecule the diffraction patterns of which could be distinguished. Astbury and Bell never obtained the crystalline form without admixture of the "wet" form. The result was the confusion of an inter-nucleotide spacing of 3.4 Å from the "wet" form with a backbone repeat of 27 Å from the "crystalline" form. The presence of more than one structure in the fibre could be detected by an examination of the equatorial reflections:

	Wet Pattern	Crystalline Pattern	Mixture
20 Å	v. strong	v. weak	strong
11.3 Å	zero	v. strong	strong

Finally the behaviour of the fibres toward water strongly suggested that the phosphate groups must be on the outside of the molecule.

CHAPTER TWENTY

DNA as a Triple Helix

Crick was at work in the Strangeways Laboratory when he received the following letter:

> My Dear Crick,
>
> How is Cambridge? Is the cold wind blowing across the fens, frisking up the waters of the Cam, whistling through the barbed wire on college walls, rattling the chain padlocks on college gates and causing a healthy glow to appear in the faces of bedmakers and undergraduates hurrying across the cobbles to the college bathroom? Is it blowing in under the door of the Strangeways, congealing the culture media and causing all honest amphibians to hibernate?
>
> And when you come to town next send me a card in advance and I can reply by phone and suggest a date for dinner. I have made some very good dinners lately and am getting in a barrel of cider.
>
> Do let me know, won't you?
>
> Yours,
>
> Maurice Wilkins (Wilkins, 1948?)

The friendship between these two scientists played an important part in the events leading to the discovery of the double helix. How did they come to know one another? Although born in New Zealand, Wilkins had lived in Birmingham from the age of eight. Crick came from Northampton. For their university training Crick went to London, Wilkins to Cambridge. Both men began research before the outbreak of war, Crick in London and Wilkins in Birmingham. When the war claimed their services Crick worked in Britain and Wilkins joined the Manhattan project in Berkeley. It was almost as if they were destined not to meet. But both men had at different times worked under Harrie Massey (now Sir Harrie) and when they thought of going into biology first Wilkins and then Crick went to Massey for advice. Massey suggested to Crick that he should meet Wilkins. Crick recalled that he "didn't know many biophysicists in those days. I might have just met Randall, but Wilkins I knew personally . . . (Crick, 1968/72).

Watson met Wilkins for the first time in Naples in the spring of 1951, when he tried to chat him up on an excursion to the Greek temples at Paestum. But Wilkins was reserved. Indeed, R. D. Preston recalled one of those at the temples using him as a shield from Watson's searching eye (personal communication). Very likely it was Wilkins!

The Meeting between Watson and Crick

When Watson came to work in Cambridge he found himself sharing an office with Robert Parish and Francis Crick. Because of the difficulties with Paul Weiss over the transfer of his fellowship to Cambridge, Watson had described himself as working on the structure of proteins and plant viruses under Roy Markham, while in fact he was attempting to crystallize myoglobin for X-ray diffraction studies under Kendrew at the Cavendish. As his efforts in this direction were unsuccessful he was able to spend several hours every day talking to Crick whom he described to Delbrück as "no doubt the brightest person I have ever worked with and the nearest approach to Pauling I've ever seen . . . He never stops talking or thinking . . ." (Watson, 1951b). Long after the event Watson said that in Crick he had found someone "who knew that DNA was more important than proteins" (*Helix*, 48). Crick is by no means so sure that this was the case, and commented

> You ask me what I knew about DNA and chromosomes when I first met Watson and the answer is I really do not remember . . . I must have known about DNA. When I was at the Strangeways I was asked to give a lecture and it was on, or I chose, DNA as a subject. I didn't know much about it and I remember pointing out that radiation did reduce the viscosity in a very strange way suggesting that the molecules were long or something. So clearly I must have been interested but I think, in present-day terms, I was not convinced of the overwhelming importance of DNA to suggest that I did experimental work on it rather than on proteins, because I thought proteins were also important and here was the work going on. I suspect very much that if I could find the notes of that lecture they would reveal what I thought before I met Jim Watson. I am fairly sure that the whole theme of the lecture was that the central problem of biology was to explain how genes replicated and how they acted and to point out that the main thing they were likely to do was to control the synthesis of proteins, in particular their amino acid sequences. Unfortunately I have no documents to support this recollection.
>
> (Crick, 1968/72; 1973)

As a test of Crick's attitude to DNA before he met Watson we may inquire as to his attitude to Wilkins' work. Wilkins' colleagues recall how Crick teased his friend when the difficulties of working with DNA were brought up. Once in the Embankment Gardens over tea Crick quipped: "What you ought to do is get yourself a good protein" (G. Brown, 1969). Since Geoffrey Brown was in Stockholm from October, 1950 to September, 1951 we may suggest that this occasion was in the summer of 1950, when admittedly the DNA pictures looked extremely unpromising to a protein crystallographer.

Crick is convinced that he was present at Perutz' Protein Conference in the Cavendish, yet he cannot recall Wilkins talking about DNA. If this was the 1950 meeting, then, as Wilkins recalled, there was little on the X-ray data for DNA (Wilkins, 1972). We should remember, too, that according to Wilkins' collaborator, Bill Seeds, Crick was in active conversation at the back of the hall during the proceedings (Seeds, personal communication).

Nor should we assume that Wilkins was convinced that DNA was the genetic substance when he got those first X-ray pictures of the Signer fibres in June, 1950. That August he confided in Markham: "What we would really like to do, of course, is to find what nucleic acid is in cells *for*. I rather hope my new ultraviolet microscope used on living cells may help by telling us more clearly just where (and when) the nucleic acid occurs in cells" (Wilkins, 1950d). By the following May he was much more confident: "the study of crystalline nucleoproteins in living cells may help one to approach more closely the problem of gene structure" (1951a, 105). Before Watson heard these words in Naples his attitude to nucleic acids seems to have been somewhat ill-defined. He wrote of his daydreams "about discovering the secret of the gene, but not once did I have the faintest trace of a respectable idea" (*Helix*, 30). To be sure, he had come to Copenhagen to learn nucleic acid chemistry, DNA looked like the stuff of the gene, but whether DNA held the key to the *duplication* of the gene remained to be seen. The immediate problem was how to get at the structure of the genetic material. Chromosomes were no good. Nucleoproteins were messy, but Wilkins taught Watson in Naples that the approach to the structure of the gene could be made through the structure of oriented fibres of the sodium salt of DNA.

What can we guess as to the nature of the discussions which took place in October, 1951 between Watson and Crick? I suggest the following: The important feature of the genetic map was that it represented a *linear sequence* of units. These units were divisible into a linear sequence of sub-units right down to the molecular level. To this sequence there must correspond a chemical sequence, presumably of nucleotides. But what of the magical property of self-duplication possessed by chromosomes and viruses? From Caspersson's work it was accepted that DNA was necessary for chromosome duplication, but it was by no means clear that DNA could duplicate itself without the aid of protein. Nor can it. For Watson and Crick in 1951 there was also the corollary: the possibility that the nature of this duplication would not be discovered from the structure of DNA. They might indeed work out this structure only to find that the duplication of the gene remained shrouded in mystery. This was the case with their first model and with Pauling's 1953 model but the immediate question was how to tackle the structure? Here Crick had no doubt whatever as to the approach required. They must use the method Pauling had adopted for the α-helix.

As we saw, Pauling and Corey built up the polypeptide chain from a detailed knowledge of its sub-units. There is no evidence that Watson and Crick did the same for DNA in 1951. Did they read Furberg's paper in *Acta Chemica Scandinavica* for 1950 on the stereochemistry of the nucleosides at this time? There is no hint in surviving documents to show that they did. Astbury's well-known paper of 1947 was known to them; it contained the right information on the linkages in the chain, but where in that paper would they have found any information on the precise bond angles and inter-atomic distances in the sub-units? Such information only became available later (Pitt, 1948; Clews and Cochran, 1949; Broomhead, 1950).

Now it is surely inconceivable that Watson and Crick were not made aware at this time of the work of Cochran and Broomhead in the Cavendish, and of Todd's work on the linkages in the polynucleotide chain. Before Watson's arrival Crick and Cochran had shared the laboratory on the top floor of the Old Cavendish building. Now Cochran's paper on the crystal structure of adenine hydrochloride, in which the positions of the protons near to the rings had been determined (see Fig. 20.1), was published in 1951.

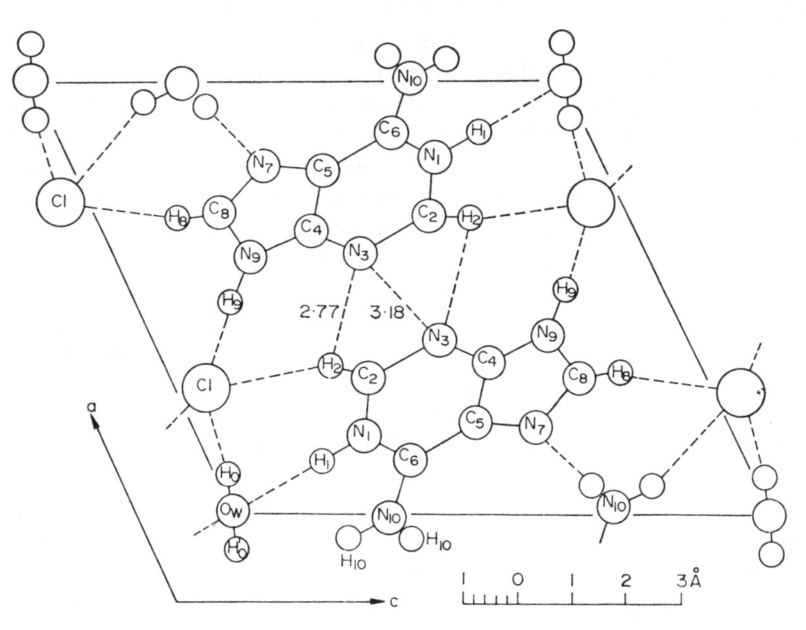

Fig. 20.1. The structure of adenine hydrochloride (from Cochran, 1951, 88).

Then surely it was established that the protons do have fixed locations, and Crick ought to have accepted this in his model building. Obviously this was the *simplest* line to take, but was it justified? Crick thought it was not, because there might well be all the world of difference between adenine in DNA or in the free state, on the one hand, and the *hydrochloride*, with an extra hydrogen atom, on the other.

Crick preferred to leave unsure restrictions well alone and starting with the minimum number of postulates to use model building. By this technique many possibilities could be eliminated and then all possible data could be used to select from those structures which remained the most probable. "All we had to do", recalled Watson, "was to construct a set of molecular models and begin to play—with luck, the structure would be a helix" (*Helix*, 50–51). But this exercise would be useless if the end product failed to match the diffraction pattern produced by DNA. In Astbury's 1947 paper there was a half an X-ray diagram of a stretched film of DNA taken by Bell before the war, which showed the 27 Å repeat, the 3.4 Å meridional arc and the 16.2 Å side spacing along the equator. But the diffraction maxima were far from sharp. For more reliable data they would have to turn to Crick's friend, Wilkins, who was by this time back from the United States. So Wilkins was invited to Cambridge for the week-end. Then they learnt that he had obtained no improved photographs of crystalline DNA since the summer of 1950. However, Wilkins was able to draw their attention to the evidence he had given for a helical DNA structure at Perutz' second Protein Conference (see p. 341). They also learnt nothing about Franklin's latest results since Wilkins had been excluded from her circle of activities. So there was nothing for it but to come to the colloquium at King's at which she was to present a report on her work.

Watson's Visit to King's

It was probably because Crick had already made other plans that Watson was the one to attend the colloquium. By this time he had had six weeks in Cambridge during which he could get to grips with the principles of X-ray crystallography and learn from Crick. So he was sent instead. Unfortunately he had not learnt enough, and when Crick pumped him for details Watson's answers were vague. Yet scarcely a week passed by before Watson and Crick had produced a model and Crick had written a summary of the way they tackled the problem.

> *The Structure of Sodium Thymonucleate. A Possible Approach: Summary*
> *Introduction*
> Stimulated by the results presented by the workers at King's College, London, at a colloquium given on 21st November, 1951, we have attempted to see if we can find any general principles on which the structure of DNA might be based. We have tried, in this approach, to incorporate the *minimum* number of experimental facts, although certain results have suggested ideas to us. Among these we may include the probable helical nature of the structure, the dimensions of the unit cell, the number of residues per lattice point, and the water content. Having arrived at a tentative structure in this way, we have generalized what we regard as the important features and now present these as postulates.
>
> *Postulates*
> 1. That all residues are equivalent. As Pauling has pointed out, this necessarily leads to helical structures. We do not feel absolutely bound by this postulate, which can be relaxed in two ways: (a) while certain parts of each residue may be equivalent, others may not be equivalent. In the strict sense this is obvious, since the basic rings are different in different residues. (b) the residues may fall into a *small* number of groups, within which all residues are equivalent.
> For the moment we favour structures in which the phosphate link and the sugar ring are all equivalent.

2. That the structure will be dominated by the charged atoms. There are no atoms which can *donate* hydrogen bonds except in the basic rings and the water. Thus hydrogen bonding is unlikely to play the dominant part in the structure that it does in the polypeptide α-helix. Electrostatic forces are so big relative to van der Waals forces that we may be confident that the Na^+ and the PO_2^- will mainly decide the arrangement.

As the result of our first two postulates we surmise a third, rather less certain one:

3. The Na^+ and the PO_2^- will all lie very close to the surface of a single cylinder.

If the structure is based strictly on a helix all Na^+ ions will lie on the surface of one cylinder, and all PO_2^- ions on the surface of a second one. We guess that the minimum local neutralization will occur when the radii of these two cylinders are very nearly the same. It may be possible to check this idea by calculations. (In conversation with Perutz and Kendrew, it was pointed out that the above is not the most general condition, which would be that the Na^+ radius was approximately the mean of the two O- radii.)

A Systematic Approach

Starting from these postulates we can now attack the problem of possible structures systematically. It is first necessary to assume a co-ordination number in the cylindrical surface for the Na^+. This is likely to be an even number, because of the half charge on the oxygens. Two is too small to be very probable, and 8 is too big. This leaves 4 or 6. We have so far concentrated on 4, but 6 may not be impossible, and should be investigated. Here again some simple electrostatics may help.

Having chosen a co-ordination number the next problem is to write down systematically all the possible topological combinations. This is done by writing down all the possible linkage schemes, which satisfy the first postulate, for an infinite plane surface. A strip of this surface is then isolated and folded round into an infinite cylindrical sheet. The number of ways of doing this can then be explored systematically.

Having thus obtained all possible linkage schemes, the next step is to build models of them all. Up to this point one has only to consider the Na^+ and PO_2^- in isolation, and the model is built in this way. The model is then subjected to a series of tests, in which the experimental data are increasingly used. For example, it may prove impossible to link any two phosphorus atoms satisfactorily with the sugar, because they are too far apart. Then it may be possible to link them, but only in such a way that the normal van der Waals conditions are violated. The number of residues per 27 Å length of helix may be quite wrong, and so on. In doing this it is particularly important not to use as a basis of rejection rather vague criteria, such as not quite seeing how it fits in with possible preconceived ideas about the X-ray pattern, and not quite seeing how it would stretch. If great care is not taken the right model may be rejected because of some difficulty which will sort itself out at a later stage.

An Example

To make matters a little clearer we give here an account of the line we have mainly followed so far.

We start off with co-ordination 4, and the basic net

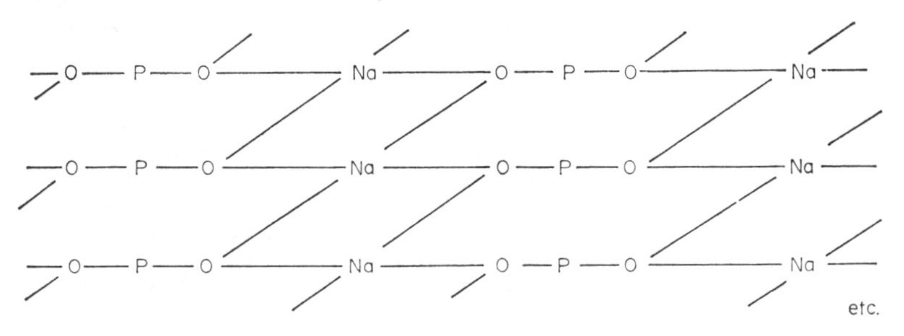

This is a topological scheme only; the angles are not supposed to be anything like those drawn. It is easily found that any structure can be defined precisely by

(a) the number of sodium ions passed in going once round the cylinder.

(b) the number of sodium ions up or down from the start after going once round.

These rules can be made precise by giving an unambiguous rule for traversing the structure. The one we have used is "at any junction take the diagonally opposite path". For this purpose the O, P and O are considered as being topologically at a single point.

We have roughly built some of the 2 Na structures. They tend to give about 16 residues in 27 Å, so we have temporarily put them on one side, though they may be appropriate for stretched DNA. The 3 Na structures are more interesting, as they give about 24 residues in 27 Å and have a hole in the middle which is about the right size to accommodate about 8 water molecules. We are now about to explore these systematically.

Finally we note a few isolated points which have emerged so far in this study.

1. If a helix consists of 3 separate polynucleotide chains intertwined, and the X-ray pattern repcats after 27 Å, it does *not* follow that the number of residues in 27 Å must be a multiple of 3, since the repeat may be from one chain to another.

2. For a helix to fit in with a neighbouring identical helix, with an identical orientation, it is likely that a bump on one will be opposite a hole in the other. For any given helical scheme one can thus predict very roughly the monoclinic angle β.

3. In *ribose* nucleic acid the extra oxygen is likely to be in such a position as to make the DNA structure impossible. In other words its stearic effects are likely to be more important than its power to form hydrogen bonds.

(Crick and Watson, 1951)

This is a revealing document. It shows how Crick was living under the shadow of the α-helix. Pauling had defied the X-ray evidence from α-keratin when he first built the α-helix. The Cavendish group had used the α-keratin data to rule out non-integral helices. Astbury had continued in his rejection of helical folds in α-keratin because he could not see how inter-chain sulphydral bonds could remain intact on stretching if the chains were uncoiling. So Crick emphasized the need for a systematic and general approach coupled with reluctance to use limiting empirical data until later in the work.

Since Pauling, as well as Watson and Crick, came up with a three-strand model with the backbones on the inside one suspects that there was a good reason for building the molecule this way. The Cambridge pair argued along these lines. The bases are of different shapes and sizes. Their sequence along the polynucleotide chain is probably highly irregular. Yet DNA forms an ordered aggregate which gives a crystalline diffraction pattern. The order must be due to those features of the molecule which are regular. It cannot therefore concern the bases. It must concern the sugar–phosphate backbone. Therefore let us build models in which we neglect the bases.

The natural tendency, having made this decision, was to assume that the bases faced outwards in which position they did not enter into the consideration of how to pack the atoms of the backbone structure to yield a plausible model. As they later commented on the above line of reasoning, it seemed "that the structure was based upon features common to all nucleotides. This suggested that in the first instance one should consider mainly the configuration of the phosphate sugar chain, with an 'average' base attached to each sugar. In other words, an idealized polynucleotide with all the monomers the same" (Crick and Watson, 1954, 83).

There was also another reason for the backbones being placed on the inside of the molecule. This, ironically enough, was due to a restriction of a chemical nature which Crick introduced. This concerned the bases, which were described as having several tautomeric forms. This term implies that one form can change into another by the migration of a proton from one location on the molecule to another location (see Fig. 20.2). Therefore he ruled out hydrogen-bonding between bases as the source of the forces holding the long

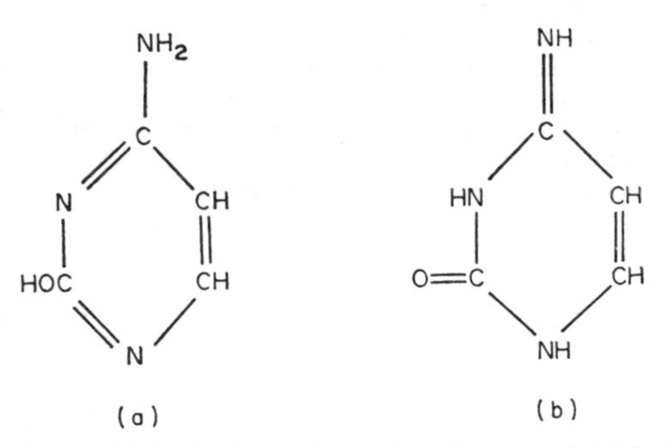

Fig. 20.2. The cytosine model (a) in the amino and enol form and (b) in the imino and keto form.

chains together in a rigid helical rod. Crick explained the drift of his argument as follows:

> We knew that hydrogen bonding was important in the α-helix. Where I made a complete mistake was assuming that each base actually existed in more than a single tautomeric form. Even if the less correct tautomer of each base only existed for 5 or 10 per cent of the time this would make it difficult, I thought, to use the hydrogen bonds of the bases to build a regular structure. I therefore concluded, quite incorrectly, that hydrogen bonds could play little or no part in forming the structure.
>
> (Crick, 1968/72)

Nor did he reckon the structure could be held together by hydrogen-bonds with water molecules, since according to Watson there was so little. This absence of water meant that local charges could not be neutralized by the presence of water shells around them. So the charged groups, Na^+ and PO_2^-, had to be close together. The obvious conclusion that followed was that the structure was held together by the attraction between these oppositely charged constituents of the sodium salt of DNA.

> We tried to make a model which balanced the positive and negative charges. This was completely wrong because in fact there is a lot of water there. I did not know enough chemistry to know that things like sodium are highly likely to be hydrated anyway. Otherwise I would have picked up the mistake. It was not only that Watson made the mistake but that I did not notice it, which is equally blameworthy. So it was a complete

waste of time. Watson gives a pretty accurate account but it was a bit more systematic than he describes it. We made several nets. I formalized the problem and made several geometrical theorems, but they didn't do us any good!

(Crick, 1968/72)

At the time Watson and Crick were highly pleased with this three-strand helix held together by the charged groups of the phosphate backbones (see Fig. 20.3a). The latter may have looked a trifle awkward, but the bases stood out nicely, ready for interaction with protein. Explaining their naivety later Crick said: "Well you see, we had never built any models before and if I had the slightest experience of model building I would have realized that we

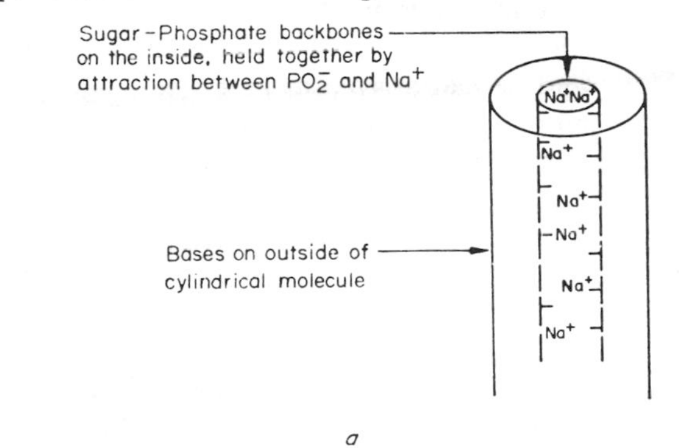

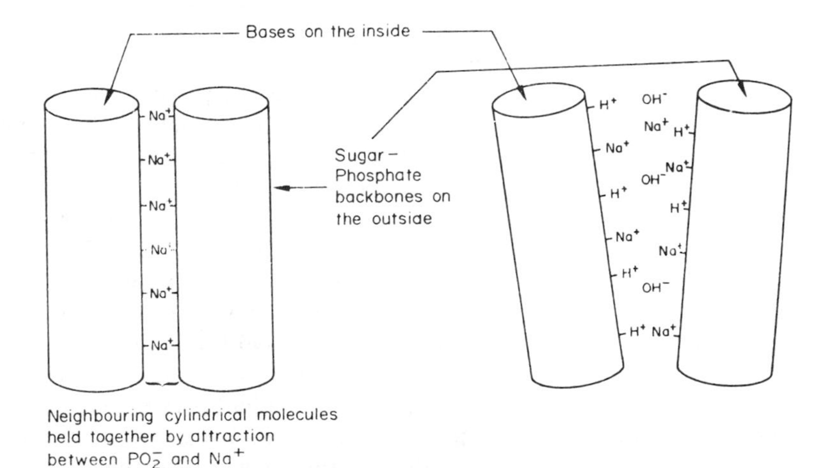

Fig. 20.3. (a) Watson and Crick's scheme in 1951 (b and c) Franklin's scheme in 1951. It should be noted that no diagrams were drawn in 1951; these are the author's based on the manuscript descriptions.

were not anywhere near anything good, as of course we gradually began to see. Possibly, we would not have shown it to the King's people quite so soon if our colleagues, like John Kendrew, had not said we could not go on doing things without telling them. What we should really have done was to have another week or so and settle down and do it. That was part of the reason for the fiasco. Looking back on it, it was no good at all. We never built it properly. As Watson says it is very difficult to build models like that" (*Ibid.*).

Obedient to their colleagues' advice they telephoned Wilkins. The next day Wilkins, Seeds, Franklin and Gosling made a special trip to Cambridge to view the model. Watson has described the awful anticlimax that followed Crick's enthusiastic peroration on the power of the helical theory which Cochran, Vand and he had arrived at four weeks ago. But before Watson and he could go on to show how their model was compatible with the X-ray data Wilkins was complaining that his colleague, Stokes, had already worked out this theory back in the summer, and Franklin was positively snorting at all this talk about helices. As the X-ray crystallographer on the job she knew they had a long way to go before the X-ray data was clear enough even to discuss possible structures. The molecule might well be helical but "there was not a shred of evidence" that it was! (*Helix*, 94). And as for the role of cations, why, these must be surrounded by shells of water, which meant that their charges could be neutralized at a distance, so they were hardly likely to be tightly bound to the phosphate groups. Then the awful truth came out. Watson had got the water content wrong. It was not four molecules per lattice point as Watson had reported it but eight per nucleotide! (Franklin, 1951b). All sorts of structures now became possible. The argument in favour of their particular model no longer held.

Watson and Crick tried to save the day by suggesting a future course of action in which the two groups would collaborate. But Franklin and Gosling very understandably would have nothing to do with such a suggestion. They had witnessed two clowns up to pranks. Why should they condone their behaviour by joining forces with them? And so back went the four King's scientists to London leaving Watson and Crick deflated in Cambridge.

Next time—if there was to be a next time—Watson and Crick would be more cautious. No pleading by their colleagues that Wilkins must be told would sway them a second time. This lesson they had learnt the hard way. Franklin's lecture on the relation of the DNA structure to water, however, did not apparently go home in Watson's case. She had pointed out that a structure which so readily took up water, despite its large content of non-polar groups (the bases), could only do so if the polar groups (phosphates) were *on the outside* of the molecule. The "crystalline" form was well ordered because the exposed phosphates kept close to the cations *between* neighbouring molecules. The long rod-shaped molecules were therefore held tightly "back-to-back". Drying intensified this ordering effect. Moistening progressively destroyed it (see Fig. 20.3b, c). Although Watson and Crick did not know it there was also a good body of evidence from physical chemistry in support of Franklin's assertion.

Electrometric Titration

John Masson Gulland had begun the electrometric titration of calf thymus nucleic acid shortly after his return to University College Nottingham (now the University of Nottingham) from the Ministry of Supply. At the time, changes in the viscosity of nucleic acid solutions which resulted from the addition of acid or alkali were attributed to depolymerization by rupture of covalent bonds. Gulland and his colleagues came to favour another explanation which they formulated by analogy with the proteins. What brought about these changes in viscosity, they argued, was to be likened to the denaturation of a protein, in which hydrogen bonds were broken. Now at pHs of 2 and upwards all the primary phosphoryl groups were ionized. But titration above pH 2 revealed other dissociations. These occurred in two places on the scale of pH (4–5, 11–12) separated by a plateau (5–11) where no groups were liberated (see Fig. 20.4). There was a hysterisis so that in the back-titration the plateau was a shorter one indicating that in the environment of the fragmented micelle the PK values of the nucleotides were different. They argued that the amino groups of guanine, adenine and cytosine were liberated at acid pH and the "enolic hydroxyl (we would say keto) groups" of guanine and thymine at alkaline pH. Around neutrality these groups were evidently linked together, and the most likely type of bond was a hydrogen bond. The fact that the viscosity of DNA was greatly reduced at just those pHs at which the amino and enolic groups were made available for titration suggested that the drop in viscosity was due to the falling apart of a micelle originally consisting of several hydrogen bonded chains. ". . . the micelle is considered to disaggregate into smaller molecular units by rupture of the linkage between the amino and hydroxyl groups . . ." (Gulland and Jordan, 1947, 60).

In 1951 D. O. Jordan described this work in Randall's new journal *Progress in Biophysics*. The King's group could hardly not be aware of it. Jordan was then able to refer to the evidence for hydrogen bond formation between bases in the crystal state. These bonds were pictured by Gulland and Jordan as forming between nucleotides on adjacent chains. Jordan was clearly not in sympathy with the idea that DNA was a helical molecule. In solutions free from electrolytes Jordan knew that the DNA molecule was stretched to form a stiff rod of high asymmetry. He suggested that the repulsion between the negatively charged phosphoryl groups was responsible. On addition of an electrolyte these charges were neutralized and the molecule

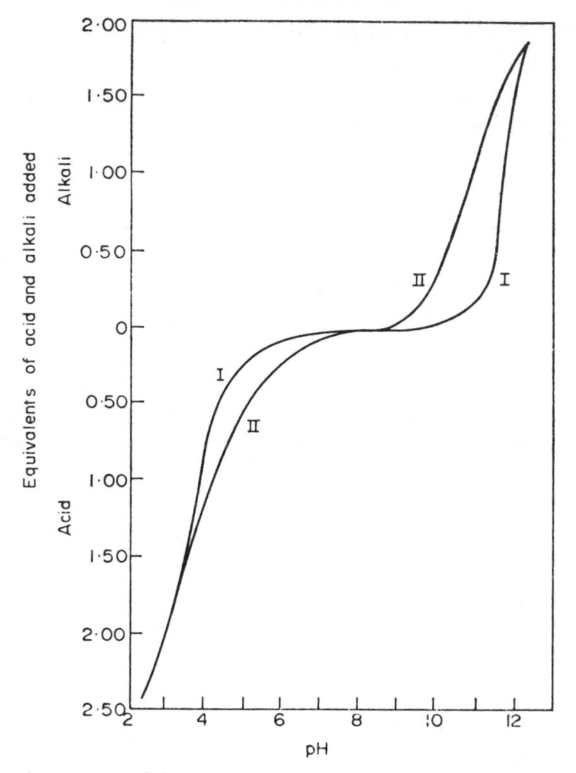

Fig. 20.4. Titration curves of deoxypentosenucleic acid of calf thymus. I is the forward titration curve, II back titration curve. Data of Gulland, Jordan and Taylor, 1947 (graph from Jordan, 1952, 70).

could roll up as in Fig. 20.5. Hence there was no suggestion that the hydrogen bonded bases on neighbouring chains were protected by a helical sugar–phosphate coil around them. This was Franklin's suggestion in November, 1951.

The Ban on Watson, Crick and Wilkins

The result of the King's visit to see the 3-strand Watson-Crick model was that Randall and Bragg had a discussion and agreed that the structure of DNA was to be left to Randall's group. Crick was to concentrate on his thesis, "Polypeptides and Proteins: X-ray Studies", and it seemed logical that Watson, who had been formally assigned to the plant virologist, Roy Markham, should work on TMV. Watson felt that TMV might yield the clue as to the structure of DNA, and by June he did succeed in getting a very good X-ray picture with helical characteristics, but the pattern was too complex to provide the sort of clue he wanted.

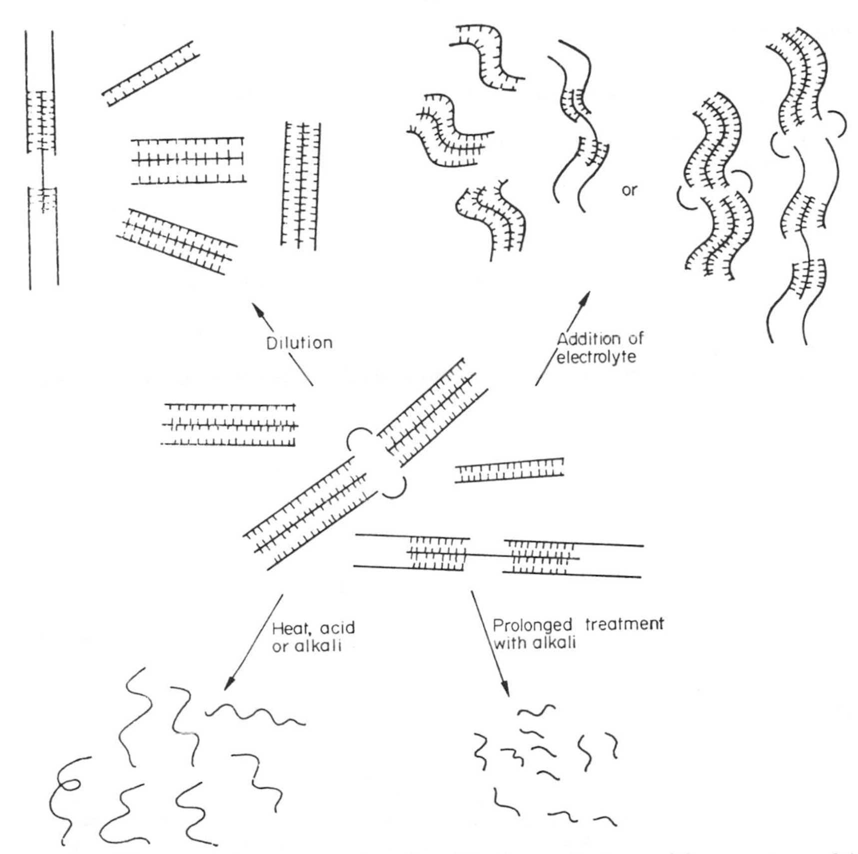

Fig. 20.5. Diagrammatic representation of possible changes in shape of deoxypentosenucleic acid molecule after addition of various reagents (from Jordon, 1952).

Meanwhile Wilkins' work suffered a set-back. On his return from the United States he had proposed that Franklin have the Signer DNA. He modified the single-crystal camera that he had used in 1950 "to give better collimation and, in December, 1951, observed layer lines and equatorial reflections in B patterns from Chargaff DNA from *E. coli,* wheatgerm and pig" (Wilkins, 1968). Against this success there were two failures. He could not get the Chargaff DNA samples to undergo the A/B transformation. Wilkins could not get the A pattern at all. Nor was the resolution of the spots as good as Franklin achieved in 1952 with the Signer DNA. Further progress with DNA, it seemed, was impossible.

Fortunately Wilkins could still lay claim to the structure of nucleoproteins. So with the aid of his improved camera he had another go at getting a diffraction pattern from *Sepia* sperm. Before he left for a combined holiday and work trip on the Continent in the spring of 1952 he reported his success to Crick.

> I have got *much* better X-ray pictures of the sperm Squid which show very nicely a whole series of helical layer lines and one inter micelle spacing. These spots will not overlap when a disoriented specimen is used and I want to do it on *living sperm* in glass tubes. I believe your friend in Rome offered me use of an X-ray tube . . .
>
> I did the transform of the micelle of Na projected on a plane at right angles to the axis and have done more on the layer line intensities also repeated Oster and Riley on the long distances on wet gels *oriented.* I have found several of your suggestions very valuable but am fairly convinced for many reasons the phosphates must be on the outside. There is just one part of the picture that still puzzles me but if helices are right we must hit on the explanation soon. I am really getting down to the job myself but haven't done anything on models and chemistry as I think the picture holds a key not yet recognized which will then more or less directly give us the model. I now take my own X-ray pictures and have made new cameras (hence better sperm pictures). But I believe as you that the key to nucleoprotein lies in the crystalline picture. The exact crystallinity is of interest in that it gives us definitely the *general* micelle arrangement if you follow me. Franklin barks often but doesn't succeed in biting me. Since I reorganized my time so that I can concentrate on the job, she no longer gets under my skin. I was in a bad way about it all when I last saw you.
>
> When I saw Jim last before Christmas he said you people couldn't build a model along your lines. If that is so it simplifies things a bit. Anyway I am pretty certain the phosphates are on the outside. I won't start making any references to the "business" between you people and us over na* but look forward to discussing all our latest ideas and results with you again. Why don't you come and have lunch with me when you are next in town?
>
> (Wilkins, 1952)

He then went on to outline his plans for getting diffraction patterns from living nuclei or whole cells to match against those from disoriented nucleoprotein as isolated by biochemists.

> Thus we may be able to prove the helical idea for ordinary cells as well as special low water content cells such as sperm. But apart from little excursions of a day or so this *must wait.* Did you hear I *have* got a Rio invitation? Isn't it grand! In some ways I feel I am a very lucky fellow. And there I hope to collect sperm bundles which contain separate chromosomes and do X-ray pictures of each of the two chromosomes in the cell spiralized and unspiralized etc. This is rather gilding the lily but may be very useful in a year or 2's time to link the X-ray business more with *real chromosomes.* I hope I don't go to the devil with all these trips but so far I think they are well worthwhile from the work standpoint.
>
> (*Ibid.*)

From this letter I suggest that Wilkins' sole contact with Watson and Crick since the November visit to Cambridge had been a brief visit from Watson just before Christmas. Wilkins did not yet want to start model building. The picture, he confided "holds a key not yet recognized which will then more or less directly give us the model". Also clear from this letter is the fact that Wilkins was eager to share his results with Crick.

Later that year Wilkins obtained trout sperm heads, with tails removed, from A. Felix at Frankfurt. Since they were unfixed and undried Wilkins expected "their structure to be similar to that of sperm in the living state".

* *na* = nucleic acid.

Because they were optically isotropic they corresponded more closely to ordinary cells.

> The photograph is similar to *Sepia* except that each arc has become a complete ring. This suggests that both the isotropic and anisotropic sperm heads are composed of similar crystalline bundles of long-chain nucleoprotein molecules.
>
> (Wilkins and Randall, 1953, 192)

Whilst admitting the importance of this demonstration that extracted and *in situ* DNA showed the same structural features one wants to know whether Wilkins stated to which structure—"wet" or "crystalline"—the sperm pattern approximated. Neither in his letter to Crick in 1952 nor in his report with Randall (received by *Biochimica et Biophysica Acta*, November 27, 1952) did he discuss this point. Recently Klug drew attention to this fact: ". . . the periodicities were not sharply defined and no assignment to one of the two known—but as yet unpublished—forms was reported" (Klug, 1968, 844). One must remember, however, that in 1951–52 Wilkins had been unable to produce the A form since he did not have access to the Signer DNA, which alone gave it. The drawing which Wilkins put in his letter to Crick shows the unmistakable cross-ways made up of intensities on layer lines which Wilkins numbered 1, 2 and 3. Whilst admitting that there were doubts, can one in all honesty say that this *Sepia* pattern looks no more like a B pattern than like an A pattern? Much later Wilkins summarized the situation thus:

> By December, 1951, equatorial reflections on all the patterns gave roughly the outer diameter of the DNA helix, and the helix pitch was defined for the *A* and *B* structures and by February, 1952, the pitch of the helix from sperm heads had also been observed. . . . I was visiting the Cricks fairly frequently during, I think, 1951 and 1952. I therefore was able to pass to Crick and Watson a general picture of the DNA molecule, approximate number of chains, pitch and diameter of the helix and base stacking . . .
>
> (Wilkins, 1968)

This is partly born out by a long letter which Watson wrote to Delbrück in May, 1952. On DNA he reported the following:

> *The Structure of DNA:* Recently Maurice Wilkins of King's College London has obtained extremely excellent X-ray diffraction photographs of orientated Na^+ thymonucleic acid. The spots are extremely sharp indicating very good orientation in three dimensions and should provide sufficient detail for a rigorous interpretation of the pattern. The pattern has been indexed on the basis of a unit cell with the following shape (Fig. 20.6). From

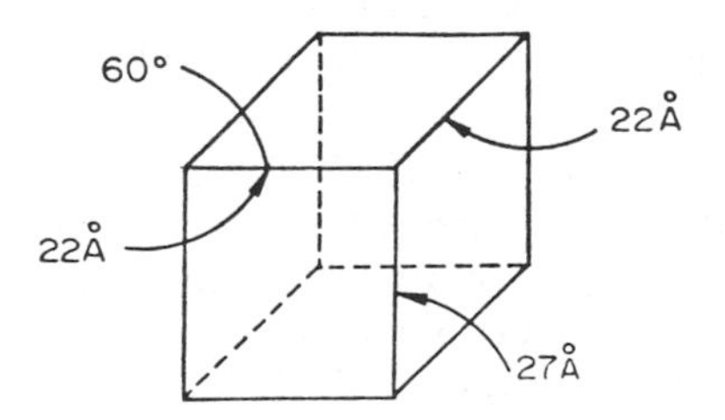

Fig. 20.6. The unit cell in Watson's letter to Delbrück.

> density arguments it is probable, that 14–17 nucleotides are contained in this cell. The King's people believe that the nucleotides are arranged in a helical manner, but as yet the X-ray evidence is far less convincing than for synthetic polypeptides.
>
> It is quite obvious that a great deal of work should go into elucidating this structure. However, the people at King's are involved in a fight among themselves and so at present no real effort is being made to solve the structure. We have attempted to interest them in a Pauling approach of model building and in fact this winter we did spend several weeks attempting to build plausible models. However, we have temporarily stopped for the political reason of not working on the problem of a close friend. If, however, the King's people persist in doing nothing, we shall again try out our luck.
>
> Though the King's group has made little progress in the main problem, they have established the following facts of interest:
>
> 1. The Na^+ salt of purified DNA from calf thymus, herring sperm, and *E. coli* gives identical X-ray photographs, that is a repeat every 27 Å along the axis—it thus appears that a basic structural pattern for DNA exists (at least for the Na^+ salt).
>
> 2. X-ray photographs of orientated intact squid (*Loligo*) sperm gives the same DNA pattern as purified DNA!!! thus showing that the DNA within the sperm is present in an orientated arrangement. "Powder" photographs of orientated squid sperm give a powder diagram with rings corresponding to the strong spots on orientated DNA photographs. It is thus possible to demonstrate the presence of orientated DNA within specimens which *cannot be orientated.* Maurice and I are now beginning to investigate T2 bacteriophage from this viewpoint.
>
> (Watson, 1952a)

With reference to this document Wilkins wrote:

> Up to that time optical measurements were the only indications that the structure of the DNA that chemists extracted from cells was the same as that in intact chromosomes. The X-ray work provided direct evidence of detailed similarity and showed that X-ray crystallographers studying crystalline DNA were not studying a chemist's artefact but a structure directly relevant to the nature of the gene.
>
> (Wilkins, 1972)

But what of the crystallographic data in Watson's letter? From the dimensions given by Watson in Fig. 20.6 it seems that he was referring to the primitive cell rather than to the unit cell. Even so, the reference to 14 to 17 nucleotides is odd. It is true that Wilkins had earlier suggested a single helix in the unit cell (see Chapter 19). Watson, it seems was relying on earlier statements of Wilkins. Also noteworthy is the absence of any reference to the existence of two forms of DNA, or to the 34 Å repeat in the picture obtained at high humidity. Was Watson following Wilkins here also? It seems very likely, since neither in Franklin's November, 1951 colloquium notes, nor in her first Fellowship report can this datum be found. This letter therefore serves to show how little Watson, Crick and Wilkins knew of Franklin's and Gosling's recent data. Consequently Watson was assuming that Wilkins' new pictures of herring sperm and *E. coli* DNA were of the crystalline [A] form. This confusion of intensity data from A and B forms was apparently inherited not only by Pauling, but by Watson and Crick in 1952. This conclusion serves to underline the crucial contribution made by Franklin and Gosling when they determined the location of the different meridional arcs from their strikingly distinct X-ray pictures.

Why was this? Wilkins' comment was that "it certainly could have been measured. The layer line streaks were clearly visible on the first Franklin B patterns. Both the A and B periods were 'around 30 Å' and I suggest that the B patterns had simply not been measured accurately" (Wilkins, 1972). This was probably because they were not required for making the distinction between the two forms, which could be characterized by other criteria: (1) The sharp spot at ~22 Å on the equator of the B form in place of the 16.5 Å spot in the case of the A form. (2) The prominent meridional arc at 3.4 Å in the B form, absent from the A form. (3) The lesser crystallinity of the B form. Sometime during December, 1951 and January, 1952 Wilkins obtained a much improved pattern from sepia sperm in which the arcs on the first four layer lines were clearly marked. In the spring he described this pattern in a letter to Crick (see p. 366). Although Wilkins did not tell Crick so, this pattern was clearly of the B type, and the 34 Å repeat could have been calculated from it, but Wilkins "had made no accurate measurement at that time" (Wilkins, 1972). Hence it came as a surprise to Watson when he saw Franklin's 1952 B pattern for the first time in January, 1953, and learnt that its repeat distance was 34 Å.

Franklin's New Data

In the spring of 1952 Franklin and Gosling were busy taking trial short exposure photographs of DNA at 75 per cent R.H. with the Phillips microcamera. They also tried out the special tilting stand designed to search for wide angle repeats along the fibre axis.

On the 18th April, Franklin recorded in her notebook the details of a single fibre photograph of the crystalline form which showed some double orientation. She underlined the last two words. She had no record of the exposure time as this had been interrupted by her absence over Easter, but she seemed confident that this photo had much significance, and when Crick and Watson attended the one-day protein conference which the Royal Society held on the 1st May they learnt that Franklin "was insisting that her data told her DNA was *not* a helix" (*Helix*, 122). The irony of the situation is all the more remarkable when we find that on the very same night, and on the following one, Franklin was taking an X-ray picture of a single fibre which had once given the crystalline pattern but had since changed irreversibly so as to give the "wet" or B pattern even at 75 per cent R.H. This further exposure at 75 per cent R.H. at the beginning of May now gave not a "good 'wet' photo" but a "very good 'wet' photo". This, it appears, was no other than the dramatic picture of the B form which was to startle Watson some nine months later! Another good photograph of the B form at this time enabled her to resolve the very strong equatorial reflection at about 22 Å into a doublet composed of two spots 2.5 Å apart, equivalent to a layer of water one molecule thick. "This suggests", she wrote, the "coexistence of two phases differing only by one moleculer layer of water separating [the] chain units" (Franklin, 1952a). Wilkins regarded this observation as "almost certainly fortuitous". It has never been observed in any of the extensive further studies of DNA, and he attributed it to two different swollen states (Wilkins, 1972).

Franklin also calculated the spacing of the first layer line. It was between 33 and 34 Å. So the very prominent arc on the meridian at 3.4 Å must, she concluded, be on the tenth layer line. That month she wrote in her notebook that the clarity of the equatorial spots:

> together with the 3.4 arc lying on a layer-line, suggests that there *is* an integral number (or single fractional number) of residues per turn of helix (if there is a helix) even in the "wet" state.
>
> In passing from "crystalline" to "wet" the predominent equatorial spacing is approximately doubled, and the fibre-axis period extended by 25 per cent (27 Å to 34 Å).

Franklin's Anti-helical Data

It might seem obvious that if the DNA fibre consists of cylindrically-shaped helices packed parallel to the fibre axis, the indexing of spots in left and right hand sectors of the diagram should confirm radial symmetry. But when Franklin and Gosling started to index the A pattern they found ambiguities which were in due course to lead them to adopt an anti-helical position. In April, a case of double orientation was discovered which they used to check the reliability of their indexing. Double orientation is not in itself incompatible with a helical structure, but when indexed the results seemed to refute the assumed radial symmetry of intensities. Spots of equivalent intensity in left (L) and right (R) hand quadrants were indexed. "It was then found that all reflections labelled L had been allocated indices $hk\bar{l}$ whereas all R reflections had indices hkl" (Franklin, 1953, 7). In August she tried to repeat the production of double orientation, but failed. She concluded that it was due either to a mechanical accident or to preferential orientation of the crystallites near the surface. "The anti-helix case", recalled Wilkins, "described at a meeting of those in our laboratory concerned with DNA by Franklin was not based on the double orientation but on the claim that $hk\bar{l}$ reflections had a marked tendency to be more intense than hkl reflections. As presented, the data could not be compatible with a helical structure" (Wilkins, 1972).

Now, as Crick has explained to me, the fibre diagram of helical molecules in parallel array can give the appearance of radial asymmetry because the molecules are packed in monoclinic cells, so that the view of the crystal lattice can differ from one vantage point to another. The transform can alter rapidly as the sampling point changes, with the result that the pattern of overlapping intensities differs in different sectors of the diagram. This possibility neither Franklin nor Gosling appreciated at the time. It is all the more ironic, therefore, that initially their anti-helical stand rested on the indexing.

Franklin and Gosling went on to calculate the cylindrically averaged Patterson of the A form in order to resolve the ambiguities in indexing and thus derive reliable parameters for the unit cell. This fine work was completed during the summer of 1952 and in the report which they wrote at the beginning of September, 1952 for the visit of a committee of the MRC they gave the following very important information:

> On the Patterson function obtained, the lattice translations could be readily identified. On the basis of a unit cell defined by $a = 22.0$ Å, $b = 39.8$ Å, $c = 28.1$ Å, $\beta = 96.5°$, the 66 independent reflections observed could all be indexed with an error of less than 1 per cent.
>
> (Randall, 1952)

The extraordinary fact is that Franklin and Gosling made no reference whatever to her anti-helical views in this report, whereas Wilkins reported these views in a tone suggesting reluctant acceptance of them.

> Using two-dimensional data, the most reasonable interpretation was in terms of a helical structure and the experimental evidence for such helices was much clearer than that obtained for any protein. The crystalline material [A form] gives an X-ray picture with considerable elements of simplicity which could be accounted for by the helical ideas, but three-dimensional data shows apparently that the basic physical explanation of the simplicity of the picture lies in some quite different and, *a priori*, much less likely structural characteristic. The 20 Å units, while roughly round in cross-section, appear to have highly asymmetric internal structure.
>
> (Randall, 1952)

In Wilkins' defence it should be pointed out that he did not *see* any of Franklin's anti-helical data until much later. He was merely stating Franklin's oral opinion on the evidence from the crystalline A form. Franklin, for her part, never published any directly anti-helical views and when she described the cylindrical Patterson function in *Acta Crystallographica* in 1953 (see Fig. 20.7) she merely wrote:

> It will be observed that the C-face-centred monoclinic unit cell is near-hexagonal in projection. Nevertheless, there is evidence to suggest that the symmetry of the crystallite itself is far from cylindrical. Although the nature of the accident which produced double orientation of the crystallites in a fibre is unknown, it seems unlikely that it could have occurred at all if the individual crystallites had a high degree of symmetry about the fibre axis.
>
> (Franklin and Gosling, 1953e, 684)

When we turn back to Franklin's typescript of this paper we find one word is different. "There is evidence to suggest that the symmetry of the structural unit . . ." Here she was referring to the molecule or group of molecules which form the structural unit in the unit cell. If we go back to an earlier handwritten draft we find the phrase: "*either* the structural unit itself *or* the way in which the units are packed together is highly asymmetric". But what do

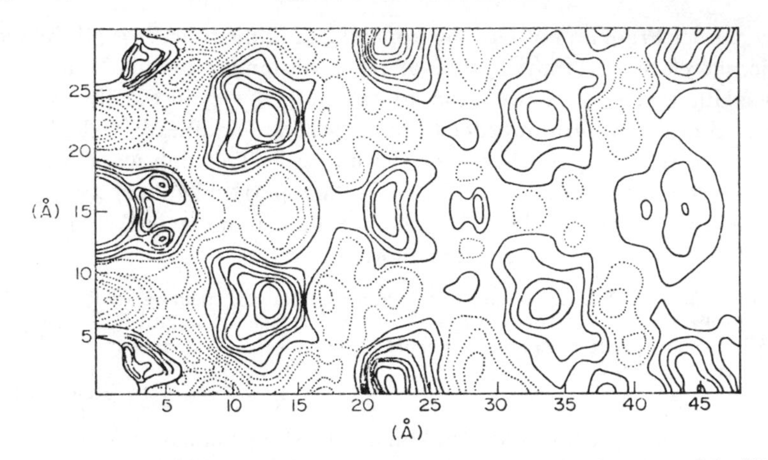

Fig. 20.7. Cylindrical Patterson function of NaDNA as prepared before news of the Watson, Crick model reached the authors (from Franklin and Gosling, 1953e, 682).

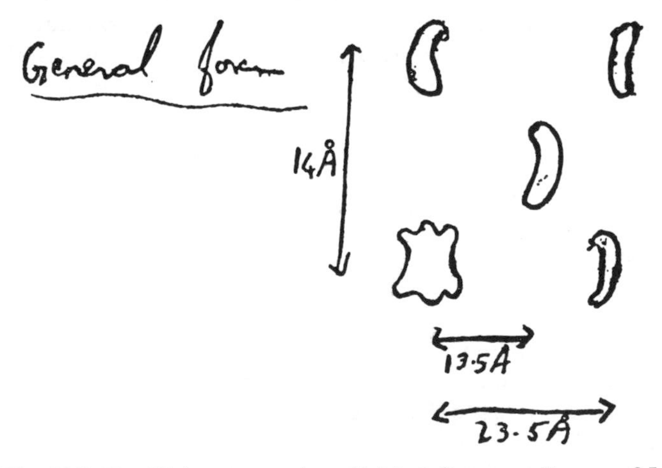

Fig. 20.8. Franklin's notes on the cylindrical Patterson diagram of July, 1952 (from her research notebook).

we find in her notebook for July, 1952? First she drew the general form of the Patterson function (see Fig. 20.8), then wrote:

> There is no indication of a helix of diameter 11 Å. The central banana-shaped peak fits curve calculated for helix of diameter 13.5 Å having 2 turns/unit cell. If a helix, there is only one strand (2-strand helix would give θ).
>
> Helix does not explain vertical short vector ~ 4 Å (this is superimposed on peak at origin, so represents true distance of > 4 Å, i.e., *not* 3.4 Å). But if there is a *flat* banana-like unit in [the] structure, with banana axis parallel to fibre axis, this vector is explained . . .
>
> Again, the dimensions of this cell make a helical structure improbable—if helices [are] of 13.5 Å *diameter* how is the remaining space in the long period filled?
>
> Suggests rather a double-sheet structure

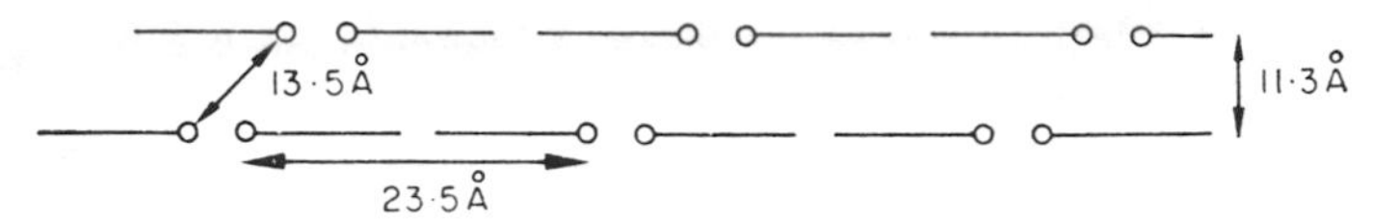

> Also a structure of this kind seems necessary to explain double orientation in diagram 45. Effect possibly due to preferential orientation of sheets at right angles [to] surface of fibre, and not all of fibre in beam (though nearly all of fibre *was* in beam).

This, to our ears, extraordinary piece of interpretation shows, according to Crick and Klug, that Franklin was considering three separate helices of diameter 11 Å running side-by-side, the centres of the groups of three being about 20 Å apart. The *self* Patterson of any one such helix should give phosphate–phosphate vectors at 11 Å. Instead, Franklin found it was at 13½ Å. Therefore she rejected helices. Now for some reason—anti-helical prejudice or just a natural failure to entertain the idea—she did not pursue *other* helical possibilities. For instance, she mentioned the possibility of two coaxial chains (a double helix) but did not realize, incredible as it may now appear, that such a structure would account for the peak at $c = 14$ Å in terms of the halving of the 28 Å repeat. The central banana-like peak would then result from the intersection of the two strands; the 5 Å peak from the first phosphate vector between neighbouring phosphates on a strand, and the trough of low electron density at $x = 6.7$ Å was merely the fall one gets when moving out from the origin. It is, of course, not zero intensity, as the dotted contours on Franklin's and Gosling's 1953 paper show. By making an arbitrary selection of the density at which continuous lines were used the impression of discontinuity was heightened.

Franklin's Attitude to the Helix

How was it possible for Franklin to go off on this anti-helical tangent in May, 1952 when she had written in the previous February: "The results suggest a helical structure (which must be very closely packed) containing probably 2, 3 or 4 coaxial nucleic acid chains per helical unit and having the phosphate groups near the outside" (Franklin, 1952, 4). This question has already been discussed (Klug, 1968), so all we need do here is to enumerate possible reasons.

Franklin was a professional structural crystallographer who distrusted intuitive guessing, and who wanted to solve the structure by direct methods, i.e., without introducing assumptions in the form of hypothetical structures. She was not against helices as such, but against *assuming* helices when the evidence, in her opinion, was inadequate. This attitude was dictated by her own inclination as a research student and by her admiration of Bernal, whose lecture, "Considerations on the Present State of Protein Structure", she atsended at the international congress on crystallography in 1951. The notes the made of this lecture show that Bernal's message was not lost on her. Bernal had described Pauling's approach as that of using chemical knowledge to build structures, then using X-ray data as a test. This was deductive and speculative. In the early stages it was justified, but, wrote Franklin, "time has come to review evidence and assumptions—have we found *the* solution or *a* solution." Bernal contrasted the deductive Pauling method with the inductive empirical method of deriving chain types from Patterson sections. "Real test requires scattering far enough out in crystal diagrams and final structure by Patterson sections (<2 Å)." Undoubtedly this was the line she followed a year later when she embarked first on a cylindrical Patterson and then on a 3-dimensional Patterson analysis. A structure like DNA might seem ideally suited to such a treatment, since it contains one type of atom—phosphorus—with a much stronger scattering power than any other. Certainly the Patterson diagram was much simpler to interpret in terms of a helical conformation (see Fig. 20.9), but even then it proved misleading, for as Wilkins recalled "the Patterson study led to the idea that the PO_4 chains were separated by $c/2$ whereas they are really $\sim c/3$ apart. To take the Patterson peaks as indicating PO_4 positions was an oversimplification" (Wilkins, 1972). The discussion of this subject after Klug's paper "Rosalind Franklin and the Discovery of the Structure of DNA" appeared resulted in the following statement by him:

> . . . although the cylindrical Patterson function calculated by Franklin and Gosling unquestionably shows the existence of a double helix in the A form, the resolution of the map is only about 5 Å, so that the details of the model fitted are only approximate. The later definitive work of the King's group (Fuller, Wilkins, Wilson and Hamilton, 1965) shows that the two chains are far from equally spaced. Likewise, it is doubtful whether the resolution in the three-dimensional Patterson map was high enough to enable the orientation of the helical molecules in the crystal cell to be deduced without ambiguity. It is the opinion of Fuller *et al.* that the X-ray data used were too sparse to settle this problem, and even their more comprehensive data to 3 Å resolution were "no more than sufficient for the purpose".
>
> (Klug, 1968, 880)

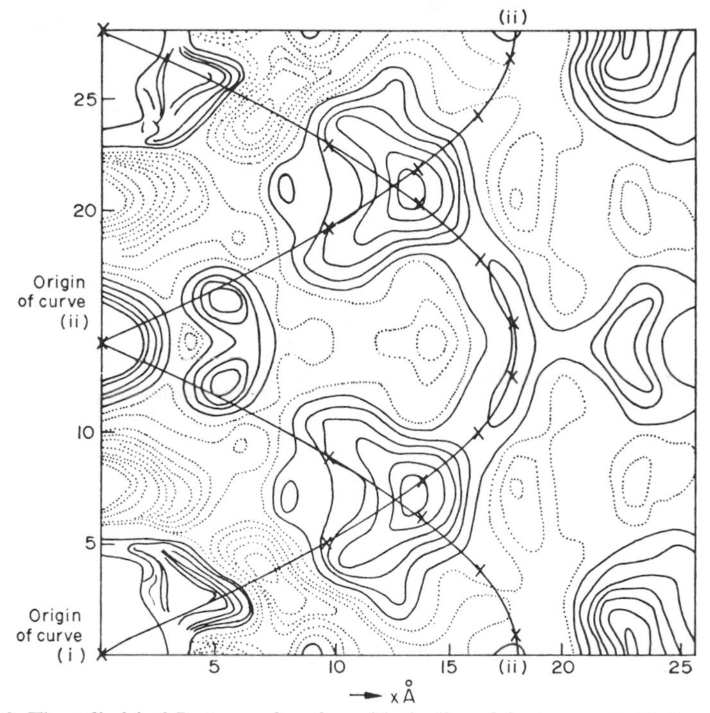

Fig. 20.9. The cylindrical Patterson function with the line of the two coaxial helices drawn through it, as drawn by the authors after learning about the Watson–Crick model (from Franklin and Gosling, 1953d, 156).

Franklin and Gosling were, of course, very concerned about the interpretation of their cylindrical Patterson diagram. Therefore they went to see C. H. Carlisle at Birkbeck College, and asked him: "Do you think that a Patterson of this type would be consistent with the phosphates being on the outside of a helix?" (Carlisle, 1968). Although Carlisle replied in the affirmative they rejected this interpretation.

Now precisely because Franklin *wanted* to use Patterson analysis she concentrated on the A form and left that magnificent picture of the B form obtained at the beginning of May, 1952 until February, 1953. Her notebook shows that she turned to the B form on the tenth of the month, a Tuesday. Little more than a week had elapsed since Watson had come to King's bringing with him Pauling's manuscript on his triple *helical* model! Was her decision to return to the B form due to Watson's visit, and to Pauling's advocacy of helices for DNA, or had she simply exhausted the possibilities of the A form for the time being? Crick and Klug incline to the latter reason. In the winter of 1952/3 Franklin had been at work on the interpetration of the 3-dimensional Patterson diagram; after the sheet structure composed of antiparallel rods in pairs, back-to-back, she had gone on to consider systems of diagonally-packed rods such as would simulate the diffraction pattern of a helix. In January, 1953 she started model building to explore the possible orientation and dimensions of the sugar-phosphate backbone. Still rejecting helices, she investigated a figure-of-eight structure in which a single chain formed a column of repeating eights. This was aimed at accounting for the halving of the unit cell in the fibre direction and for the close packing of a structure which on addition of water could open out to give structure B.

There was, of course, a particular reason why Franklin allowed herself to be misled by the Patterson diagram. Before she got it she had obtained the X-ray photo showing double orientation. After her first attempt to interpret the Patterson diagram she went back to taking more X-ray pictures in an effort to repeat the double orientation (August, 1952). Aaron Klug has implied that Franklin was consistent in advocating a helix for the B form, but that in 1952 she opted for a non-helical structure for the A form (Klug, 1968). Wilkins had this to say:

> My own view remains that her original intensity data showed considerable asymmetry and was incompatible with a helix, but that critical examination of her measurements showed that the intensities and indexing were not sufficiently accurate to justify the conclusion that there was real asymmetry not compatible with a helix.
>
> You [Klug] may well be right that Rosalind fairly consistently thought the B form was helical. If she did, however, she kept this view to herself and my own view was that, if A was non-helical, it was quite unreasonable to suppose that B could be helical.
>
> (Wilkins, 1969)

I would go further and suggest that from May, 1952 until Watson's visit in January, 1953 Franklin would gladly have buried helical DNA for the B form as well as for the A form. This would account for the mock funeral of the helix which was jokingly held at King's in 1952, for the skull and crossbones beside Gosling's conclusion, from a model building, that DNA must be a helix (undated manuscript, but probably February, 1953), and for the anti-helical impression Franklin gave to her colleagues and to Crick who, in the summer of 1952, assured her to no avail that the double-orientation photo was misleading. Did she revert to her February, 1952 position when she heard about Pauling's helical model? It could well be.

Pauling's Triple Helix

When asked how he became interested in DNA, Pauling replied that he knew Avery, Macleod and McCarty and their work with pneumococcus.

> During the war I was doing some work with pneumococcus polysaccharide and antibodies to pneumococcus polysaccharide and I knew the contention that DNA was the hereditary material. But I didn't accept it, I was so pleased with proteins you know that I thought that proteins probably are the hereditary material, rather than nucleic acid,

but of course that nucleic acid played a part. And in whatever I wrote about nucleic acid, I mentioned nucleoproteins, and I was thinking more of the protein than of the nucleic acid at this time.

(Pauling, L., 1968)

Pauling went on to say that he had worked on the structure of a number of polymers, such as the polysaccharides, without publishing anything. The structure of such compounds intrigued him as did that of DNA, so he "thought it would be worthwhile to analyse Astbury's photographs of DNA" although he was not yet convinced that DNA was the hereditary material. When Peter Pauling was asked the same question he took the view that his father was interested "in nucleic acid because nucleic acid is a chemical compound, not because it is the genetic material..." (Pauling, P., 1967).

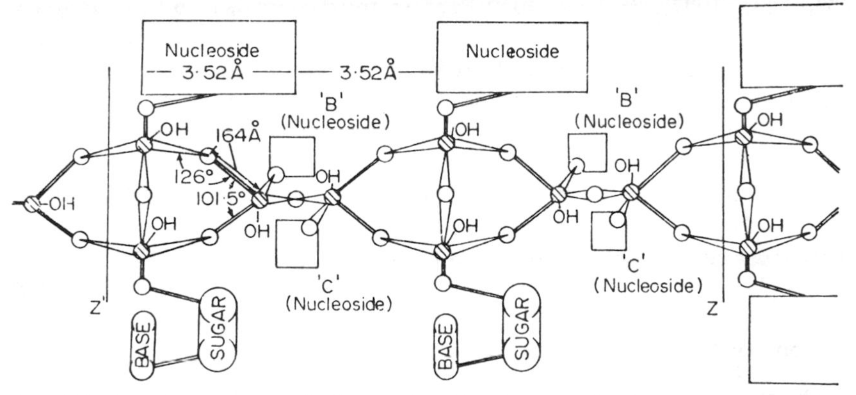

Fig. 20.10. Top view of a unit cell of a nucleic acid polymer chain: ○ = oxygen atom ⊘ = phosphorus atom. In the repeat distance Z — Z′ (14 Å) there are eight residues (from Ronwin, 1951, 5142).

Then why did Pauling set to work on DNA in 1952 at just the time when the King's work was progressing? The reason, I believe, has nothing to do with King's but with an extraordinary paper by Edward Ronwin which appeared in the *Journal of the American Chemical Society* in November, 1951. It was entitled "A Phospho-tri-anhydride Formula for the Nucleic Acids". Its author argued that a careful study of the data on the structure and behaviour of the nucleic acids led him to conclude, in defiance of the classical work of Levene, that there was but one phospho–sugar ester bond per nucleotide. So he built a phosphate backbone instead of a sugar–phosphate backbone. The core of his molecule therefore consisted of phospho–anhydride links, each phosphorus atom being bound to five oxygen atoms (see Fig. 20.10). He admitted that P_2O_5 "reacts with water in a violent manner", but polymeric forms behaved in a milder fashion. Todd's comment to Markham in Cambridge was that if DNA fizzed when you put it in water then the Ronwin structure could be right!

Pauling did not joke about the structure, he "was offended that it was nonsense" (Schomaker, 1968) and got Schomaker to join him in denouncing it to the *Journal.*

> The proposer of this extraordinary formula for the nucleic acids has not quoted any significant evidence in support of it. The ligation of five oxygen atoms about each phosphorus atom is such an unlikely structural feature that the proposed phospho-tri-anhydride formula for the nucleic acids deserves no serious consideration.
>
> (Pauling and Schomaker, 1952a)

That spring Ronwin wrote to Pauling drawing his attention to the four compounds synthesized by Anschütz for which this chemist had proposed structures in which the phosphorus atoms were ligated with five oxgen atoms (Anschütz, 1927). In June, 1952 Pauling and Schomaker had to withdraw their statement that there was no precedent for such structures. But they drew attention to the extreme sensitivity of Anschütz' compounds to water, so sensitive "as to make it impossible to determine, except roughly, their melting points and other physical properties" (Pauling and Schomaker, 1952b, 3713).

As the months of 1952 rolled by Pauling's interest in DNA grew. In June he was shown Watson's long letter to Delbrück about the work on DNA and TMV in England (see p. 367). This letter was in support of Watson's wish to spend his first year as a Fellow of the National Foundation for Infantile Paralysis in Cambridge rather than in Caltech. Delbrück wrote to the National Foundation saying:

> Dr Pauling and I have gone over Dr Watson's letter in detail and we both wish to endorse his program and the reasons he gives for continuing it for one year in Cambridge. The technical set-up in Cambridge is indeed much superior for the particular problem Dr Watson wants to tackle. This is true at least for this coming year. The following year Dr Pauling hopes to have equivalent facilities here.
>
> (Delbrück, 1952b)

Pauling therefore knew that Wilkins had obtained "extremely excellent X-ray diffraction photographs". Indeed he would very likely have been able to see the King's pictures that year had the U.S. government not refused him a passport, thus preventing him from attending the one-day Royal Society protein meeting in London.* His colleague, Robert Corey, did come and was shown DNA pictures by Rosalind Franklin. Corey reported back to Pauling that the King's group had got good pictures, but when Pauling wrote to Wilkins asking for details, the latter replied that he had not yet reached the stage when he wished to release them.

Although Pauling had fresh pictures of DNA taken at Caltech that year they were "inferior to those of Astbury and Bell" (Pauling and Corey, 1953, 85). Not until after the Pauling structure was published did Alexander Rich, then a postdoctoral fellow at Caltech, obtain really good pictures. Little did

*Held May 1, 1952.

Pauling know that E. Beighton, in Astbury's department at Leeds, had obtained beautiful B patterns in 1951 about which nothing had been done!

Astbury had put Beighton on to the X-ray analysis of calf thymus DNA, sent by Chargaff at Astbury's request, in the spring of 1951. Beighton took 15 to 20 minute exposures using the powerful rotating anode X-ray tube in the department. He stretched the DNA material by 300 per cent, dripped water on to it the while, and obtained very good patterns of the type now known as B. The same specimen left at room humidity gave him the crystalline pattern already discovered by Wilkins. Strangely enough, Astbury was disappointed by these results and both Beighton and he obviously thought that the 1938–39 Bell pictures represented pure forms, whereas Beighton's B form was a mixture—with fewer X-ray spots of course!

So there was Pauling in Pasadena with only the pre-war Astbury–Bell pictures as his check on whether the structure he built up by his stochastic method was right.

Pauling's Strategy

Pauling began with the stereochemistry of the nucleotides. Todd had just cleared up the question of the sugar phosphate linkages with his discovery of cyclization (Brown and Todd, 1952). In 1951 his formation of a cyclonucleoside gave evidence for the β configuration of the glycosidic link (Clark, Todd and Zussman, 1951). Pauling also knew Furberg's work on cytidine which gave the correct orientation of sugar to base. So he built polynucleotide chains incorporating all these features.

Next he argued that, as in the case of the polypeptides, so in that of the polynucleotides, the general operation for converting the asymmetrical subunits (nucleotides) into a symmetrical structure was a rotation-translation leading to a helix.

Then came the troubling problem of density, which as we saw Astbury and Bell did not analyse adequately. Doing the same type of calculation as they had done, Pauling arrived at the conclusion that there must be three nucleotides every 3.4 Å along the molecule, and therefore three chains. He calculated the volume of a nucleotide from the dry DNA density (given by Astbury and Bell) as

$$\frac{330}{1.62} \times 1.675 = 338 \text{ Å}^3,$$

and the cross-sectional area per residue as 303 Å^2. The length per residue thus amounted to

$$\frac{338}{303} = 1.12 \text{ Å},$$

or three residues every 3.4 Å. "Accordingly, the reflection is to be attributed to a unit consisting of three residues" (Pauling and Corey, 1953, 86). Therefore there must be three chains.

It is not clear to me how Pauling arrived at the cross-sectional area per nucleotide. He had Robley Williams' evidence from electron micrographs that the DNA molecular fibrils had a diameter between 15 and 20 Å (Williams, 1952). The equatorial spots on the Astbury–Bell pictures gave a minimum side spacing at 16.2 Å, which Pauling reckoned corresponded to a molecular diameter of 18.7 Å. If we take this figure for the calculation of the cross-sectional area *of the molecule* we get 235 Å^2. Thus the volume of a disc 3.4 Å in height is 935 Å^3. The number of nucleotides in this disc is therefore 935/338=2.8, and if we take the density of DNA at ordinary humidity as given in Bell's thesis we get 2.57 nucleotides. This shows that if Pauling had had all the 1939 Leeds data he would perhaps have continued with two-chain models!

As his notes show, Pauling had begun experimenting with a two-chain structure in the unit volume, but this gave him "a lot of extra space, things rattle around because the density as calculated, . . . permitted three chains in this unit volume, and with two chains they would fill up only two thirds of the space..." (Pauling, L., 1968). Pauling was therefore forced to abandon a double helix and build a triple helix. Fifteen years later he commented that he "would have preferred two chains, because I had decided several years earlier that the gene involved two complementary molecules, each of which could act as a template for the synthesis of a replica of the other" (Pauling, L., 1967).

Then he asked what was the nature of the *core* of the molecule, because a good fit here would give a stable helix. Packing of atoms at the core was a more difficult problem to solve than packing at the periphery. The lesson he had learnt from the polypeptides was that the structure "in which the atoms are packed in a satisfactorily close manner about the axis" is more likely to be right (α-helix right, γ-helix wrong). Then came the elimination of alternatives:

> It is found by trial that, because of their varied nature, the purine–pyrimidine groups cannot be packed along the axis of the helix in such a way that suitable bonds can be formed between the sugar residues and the phosphate groups; this choice is accordingly eliminated. It is also unlikely that the sugar groups constitute the core of the molecule; the shape of the ribofuranose group and the deoxyribofuranose group is such that close packing of these groups along a helical axis is difficult, and no satisfactory way of packing them has been found. An example that shows the difficulty of achieving close packing is provided by the polysaccharide starch, which forms helixes with a hole along the axis, into which iodine molecules can fit. We conclude that the core of the molecule is probably formed of the phosphate groups.
>
> (Pauling and Corey, 1953, 89)

What holds the Structure Together?

Pauling's attitude on this point was that if the structure packed together well it was partly because van der Waals distances were satisfied and partly because valency forces, weaker than covalent bonds, were present. Unlike Watson and Crick in their 1951 structure, and Astbury in his pile-of-pennies model, Pauling paid no attention to electrostatic attraction between bases. Instead he introduced hydrogen bonding between the phosphate groups. At any one level the phosphate groups of the three nucleotides were linked to each other by hydrogen bonds with dimensions as in potassium dihydrogen phosphate. The presence of the protons necessary for such a scheme would depend upon the primary phosphoryl group of each phosphate being unionized. Pauling felt that although the extracted DNA from which the Astbury–Bell pictures were taken was known as the sodium salt, the Na^+ ions were in all probability not adjacent to the primary phosphoryl groups. Later he recalled that he had "analysed some reported titration curves myself, in the effort to decide the extent to which nucleic acid would be ionized, and I concluded that a good number of the phosphate groups would not be ionized at physiological pHs" (Pauling, L., 1968). This is a surprising statement in the light of the fact that no discussion of the subject is to be found in the 1953 paper, and that, as we have seen, the titration curves of DNA concerned, not the dissociation of phosphate groups, but that of amino and keto groups in the bases. The only obvious conclusion to which one is forced by the facts at present available is that Pauling treated the structure of DNA chiefly as a problem in phosphate chemistry. In cutting a way through to what he conceived as the heart of the problem—packing the phosphate tetrahedra together—he passed over other features, such as the stacking forces between bases, and having succeeded in getting the sugar rings to fit, albeit not in the manner suggested by Furberg that year, he was happy to put the bases on the outside of the structure (see Fig. 20.11). Each of the three chains repeated (returned to the equivalent position) after executing seven turns involving 24 residues. Because there were three chains in the helix there was a crystallographic repeat after 8 residues and at a distance along the helix of 27 Å. And in the 1953 paper sodium ions were not even mentioned.

What mattered to Pauling was that the structure was stereochemically feasible, it was in tolerably good agreement with the Astbury–Bell pictures and with the observed reflections which Astbury and Bell had published in *Tabulae Biologicae* (1939, 95). The predictions from the helical theory of Cochran, Crick and Vand likewise agreed satisfactorily with layer-line intensities. He had begun the paper in optimistic tones. It was a "promising structure . . . not a vague one, but is precisely predicted; atomic co-ordinates for the principal atoms are given . . . This is the first precisely described structure for the nucleic acids that has been suggested by any investigator"

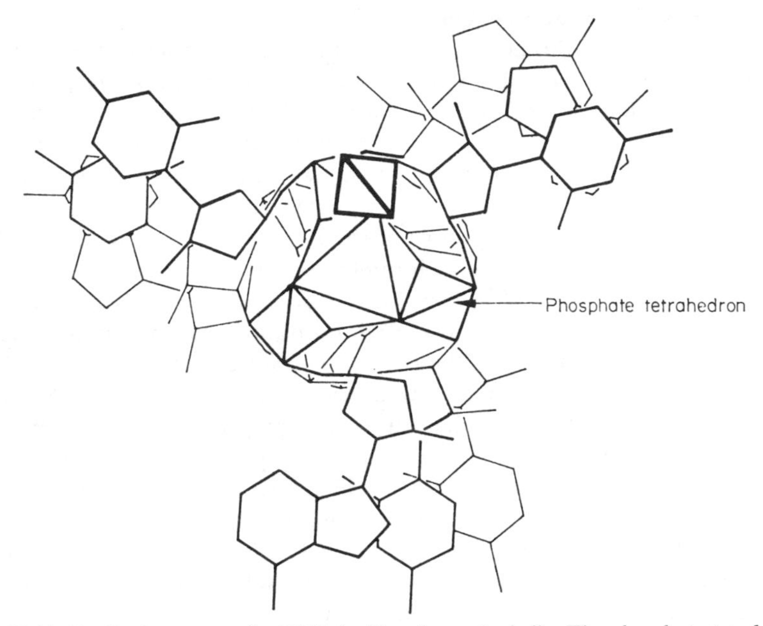

Fig. 20.11. Pauling's structure for DNA looking down the helix. The phosphate tetrahedra are hydrogen bonded to one another across the hole in the centre (from Pauling and Corey, 1953, 92).

(Pauling and Corey, 1953, 84). He closed with a comment on biological function:

> It is interesting to note that the purine and pyrimidine groups, on the periphery of the molecule, occupy positions such that their hydrogen-bond forming groups are directed radially. This would permit the nucleic acid molecule to interact vigorously with other molecules. Moreover, there is enough room in the region of each nitrogen base to permit the arbitrary choice of any one of the alternative groups; steric hindrance would not interfere with the arbitrary ordering of the residues. The proposed structure accordingly permits the maximum number of nucleic acids to be constructed, providing the possibility of high specificity. As Astbury has pointed out, the 3.4 Å X-ray reflection, indicating a similar distance along the axis of the molecule, is approximately the length per residue in a nearly extended polypeptide chain, and accordingly the nucleic acids are, with respect to this dimension, well suited to the ordering of amino acid residues in a protein. The positions of the amino acid residues might well be at the centers of the parallelograms of which the corners are occupied by four nitrogen bases. The 256 different kinds of parallelograms (neglecting the possibility of two different orientations of each nitrogen base) would permit considerable power of selection for each position.
>
> (Pauling and Corey, 1953, 96)

Pauling had sent the manuscript off at the end of December, 1952, but by the following February it was clear that the structure needed modifying. On February 4 Pauling wrote to his son:

> . . . we have found that the atomic co-ordinates in our nucleic acid structure need to be changed a bit. One of the van der Waals contacts is too close—between carbon atom 5′

and an inner oxygen atom of the phosphate group to which it is bonded. I judge that I overlooked this contact when I was making the final calculations—in the preliminary structure there were several van der Waals contacts that were pretty small, and I hunted around for parameters that would increase all of them until they were at least acceptable. Apparently I forgot to include this contact in the final process. It will be a few weeks before we have finished the job of checking over the parameters again, but I expect them to come out all right—at any rate I hope so.

(Pauling, L., 1953b)

Two weeks later he wrote: "I am checking over the nucleic acid structure again, trying to refine the parameters a bit. I think that the original parameters are not exactly right. It is evident that the structure involves a tight squeeze for nearly all the atoms" (Pauling, L., 1953c).

Evidently the work of checking the structure had been done very quickly. As Schomaker remarked: "The typical thing with Pauling is that he works and thinks enormously hard on something and makes a breakthrough on it. Then he works enormously fast and prepares a marvellous seminar that tells all about it..." (Schomaker, 1968). When it became clear that some modification to the structure was needed, Pauling consulted Schomaker. He had visited him frequently to talk about DNA. It was clear from these visits that the structure of DNA had become for Pauling "so important; it's so wonderful, and I think I am understanding something about it... He was just captivated with the thing" (Schomaker, 1968).

About three weeks after the error over the van der Waals distances had been spotted, Schomaker came up with an improvement. He suggested rotating the phosphate groups through 45° about their vertical axes. In his note of this modification, dated February 27, Pauling wrote "O_{III} — $O_{II'}$ = 4.26 Å OK". Schomaker, who had earlier been doubtful about the structure, now felt convinced. It just fitted together so beautifully that it had to be right.

The Fraser Helix

One more attempt had been made to build a triple-stranded helix. This was the work of Fraser, the infra-red expert, who in 1951 had built a modified form of the backbone-central single-stranded model of Furberg. In 1952 he built a model composed of three strands, the bases being hydrogen-bonded to one another and on the inside of the helix. This model has not survived and was not published in 1953 because Crick was against having what was clearly the wrong structure appearing along with their own. Hence even at King's, where the latest results originated, the verdict in 1952 was in favour of a three-strand model. This model did not account for the symmetry of the fibre nor did it incorporate the Chargaff ratios. Nevertheless, it was the closest approach to the correct structure in the winter of 1952.

CHAPTER TWENTY-ONE

DNA as a Double Helix

Before 1953 all attempts to build models for DNA were confined either to single or to triple helices. We have seen that Pauling had hoped to build a double helix in 1952, but had opted for a triple helix on the basis of density. When we turn to the important paper published by Furberg in the summer of 1952 we are tempted to see in it a very close approach to the model subsequently built by Watson and Crick. But his base-inside model (I) was not a double helix and he suggested it as one of two alternatives. Yet the two strands in the Watson–Crick model surely have roughly the same conformation as the single strand in Furberg's model I? As he later remarked (1967), "nucleic acids do occur in single-stranded forms, and these are important too; although their structure is unknown, they might well be single-helical structures like my model" (Furberg, 1967). After all, Furberg concluded from a comparison of densities that the nucleotides in DNA and in crystals of the nucleotides would be packed "so as to be roughly equally close in the two cases" (Furberg, 1952, 634). From the study of nucleoside and nucleotide crystals he recognized that: "Extensive systems of hydrogen bonds are found in all the crystals studied, and such bonds must be expected to occur also in the nucleic acids, probably both within and between the molecules" (*Ibid.*). He added that titration curves had already indicated the presence of hydrogen bonding.

If Furberg was so near to the solution of the structure, the more is the pity that Bernal could not have arranged for him to stay on, or to continue the work in Oslo in collaboration with Birkbeck. Certainly Furberg was keen to continue working on nucleic acids, but whether he would have taken to fibre work instead of single crystal analysis, is not so clear, but clearly, he wanted to establish the structure as far as possible on the basis of single crystal work. In this chapter we shall see how different was the approach of Watson and Crick and what a significant advance they made upon the important but incomplete picture achieved by Furberg. In truth, there was all the difference in the world between a single and a double helix.

The Role of Cations

Although Franklin had assured Watson and Crick that the sodium ions in DNA must be surrounded by water shells, and on the outside of the molecule,

Watson was still keen on giving cations a crucial role in the structure. He favoured magnesium ions for the following reasons: Lwoff had evidence suggesting that cations were necessary for the induction of phage reproduction. The role of divalent metal ions, he asserted, had been "neglected in theories concerned with replication" (Lwoff, 1952, 167). This remark prompted Watson to suggest that such cations neutralized the negative charges on the DNA chain. Knowing from Wilkins that the diameter of this chain was not >22 Å, and from Hershey that all the DNA in phage was packed into a core, the diameter of which Watson reckoned at 220 Å, it was clear that something other than the basic amino acids in the phage protein must bring about this neutralization. Surely it was metal ions. Therefore it seemed probable that phage DNA was present "in the form of a metallic salt" (Watson, 1952b, 171). Dr Bergold, he added, had told him about significant quantities of Mg^{++} in a polyhedral virus. Unfortunately no search for metallic cations had yet been made in DNA. In his research report to Delbrück, a month after he had made this suggestion, Watson spoke of his future plan to work on the structure of RNA.

> So far no one has obtained an oriented photograph of RNA. I wonder if it is not because everyone makes up the RNA in the form of an Na^+ salt while its "natural" salt may be a different cation such as Mg^{++} or Ca^{++}. Maurice Wilkins and I may have a preliminary try from this viewpoint sometime in the summer.
>
> (Watson, 1952a)

By the autumn of 1952 Watson had obtained a sample of phage DNA from Maaløe in Copenhagen and this failed to show the presence of magnesium in Ramsay's flame photometer in Cambridge. The cation idea no longer looked promising.

Base Pairing

Watson's enthusiasm for divalent metal ions did not appeal to Crick, who argued that since there was plenty of water in the fibres the demand for local neutrality was satisfied without bringing these cations into intimate association with the structure. They were in any case surrounded by water shells. Was it not wiser to look to the bases for the forces holding the several helices together?

> As hydrogen bonding still seemed unlikely I considered electrostatic attraction between the dipoles which would be produced if there were a net shift of charge over the individual bases, and I asked Professor Kemmer if he knew of someone who could calculate what attractive forces might be generated by the dipoles of two bases lying one over the other.
>
> (Crick, 1968/72)

Kemmer put Crick on to the young Cambridge mathematician, John Griffith, nephew of Fred Griffith the discoverer of bacterial transformation. Their collaboration began after a talk in the old Cavendish Building by Tommy Gold on "The Perfect Cosmological Principle".

Griffith and I went after a talk by Tommy Gold to the Bun Shop and had a drink. We discussed "What is the perfect biological principle?" I said: "Since the bases are flat, perhaps that is so that they can stack on top of one another and attract. Why not work out if adenine attracts adenine, and so on?"

(Crick, 1968/72)

The idea was, of course, that the bases belonging to different chains of the same molecule are interleaved. Later Crick spotted Griffith in the tea queue and he asked:

"Have you done the calculations?" and Griffith said: "Yes, and I find that adenine attracts thymine and guanine attracts cystosine". Now I did not say: "This explains Chargaff's rules," which any sensible person would have done if they had ever heard of them. What I said immediately, before he had time to say anything, was: "Well that is all right, that is perfectly O.K., *A* goes and makes *B* and *B* goes and makes *A*, you just have complementary replication." Now I said it, he did not say it, as I remember very clearly. Nor did I say (and I remember this, because much later on I thought how stupid I had been not to say it), "Of course, if you have complementary replication according to your

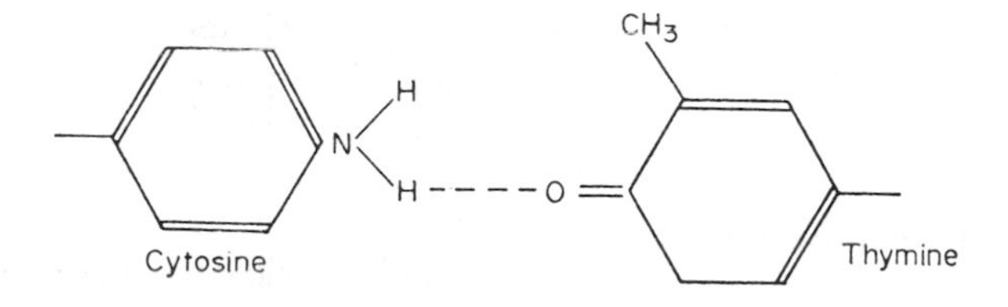

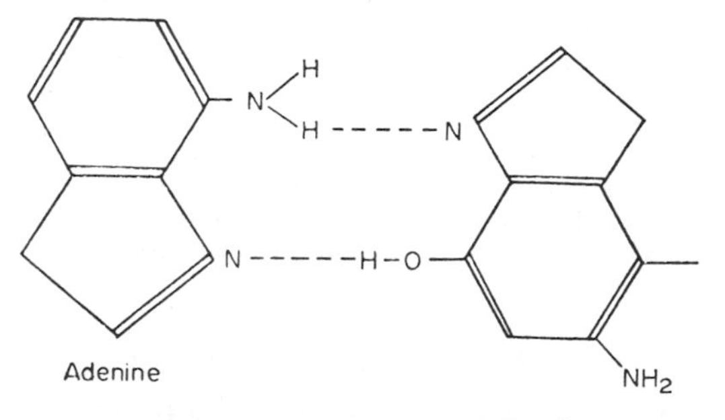

Fig. 21.1 Probable nature of Griffith's schemes for base-pairing in 1951 (from Olby, 1970, 957).

> scheme you should find $A = T$ and $G = C$." Because I did *not* make that remark I am confident that I was not aware of Chargaff's rules at the time, even if Jim had, as he claimed, mentioned them to me earlier. If he had told me I had simply forgotten them.
>
> (Crick, 1968/72)

At this stage Crick had realized, and Griffith too, that preferential attraction between adenine and thymine, and between guanine and cytosine would give rise to complementary replication. Neither had yet read Pauling's and Delbrück's 1940 paper stating a preference for this type of replication rather than like-with-like schemes, but Griffith had already, unknown to Crick, worked out a scheme of hydrogen-bonded complementary base pairs (pyrimidine with pyrimidine and purine with purine) as in Fig. 21.1, which he never published. That he did not mention this to Crick is understandable, for he was a young mathematics graduate taking an undergraduate course in biochemistry, introverted and unassuming; whereas Crick was older, extroverted, and in Griffith's eyes an established scientist. As Crick's questions all related to stacking pairing schemes, not planar ones, there was no purpose in Griffith's bringing up his hydrogen-bonded scheme.

Griffith had become interested in biochemical genetics from taking the Part II biochemistry course in the Cambridge Tripos in 1950–51. It was probably around the spring of 1951 that he began to puzzle over gene replication. At first he had considered amino acid matching for protein replication, but on learning about the possible role of nucleic acids in the process, he considered base-pairing schemes. From what source he learned of the hereditary role of DNA he cannot now recall, though he thinks that he probably read a review article.

Shortly after Crick's meeting with John Griffith in the tea queue, John Kendrew introduced Watson and Crick to Erwin Chargaff who was visiting Cambridge that July. Crick recalled the occasion as follows:

> We were saying to him as protein boys: "Well what has all this work on nucleic acid led to; it has not told us anything we want to know." Chargaff, slightly on the defensive, "Well of course there is the 1:1 ratios." So I said: "What is that?" So he said: "Well it is all published!" Of course I had never read the literature, so I would not know. Then he told me, and the effect was electric. That is why I remember it. I suddenly thought: "Why, my God, if you have complementary pairing, you are bound to get a one to one ratio." By this stage I had forgotten what Griffith had told me. I did not remember the names of the bases. Then I went to see Griffith and I asked him which his bases were and wrote them down. Then I had forgotten what Chargaff had told me, so I had to go back and look at the literature. And to my astonishment the pairs that Griffith said were the pairs that Chargaff said.
>
> (Crick, 1968/72)

The strange aspect of this story is the fact that Griffith was able to arrive at the correct pairs of bases using calculations which he knew and which were generally known to be too approximate to give reliable answers except for the simplest of compounds. It is clear that he examined unlike pairs because of the likelihood that replication is complementary rather than identical. On the other hand, he would hardly have omitted telling Crick of Chargaff's data had he known it at this stage. The correlation would have been so exciting as to cause the quiet and calm young Griffith to proclaim it. The correct pairing must have been a fluke. At the time, Griffith told Watson and Crick that π orbital interaction might account for the adenine–thymine and guanine–cytosine preferences. In fact, specific pairing is brought about not by electrostatic attraction in the stacking direction—though such forces help to hold the molecule together—but by hydrogen bonding perpendicular to the fibre axis. The *Double Helix*, according to Crick, "gives the impression that we knew all about Chargaff's rules before we talked to John Griffith. Now that may be formally so, but it does not correspond with what was actually going on in our minds" (Crick, 1968/72).

On the other hand it is very unlikely that Watson did not already know Chargaff's work before these events took place. Roy Markham thought that the Harvard molecular biologist, Paul Doty, who met Watson and Crick in the Bun Shop in the winter of 1951/2 must have told Watson about Chargaff's work (Markham, personal communication), but if Watson told Crick, as he claims he did (*Helix*, 126), Crick forgot. Watson certainly gave Chargaff no indication that he already knew his work. He had, it appeared, already suffered from Chargaff's scorn before the conversation turned to DNA. He would have been foolish to admit any knowledge which Chargaff could test out. Crick was more daring, with the result that he was led by Chargaff into "admitting that he did not remember the chemical differences among the four bases. The *faux pas* slipped out when Francis mentioned Griffith's calculations. Not remembering which of the bases had amino groups, he could not qualitatively describe the quantum mechanical argument until he asked Chargaff to write out their formulae" (*Helix*, 130). It is not surprising, therefore, that Chargaff could recall the Watson and Crick of 1953 as a strangely ignorant couple whom he likened to "Pitchmen"—i.e., salesmen or ad men shooting a line.

> When I met them first in Cambridge in the spring of 1952, they did not seem to know much or anything of my work, nor even of the structure and chemistry of the purines and pyrimidines. They were, however, extremely eager to match Pauling's α-helix by something similar in a polynucleotide, and they talked so constantly about "pitch" that I wrote in a little notebook of mine: "Two pitchmen in search of a helix." So far as I can recollect it was I who first explained to Watson and Crick, at a luncheon arranged by Kendrew, our observations on the DNA regularities and told them that whatever structure they came up with would have to take account of our complementary relationships. I believe they asked me about possible fits between the bases and I said: "Adenine with thymine, guanine with cytosine, purines with pyrimidines."
>
> (Chargaff, 1970)

Chargaff, it seems, had underestimated the two "Pitchmen". They had realized that the base pairing (although not transverse, hydrogen-bonded pairing) could be the cause of the Chargaff rules. This simple inference, startling in its possibilities, left Chargaff unmoved, but excited Crick to fever pitch. It escaped everyone except Wyatt and Cohen who first thought of it in March, 1953 (Wyatt and Cohen, 1953, 780; Cohen, 1968, 28). It reminds one of the explanation of Mendelian ratios in terms of germinal segregation. Of the so-called three rediscoverers and the several other plant breeders who published Mendelian ratios in and around 1900 there are good grounds for believing that only one established the connexion independent of reading Mendel's paper—Carl Correns. What seems an obvious step with the advantage of hindsight is rarely so without it. Whether Crick would have realized that Chargaff's ratios meant base pairing had he not known the outcome of Griffith's calculations cannot be decided, but Griffith's work made sure he did.

We are now faced with an even stranger episode in the story. Watson and Crick were now fully aware of Chargaff's rules and—on the basis of Griffith's calculations—of the possibility that a structural relationship between pairs of bases might explain them. Why did they not at once set to work to build a model incorporating base pairing? Here it is easy to telescope events and to imagine that they had in mind hydrogen-bonded base pairing in the plane of the bases, whereas in fact not once in 1952 was any attempt made to figure out such schemes, for they were thinking in terms of stacking pairing of interleaved bases. Hydrogen bonding was, they thought, too unreliable as a mechanism of replication. Here Crick made the mistake of assuming that both tautomeric forms of the bases exist in the same DNA molecule, that the proton could shift from one position to another, thus altering the sites for hydrogen bond formation. Now if these sites varied in the same base, how could one get constancy of matching in replication? The answer, it appeared, was by preferential "stacking pairing" of the bases, which could take place in a variety of ways. The bases could project out from the molecule and interleaved pairing with bases of the nascent daughter polynucleotide strand could occur as in Fig. 21.2, and this pairing need have nothing to do with the structural features of the original parental DNA molecule. In other words, *how the several strands of the helix were packed together in the cylindrical molecule had no necessary connexion with the mechanism of its replication.*

Stacking pairing schemes, even if they were incorporated into the structure of the helix, had to be treated with caution. Both Crick and Griffith knew how unreliable are conclusions based on quantum mechanical arguments. Further, the Chargaff rules might yet prove to be a red herring; they might be the product of some feature of the code or of the metabolism of the bases. Crick hated using more than the minimum number of assumptions, for by so doing, it would be easy to exclude the correct approach. They had yet to find some direct evidence for base pairing.

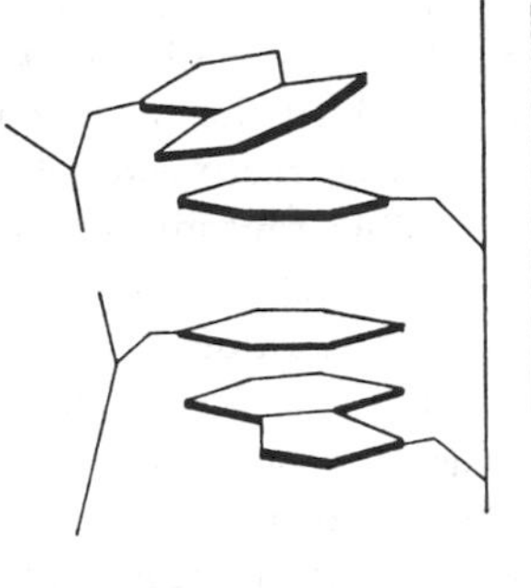

Fig. 21.2. Complementary replication A to T and C to G by specific 'stacking' attraction. (No such diagrams are to be found in manuscripts of the period.)

At the end of July, 1952 Crick tried to get such evidence by determining the ultraviolet absorption of mixed and pure bases and nucleosides (Crick, 1952). If the mixtures entered into pairing relationships, they would absorb less ultraviolet radiation than would the sum of their constituents irradiated separately. These experiments failed to show any difference, because Crick lacked small enough cells to achieve the required concentrations without bringing on opacity, and as he was in the last year of his thesis, he dropped this unpromising attempt. But he continued to feel the need for more evidence.

The Tautomeric Forms

We come now to the most ironical part of the story. Watson and Crick had discarded hydrogen bonding as a means for holding the structure of the DNA molecule together in 1951. Now that they had got *the* clue to the structure—base ratios means base pairing—they still showed no sign of reconsidering hydrogen bonding. How could this be? Because of the tautomeric forms.

There was really only one piece of experimental evidence in favour of the existence of tautomeric forms in purines and pyrimidines, and that was the shift towards the red end of the ultraviolet absorption spectrum caused by raising the pH of their solutions. Work done at Siena Heights College (Stimson and Reuter, 1943) showed such a shift and these workers attributed it to "enolization"—the change from the keto to the enol form.

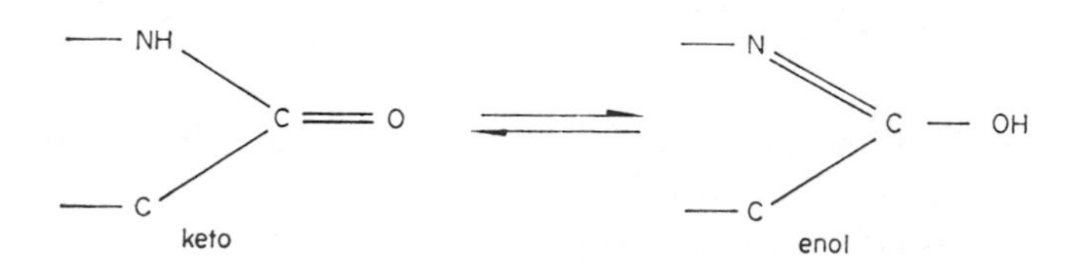

Subsequent work at the Sloan Kettering Institute did not alter this conclusion (Cavalieri *et al.*, 1948), and although D. O. Jordan, in his paper in Randall's *Progress in Biophysics* represented all the bases which are found in DNA in what we now know to be their correct forms—keto and amino—he did not conclude that the other possible tautomers did not exist. On theoretical grounds they were expected to exist somewhere, some of the time. Jordan's attitude was that one cannot unquestioningly apply to solutions of the bases what is found in their crystals, much less in the crystals of their derivatives. And it was in their solutions that the "enolization" appeared to occur.

Even if one assumed that what applied to the crystals of the bases also applied to their solutions, X-ray diffraction studies failed to yield a decisive answer. Such a programme was initiated in 1945 by Todd, Clews and Cochran at Cambridge, with the aim of establishing the structures of the bases and locating the positions of the protons. Similar work was begun in Bernal's laboratory where G. J. Pitt tried to locate protons in a pyrimidine but failed owing to the complications caused by the presence of water of crystallization (Pitt, 1948). June Broomhead continued the Cambridge programme (Broomhead, 1948, 1950, 1951; Clews and Cochran, 1948, 1949) but unfortunately fell for L. Hunter's theory of "mesohydric tautomerism" (Hunter, 1945), according to which proton migration from one site to another took place in compounds with structures like the purines and pyrimidines with such ease that it was "meaningless" to attempt to distinguish the different tautomers. When Broomhead found a curious electron density distribution near the amino groups in adenine hydrochloride she thought it might well mean that a Hunter-type proton interchange was taking place between the amino group of C_6 and the nitrogen of N_7.

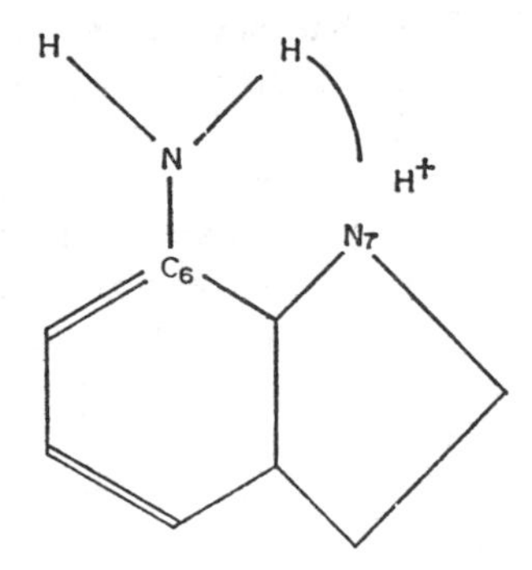

Even when Cochran localized all the protons in adenine hydrochloride (Cochran, 1951) Broomhead held to her opinion, and the question of the keto or enol form remained open since this group is not present in adenine.

The general uncertainty in the '40s over the tautomers found its expression in the textbooks of the day. Here the enol form was depicted, there the keto, here the imino, there the amino. Until this confusion was sorted out Watson and Crick were doomed to fail in their quest for the structure of DNA. Little did they know that the American crystallographer, Jerry Donohue, had discussed Broomhead's data, and from a consideration of the hydrogen bonding scheme had come to the conclusion that for adenine hydrochloride there was in the crystalline state *only one position for the protons, although there were four resonance structures.* His criteria were acceptable distances and orientations for hydrogen bonding. He was left with:

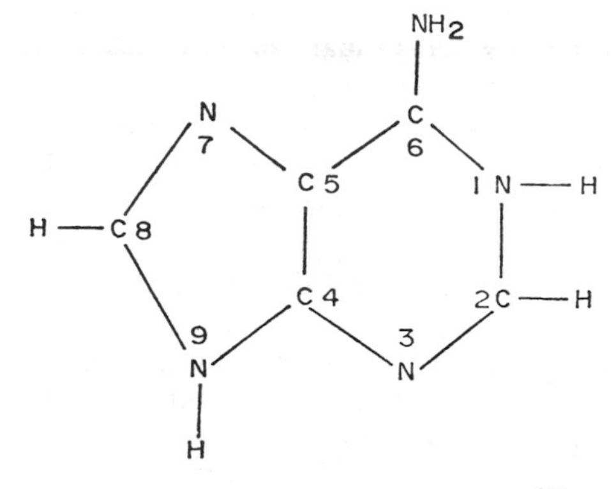

(Donohue, 1952, 504)

Nor did Watson and Crick know in 1952 what Marshall and Walker had concluded a year before, namely that the shift in the ultraviolet absorption spectra of the bases was an ionic phenomenon and not due to enolization (Jordan, 1955, 450). We shall see later how Watson and Crick used tautomerism to account for mutation in their successful model, and Donohue was later to suggest that "a guanine molecule will assume the tautomeric form which will permit it to form the best hydrogen bonds in a given environment" (1955, 735). Really, as it later transpired, tautomerism was a misnomer; for no evidence pointed to its existence in the bases, either in or out of DNA! Crick can surely be excused for assuming that because the bases were said to exist in tautomeric forms, that both forms existed in significant proportions in DNA.

News of the Pauling Structure

At the beginning of 1953 Watson had more or less resigned himself to the opinion that Pauling had solved the structure of DNA. In December, 1952 Peter Pauling had shown them a letter from his father stating that Corey and he had a structure for DNA. Careful as ever, Linus Pauling gave no details.

As Watson and Crick passed the letter to and fro their frustration grew. Crick paced "up and down the room thinking aloud, hoping that in a great intellectual fervour he could reconstruct what Linus might have done" (*Helix*, 156). Perhaps he had got it wrong. Surely he had not seen the King's pictures. But Watson's heart told him that Pauling had got it right. Crick, on the other hand, was not so sure. Pauling was sometimes right and sometimes wrong. "Naturally we were apprehensive, but I was more stoical, Jim was more nervy about things" (Crick, 1968/72). And so, while Peter Pauling spent Christmas in the beautiful Erzgebirge, Watson went ahead with his holiday plans and by the New Year he was in Zermatt skiing with his sister Betty. From there his current interest in microbial genetics took him to Milan to see Cevalli-Sforza, and by mid-January he was back in Cambridge, having settled Betty at Camille Prior's establishment. The next task was to find out the latest on Pauling's structure for DNA. Peter revealed that his father had made one reference to DNA in his letter of December 31. It read: "Corey and I have sent off a paper on the structure of nucleic acids. Would you like to have a copy?" The worst had happened: the structure was soon to be published. Anxiously Watson awaited the arrival of the promised copy. True to his word, on hearing from his son, Pauling put the MS in the post, and in his covering letter of January 21 he wrote "I was interested to learn what is said about us in Cambridge". What, one wonders, had his son let on? This letter also supplied an important piece of news—the DNA paper was to appear in the February issue of the *Proceedings of the National Academy of Sciences*.

Only those who have experienced the trauma of competition in research will know how Watson and Crick must have felt with Peter Pauling in the room, the manuscript from his father sticking out of his pocket. Had Pauling really discovered the secret? Did he know all? Was this, as they feared, just a repetition of the α-helix affair, this time with DNA? Many years later Watson had not forgotten how his "stomach sank in apprehension at learning that all was lost". Peter had that important look on his face, and in his forthright manner he told them that the structure was a three-chain helix with the phosphates on the inside. So, thought Watson, they had been on the right path in 1951 after all! When Peter Pauling handed Watson the manuscript the latter scanned the summary and introduction, then studied the figures giving the locations of the atoms. Something looked amiss, and when he examined the illustrations the truth dawned on him that the phosphate groups were *not* ionized. It was unbelievable! If this feature of the structure had been an incidental one it would not have been worth pursuing, but it was not. The phosphates had to be un-ionized if the structure was to hold together, for the only links between the three chains were hydrogen bonds. The phosphate group has three ionizable groups. In the polynucleotide chain two of these are involved in covalent links with the sugar residues. Except at the end of the chain, therefore, there is but one hydroxyl group per phosphate, which by loss of its hydrogen becomes negatively charged. In Pauling's structure each "primary" group retained its hydrogen ion, and this could form a hydrogen bridge over to the P = O group of a neighbouring chain. Granted that these ionization constants would be different for different structures of DNA, the fact remains that the evidence from the substance itself as extracted from calf thymus and herring sperm pointed to 50 per cent dissociation of the primary phosphoryl groups at a pH *below* 2 (Gulland *et al*, 1947*). Under normal physiological conditions, therefore, one would expect all the phosphoryl groups to be dissociated. No doubt Watson took an extreme position when he declared that everything he knew "about nucleic acid chemistry indicated that the phosphate groups never contained bound hydrogen atoms" (*Helix*, 160). All the same, one naturally expected the evidence for the total or partial un-ionized state of NaDNA to be cited. Earlier analysis of the sodium content of the extracted material had shown molar equality between Na and P (Gulland, Jordan and Taylor, 1947), from which Jordan argued that "in view of the fact that the amount of secondary phosphoryl dissociation is small these [Na^+ ions] must be combined largely or entirely with the primary phosphoryl dissociations" (Jordan, 1951, 71).

Watson lost no time confirming his suspicion. He raced over to get assurances from Roy Markham and from the organic chemists. He reckoned that by mid-March the structure which Pauling and Corey had produced would be subject to international discussion. By that time the February issue of the *PNAS* would be everywhere. So there was no time to be lost. There would be six weeks at the most before Pauling would be back in full pursuit of the structure, his error having been exposed. Crick was likewise amazed at Pauling's error, and agreed that something must be done to get the correct structure before Pauling realized his mistake. Over him was hanging the awful memory of the planar peptide bond, and Pauling's α-helix. The opportunity which DNA presented just must not be thrown away. It was agreed, therefore, that Watson should take the Pauling MS to Wilkins, but Wilkins first saw it at Crick's house in Cambridge.

Watson's visit to King's

In the *Double Helix* Watson has described how his news was taken by Rosalind Franklin. "Coolly she pointed out that not a shred of evidence permitted Linus, or anyone else, to postulate a helical structure for DNA" (*Helix*, 165). So insulting did she find Watson's impudent attempt to tell her

* On p. 1138 Gulland, Jordan and Taylor wrote: "At the isoelectric point, therefore, nucleic acids will exist almost entirely in the zwitterionic form".

how to interpret X-ray patterns that she all but lost her temper. One suspects that her resulting threatening posture was as much due to the fright of having her pride so injured as to aggressive anger. Little did Watson know that since Corey's visit to King's in May, 1952, Franklin had been in correspondence with him. Was it through Watson or through Corey that Franklin knew about Pauling's model? We cannot be sure, but she wrote to Corey asking for details in early 1953. Corey replied:

> I did not mean to neglect your letter of January 6 asking for information about the structure which Professor Pauling and I were publishing in the *Proceedings of the National Academy of Sciences.* I might say that we are already working on a revised form of this structure in which the phosphate tetrahedra are turned through 45° with a result that the van der Waals packing and some other features of this structure appear to be considerably improved.
>
> . . . I have always been grateful to you for showing me your splendid X-ray photographs of nucleic acid fibers. I hope that data from them will soon be available in published form for use in testing the structures of nucleic acids which I am sure will be forthcoming from many laboratories.
>
> (Corey, 1953)

How infuriating it must have been for Franklin, who *was* actually working on the X-ray diffraction of DNA fibres, to find Watson, who was not, at every turn of her path. Here he was with a copy of the MS. Corey had not sent her one!

Wilkins was more receptive to Watson. He had recommended his diffraction studies of DNA. The fine diffraction pattern of the B form which he showed to Watson, however, was Franklin's famous B pattern of May, 1952. Watson did not even know of the existence of this form of DNA until that last Friday in January, although Franklin had reported its existence and showed a rather poor X-ray picture of it in 1951 in Watson's presence. This 1952 photograph offered the simplest and most striking example of a helical diffraction pattern that had ever been seen. Because the repeat was integral, unlike the polypeptides, the intense meridional spot came bang on a layer line. Because there were ten residues per chain to the pitch this spot came on the tenth layer line. The diamond-shaped absence of reflections between the zero and tenth layer lines along the meridian spoke eloquently of a helical conformation, the slope of the helix being equal to the angle which the prominent "cross-ways" made with the meridian. No wonder Watson's mouth "fell open" and his "pulse began to race". The empiricist might argue about the significance of these features in the A form but no one in their right mind, not even Franklin in private as opposed to in public, could doubt that here in the B form one had to do with a helix. But then Franklin, professional that she was, had concentrated her attention on the A form because this form gave more of the information that the crystallographer usually requires. Watson and Crick, as we have seen, had very good reasons for concentrating their attention on the B form. "We had no hesitation in choosing the latter", they wrote, "mainly because of its extremely strong 3.4 Å meridional reflexion . . . , since this gives information which can be of direct help in building models" (Crick and Watson, 1954, 84).

How many Chains?

Three lines of evidence could be invoked to argue for the number of polynucleotide chains in the molecule being either two or three. Less than two or more than three chains were ruled out by the density of the sodium salt of DNA which Astbury had evaluated as 1.65 g/cc. when dried over P_2O_5. Biological, crystallographic and density data could then be applied to resolve the difficult decision between two and three. By this time Wilkins had realized that the earlier decision in favour of three chains had been hasty. Through Franklin's work it was known that the A form was monoclinic, and that its water content (at 75 per cent humidity) was "about 40 per cent". This estimate comes from Franklin's Interim Annual Report for the period January, 1952 to January, 1953. Franklin rightly pointed out the difficulty of being sure that all this water was actually held in the crystallites which contributed to the X-ray patterns. The A form was not a perfect three-dimensional crystal by any means. She went on:

> However, calculation shows that if the water content of the crystallite is assumed to be between 22 per cent and 50 per cent of the dry weight, then the number of nucleotides per face-centred unit cell lies between 56 and 44.
>
> It was mentioned above that the occurrence of a near-meridional reflection on the eleventh layer line suggests that there are 11 nucleotides per chain in the unit cell. The density measurements therefore suggest that there are four chains passing through the face-centred unit cell (or two chains associated with each lattice pair).
>
> (Franklin, 1953a, 9)

But how could Watson and Crick know what was in this report? Franklin had not yet written it!

The answer, I believe, is that when Watson returned from King's on that last Friday in January he did not know the contents of this report. All that he learnt must have come from Wilkins. Watson had asked Wilkins why the King's group favoured three chains, and Wilkins had answered that it was because of the density. In 1951 Franklin had put the water content of the A form at 42 per cent. This and Astbury's estimate of the density at 1.63 g/cc. gave her $\sim$46 nucleotides in the unit cell (see p. 349). Her later work had shown how variable was the water content of their crystalline specimens, so the evidence was by no means satisfactory. As for the relation between the two forms, Wilkins was not able to help Watson. In his work of 1951 Wilkins did not conclude that changes in fibre length represented changes in molecular length, and although Franklin had set down the data from which this could be concluded, she had not done so herself. Hence it was her report containing this information, and not any hints from Wilkins, that was crucial for Watson (Wilkins, 1986). This was all Watson needed to decide between two and three chains. On the train that night he sketched on the blank margin of his newspaper all the details of the splendid B pattern that he could remember. Then he puzzled over the question of the number of chains until, by the time he had got back to his college, he had decided on two. "Francis would have to agree. Even though he was a physicist, he knew that important biological objects came in pairs" (*Helix*, 171).

This famous sentence has led to some exaggeration of the role of biological thinking in the discovery of the structure. Granted that two chains appealed to Watson as a biologist and led him to ask Wilkins about the data (there were other reasons of course—a three-chain model is more difficult to build, especially when the bases are on the inside) the fact remains that Watson's argument for two strands was a purely physical one which led to a value for the number of residues on a single chain in the form for which Franklin had calculated the number of nucleotides in the unit cell. His argument, a similar version of which will be found in Crick and Watson's Royal Society paper (1954, 85), would appear to have gone as follows:

> In the transformation B ⇌ A the axial periodicity (= helix pitch) changes from 34 Å to 28 Å. This represents a shortening of nearly 20 per cent. The observed length decrease of the fibre when going from B to A is also about 20 per cent, which can thus be attributed to the packing down of a helix. The residue separation in the B form is 3.4 Å. Therefore it must be $3.4 \times 0.8 = 2.72$ Å in the A form, and the number of residues in the repeat of a single chain in the A form must be $28/2.72 = 10$ [actually it is 11].
>
> If we have to do with a triple helix the number of nucleotides in the unit cell is likely to be 30, 60 or 90, if a double helix, 20, 40 or 80. The number is said to be 46, which is closer to 40 than it is to 30 or 60. Therefore a double helix is more likely than a triple helix.
>
> (Author's wording)

The next day was a Saturday. According to Watson, Crick had a slight hangover from the excellent dinner party of the night before, so that when Watson declared that he, a former bird-watcher, could now solve the structure of DNA, Crick was not amused. He soon changed his tune when Watson gave him the details of the B form as he had sketched them on his newspaper —Crick never saw the picture. The heavy spot on the equator at about 20 Å assured them that the diameter of the helix was in agreement with EM pictures, about 20 Å. Of course they had known about the 3.4 reflection long before, but was it going to prove like the 5.1 Å reflection of α-keratin, an off-meridional doublet? Watson emphasized the fact that of all the reflections on the B diagram this meridional 3.4 was far the most intense. This assured them that Astbury's and their assumption that this spot represented the distance separating neighbouring bases in the axial direction was justified. So they had the 3.4 Å, 34 Å, 20 Å, helix slope about 40°*, ten residues per chain

* According to Crick, Watson's recollection of the slope was too vague to give more than the very roughest idea.

repeat and the argument from the density of the A form that two chains was more likely than three.

According to Watson, Crick "drew the line against accepting my assertion that the repeated finding of twoness in biological systems told us to build two chain models. The way to get on, in his opinion, was to reject any argument which did not arise from the chemistry of nucleic acid chains" (*Helix*, 175). When asked about this many years later Crick recalled:

> I said, "two chains I agree is the most likely, but we must not prejudge the issue; let us go on the data as to whether it is two or three. After all you might have two chains for replication and another chain for the cytoplasm or something." So it is all very well for him to say that any biologist would know you have two, but that is exactly the type of argument that I regard as terribly dangerous. That is how you make the mistakes . . . I went along with Jim doing the building of two but if two had gone wrong I would have said: "Let's try three." But once we had got the C_2 symmetry and the base pairing there was no point in looking for three.
>
> (Crick, 1968/72)

The Position of the Sugar–Phosphate Backbones

In November, 1951 Franklin had given her reasons for believing that the phosphate groups must be exposed, and therefore on the outside of the cylindrically shaped molecule of DNA. We saw that in their first attempt to build a model Watson and Crick did not use this vital piece of information—presumably because Watson failed to digest it when he attended Franklin's seminar. Then came Franklin's criticism of their first attempt which Crick recalled as follows:

> The phosphates are always hydrated. Therefore our first model was wrong. If they are hydrated, therefore they are on the outside. But that second argument was much weaker. They could be hydrated and on the inside. Then the water would have to form part of the structure. If on the outside—it could wobble around just hydrating. That she told us when she criticized our model.
>
> (Crick, 1968/72)

Crick did not recall the other evidence in support of the exposed position of the phosphates—titration data and the Franklin model for explaining the $A \rightleftharpoons B$ transformation—given at the November colloquium. Presumably Watson never told him. Nor did Crick recall the details of Franklin's cylindrical Patterson map, because he did not see it until after the Watson–Crick model had been published. A copy of the typescript of the paper containing the cylindrical Patterson was lent by Franklin to Watson only later, and he passed it on to Crick. This map could be interpreted as showing the powerfully-scattering phosphorus atoms at a diameter of 18 Å and spaced 5.7 Å apart along the chains. So Watson was partly right when he said that Franklin's "evidence was still out of reach of Francis and me" (*Helix*, 169). And so for one-and-a-half days Watson worked on two-chain models, their backbones in the centre, for to anyone who thought along biological lines it was obvious that the bases had to be on the *outside*, otherwise how would the encoded information which they contained be read off? There was, too, the chemical fact that under physiological conditions DNA was normally associated with histone, with which it formed a salt-like combination. How could the basic arginine side chains come into close proximity with the negatively charged phosphate groups without completely covering up the way to the bases unless the bases were on the outside? The idea of interleaved arginine side chains going in between the bases to the central phosphates had attracted K. G. Stern in 1951 and Watson had fallen for the idea also in 1951.

There was another and more pressing reason for building backbone-inside models. "As long as the bases were on the outside one didn't have to worry about how to pack them in", said Crick. "When I said: 'Why not put them on the outside?' Jim replied: 'That would be too easy!' This prompted the retort: 'Then why don't you do it?'" (Crick, 1968/72). So awful did the two-chain-inside models look, even worse than the old three-chain models, that Watson gave up in despair and went out to play tennis in the lovely early spring of that year. Finally, over coffee, wrote Watson,

> I admitted that my reluctance to place the bases inside partially arose from the suspicion that it would be possible to build an almost infinite number of models of this type. Then we would have the impossible task of deciding whether one was right. But the real stumbling block was the bases. As long as they were outside, we did not have to consider them. If they were pushed inside, the frightful problem existed of how to pack together two or more chains with irregular sequence of bases. Here Francis had to admit that he saw not the slightest ray of light.
>
> (*Helix*, 177–178)

The real problem, then, which in the end forced Watson to build backbone-outside models, was that of packing. As they said in 1954:

> No difficulty was found in obtaining repeat distances of 3.4 Å in the fibre direction as long as we considered only the charged groups. When, however, we attempted the next step of joining up the phosphate groups with the sugar groups we ran into difficulty. The phosphate groups tended to be either too far apart for the sugars to reach between them, or to be so close together that the sugars would only fit in by grossly violating van der Waals contacts. At first this seemed surprising, as the sugar–phosphate backbone contains, per residue, five single bonds about all of which free rotation is possible. It might be thought that such a backbone would be very flexible and compliant. On the contrary, we came to realize that because of the awkward shape of the sugar, there are relatively few configurations which the backbone can assume. It therefore seemed that our initial approach would lead nowhere and that we should give up our attempt to place the phosphate groups in the centre. Instead, we believe it most likely that the bases form the central core and that the regular sugar–phosphate backbone forms the circumference.
>
> (Crick and Watson, 1954, 85)

Wilkins' Visit to Cambridge

At this juncture Wilkins accepted an invitation from Crick to spend the week-end in Cambridge at his newly acquired and charming old house, 19 Portugal Place, now appropriately called the "Golden Helix". On the Thursday he wrote (presumably February 5) saying:

> I should have replied before but was feeling lousy and thought I might be too depressing and should keep away from humanity. (I feel alright to-day.) Rosie's [Franklin's] colloquium made me a bit sicker. God knows what will become of all this business. They talked for 1¾ hours non-stop and she effectively refused to answer questions. They had a unit cell big enough to sit in (but nothing in it) . . . It is very nice of you to get Pauling's paper and I will tell you all I can remember and scribble down from Rosie.
>
> (Wilkins, 1953a)

We learn from Watson that "almost as soon as he [Wilkins] arrived from the station Francis started to probe him for fuller details of the B pattern. But by the end of lunch Francis knew no more than I had picked up the week before" (*Helix*, 179). Wilkins stated that he visited the Cricks, as far as he could remember, "fairly frequently" during 1951 and '52. "I therefore was able to pass to Crick and Watson a general picture of the DNA molecule, approximate number of chains, pitch and diameter of helix and base stacking . . ." (Wilkins, 1968). But Wilkins brought no X-ray pictures with him to Cambridge, and earlier when he wrote to Crick in 1952 describing his much improved pictures of *Sepia* sperm heads which showed "very nicely a whole series of helical layer lines and one inter-micellar spacing" he gave no dimensions and in the rough drawing showing the helical dashes on layer lines 1, 2, and 3 and the meridional arc there was no indication as to whether the latter fell on a layer line or off it, and just how far out it was one could not tell. In this instance he had very likely not yet measured it; after all, in 1952 he had been virtually forbidden to work on DNA. Hence his interest in the nucleoproteins, and his work in November and December with H. G. Davies on a technique for weighing chromosomes. Of course he was sore about DNA, but his reply to Crick's invitation was cheerful: "It is very nice of you to get Pauling's paper and I will tell you all I can remember and scribble down from Rosie" (Wilkins, 1953a). Maybe Franklin refused to divulge any of her results. Perhaps Wilkins, who had at last begun to work on DNA again, began to feel some reticence about telling his Cambridge friends all he knew. Their eagerness to learn the latest news did not help. Crick had clearly been very anxious. At first Wilkins did not reply to the invitation and this led to a "shoal of daily letters" from Crick (Wilkins, 1953a). But when he came, Watson and Crick failed in their attempt to open him up. They had instead to be satisfied with Wilkins' assent to their starting model building again, although even this "was given slightly reluctantly, there was no doubt about it; I remember it quite vividly, and I think it was in our dining-room at Portugal Place", said Crick. "I remember the scene because we were somewhat edgy about it" (Crick, 1968/72).

Peter Pauling was also around at that time and recalls having tried to persuade Wilkins:

> putting in arguments that if Maurice didn't build models my father would, and he would get it right next time. One of the rules of the stochastic method is that you can't

> guess more than once, but since my father says there are two of them—Pauling and Corey—they can guess twice! This was the chief argument we were all using on Maurice—that if he didn't do it my father would. Maurice said he wasn't going to build models until Rosie Franklin left in four weeks time.
>
> (Pauling, P., 1967)

What would Wilkins' reaction have been had he known that during the whole of the time since Watson's visit to King's a week ago our American research fellow had been trying to build models, apart that is from the times when he was playing tennis or watching films. And there was Wilkins, sitting in Crick's house, dreaming of the happy day when the cloud hanging over him at King's would lift. Already he was marshalling his team. H. R. Wilson had joined him, and together they had produced a good B pattern of *E. coli* DNA, though not as good as Franklin's best picture with the Signer DNA. Leonard Hamilton was soon to send DNA from New York. How would he have replied to his friends' questions had he known what they would do during the next four weeks, and would he have told them more about his results had he known that they had access to a copy of a report from Randall's unit, which included a summary of Franklin's results?

The MRC Report

It must have been during the second week of February that Perutz passed on to Crick his copy of the report which Sir John Randall had circulated to all the members of the Biophysics Research Committee who came to see round his Unit on December 15, 1952.

How did this report help Watson and Crick? It confirmed Watson's recollections of the data on the water content and axial repeat in the A and B forms:

> The crystalline form of calf thymus DNA is obtained at about 75 per cent RH and contains about 20 per cent by weight of water.
>
> Increasing the water content leads to the formation of a different structural modification which is less highly ordered. The water content of this form is ill-defined.
>
> The change from the first to the second structure is accompanied by a change in the fibre-axis repeat period of 28 Å to 34 Å and a corresponding microscopic length-change of the fibre of about 20 per cent.
>
> (Randall, 1952)

Her argument for the exposed position of the phosphorus atoms from the behaviour of DNA to water then followed:

> The phosphate groups, being the most polar part of the structure, would be expected to associate with one another and also with the water molecules. Phosphate–phosphate bonds are considered to be responsible for intermolecular linking in the crystalline structure. The water molecules are grouped around these bonds (approximately four water molecules per phosphorus atom). Increased water content weakens these bonds and leads, first, to a less highly ordered structure and, ultimately, to gel formation and solution.

But Watson and Crick also learnt something which was new to them:

On the Patterson function obtained, the lattice translations could be readily identified. On the basis of a unit cell defined by

$$a = 22.0 \text{ Å}$$
$$b = 39.8 \text{ Å}$$
$$c = 28.1 \text{ Å}$$
$$\beta = 96.5°$$

the 66 independent reflections observed could all be indexed with an error of less than 1 per cent.

So the crystalline form of DNA was monoclinic C2. Now as Crick recalled, Watson knew nothing about space groups,

And I knew very little. But it so happened that the space group of horse haemoglobin, on which I had been working for some years, is C2. Therefore it happened to be one with which I was familiar. I knew immediately that the space group C2 has diads at the side. Therefore if these conclusions were correct, and there are ways of weazeling out of them, . . . the naive and as it turns out correct interpretation is that there are diad axes at the side. Jim never liked this argument, I don't think he could quite follow it.

(Crick, 1968/72)

It was not merely that the cell was monoclinic, for as Crick pointed out:

A true monoclinic cell must have diads, *or* screw diads, *or* both. I knew [before seeing the MRC report] that at King's they thought the unit cell was monoclinic, but this could have implied only screw diads (that is, *between* molecules) which would have told us nothing about the molecule itself. But C2 implies that there were not only screw diads but diads as well. This was the crucial fact. Moreover the cell dimensions showed that the diads were not perpendicular to the plane of hexagonal close packing, but *in* that plane. Examination of the space group implies also that the diads are likely to relate elements *within* the same molecule. Hence one chain must run up and the other run down. Notice that the symmetry does not apply to the base sequence, but we appreciated that, at that resolution, one base-pair looked very like any other.

(Crick, 1973)

Because the implications of this crystallographic argument were not clear to Watson, he went on trying to build the model with the sugars too close—they were only separated by 18°. One day when he left the model to play tennis he said to Crick: "You try it." And while Watson was out Crick, who was convinced that Watson's chain was too tight, built one with a 36° rotation per residue. "I left a little note for him saying: This is it—36° rotation" (Crick, 1968/72).

The argument behind this decision went as follows:

The sugar–phosphate chain is polar—it has a direction dictated by the 3′, 5′—phosphate-di-ester linkage—and in a double helix the two chains are "parallel" if they run in the same direction. In such a structure the upper and lower halves of the helix would be identical, and the repeat of the structure in the axial direction would be halved. Therefore the repeat at 34 Å would represent half the pitch, it would correspond to a rotation of the chain around the axis of 180°. But there are ten residues in this distance, therefore the rotation per residue would be 18°. If, on the other hand, the two chains run in opposite directions—are anti-parallel—then the pitch is not halved, and 34 Å represents a rotation of 360°. Therefore the rotation per residue is 36°.

How was it possible to decide between these two alternatives? Crick argued that if the chains were parallel the axis of symmetry would be *along* the molecule in the *c* direction. Rotation of the cylindrical molecule about its own axis through 180° would place all the atoms of the sugar–phosphate backbones in equivalent positions.

In the case of anti-parallel chains, the atoms would only be placed in equivalent positions by rotation about an axis at right angles to the axis of the DNA cylinder. If the symmetry of the crystal was dictated by the symmetry of the double helices then monoclinic C2 obviously meant that the chains were anti-parallel. As Crick said much later: "I don't think I would have thought of running them in the other direction but for the clue of the C2 symmetry . . . It was absolutely crucial to have this idea . . . In my view it is an obvious enough idea for a trained crystallographer, but for someone like myself, who was a beginner, it needed a clue to get to that point, and the clue was Rosalind Franklin's data. It was very difficult for Jim to grasp the fact that the chains were running in the opposite directions. For example, when we got the chains the other way [anti-parallel] he could not build the second chain the other way up. It was just too difficult! Somehow his mind didn't work that way" (Crick, 1968/72).

Now why was it only Crick who saw the meaning of the C2 symmetry? Was it just a question of being a quicker thinker? I think not. Watson did not know the meaning of monoclinic C2. Franklin was not in favour of helices for the crystalline form until after Watson's visit at the end of January, 1953. Before that date she had attempted to account for the symmetry of the A form in terms of zig-zag chains placed back-to-back in which like stretches of base sequences alternated from one chain to the other. Wilkins, I suggest, did not think of anti-parallel chains because he was not attempting to build a *double* helix, instead he was playing with four helically wound strands. Franklin's February analysis of the B form had led her to favour either two non-equivalent strands, or a single strand. The symmetry suggested an even number of strands, but only if the sugar-phosphate backbone were polar. Brown and Todd had shown that they were consistently 3′,5′-phosphate links, but when Franklin was shown the double helix she asked Watson and Crick whether it was certain that the linkages in the backbone went 3′,5′-, 3′,5′-, 3′5′-,

> as opposed to 3′,5′-, 5′,3′-, 3′,5′-, 5′,3′-, which would have enabled you to put a diad axis in on one chain, because it had no polarity. With gimmicks like that you can get three chains in the asymmetric unit, because you can get down to one chain. Once you have got one chain you can multiply up to three. Donohue had another argument: possibly the symmetry was false, and only reflected the arrangement of the phosphates. It

didn't show the two chains running in opposite directions . . . Only the phosphates showed [in the symmetry] and they were not polar. Only the sugars were polar, and then you can do the same thing over again.

So the argument is not completely foolproof. The density supported two chains, the shape of the unit cell suggested the diads were at the side. Any reasonable person would have said: "That suggests two chains running in opposite directions." That was the naive interpretation. In any argument there is almost always a gimmick by which you can get round it. You just have to decide whether to go along with the argument, and ignore the let-out, or alternatively to ignore the argument because it is not completely fool-proof.

(Crick, 1968/72)

Thus, by the beginning of the third week of February Watson and Crick had in the office, which they shared with Jerry Donohue and Bob Parish, an acceptable model of one sugar–phosphate backbone for a double helix—it was not necessary to build both strands at this stage. Franklin had given the argument for a backbone-outside molecule, Watson the argument for a double helix, and Crick the argument for anti-parallel chains and a residue separation of 36°. Since this model was in agreement with all the data in the MRC report Crick was very enthusiastic that Watson should press on with its refinement. Watson was not enthusiastic; for the problem that now faced him was what to do with the bases.

Base Pairing

Since no models of the bases were available in the Cavendish—sometime during February, Watson had asked for them to be constructed—he turned to J. N. Davidson's Methuen Monograph, *The Biochemistry of Nucleic Acids*, first edition 1950, and copied out the structural formulae of the bases (see Fig. 21.3). These reflected the "voting agreement" in favour of the amino and enol tautomeric forms. Davidson had based his chapter on the purines and pyrimidines on B. Lythgoe's recent review article (Lythgoe, 1949). "My aim", wrote Watson, "was somehow to arrange the centrally located bases in such a way that the backbones on the outside were completely regular . . ." (*Helix*, 182). Unfortunately the facts of genetics demanded that DNA must show the maximum variety of encoded information. By making this a condition of his model Watson found himself in a hopeless mess. "In some places the bigger bases must touch each other, while in other regions, where the smaller bases lie opposite each other, there must exist a gap or else their backbone regions must buckle" (*Helix*, 182–183). How the demands of biology were making the life of the model builder difficult!

This was not all, the structure had to be held together; the helices had to intertwine in order to produce a very rigid rod with constant diameter. So far Watson had progressed haltingly, depending on how much help Crick gave him, or how much of Crick's advice he chose to follow. Now Watson struck out on his own and reversing the earlier decision to reject hydrogen bonding as a significant feature of the structure of DNA he began a literature search. When he turned to the acid-base titration work of Gulland and Jordan, which we have described in Chapter 18, he found these authors suggesting that their results indicated the presence of hydrogen bonding between the bases of neighbouring molecules in a micelle. But Watson perceived the significant fact that such bonding seemed to be present even at

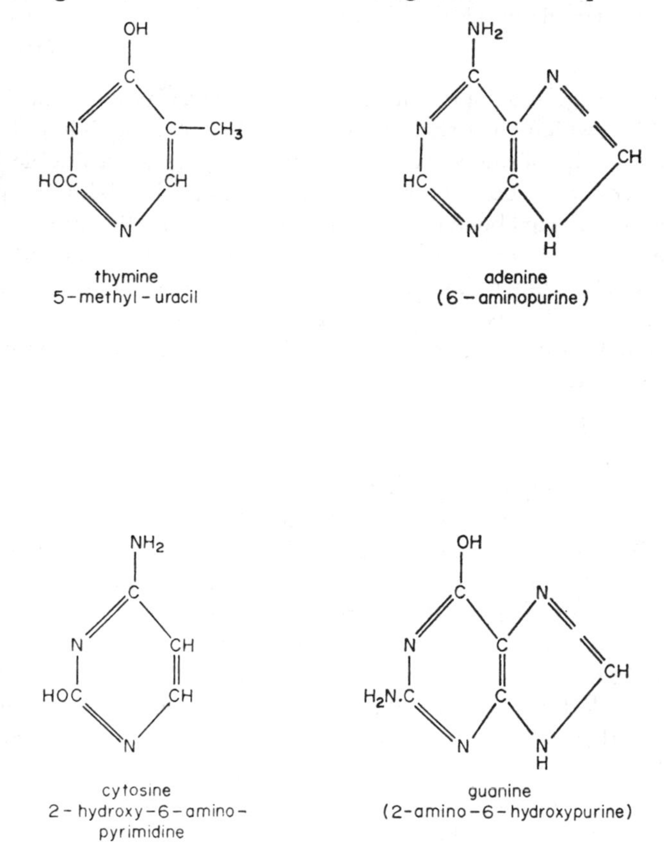

Fig. 21.3. The structure of the bases in DNA according to the standard textbook (from Davidson, 1953, 7–8).

very low concentrations of DNA, a hint that "the bonds linked together bases in the same molecule. There was in addition the X-ray crystallographic result that each pure base so far examined formed as many irregular hydrogen bonds as stereochemically possible. Thus, conceivably the crux of the matter was a rule governing bonding between bases" (*Helix*, 183).

It was then that he consulted June Broomhead's thesis. There to his amazement he found a regular pattern of hydrogen bonds between neighbouring adenine and guanine bases at just the right order of distance for packing inside his double helix (see Fig. 21.4). All Watson had to do was to read the accompanying text to find out which of the several dotted lines in Broomhead's diagrams represented in her opinion hydrogen bonds. He could then isolate two neighbouring, hydrogen-bonded adenine hydrochloride molecules from Broomhead's diagram, and there was a base pair! Similarly, her diagram of guanine hydrochloride gave him hydrogen-bonded pairs of this purine. It took little effort to construct similar pairs of pyrimidine molecules using the structural formulae in Davidson's book. The "potentially

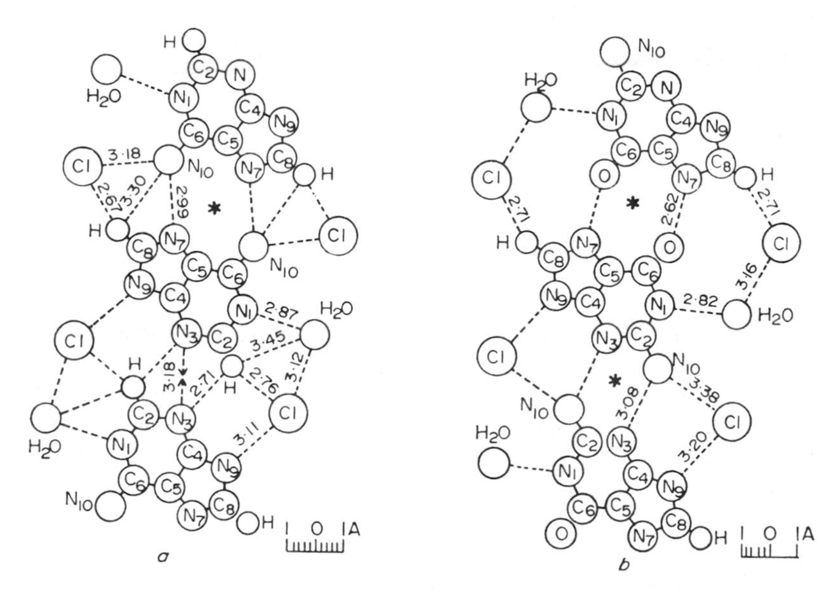

Fig. 21.4. (a) A section of the structure of adenine hydrochloride hemihydrate in the plane of one of the layers (b) a section of the structure of guanine hydrochloride monohydrate in the plane of one of the layers (from Broomhead, 1951, 98).

profound implications" of this idea of a like-with-like pairing of the bases inside the double helix struck Watson with all the force of a revelation (see Fig. 21.5). He did not attempt to fit the base pairs into his model of the backbone at this stage because he still lacked the precisely cut representations of them in metal. To be sure he worried about how on earth he would ever be able to keep the backbone regular and fit two pyrimidines above two purines. He also worried about the frequency of pairing mistakes. If these were high then base pairing could not be the basis of gene replication, for the facts of genetics told him that such mistakes must be infrequent. If only base pairings were the basis of replication, then he would be really happy. The

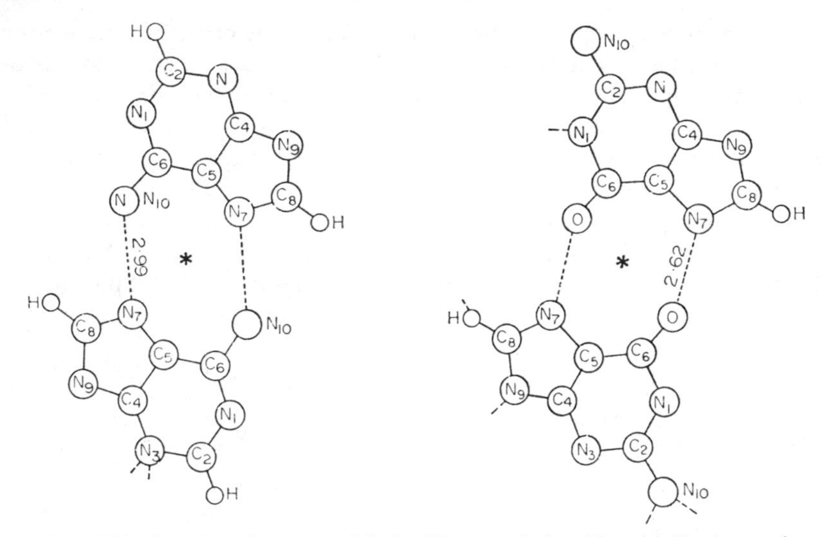

Fig. 21.5. This shows how it was possible for Watson to isolate like-with-like base pairs and to select suitable sites between which hydrogen bonding might exist. The protons involved in hydrogen bonding along the dotted lines are not shown.

structure would not be just a nice fit, with a pretty pattern of hydrogen bonds. It would be highly suggestive to the biologist. To his delight and amazement "the answer was turning out to be profoundly interesting" (*Helix*, 188). As he lay awake, eyes closed, adenine residues whirled in front of him. "Only for brief moments did the fear shoot through me that an idea this good could be wrong" (*Ibid.*).

News Reaches Caltech

The next day was Friday, February 20. Watson hurried over his breakfast, then returned to his College to write a covering letter to send off with the MS which he and Bill Hayes had recently written on the genetics of K12. This manuscript had been the chief purpose of Watson's recent visit to London when he also took the Pauling manuscript on a structure for DNA to King's. This covering letter shows not only that Watson was extremely pleased with his like-with-like base pairing scheme, but that he was every bit Delbrück's equal when it came to an argument. It goes:

> Dear Max,
>
> I am enclosing a MS about K12. To a very large extent it is identical to the formal genetics which I sent in late September. If possible I would like you to submit it to the *Proceedings of the National Academy*. In this way it would be published before the Cold Spring Harbor symposium. . . .
>
> I hope that you shall find it clear, despite my inability to write in a consistent style. If you believe that the MS would be improved by minor alterations please make them. It

would be nice if we could avoid a repetition of the almost comical revisions of revisions of revisions which the transfer manuscript of Ole and I went through.

I am now extremely busy, largely working on the DNA structure. I believe we are close to the solution. We have seen Pauling's paper on nucleic acid. Have you? It contains some very bad mistakes. In addition we suspect he has chosen the wrong type of model. All in all, however, Pauling's paper is at least in the proper mood and the type of approach which the people at King's College, London, should be taking instead of being pure crystallographers. I had started on DNA when I first arrived in Cambridge but had stopped because the King's group did not like competition or cooperation. However since Pauling is now working on it, I believe the field is open to anybody. I thus intend to work on it until the solution is out. To-day I am very optimistic since I believe I have a very pretty model, which is so pretty I am surprised that no one has thought of it before. When I have the proper coordinates worked out, I shall send a note to *Nature* since it accounts for the X-ray data, and even if wrong, is a marked improvement on the Pauling model. I shall send you a copy of the note.

(Watson, 1953a)

For a moment or two Watson considered giving further details of his like-with-like base pairing scheme, but as he was in a rush to get to the laboratory he ended the letter and popped it in the box.

Meanwhile, as we saw in the last chapter, Pauling was having difficulty with some of the van der Waals distances near the centre of his model. Peter Pauling learnt about this in two letters from his father (sent February 4 and 18). Added to this, Linus had to contend with the criticism of his model from Watson and Crick which Peter had passed on to him.

I am glad to have news from you about the nucleic acids. I have not understood why the Cambridge people should object to our structure as a structure for nucleate ion as well as nucleic acids. . . .

I heard a rumour that Jim Watson and Crick had formulated this structure already sometime back, but had not done anything about it. Probably the rumour is exaggerated. We are glad to learn about what is going on in London. We had heard last summer that Miss Franklin was leaving, and then had heard that she was still there. Your letter tells us that she will be gone in a couple of months.

(Pauling, L., 1953b)

Two weeks passed and then Pauling wrote yet another letter from Pasadena, this time to Watson, whom he wanted to be present at the protein conference which he was organizing for the coming September. He went on:

I think that we might as well discuss the structure of nucleic acids, as well as of proteins, at the conference, and I hope accordingly that you will arrive in time. Randall is going to be here from King's College.

Also, Delbrück said that you had found a beautiful new structure for the nucleic acids, but had not yet worked it out in detail. He could not tell whether it was significantly different from our structure, with three polynucleotide chains twisted around one another, and the phosphate groups near the axis of the molecule, but the wording, as a beautiful new structure (this is his memory of what was in the letter) suggests that it is not a variant of the old one. I trust that you will continue to work on the problem—Professor Corey and I do not feel that our structure has been proved to be right, although we incline to think that it is.

I may mention that we have made a small revision in it. I had not had very good success in trying to make small changes in the parameters, in order to increase some of the

> interatomic distances corresponding to van der Waals contacts. Professor Verner Schomaker then pointed out a significant improvement could be effected by rotating the phosphate tetrahedra about their vertical axes through 45° . . .
>
> (Pauling, L., 1953c)

Little did Pauling know when he wrote this letter that Watson's "very pretty model" was a set of base pairs in which all the tautomeric forms were wrong save that for adenine!

Four days after Pauling had written this letter to Watson he drove over to Riverside, as he told his son,

> . . . to look over a collection of 800 organic phosphates that they have there, in the Citrus Experiment Station. I looked them over, and selected four substances for X-ray investigation. Three of them are phosphate diesters. I want very much to find out what the structure of a substance is in which two of the hydrogen atoms of phosphoric acid have been replaced by hydrocarbon groups—presumably the phosphates in nucleic acid are of this type, and so far as I know no one has published a structure determination of any such substance. I think that these substances are more interesting in relation to nucleic acid than the nucleotides themselves, at any rate so far as the phosphate group goes. We have a man, Dr Rollet, due here next week from Leeds—he is one of Cox's men—whom we propose to get started on the precise structure determination of one or more of these crystals.
>
> (Pauling, L., 1953d)

Rejection of Watson's Pretty Model

As soon as Watson had posted his letter to Delbrück on the morning of Friday, February 20, he went off to the Cavendish. According to Watson his pretty scheme was promptly torn to shreds. This makes a good story but seems unlikely. The letter in which he confided to Delbrück that he had a very pretty model was the same letter in which he asked Delbrück to submit the K12 manuscript to the PNAS. It must have been a week later that Watson received Delbrück's reply saying that he would do as requested. When published Watson's and Hayes' paper bore the note "communicated by M. Delbrück, February 27". It must have been after receiving *this* letter that Donohue assured Watson that he was using the wrong structural formulae for guanine, thymine and cytosine. They did not exist in the enol form but in the keto (see Fig. 21.6). Had he not already sought to demonstrate this in the case of Broomhead's data (Donohue, 1952)? On Watson he urged the analogy between the partial double bond of the keto group in these bases and in diketopiperazine, and in the latter it was unquestionably in the keto form. "I was thus firmly urged not to waste more time with my hare-brained scheme" (*Helix*, 192). Since Donohue recalls having discussed the tautomeric forms of the bases with Watson at this time "more than once" (Donohue, 1969b) it would seem that Watson did not give up his like-with-like scheme as quickly as he later recalled. Donohue must have had to attack it on at least two occasions, the last being towards the end of February, probably Friday the 27th.

Watson had many reasons for not giving up his scheme. Although Pauling and Delbrück (1940) had favoured a complementary scheme for the replication of the gene, the majority of geneticists were not sold on it, their hero—J. H. Muller of Indiana University, where Watson was a graduate student—favoured a like-with-like scheme. He thought, too, that the enzyme involved

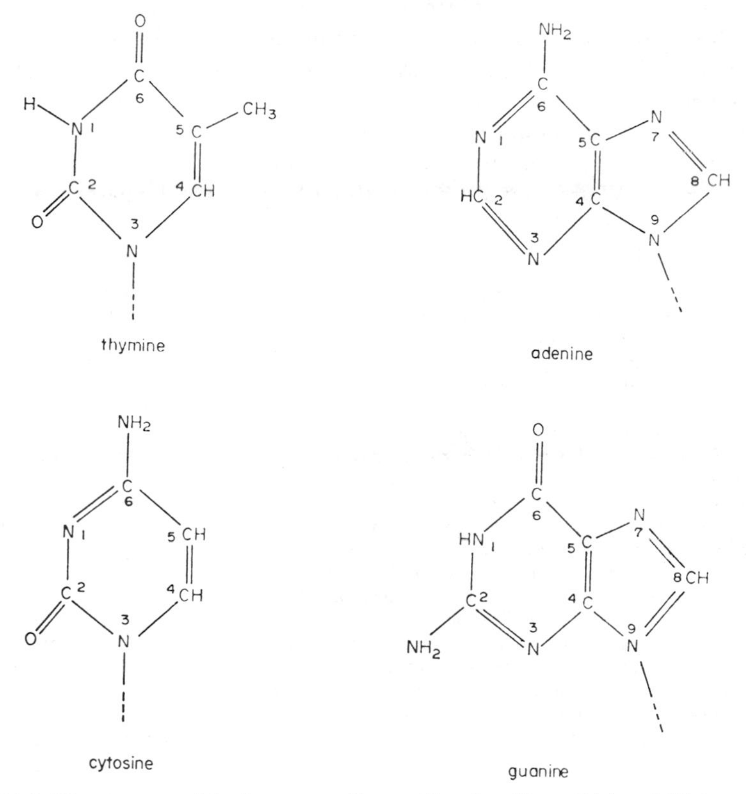

Fig. 21.6. The structure of the bases according to Donohue (from Crick and Watson, 1954, 87).

in replication would work best if it recognized two adenines, another recognizing two guanines etc. Crick pointed out that the scheme did not account for the C2 symmetry and Chargaff's base ratios. On the 27th of February then, Watson's pretty scheme was shattered. Crick recalled this dramatic occasion as follows:

> The crucial point was when we told Donohue about the tautomeric forms, because then you could put the hydrogen bonds together. At that stage, and I remember this very clearly, Jerry and Jim were by the blackboard and I was by my desk, and we suddenly thought, "Well perhaps we could explain 1:1 ratios by pairing the bases." It seemed too

> good to be true. So at that point all three of us were in possession of the idea we should put the bases together and do the hydrogen bonding and it was the next day that Jim came in and did it.
>
> (Crick, 1968/72)

On this crucial incident Watson's and Crick's recollections do not agree. I do not find this surprising for, as Watson remarked, Crick was always trying to organize the others in the office, so they had a built-in tendency to think of an alternative plan of action to the one he suggested. He talked so much that Watson did not necessarily take note of the important bits (Watson, personal communication). What Crick recalled as a precise programme—to pair the bases in accordance with Chargaff's 1:1 ratios—Watson recalled merely as a rejection of his like-with-like scheme.

The next day Watson got out his cardboard replicas of the bases again and tried once more to fit like pairs using the keto forms, but saw all too clearly "they led nowhere".

> When Jerry [Donohue] came in I looked up, saw that it was not Francis, and began shifting the bases in and out of various other pairing possibilities. Suddenly I became aware that an adenine–thymine pair held together by two hydrogen bonds was identical in shape to a guanine–cytosine pair held together by at least two hydrogen bonds (see Fig. 21.7). All the hydrogen bonds seemed to form naturally; no fudging was required to make the two types of base pairs identical in shape. Quickly I called Jerry over to ask him whether this time he had any objection to my new base pairs.
>
> When he said no, my morale skyrocketed, for I suspected that we now had the answer to the riddle of why the number of purine residues exactly equalled the number of pyrimidine residues. Two irregular sequences of bases could be regularly packed in the centre of a helix if a purine was always hydrogen-bonded to a pyrimidine. Furthermore, the hydrogen-bonding requirement meant that adenine would always pair with thymine, while guanine could pair only with cytosine. Chargaff's rules then suddenly stood out as a consequence of a double-helical structure for DNA. Even more exciting, this type of double helix suggested a replication scheme much more satisfactory than my briefly considered like-with-like pairing. Always pairing adenine with thymine and guanine with cytosine meant that the base sequences of the two intertwined chains were complementary to each other. Given the base sequence of one chain, that of its partner was automatically determined. Conceptually, it was thus very easy to visualize how a single chain could be the template for the synthesis of a chain with the complementary sequence.
>
> (*Helix*, 195–196)

When Crick came in, the characteristic scepticism of the collaborator was quickly dispelled. The similar shape of the two base pairs was compelling. But was there another way of pairing the bases so as to give the Chargaff ratios? Shifting the bases into other positions did not reveal one. Then he was struck by the symmetry of the glycosidic bonds which joined the bases to the respective backbones. They were related by a diad axis in the plane of the base pairs, which could be rotated through 180° to give the same orientation of these bonds. Knowing from that MRC report that crystalline DNA had C2 symmetry again proved crucial "because as soon as I saw Watson's base pairing I said: "Look, it's got the right symmetry" (Crick, 1968/72). This, together with the density, dominant features of the B diffraction pattern,

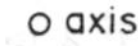

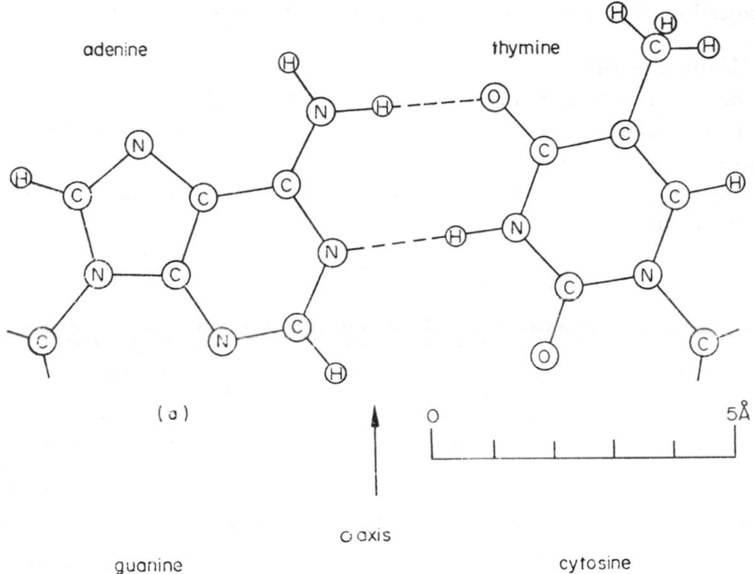

Fig. 21.7. * Complementary base pairing (a) adenine with thymine (b) guanine with cytosine. (from Crick and Watson, 1954, 88–89).

and Chargaff's ratios could all be accounted for by their model. In principle, the biological functions of encoding genetic information, replicating and expressing it, could also be envisaged. This was no false trail. This was

* Note the absence of the third hydrogen bond in the G-C pair. This was deliberate.

not a case of anti-climax in which a boring structure was emerging with no suggestions for biology. It was just so good it had got to be true!

Building the Model

It was now lunch time. Crick "winged into the Eagle to tell everyone within hearing distance that we have found the secret of life" but Watson felt "slightly queasy" (*Helix*, 197). Indeed he remained highly nervous about letting the news out, and two weeks passed before Wilkins and Delbrück were informed. First, Watson and Crick had had to wait for the metal bases. Then they had to build a very small piece of the molecule to see if it really was stereochemically satisfactory.

> It was not done on a base pair with the two residues on either side. It was done on one residue and the appropriate base, because I worked out a theorem, which now escapes me, which expressed the geometrical restraint of the other base [not built] on this one [built]. I knew that as long as I obeyed this geometrical rule, some projection of something on something else, one could do it. And so all the model building was done on a sugar, a phosphate, and a base and just the first atom of the next sugar which you had to put in the right position.
>
> (Crick, 1968/72)

In that first week of March the excitement mounted. They could talk and think of nothing else but the structure of DNA. At 19 Portugal Place it was still DNA. At mealtimes, before mealtimes and after mealtimes it was DNA. What a week! The model building, recalled Crick, "started about Wednesday and finished on a Saturday morning, by which time I was so tired, I just went straight home and to bed" (Crick, 1968/72).

Calculating the Co-ordinates

By this time Crick felt they were home and dry—the model could be built and the urgency was over. So they could work more leisurely during the second week of March. On Monday a letter arrived from Wilkins, in which he said:

> My dear Francis,
> Thank you for your letter on the polypeptides. I think you will be interested to know that our dark lady leaves us next week and much of the 3 dimensional data is already in our hands. I am now reasonably clear of other commitments and have started up a general offensive on Nature's secret strongholds on all fronts: models, theoretical chemistry and interpretation of data, crystalline and comparative. At last the decks are clear and we can put all hands to the pumps!
> It won't be long now.
> Regards to all
> Yours ever
> M.
>
> (Wilkins, 1953b)

Presumably Wilkins was told about their new model that week (March 9–14) and Watson wrote to Delbrück on the Thursday a long letter giving full details.

Dear Max,

Thank you very much for your recent letters. We were quite interested in your account of the Pauling Seminar. The day following the arrival of your letter, I received a note from Pauling, mentioning that their model had been revised, and indicating interest in our model. We shall thus have to write him in the near future as to what we are doing. Until now we preferred not to write him since we did not want to commit ourselves until we were completely sure that all of the van der Waals contacts were correct and that all aspects of our structure were stereochemically feasible. I believe now that we have made sure that our structure can be built and today we are laboriously calculating out exact atomic coordinates.

Our model (a joint project of Francis Crick and myself) bears no relationship to either the original or to the revised Pauling–Corey–Schomaker models. It is a strange model and embodies several unusual features. However, since DNA is an unusual substance we are not hesitant in being bold. The main features of the model are (1) The basic structure is helical—it consists of two intertwining helices—the core of the helix is occupied by the purine and pyrimidine bases—the phosphate groups are on the outside. (2) The helices are not identical but complementary so that if one helix contains a purine base the other helix contains a pyrimidine—this feature is a result of our attempt to make the residues equivalent and at the same time put the purine and pyrimidine bases in the centre. The pairing of the purine[s] with pyrimidines is very exact and dictated by their desire to form hydrogen bonds—adenine will pair with thymine while guanine will always pair with cytosine. For example

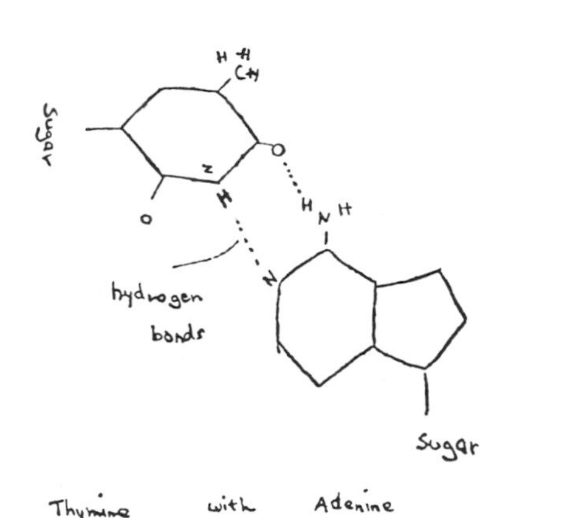

or

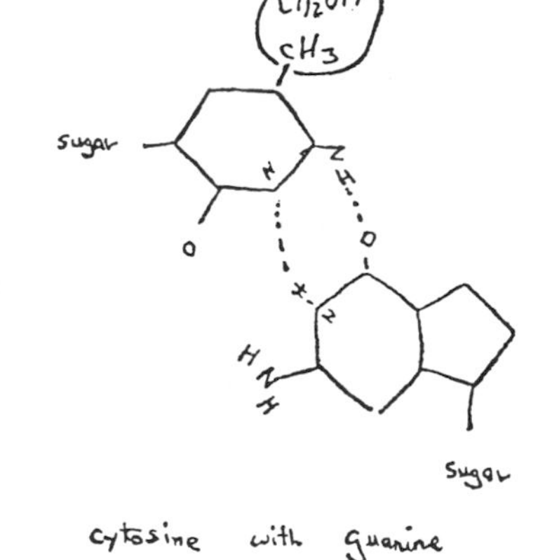

Fig. 21.8. The base-pairs as Watson drew them in his letter to Delbruck of March 12 (as published in *The Double Helix*).

While my diagram (see Figure 21.8) is crude, in fact these pairs form two very nice hydrogen bonds in which all of the angles are exactly right. This pairing is based on the effective existence of only one out of the two possible tautomeric forms—in all cases we prefer the keto form over the enol, and the amino over the imino. This is definitely an *assumption* but Jerry Donohue and Bill Cochran tell us that for all organic molecules so far examined, the keto and amino forms are present in preference to the enol and imino possibilities.

The model has been derived almost entirely from stereochemical considerations with the only X-ray consideration being the spacing between the pairs of bases 3.4 Å which was originally found by Astbury. It tends to build itself with approximately 10 residues per turn in 34 Å. The screw is right-handed.

The X-ray pattern approximately agreed with the model, but since the photographs available to us are poor and meagre (we have no photographs of our own and like Pauling must use Astbury's photographs), this agreement in no way constitutes a proof of our model. We are certainly a long way from proving its correctness. To do this we must obtain collaboration from the group at King's College, London who possess very excellent photographs of a crystalline phase in addition to rather good photographs of a paracrystalline phase. Our model has been made in reference to the paracrystalline form, and as yet we have no clear idea as to how these helices can pack together to form the crystalline phase.

In the next day or so Crick and I shall send a note to *Nature* proposing our structure as a possible model, at the same time emphasizing its provisional nature and the lack of proof in its favour. Even if wrong I believe it to be interesting since it provides a concrete example of a structure composed of complementary chains. If by chance it is right then I suspect we may be making a slight dent in the manner in which DNA can reproduce itself. For these reasons (in addition to many others) I prefer this type of model over Pauling's which if true would tell us next to nothing about the manner of DNA reproduction.

P.S. We would prefer your not mentioning this letter to Pauling. When our letter to *Nature* is completed we shall send him a copy. We should like to send him coordinates.

(Watson, 1953b)

In their effort to get the neatest possible conformation of the backbones Watson and Crick took co-ordinates on two slightly different chains. Using a plumb line the positions and heights of the atoms were measured and recorded one by one. They took trouble over the bond lengths which they calculated but the van der Waals distances they checked only roughly with a ruler. As they lacked a gauge they did not check the angles, for as Crick explained:

All we wanted to do was to build any model which had reasonably acceptable co-ordinates. And even then, you know, it is not all that good, because when you get the co-ordinates, you should check them. Nowadays it is done by putting them into a computer which will give you all the bond distances and angles of the chemical bonds, which shows you whether you have done your model building carefully. Well, we did not have a computer programme in those days. I did in fact check all the distances and alter the co-ordinates a little if there was one distance a bit too short or too long. But being a bit lazy and not having at my finger tips the formula for an angle between three points, I never checked the angles. So you will find that the distances are pretty good but some of the angles are really a bit off.

(Crick, 1968/72)

The result of their concentration on getting a neat backbone was that the bases were not far enough in to the centre of the molecule. The phosphate bridges (including the bond C'_4—C'_5) were all in the plane of the bases, and when the latter were viewed from above, the bonds of the phosphate bridge made a straight line (O_{IV} P O_{III} C'_5 C'_4) (see Fig. 21.9). Crick reckoned that if they had realized "how much the stacking energy was, we would probably have tended to build models with the bases nearer in."

I think in the real model the bases like to overlap. Consequently what I think happens is that the bases, as it were, pull in towards the middle, distorting the backbone slightly to do so, which I think is the reason why our model has the bigger radius.

(Crick, 1968/72)

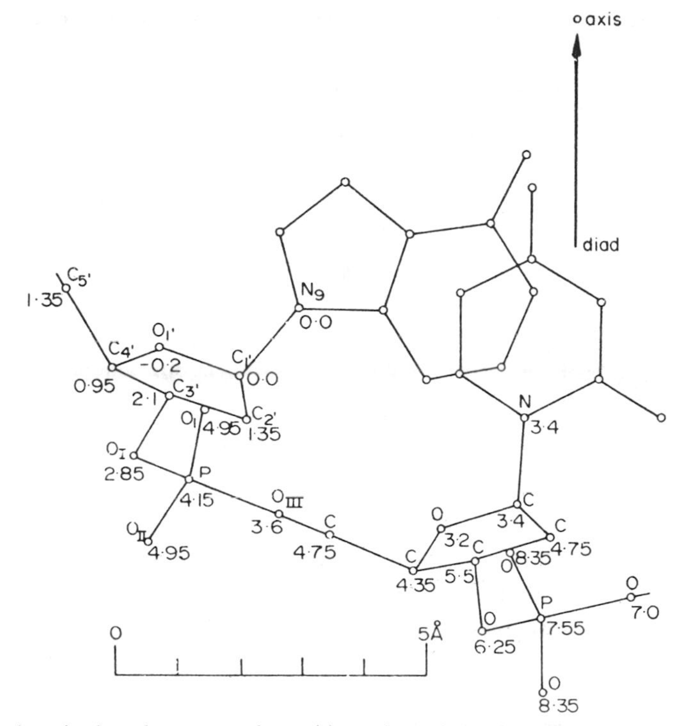

Fig. 21.9. A projection of two successive residues of one chain of the structure. The direction of projection is parallel to the fibre axis. The figures show the height of each atom above the level of the lower base (from Crick and Watson, 1954, 91).

Acceptance of the Model at Kings'

Wilkins must have received a copy of the paper which Watson and Crick had been typing very soon after they had completed checking the co-ordinates of the model for, on the 18th of March, he wrote:

> I think you're a couple of old rogues but you may well have something. I like the idea. Thanks for the MSS. I was a bit peeved because I was convinced that 1:1 purine pyrimidine ratio was significant and had a 4 planar group sketch and was going to look into it and as I was back again on helical schemes I might, given a little time, have got it. But there is *no good grousing*—I think its a very exciting notion and who the hell got it isn't what matters.
>
> But I hope you won't mind me being a trifle awkward on 2 points. First I think Frazer should have the opportunity of publishing his 3 chain model, we rather stopped him doing so a year ago because we thought we would do better and didn't. I have sent him a cable. I don't think Frazer's thing anything like as nice as yours but its a hell of [a] sight better than Pauling's.
>
> Second, since we got the 3 dimensional data we saw very soon that there was no conclusive evidence against helices and R.F's [Franklin's] dogmatism was—anyway. And I have recently removed several of the difficulties of interpreting the paracrystalline picture as a helix. We should like to publish a brief note with a picture showing the general

helical case alongside your model publication. Frazer's thing can be tucked away anywhere just for record purposes.

I can have the whole thing ready in a few days. I think the two publications would look nice side by side. We would not have time to do a detailed check on your model of course. Just quite general.

I think this is quite a reasonable suggestion and hope you won't mind the slight delay that may result in your own publication.

Just heard this moment of a new entrant in the helical rat-race. R. F. & G. [Gosling] have served up a rehash of our ideas of 12 months ago. It seems they should publish something too (they have it all written). So at least 3 short articles in *Nature.*

As one rat to another good racing.

M.

Suggested Modification to your MSS

Could you delete the sentence "It is known that there is much unpublished experimental material". (This reads a bit ironical.) Simply say "The structure must of course be regarded as unproved until it has been checked with fuller experimental material e.g. [8] [i.e., reference 8 in the Watson–Crick paper.]

Delete *very beautiful* and say "We have been stimulated by the work of the group at King's or something."

M.

(Wilkins, 1953c)

Franklin had by this time moved to Birkbeck College, where she was to start work on the structure of TMV. We saw that in February she had returned to the B pattern and asked the question: "Evidence for 2-chain (or 1-chain helix)?" By the end of the month she was studying the layer line intensities in detail and concluded:

> Structure B does not fit single helical theory, even for low layer-lines. ξ-values of first maxima are too small for single-strand helix, and even more so for multi-strand . . . Structure B *if* helix, is single-strand helix since 2-strand, fitting fifth layer-line maxima to $J_{10}(x)$, would require $\gamma = 17$ Å which is much too big.

She then examined the case for an outer helix of radius 8.5 Å and came to two important conclusions:

> If single-strand helix as above is basis of structure B, then structure A is probably similar, with P–P distances along fibre axis < 3.4 Å, probably 2 to 2.5 Å (c.f. 2 Å indicated by position of P–P peaks in Patterson, and 2.5 Å indicated by position of 11th layer line reflection.)

She now realized that her earlier back-to-back chain structure for the A form was impossible "Single-strand helix for structure B rules out atom-pair theory for structure A (with 7 pairs in fibre period) . . ." Second, she appreciated that if the member chains of a helix were *not* placed equivalently, then the B pattern *was consistent with more than one strand.* Taking the radius of the helix in structure B as 8.5 Å and the volume of a nucleotide as 336 Å^3, she calculated the number of nucleotides in the repeat as 23. This suggested the presence of two rather than three strands.

When Franklin moved to Birkbeck she wrote a paper with Gosling summarizing this analysis and concluded:

> The [B] structure does not contain more than one equivalent coaxial chain, but the possibility of non-equivalent co-axial chains is not eliminated.
>
> The total absence of an inner maximum on the fourth layer line suggests that if there are two non-equivalent co-axial chains, these are separated by 3/8 of the fibre-axis period, that is by 1.3 Å in the fibre-axis direction.
>
> (Franklin and Gosling, 1953a)

One might say that Franklin and Gosling were cautiously moving in the right direction, so that by the time they were shown the Watson–Crick typescript a few days later it required little effort on their part to adapt what they had set down on the 17th of March in support of the Watson–Crick model (Franklin and Gosling, 1953c).* Their analysis of the A form followed later that year (Franklin and Gosling, 1953d).

Delbrück's Reception of the Model

While Watson and Crick had been feverishly working on their model to see if it was stereochemically feasible there had also been activity in Pasadena. Delbrück had attended a seminar which Pauling gave on Thursday, March 5, to describe his model, and to explain the modification introduced by Schomaker (see p. 383). Sinsheimer was also present. After the seminar was over Delbrück and he went to see Pauling. Sinsheimer pointed out that his sequence analysis of DNA (Sinsheimer, 1952) ruled out the possibility of attributing the Chargaff ratios to the nucleotide sequence. Some *structural* feature must be involved, but Pauling's model did not offer any suggestions. They also expressed the view that the phosphates were charged and would therefore repel one another. But Pauling naturally defended his model and emphasized the crystallographic evidence in its support (Sinsheimer, personal communication).

Two weeks later Delbrück invited Pauling over to his office to see Watson's letter of March 12. When Pauling had read to the end he said to Delbrück with a twinkle in his eye: "You didn't read what was at the end of the letter did you?" As Delbrück explained to Watson:

> I decided to ignore your postscript saying you preferred I shouldn't show your letter to Pauling, because I had promised to let him know when I heard more from you, and because I wanted to tell others here in the laboratory about it, so he would have heard anyhow, and because I know he was dying to hear what you had cooked up, and because you said you would tell him anyhow in a few days, and last not least, because, as you know, I hate secrecy in scientific matters.
>
> He was quite thrilled by your model and so was Sinsheimer, and so am I. Pauling said he is in hopes of getting a copy of the King's people's data very soon, and they should certainly settle the major issues.
>
> (Delbrück, 1953b)

In the third week of March Watson had rushed off to Paris to do an experiment on the inactivation of transforming principle with Harriet Ephrussi-Taylor. On his return he wrote a letter to Pauling which Crick and he signed. They enclosed a copy of their letter to *Nature* entitled: "A Structure

*But see further discussion of this point in the Postscript, p. 445.

for Desoxyribose Nucleic Acid", and told him that the King's College workers were also going to publish some of their experimental data. By this time they knew from Peter Pauling that his father was to visit Cambridge that Easter, on his way to the Solvay Conference in Belgium. "We are looking forward very much" they wrote, "to your visit and for the opportunity of a full discussion about DNA. Would you mind treating this [paper] as confidential for a few days as Professor Bragg has still not been able to hear about it" (Watson and Crick, 1953a). Bragg was still down with flu.

Watson also sent a copy of their *Nature* letter to Delbrück. From his covering letter we learn that Watson and Crick had come to an agreement with Wilkins and the other members of the King's group according to which "all comparisons of the experimental data" with the model were to be done by the latter. Watson then added:

> Concerning our assumption as to the equivalence of (adenine and thymine) and (guanine and cytosine) several days ago, I ran into Wyatt at the Institute Pasteur. He tells me that the more he refines the analysis of the bases, the clearer he finds the 1 : 1 equivalence. This 1 to 1 ratio also holds for 5 methyl-hydroxy, cytosine which after more careful analysis seems to be equal to guanine.
>
> (Watson, 1953c)

Publication of the Model

At the end of March Wilkins was able to write to Crick:

> You will be relieved (I am) to hear that all is safe in the hands of Gale [the editor of *Nature*] who may be able to get it in April 25. There was a frantic chaos at the last moment —no typists for retyping (Pauline helped us out) and two missing figures which after a long search turned up in Randall's bag and in Rosie's room. If Pauling is interested please show him my draft MSS.
>
> (Wilkins, 1953d)

The papers did appear in the April 25 issue of *Nature*—first Watson's and Crick's, then the paper by Wilkins, Stokes and Wilson, and last Franklin's and Gosling's.

At the same time Watson and Crick were busy writing a speculative paper on "The Genetical Implications of the Structure . . ." In their first paper to *Nature* they had merely remarked: "It has not escaped our notice that the specific pairing we have postulated immediately suggests a possible copying mechanism for the genetic material" (1953b, 737). In their second paper they suggested that this process involved separation of the co-axial strands, each of which served as a template for laying down a complementary polynucleotide strand, thus generating two double helices where there had been one. The problems which such a scheme suggested were discussed, and they concluded by emphasizing how speculative was their scheme.

> Even if it is correct, it is clear from what we have said that much remains to be discovered before the picture of genetic duplication can be described in detail. What are the polynucleotide precursors? What makes the pair of chains unwind and separate? What

is the precise role of the protein? Is the chromosome one long pair of deoxyribonucleic acid chains, or does it consist of patches of the acid joined together by protein?

Despite these uncertainties we feel that our proposed structure for deoxyribonucleic acid may help to solve one of the fundamental biological problems—the molecular basis of the template needed for genetic replication. The hypothesis we are suggesting is that the template is the pattern of bases formed by one chain of the deoxy-ribonucleic acid and that the gene contains a complementary pair of such templates.

(Watson and Crick, 1953c, 966–967)

The excitement and the enthusiasm of those around Watson was almost too much for him. He began to wonder whether their structure was right after all. Earlier he had confided in Delbrück his inclination to turn his back on the all too-challenging demands of the DNA discovery. "I have a rather strange feeling about our DNA structure. If it is correct, we should obviously follow it up at a rapid rate. On the other hand it will at the same time be difficult to avoid the desire to forget completely about nucleic acid and to concentrate on other aspects of life. This mood dominated me in Paris . . ." (Watson, 1953c). A month later he described events in Cambridge and revealed his disquiet:

Crick was very much in favour of sending in the second *Nature* note. To preserve peace I have agreed to it and so it shall come out shortly since Gale (the editor of *Nature*) is very close to Bragg. It is all rather embarrassing to me since the Professor (Bragg) is frightfully keen about it and insists upon talking about it everywhere. Until we produced the model Bragg did not know what either DNA or genes were and his reaction to our original *Nature* note was "It's all Greek to me." After we had convinced him that DNA might be interesting, he then got out of control and I spent much of my time de-emphasizing it since I have not infrequent spells of seriously worrying about whether it is correct or whether it will turn out to be Watson's folly.

Bragg, however, remains cheerful as ever, and has even told the story to the press and so last Friday's *News Chronicle* carried a story on how the secret of life was discovered in Cambridge. This immediately led a reporter of *Time* to Bragg and I am dreadfully afraid that I shall see the story in glossy print when I am in the States.

(Watson, 1953d)

Later Watson was to ask Crick not to allow a programme he had given on the radio to be repeated on the third programme. ". . . I still think", wrote Watson from Pasadena, "a talk on the Third would be in bad taste. There are still those who think we pirated data and I'm of the belief that a few enemies are worse than a few admirers . . . My main concern is not to be dragged into it as I'm afraid I was in Cambridge" (Watson, 1953g). Crick emphasized that this broadcast was not a gossipy account. "It was a perfectly straightforward account of the structure" (Crick, 1968/72).

Cold Spring Harbor

Viruses formed the subject of the 1953 symposium at Cold Spring Harbor, so many members of the Phage Group were to be present. Such an opportunity for launching the Watson–Crick model was not to be missed. When the meeting opened on June 5th Delbrück was able to announce "a last minute addition to the program . . ."

> The discovery of a structure for DNA proposed by Watson and Crick a few months ago, and the obvious suggestions arising from this structure concerning replication seemed of such relevance to many of the questions to be discussed at this meeting that we thought it worthwhile to circulate copies of three letters to *Nature* concerning this structure among the participants before the meeting and to ask Dr Watson to be present . . .
>
> (Delbrück, 1953c, 2)

At last phage workers had, declared Hershey, "for the first time what Delbrück used to call a party line. For the first time, too, biologists were in possession of a chemical structure that said something about function . . ." (Hershey, 1966, 100). Cohen has claimed that: "Insight concerning the probable role of phage DNA became broadly apparent to phage workers only after the discovery of the structure of DNA by Watson and Crick; this was communicated to virologists generally in the Cold Spring Harbor Symposium in the summer of 1953" (Cohen, 1968, 25).

Then, as now, there were problems about the mechanism of DNA replication which Delbrück and Pauling appreciated, but these did not inhibit their enthusiasm. At the Solvay Conference on proteins, held in Brussels from April 6 to 14, 1953, before the publication of the first paper by Watson and Crick describing their structure, Pauling said:

> Although it is only two months since Professor Corey and I published our proposed structure for nucleic acid, I think we must admit that it is probably wrong.
>
> I have just learned, a few days ago, that Dr Wilkins, of King's College, London, and his collaborators have shown that there are two different forms of sodium thymonucleate, giving different X-ray patterns, and that our X-ray photographs, and the older ones of Astbury and Bell, show a superposition of the two patterns. We were misled in this way: we attempted to formulate a structure accounting for a feature of one pattern (the 3.4 Å meridional reflection) and a feature of the other pattern (the 27 Å repeat).
>
> We also concluded that the nitrogen bases could not be near the axis of the helical molecule. This conclusion, which Watson and Crick have shown to be unjustified, resulted from our assumption that each nucleotide residue should provide a suitable place for any nitrogen base, either purine or pyrimidine. Watson and Crick have as an extraordinary feature of their structure the occurrence of purine–pyrimidine pairs (guanine–cytosine, adenine–thymine), which lie near the axis of the helix.
>
> Although some refinement might still be made, I feel that it is very likely that the Watson–Crick structure is essentially correct. In its feature of complementariness of the two chains it suggests a mechanism for duplication of a chain by a two-step process—a molecular mechanism that may well be the mechanism of hereditary transmission of characters.
>
> I think that the formulation of their structure by Watson and Crick may turn out to be the greatest development in the field of molecular genetics in recent years.
>
> (Pauling, L., 1953e, 113)

On the 1st May, Watson addressed the Hardy Club at Peterhouse, Cambridge; characteristically he gave his talk a low-key title: "Some remarks on desoxyribonucleic acid". Back in California Delbrück wrote excitedly to Watson:

I have a feeling that if your structure is true, and if its suggestions concerning the nature of replication have any validity at all, then all hell will break loose, and theoretical biology will enter a most tumultuous phase. Only part of this will involve chemistry, analytical and structural. The more important part will consist in attempts to take a fresh view of the many problems of genetics and cytology which came to dead ends during the last forty years.

(Delbrück, 1953d)

CHAPTER TWENTY-TWO

Conclusion

> The ultimate aim of the modern movement in biology is in fact to explain *all* biology in terms of physics and chemistry. There is a very good reason for this. Since the revolution in physics in the mid-twenties, we have had a sound theoretical basis for chemistry and the relevant parts of physics . . . So far everything we have found can be explained without effort in terms of the standard bonds of chemistry—the homopolar chemical bond, the van der Waal attraction between non-bonded atoms, the all-important hydrogen bonds, and so on . . . Thus eventually one may hope to have the whole of biology "explained" in terms of the level below it, and so on right down to the atomic level. And it is the realization that our knowledge on the atomic level is secure which has led to the great influx of physicists and chemists into biology.
>
> (Crick, 1966, 10, 11, 14)

> The universal topography of atomic particles (with their velocities and forces) which, according to Laplace, offers us a universal knowledge of all things is seen to contain hardly any knowledge that is of interest. The claims made, following the discovery of [the structure of] DNA, to the effect that all study of life could be reduced eventually to molecular biology, have shown once more that the Laplacian idea of universal knowledge is still the theoretical ideal of the natural sciences; current opposition to these declarations has often seemed to confirm this ideal, by defending the study of the whole organism as being only a temporary approach. But now the analysis of the hierarchy of living things shows that to reduce this hierarchy to ultimate particulars is to wipe out our very sight of it. Such analysis proves this ideal to be false and destructive.
>
> (Polanyi, 1968, 1312)

Crick's brave statements about explaining biology in terms of chemistry and physics have served to condemn him in the eyes of those for whom cells and organisms are more than just atoms and bonds. In justice to him I should point out that he had never denied the value of studying organisms at higher levels than the molecular, nor has he supported the view that to reduce a biological entity to physics is to do away with it. Did Watson and Crick really believe in 1953 that the duplication of the gene involved nothing more than the specific pairing of bases by hydrogen bond formation? Admittedly Crick had written to his son, Michael, "we think we have found the basic mechanism by which life comes from life" (Crick, 1953a). But the passage given on p. 420 shows that they left open the possibility that the mechanism of gene duplication might involve more than just base pairing. Was their motivation in attacking the structure of DNA that of reducing the gene to chemistry and physics? Or was "reductionism" "peripheral" (to use a term of Kenneth Schaffner's) to their programme? It has been the aim of this book to show that their programme was characteristic of that of experimental biology in the period 1900 to 1953, which was the introduction into biology of the experimental techniques and fundamental theories of physics and chemistry. What marks out their programme and that of Pauling's from those of, say the experimental cytologists, is the sort of fundamental physical theory that they introduced. It seems doubtful that their aim was to do away with all the biological entities, but only with those which proved to be experimentally untestable, quantitatively inexpressible and unrelated to chemistry and physics. As Crick has pointed out in his John Danz Lectures, the reason for wanting to "explain" biology in terms of physics and chemistry was that these sciences offered a sound foundation of knowledge "to guide the discovery of new knowledge. If the foundations of any subject are weak, then they cannot help us either to spot experimental error or to suggest fruitful theories" (Crick, 1966, 11).

Men like Pauling, Donohue and Crick saw in quantum mechanics the required foundation for explaining chemistry, and thence for chemistry to explain biology. With their entry into biology there resulted a *new kind of professionalism* marked by the demand that explanations must stand up to the rigorous standards of the new quantum physics. Resonance theory had to be applied to structural problems in the chemistry of biologically important molecules; no specific inter-molecular forces were to be allowed other than those admitted by physicists; special colloidal chemical laws were not to be invoked to account for the behaviour of macromolecules, and so on. Thus in Chapter 15 we saw that Delbrück accounted for the paradoxical constancy and mutability of the gene in terms of the persistence of molecular forces which hold atoms in fixed associations, and their occasional quantum leap to a different association. In Chapter 1 Staudinger was seen to reinterpret the so-called colloidal properties of polymeric compounds in terms of orthodox (covalently bonded) macromolecules. In Chapter 9 Beadle and Tatum described the expression of the gene in terms of the synthesis of a specific enzyme catalysing a specific reaction. In addition to showing that many properties of living things could be described in physical and chemical terms these researches involved the application of precise quantitative techniques, from which has emerged the soundly-based knowledge referred to by Crick. In this sense "explain" means more than "is conformable with" or "can be explained by", for in the process of applying quantitative experimental methods more is learned about the biological entities involved. The programme which has been described in this book does not appear to me to be so aggressively and dehumanizingly reductionist as its opponents claim. Clearly it does not add up to the full scale reductionism which Schaffner—perhaps inevitably—finds peripheral to molecular biology (Schaffner, 1969) and which Polanyi regards as false and destructive.

The Explanatory Power of the Watson–Crick Model

In 1947 Crick wrote that: "The eventual goal, which is somewhat remote, is the description of these activities [of proteins in living organisms] in terms of their structure, i.e. the spatial distribution of their constituent atoms, in so far as this may prove possible. This might be called the chemical physics of biology" (Crick, 1947). What did the Double Helix contribute to this goal? It showed that a structure as regular as DNA was known to be could accommodate within it a non-repetitive sequence of the four bases. All conceivable permutations of sequences were possible on a single polynucleotide strand, so DNA could contain a genetic code adequate to account for the great variability of blueprints in nature, providing that the DNA molecule was big enough. We know that the 1953 estimate for thymus DNA of 7×10^6 is well below the value for native DNA in the chromosome, for we now have huge values for the length of a DNA molecule in a chromosome (Kavenoft and Zimm, 1973).

The Watson–Crick model offered a partial solution to the problem of how the gene duplicates itself: "If the actual order of the bases on one of the pair of chains were given, one could write down the exact order of the bases on the other one, because of the specific pairing. Thus one chain, is, as it were, the complement of the other, and it is this feature which suggests how the desoxyribonucleic acid molecule might duplicate itself" (Watson and Crick, 1953c, 966). The discoverers enlarged on this process: "From time to time the base of a free nucleotide will join up by hydrogen bonds to one of the bases of the chain already formed. We now postulate that the polymerization of these monomers to form a new chain is only possible if the resulting chain can form the proposed structure" (*Ibid.*). They recognized this as a partial solution because they could say nothing about how the chains were separated for duplication, or whether a polymerizing enzyme was involved in the formation of the new polynucleotide chains. Later they were to admit that enzymes are necessary and further that such enzymes are involved in the "recognition" of the correct base-pairs. Thus Eigen and de Maeyer, who had measured the error probability in base pairing at below 10^{-8} to 10^{-10}, regarded this as too good to be "ensured by the [H] bond stability, since there is not much difference between the H-bonds in 'true' and 'false' complexes. It is more probable that the exact complementarity is expressed in the specific geometry of the pair, and that only this is recognized by the synthesizing enzyme" (Eigen and de Maeyer, 1966, 5).

Although Watson and Crick had accepted Donohue's view that the bases in DNA were only likely to occur in the amino and keto tautomeric forms, they naturally felt that the other forms did rarely occur, in which case the pairing in replication would be different. They appreciated that this offered a mechanism for mutation by a change in the base sequence. "For example", they said, "while adenine will normally pair with thymine, if there is a tautomeric shift of one of its hydrogen atoms it can pair with cytosine (see Fig. 22.1). The next time pairing occurs, the adenine (having resumed its more usual tautomeric form) will pair with thymine, but the cytosine will pair with guanine, and so a change in the sequence of bases will have

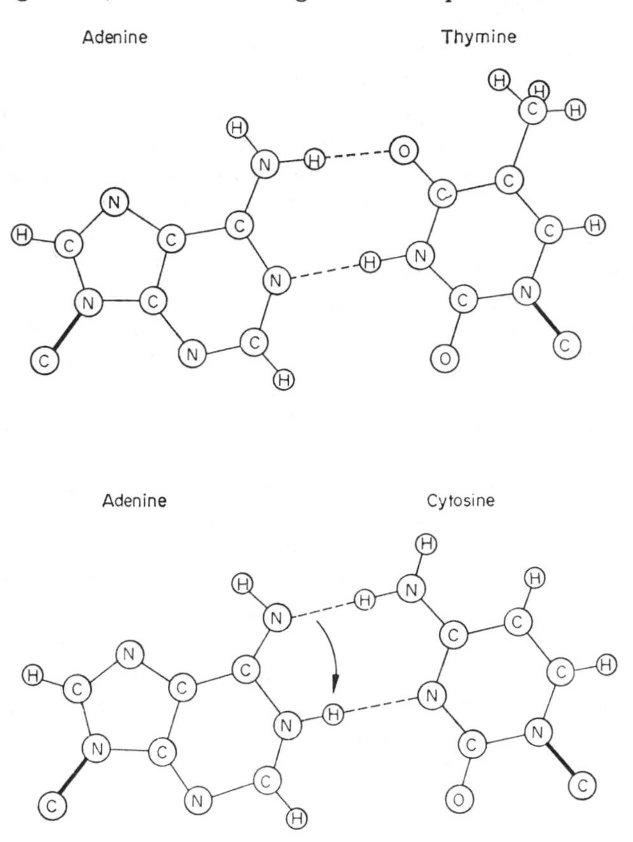

Fig. 22.1. Pairing arrangements of adenine before (above) and after (below) it has undergone a tautomeric shift (from Watson and Crick, 1953d, 130).

occurred" (Watson and Crick, 1953d, 130). There seems as yet no evidence for or against such a mechanism for spontaneous mutation (Drake, 1969); despite an intensive search for imino and enol forms nothing really positive has emerged. "The net effect of investigations by ultraviolet adsorption spectroscopy, infra-red in D_2O, nuclear magnetic resonance, and Raman spectroscopy has been negative" (Daniels, 1972). This quotation comes from a paper which Watson communicated last May for Malcolm Daniels. The latter claimed that rare tautomers could not be demonstrated by the absorption spectroscopy but that excitation spectroscopy did in fact yield evidence of the enol form in pyrimidines.

If the miraculous autocatalytic property of the gene was based upon the complementary template synthesis of DNA molecules, it followed that all information must flow from DNA. This was not stated by Watson and Crick in 1953. They wrote guardedly of the several lines of evidence indicating DNA as the "carrier of a part (if not all) of the genetic specificity of the chromosomes and thus of the gene itself" (Watson and Crick, 1953c, 964). For phage and for a Cold Spring Harbor audience they spoke more confidently: "we shall not only assume that DNA is important, but in addition that it is the carrier of the genetic specificity of the virus (for argument, see Hershey, this volume) . . ." (Watson and Crick, 1953d, 123). It is amusing to note that Hershey found the evidence in 1953 inadequate for making a scientific judgement on the genetic role of DNA. His own guess was that DNA would "not prove to be a unique determiner of genetic specificity . . ." (Hershey, 1953, 138). Although the genetic codescript could be seen very simply in the Double Helix, this model offered no obvious clues as to how the chromosomal proteins might get at the codescript and alter it. Whatever function DNA-associated protein might have, they thought, was likely to be non-specific, such as controlling the coiling and uncoiling of DNA.

Nor did their model seem to offer any obvious clues as to the mechanism for gene expression, for Watson and Crick assumed that the Double Helix remained intact during "transcription"; it did not therefore offer unpaired bases as the template for either RNA or protein synthesis. At Cold Spring Harbor Watson admitted that Crick and he could not see how DNA exerted "a highly specific influence on the cell" (Watson and Crick, 1953d, 127). Since this specificity had to come from the base sequence all they could do was to counter the obvious criticism that the bases were in the wrong place: "It should not be thought that because in our structure the bases are on the 'inside', they would be unable to come into contact with other molecules. Owing to the open nature of our structure they are in fact fairly accessible" (*Ibid.*). To be sure Watson had drawn up a scheme to express the interrelationship between DNA, RNA and proteins *before* the discovery of the Double Helix (see Fig. 22.2), in which the well-known evidence for the localization of protein synthesis in RNA-rich cytoplasmic particles was incorporated. Since, after Hershey-Chase, there was no doubt in Watson's mind that DNA had to be the hereditary material, he arrived at the sequence: DNA makes RNA makes protein. Back in Pasadena in the winter of 53/54 he pondered the role of RNA further. He claimed to have found in the published data on RNA base composition, evidence for the existence of two classes of molecules—those with non-complementary base ratios, and those with complementary ratios. This led him to picture the expression of

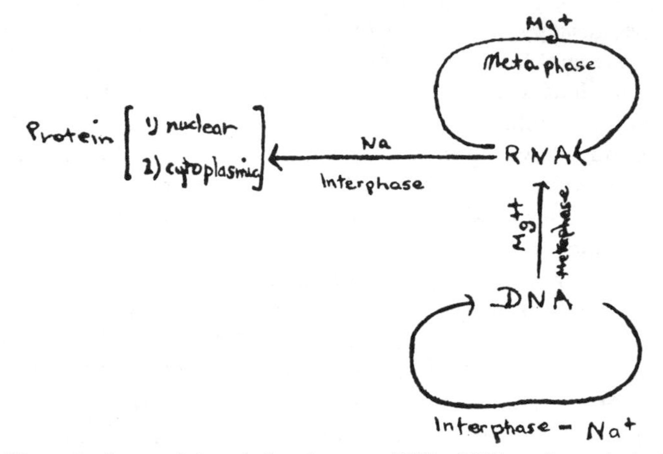

Fig. 22.2. Watson's picture of the relations between DNA, RNA and protein in 1952 (from *Helix*, p. 154).

the codescript in DNA in terms of a migration of one of the two strands of the DNA into the cytoplasm, the conversion of its sugar to ribose, which then replicated to form double-stranded RNA, and this probably acted as the template for protein synthesis (see Fig. 22.3). Fortunately Crick did not encourage him to go ahead with his proposed letter to *Nature* on this theme. Instead it was the cosmologist, George Gamow, who began to air his views on the nature of the code and on the relationship between DNA and the amino acids privately to Watson and Crick in the summer of 1953, and publicly in his note in *Nature* for February 13, 1954. He assumed that protein synthesis occurred on the DNA Double Helix, and he suggested that there existed a "lock-and-key" relationship between "various amino acids,

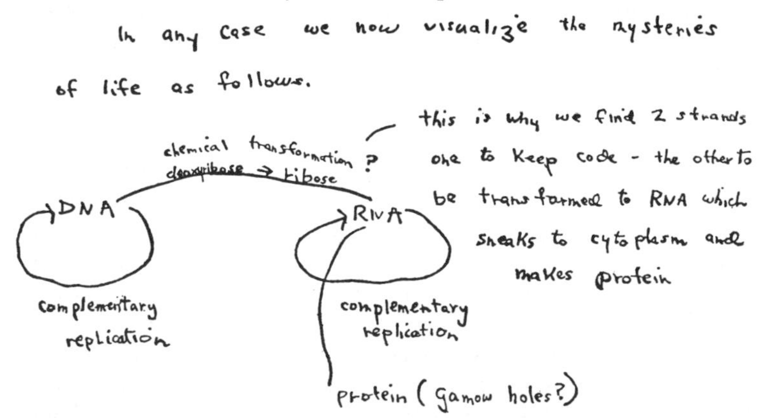

Fig. 22.3. Watson's 1954 scheme for DNA–RNA–Protein interrelationships (from Watson, 1954).

and the differently shaped 'holes' formed by various nucleotides in the desoxyribonucleic acid chain" (Gamow, 1954a). This is shown in Fig. 22.4. Like Watson and Crick, Gamow was assuming that the bases remained paired during transcription. As late as 1955 Crick still made this assumption. He also kept an open mind as to the possibility that DNA might act *directly* as the template for protein synthesis. In his criticism of Gamow's ideas he wrote: "I have tacitly dealt with DNA throughout, but the arguments would carry over to some types of RNA structure. If it turns out that DNA, in the double helix form, does not act directly as a template for protein synthesis, but that RNA does, many more families of codes are of course possible . . . In particular base pairing may be absent in RNA or take a radically different form . . . Without a structure for RNA one can only guess" (Crick, 1955, 16–17).

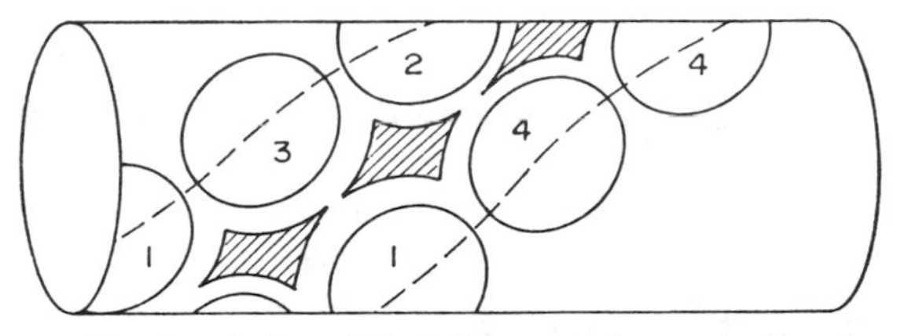

Fig. 22.4. Diagram of the rhomb-shaped "holes" between base-pairs (3 to 4 and 1 to 2) of the DNA structure (adapted from Gamow, 1954a).

Naturally attention had already shifted to RNA as the next highly significant unsolved structure. In Pasadena, Watson, Delbrück and Rich were studying the "holes" in a hypothetical model of RNA. Gamow, who visited them in 1954, wrote to Crick: "They have a model of RNA, big and nice looking, but they do not believe in it very much themselves (except Alex Rich who conceived it). It has trapezoid holes formed by two bases and two different 'sugar edges'. And there are twenty different holes" (Gamow, 1954b). In November of the following year Gamow, Delbrück and Jack Dunitz constructed a hoax letter on the structure of RNA which Delbrück sent to Crick. It contained the following news:

> Here [Pasadena] of course everybody is buzzing about Rundle's RNA structure (JACS last issue). It seems the most fantastic thing that a complete outsider from the Middle West should have hit the jackpot, and that he should have kept it to himself during all the time the paper has been in press. At first everybody was sceptical and critical of various interatomic distances but yesterday I discussed the matter with Corey who had already gone over it in great detail and seemed completely satisfied. Anyhow, since the biochemical meaning of the structure is almost more obvious than the Watson–Crick structure it would be hard to convince anybody that the structure is wrong.
>
> (Delbrück, 1955)

The latest issue of the *Journal of the American Chemical Society* was, by definition not yet available in Cambridge. Several feverish hours were therefore spent trying to build an RNA model in the Cavendish before Rich put a trans-Atlantic telephone call through to his friend Dunitz. This story serves to show how much was expected of RNA in terms of clues to protein synthesis and the nature of the code. In the event no such structure had been built, let alone by Rundle, and the clues, when they came, were not of a structural kind. As Crick already recognized, no one "looking at DNA or RNA would think of them as templates for amino acids were it not for other indirect evidence. What the DNA structure *does* show (and probably RNA will do the same) is a specific pattern of *hydrogen bonds*, and very little else. It seems to me, therefore, that we should widen our thinking to embrace this obvious fact" (Crick, 1955, 7). He went on to postulate small "adaptor" molecules which could form hydrogen-bonds with the bases in RNA and attach themselves to specific amino acids.

The Central Dogma

Between 1953 and 1957 implicit assumptions were made about the chemical codescript and its expression, which Crick made explicit in a famous lecture to the Society for Experimental Biology published in 1958. These he called "The Sequence Hypothesis" and "The Central Dogma". The former merely stated that "the specificity of a piece of nucleic acid is expressed solely by the sequence of its bases, and that this sequence is a (simple) code for the amino acid sequence of a particular protein" (Crick, 1958, 152). In order that the DNA of the gene should control the properties of the gene product he further assumed that the way a protein folded up to give its characteristic three-dimensional shape was "*simply a function of the* order of the amino acids, provided it takes place as the newly formed chain comes off the template" (Crick, 1958, 144). Thus conceived, the flow of "information" was from DNA to protein. But Crick did not stop here. He went further and in the Central Dogma asserted, albeit in a deliberately speculative context, "that once 'information' has passed into protein it *cannot get out again.* In more detail, the transfer of information from nucleic acid to nucleic acid, or from nucleic acid to protein may be possible, but transfer from protein to protein, or from protein to nucleic acid is impossible" (*Ibid.*, 153). He was later to point out that "if all possible transfers [of information] occurred it would have been almost impossible to construct useful theories. Nevertheless, such theories were part of our everyday discussions. This was because it was being tacitly assumed that certain transfers could not occur. It occurred to me that it would be wise to state these preconceptions explicitly" (Crick, 1970, 561). These were embodied in a diagram drawn but not published in 1958 (see Fig. 22.5).

By 1958 it had become evident that protein synthesis was a very complicated process. Forward translation from DNA to protein "involved very complex machinery . . . it seemed unlikely on general grounds that this machinery could easily work backwards. The only reasonable alternative was that the cell had evolved an entirely separate set of complicated machinery for backward translation, and of this there was no trace, and no reason to believe that it might be needed" (Crick, 1970, 562). In 1962, 1964 and 1968, Barry Commoner attacked the Central Dogma and the notion of DNA as "the master molecule" which was able to duplicate itself and function as the sole determiner of the amino acid sequence of proteins. The operative unit of self-duplication, he asserted, "is not DNA, but a multi-molecular system which is so complex as to require the participation of the entire living cell" (Commoner, 1964, 49). He argued that if DNA required this complex

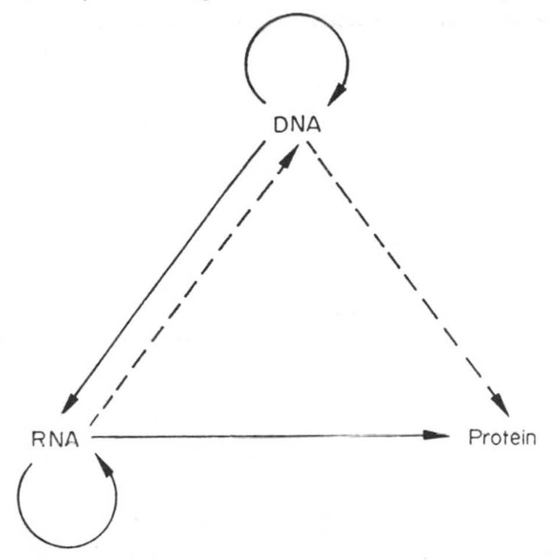

Fig. 22.5. The arrows show the situation as it seemed in 1958. Solid arrows represent probable transfers, dotted arrows possible transfers. The absent arrows represent the impossible transfers postulated by the central dogma. They are the three possible arrows starting from protein (from Crick, 1970).

machinery its action in the determination and transmission of specificity must be influenced by the several parts of that machinery, among which are proteins such as the DNA polymerases. Therefore the information in these proteins has a specific affect on the information in DNA, which cannot therefore be "the only source of protein specificity" (Commoner, 1968, 338). The forbidden transfer—protein → DNA—clearly did exist.

Crick could, as we have seen above, claim that he recognized the complex nature of the machinery referred to by Commoner. Had he not contributed to the elucidation of some of its details? He could have gone on to point out that the involvement of a protein in the duplication of DNA does not amount to a *specific* influence upon the transfer of information from the template DNA strand to the daughter strand. The specificity of this transfer, it is believed, is achieved by virtue of certain *permitted* pairing relationships of the bases which are dependent upon the formation of "acceptable" hydrogen bonds, and what Eigen and de Maeyer call "co-operativity" between just-formed and about-to-be-formed base pairs. The polymerase machinery does no more than *assure* a very high probability that only permitted pairs will be built into the structure. The action of a different DNA polymerase does not alter the *nature* of the permitted pairing relationships, *it merely lowers the fidelity of the replication* by failing to assure the incorporation of the "specificity". This was implicit in Crick's definition of information but this was missed by his critics. Nor does the discovery of a system in which RNA makes DNA (Temin, 1970) amount to a "reversal of the Central Dogma" (Nature, 1970). Crick never said that the transfer, RNA to DNA, was impossible (Crick, 1970). (Unfortunately Watson did (1965, 298).) In short it seems that the Central Dogma is here to stay, and as Crick remarked: "the discovery of just one type of present-day cell which could carry out any of the three unknown transfers would shake the whole intellectual basis of molecular biology, and it is for this reason that the Central Dogma is as important today as when it was first proposed" (Crick, 1970, 563).

Transformation of Paradigms

As stated in the Introduction an important aim of this book has been to expose to view the great historical significance of the replacement of the Protein Version of the Central Dogma by the DNA Version of the Central Dogma. In this transformation of paradigms the discovery of the structure of DNA played a crucial role. It gave confidence that DNA *could* be the sole hereditary material, and it furnished at long last a chemical barrier between the codescript (DNA) and the gene product (protein). The old confusion between genotype and phenotype which the one gene–one enzyme hypothesis failed to dispel could now be swept away (Churchill, 1973). Nor was a direct pathway for the inheritance of acquired characters a plausibility any more. At the British Association meeting in Bristol in 1955 Crick pointed this out to Darlington, the arch opponent of Lamarckism (Crick, 1973), but it was not until much later that this was stated in print (Melchers, 1965).

It is significant that the Central Dogma was introduced shortly after the work on plant virus RNA by Gierer and Schramm (1956) and by Fraenkel-Conrat (1956). Their results were encouraging to those who wished to extend the idea of DNA as the hereditary codescript to RNA. What we have called the DNA Version of the Central Dogma could then be broadened to include the RNA viruses in The Nucleic Acid Version of the Central Dogma. On the other hand it was not clear in 1958 that the folding of *all* proteins was determined simply by the one-dimensional sequence of amino acids as they emerged from the RNA template. Crick thought the γ-globulins and adaptive enzymes might prove to be exceptions (Crick, 1958, 144). The debate on this aspect of the determination of specificity continued well into the 60s. The old model of the immunologists in which the antigen acted as a mould for the folding of γ-globulins to yield an antibody with the required three-dimensional shape for binding to antigen—the instructional theory—lingered on to confuse the early attempts of Jacob and Monod to understand the control mechanism of gene expression. Today, however, the opinion that γ-globulins do not represent an exception to the rule that amino acid sequence determines the 3-dimensional shape of proteins seems to be dominant (Williamson, 1972). Antibody production in response to a novel antigen is then accounted for by the Clonal Selection Theory (McFarlane Burnet, 1959; Lederberg, 1959).

The Role of Methods

The study of the history of ideas leads all too easily to what we may call "precursor disease", a sign of which is that the adumbration of a concept may be assigned the same historical status as the subsequent explicit statement of that concept. This tendency is wrong and can be checked by studying the impact of innovations in experimental methods (including methods of interpretation). Thus we have seen in the first three chapters what a world of difference it made to the nineteenth century conception of polymers when the ultracentrifuge, fibre diffraction analysis and improved viscosimetry were introduced. Only then could a plausible and precise conception of the magnitude of macromolecules be attained. Once that had been done the science of extraction could be reviewed and the possibility explored that compounds like DNA were, in the native state, very large macromolecules which became depolymerized as a result of extraction from the nucleus. With advances in the techniques of enzyme extraction, not merely crude extracts of enzymes but purer, crystalline preparations could be prepared, their characteristics studied and methods for their inhibition found. This has been seen in the progress of work on DNase from Levene to McCarty and Kunitz. To be sure Neumann and Burian perceived the susceptibility of DNA to depolymerization long before, but it is very doubtful that they appreciated just how large was the DNA molecule—twelve nucleotides with a molecular weight of 4000 perhaps. Miescher too, recognized DNA as a slowly diffusing substance of great lability, but again it is doubtful that he had a just conception of its size—how could he? He could say that the techniques of his day were too crude, and he could look forward to better days. Likewise the suggestion that the base composition of DNA was not that of a tetranucleotide of fixed sequence had been made long before Gulland's attack of 1946, by Troland in 1917. We have seen that the data from base composition itself was strangely unconvincing. Only with the recognition of the macromolecular size of DNA combined with the application to it of the new technique of chromatography did Troland's suggestion become both conceivable and justifiable. The introduction of chromatography by Martin and Synge and its application to nucleic acids by Chargaff is therefore of great importance in the history of our subject.

In Chapter 3 we saw how the discovery of the fibre diagram justified at long last the nineteenth century tradition that birefringent fibres have a crystalline constitution, but we also saw the contrast between the model building of the 1920s, unsupported by Fourier methods, and that of the 1940s and 50s supported by those methods. In particular we noted the contrast between Astbury's and Pauling's approaches to the structure of proteins, one unsupported by single crystal studies of the amino acid and peptide building blocks, the other built on such studies. In the latter we noted the care with which important data were checked and refined using Fourier methods.

The Role of Fourier Methods in the DNA Story

Can it be said that Fourier methods played as decisive a part in the discovery of the Double Helix as in the discovery of the α-helix? Clearly no—for Crick never saw any good B patterns of DNA; Watson saw, but did not possess, Franklin's good B pattern. The only pictures they had in Cambridge to work from were those of Florence Bell as published in 1947. On the other hand they benefited from the fine 2-dimensional analysis of cytidine by Furberg, from the similar studies of the purines and pyrimidines by others, and from the crystallographic data on the unit cell and space group of DNA which Franklin and Gosling had arrived at after constructing a Patterson diagram. Watson and Crick did not see this diagram, they did no refinements of their model by Fourier synthesis, yet Crick has repeatedly emphasized the importance of the introduction of Fourier methods. Why? The answer is two-fold, for on the one hand these methods were necessary to arrive at reliable structures for the sub-units of the polynucleotide chain and on the other, Fourier theory could be applied to a very simple and regular structure, such as a continuous or discontinuous helix, in order to derive the transform thereof in reciprocal space. This gave the characteristics of the diffraction pattern, and it was at once clear to Watson and Crick that the B pattern of DNA was the most remarkable example of a helical structure in which the repeat was integral. Therefore the possible conformations of the chains and the packing of the bases must be limited. An attempt had to be made to solve it. The discovery of good B patterns of DNA in Leeds in 1951 aroused no such enthusiasm in Beighton and Astbury because they did not appreciate their significance. Now the unit cell of DNA contains nearly 2000 atoms in the A form, a fraction less in the B form. In the former, Wilkins and his colleagues observed about 120 intensity maxima (Wilkins, Seeds, Stokes and

Wilson, 1953, 759) the intensities of which they estimated with a micro-densitometer. They were then able to plot these in reciprocal space and compare them with the Fourier transform calculated for a model of the Watson–Crick type (Fig. 22.6). There was no possibility of deriving the structure directly from the A pattern, but Fourier theory and model building

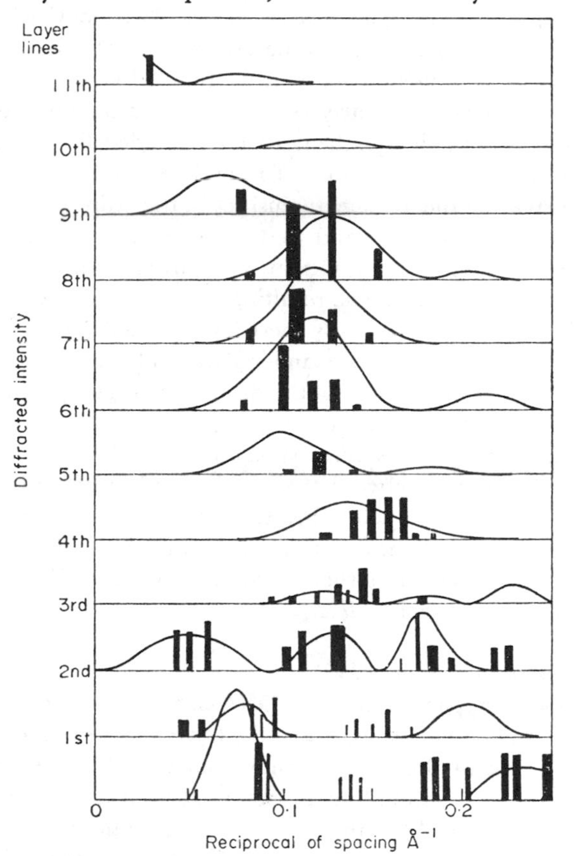

Fig. 22.6. Observed and calculated intensities of diffraction for crystalline NaDNA. The observed intensities are shown by the heights of the black rectangles and the smooth curve is calculated from the model (from Wilkins, Seeds, Stokes and Wilson, 1953, 761).

could be combined in this way to check the plausibility of the structure. It opened the way to the interpretation of fibre diagrams in a manner more rigorous than Astbury had ever imagined.

Proof of the Watson–Crick Model

A structure is said to be proven when it can be shown to follow unambiguously from the X-ray data. This demands a knowledge of the phases as well as the intensities of the spots. Lacking the phases Wilkins' group in London used the Fourier method of structure refinement (see Chapter 17). This, it will be recalled, involved using the observed intensities and the *calculated* phases, the latter being derived from the assumed model. This approach would not have come under attack if the results of single crystals studies had supported the main feature of the Watson–Crick model—the A–T and G–C base pairing. Unfortunately, crystals of mixed mononucleotides did not give complementary pairing but "Hoogsteen" like-with-like pairing (1959). Nor was it clearly established, in Donohue's opinion, that several other possible pairing schemes were excluded for DNA (Donohue, 1956). He argued that the number of intensities (1/50th of the total) employed for calculating the average intensities in the "direct method" was too small (Donohue, 1969a, 291), and that the discrepancy between observed and calculated structure amplitudes attained from Fourier refinement was too great (Donohue, 1969b). The resulting debate enlivened the columns of *Science* and *Nature,* but very recently there has been a dramatic success which promises to settle the dispute. Alexander Rich and his colleagues at M.I.T. have crystallized the sodium salt of the dinucleoside phosphate of adenosine and uracil. They found that it "forms a right-handed anti-paralled double helix in which the ribose phosphate backbones are held together by Watson–Crick hydrogen bonding between the adenine and uracil residues" (Rosenberg *et al.*, 1973, 151). They called this "the first crystal structure in which the atomic details of double helical nucleic acids can be visualized. In addition, this is the first single crystal structure showing Watson–Crick base pairing between adenine and uracil" (*Ibid.*).

The Uniqueness of Discovery

It should now be apparent that the DNA structure could have been discovered in a number of ways. It could have been deduced from single crystal studies, but in this case we may conclude that the wrong pairing scheme, i.e. Hoogsteen, would have been used until data from oligonucleotides became available. What the pure crystallographer regards as the more acceptable approach would then have led to error! Or again the traditional school of intermediary metabolism might have played the major part. Arthur Kornberg's study of nucleotide polymerizing enzymes originated in his interest in phosphorylating mechanisms. He worked upwards from nucleosides to ribonucleic acids, then to desoxyribonucleic acids. He was able to show that the direction of the chains in the synthetic DNA duplexes which he synthesized was anti-parallel, and that Chargaff's rules were obeyed in terms of base pairing. It would have been natural for him to have passed his products over to the structural chemists in the hope that they would synthesize these findings into a plausible model. Or again, the work of Rich and his colleagues on synthetic RNAs might have been extended to suggest models for DNA as well as for RNA. We may note, however, that in each of these cases the Watson–Crick model would have been discovered much later than 1953. How long would it have taken if Watson had not come to Europe and Crick had not as a result strayed from protein? One is inclined to speculate that Franklin and Gosling would have published their data, as planned, and that Pauling would therefore have been able to make a fresh start and that before the year was out, either he or Franklin or Wilkins would have solved the structure. Crick has pointed out to me the distinction between what was a chance event and what was inevitable in the work described in this book. Because of the sound foundations and long-term planning of Pauling's research it was more or less inevitable that he should discover the α-helix. The situation in the case of DNA was quite different. There was no such planning, still less sound preparation, for Watson's and Crick's work on DNA. And if Watson had come to Cambridge in 1953 or 1954 instead of in 1951 we may be fairly confident that the discovery would have been made in London or Pasadena, rather than in Cambridge. We can thus support the suggested analogy between scientific discovery and creative works of art, according to which both share in the unique (Stent, 1972).

Research Schools

One aim of this book has been to expose the contrasting approaches which resulted from the organization of scientific research into schools dedicated to different lines of attack. This has been strikingly seen in the "structural" schools of Caltech, Cambridge and Leeds. It is also evident in the tradition of experimental cytology, rooted in colloidal biophysics, and that of cytochemistry rooted in the study of fixed and stained preparations. The former flourished in the 1920s, 30s and 40s. In the work of Jacques Loeb and William Hardy it connected with the study of proteins and gave rise to the conceptions of proteins as polyelectrolytes. In the studies of Robert Chambers, Sir James Gray and J. D. Bernal it was related to the mechanics of mitosis. In John Runnström's school at the Wenner-Gren Institute in Stockholm it continued the tradition of developmental mechanics established by Driesch and Boveri. Where this biophysical–experimental approach failed was in making any meaningful contact with genetics and with the new science of macromolecular chemistry. Its practitioners were addicted to the whole cell and not to the study of extracted material, to physiological action and not to information transfers. J. D. Watson narrowly missed being drawn into this tradition when he was advised to go to Stockholm rather than to Cambridge by Paul Weiss. Crick spent two years in experimental cytology before he was allowed to turn his attention to the structural chemistry of extracted material. This was the "Old Biophysics" which has been superseded by the "New Biophysics" which emerged from the union of the "Informational" and "Structural" schools of molecular biology.

The separate development of these two schools was in part due to National Socialism and the Second World War which inhibited the development of the multi-disciplinary biophysics of inter-war Germany and caused intellectual migrations. There resulted a separate development of the "informational" school in America and of the "structural" school in England. We have seen this separation destroyed by the Watson–Crick model. The special feature of its discoverers was that in a sense they "belonged to *both* schools. "Although I was a member of the structural school", wrote Crick, "I was always keenly interested in all the genetic and biochemical work. At the same time Watson, who was a member of the Informational School, had naturally found himself turning to the structural and biochemical side" (Crick, 1973).

We have seen how the limitations of a school limited the achievements of those who worked in it; how, for instance, the British and German schools of physiological genetics failed to make the breakthrough that Beadle and Tatum achieved, how the weakness of the Cambridge and Leeds schools of molecular biologists in structural chemistry denied them the prize of the α-helix. At the same time we have noted new concepts, techniques and discoveries emerging within orthodox traditions—the X-ray crystallography of fibres was pursued by many working within the community of colloidal biophysics; the quantitative cytochemistry of the nucleus was developed by Caspersson, a student of Runnström and Hammarsten. Again it was a student of colloid science, Svedberg, who developed the ultracentrifuge. We have not therefore painted a black and white picture of good and bad traditions and of right and wrong influences, but we have observed how necessary it is to be prepared, as Watson was, to leave one tradition and enter another. This can result from external influences as when Watson wanted to gain a European experience, or from internal factors, as when Randall's group shifted the emphasis of the work at King's from experimental cytology to the structural chemistry of extracted material.

The Rockefeller Foundation

It is only to be expected that references to support from this Foundation should be frequent in this book, for it was a deliberate policy decision by the Foundation's administrators to support biophysical and biochemical work in the early 1930s. Between 1932 and 1959 the Natural Sciences section of the Foundation spent over $90 000 000, a large part of it on biological research. Muller, Delbrück, Caspersson, Hammarsten, Pauling, Corey and Svedberg all received support from the Foundation during the inter-war years. Between 1934 and 1938 Astbury received $71 000. This was in marked contrast with the 1920s when most of the funds went to physics, a subject much admired by Wickliffe Rose, President of the General and International Education Boards. There can be little doubt that without the Foundation's support through the natural sciences section of the International Education Board, there would never have been such a flowering of molecular biology in the 1930s. Why, then, did the Foundation adopt this policy?

This question has been answered by the director of the Natural Sciences section of the Foundation, Warren Weaver, from his personal involvement in the Foundation's programme. There seem to have been three factors involved: first, the wave of anti-scientific feelings against the technology of the hard sciences which was fanned by the depression, causing philanthropic organizations to consider directing their support to research of a more social and medical nature; second, the appointment of Max Mason as president of the Foundation; and third, Mason's choice of Weaver as director of the programme in the natural sciences.

Mason and Weaver were colleagues at Wisconsin where they often discussed what could be achieved in biology if the fruits of the advances in physics and chemistry were applied to physiological questions. At the time, Mason was experiencing the trauma of seeing his wife turn incurably insane, and he longed for science to give man control over such illness. Mason was a mathematical physicist, but he was not in love with quantum mechanics—far from it. Weaver recalled that he would have "plunged overboard" into this exciting field but for Mason, who was "sarcastically indifferent" (Weaver, 1970). This indifference was modified on one page (327) of the book, *The Electromagnetic Field*, which these two men published in 1929, but in the introduction we find that "the Maxwell–Lorentz theory has attained a degree of success little short of marvellous. Its triumphs have forced it into the position of an ultimate theory" (*Ibid.*, p. ix). The authors went on (p. xiii) to describe what was clearly a personal experience:

> . . . just as one's psychology is largely set before he reaches the age of three, so one's viewpoint toward electrodynamics is a fixed and settled matter by the time one has acquired the field equations and has discussed certain of their applications.

From his first incursions into the literature of biology Weaver was struck by the sparsity of really good ideas in the subject. It "lacked the kind of intellectual ferment" to be seen in the physical sciences. Now Weaver's *forte* was statistics, so it was natural that he should find sympathy for the subject of genetics, and from genetics his reading spread to biochemistry. When Mason offered him the direction of the natural science programme, Weaver was adamant that he would only consider it if "they were willing very substantially to change this programme" (Weaver, 1971, **3**, 16). They were willing; Weaver came in 1932, and at a meeting of the trustees in April, 1933, he presented a memorandum entitled, "Natural Science—programme and policy", in which he called for a "highly selective procedure" of support "if funds are not to lose significance through scattering". Instead of selection in terms of scientific leaders, as in the past, the Foundation should make the field of research play the dominant part. "Science", he declared, "had made magnificent progress in the analysis and control of inanimate forces, but it has not made equal advances in the more delicate, more difficult, and more important problems of the analysis and control of animate forces" (Weaver, 1933, 76–77). A year later Weaver could report termination of support to astronomy and meteorology and a concentration upon "experimental biology" (Weaver, 1934, 126). Beginning with the terms "physiochemical" and "experimental" biology in 1934, four years later he spoke of "molecular biology". Modern tools, wrote Weaver, were reaching deeper and deeper into the living organism, they were revealing new facts about the structure and behaviour of the "minute intercellular substances".

> And gradually there is coming into being a new branch of science—molecular biology—which is beginning to uncover many secrets concerning the ultimate units of the living cell . . .
>
> Among the studies to which the Foundation is giving support is a series in a relatively new field, which may be called molecular biology, in which delicate modern techniques are being used to investigate ever more minute details of certain life processes.
>
> (Weaver, 1938, 203–204)

A year later the terms "molecular biologist" and "molecular biology" were used by Astbury (1939b, 49) and by his research student Florence Bell (1939, 63) respectively. There may well be significance in the fact that the one who was among the first to use this term, Astbury, was in 1938 a grantee of the Foundation's programme in natural science. Neither Weaver nor Astbury, nor Bell, however, seems to have brought the term into general usage. Nor is it clear that Paul Weiss' regrouping of the National Research Council's programme in 1951 under six heads, one being Molecular Biology, had a widespread effect. The introduction of the *Journal of Molecular Biology* in 1959, however, is thought by many to have made the term fashionable.

Until 1947 when the Medical Research Council financed molecular biology in a big way in this country, it was the Rockefeller Foundation, under Weaver's guidance, that supported the subject. Just how crucial was Weaver's policy can be judged from the following comments of Bragg and Delbrück:

> Your help came at a vital time just before the war when I was trying to find some way of supporting Perutz's work and it was continued after it. This school was responsible for DNA, for the first protein structures, for the first understanding of virus structure, and for work on muscle. The extent to which the X-ray analysis of protein was pioneer work is shown by the fact that only now, 12 years after the trail was beaten at Cambridge and the Royal Institution, has any other research centre succeeded in getting a protein "out".

> I am allowing myself to put this so strongly just because I think that the Foundation's help made an outstanding difference to these advances.
>
> (Bragg, W. L., 1967a)

> As to the role of the Rockefeller Foundation's programme in these developments, I can only testify as far as I am concerned, and here very strongly and unambiguously: without the encouragement of the Rockefeller Foundation received in 1937 and their continuing support through the mid-forties I believe I would hardly have been able to make my contributions to biology.
>
> (Delbrück, 1967b)

We may conclude that Weaver's action constituted a force external to the biological sciences the effect of which was to accelerate the migration of physicists and chemists into biology and to promote the expansion of the reductionist/experimental approach. When seen against the backcloth of a broad time scale this movement is of profound significance for the history of biology. Its most dramatic and significant fruit has been the Watson–Crick model for DNA, and the Central Dogma of molecular biology.

Postscript (1994)

In the two decades since this book was published there has been not only an explosion of activity in the field, but also a dramatic evolution of the basic concepts of the molecular biology of the 1950s, alongside the introduction of an impressive range of new tools. These are now finding their place in many of the traditional specialisms of the biological sciences. How wrong were those who in the 'sixties spoke of coming to the end of a golden age![1] What a relief it was, therefore, to find, in the original Preface to this book, referring to the last chapter, the sentence: "It should convey my recognition of the limited extent of present-day knowledge, and of the possibility of further unexpected, even revolutionary, developments".

I also described the book as "a first attempt on a difficult and complex story" that would need correction and amplification in many places. I offered it "not as a definitive account" but as one which should "stimulate others to do better". Since 1974, scientists, historians and sociologists of science have taken up many of the topics sketched in this book, and have enriched our understanding of them, whilst at the same time adopting a variety of approaches. Thus Pnina Abir-Am has depicted the migration of physicists into biology as an exercise of power on their part, and their representation of the origins of molecular biology in terms of this intellectual migration she sees as an exercise in legitimation for their discipline. The claims of molecular biologists that their success represents the justification of reductionism she inclines to view instead as an interpretation of the history that suits their expansionist aspirations.[2] Lily Kay has focussed on the funding of the molecular programme at Caltech and has argued for the existence of a continuing commitment on the part of the Rockefeller Foundation and Caltech scientists to what were ultimately eugenic aims. She sees recombinant DNA technology both as the fruit of research that had its origins in the Rockefeller programme and as offering the means to achieve control of human reproduction in the interests of eugenic goals.[3] Understandably this claim has stimulated some strong objections.[4] Robert Kohler has made a very careful and illuminating analysis of Rockefeller policy in the management of scientific research. He has shown how Warren Weaver's programme grew organically out of a broadly focussed conception of instrumentation in biology by a process of symbiotic association with potential and existing scientific recipients of patronage.[5] The reader of the last three pages of Chapter 22 of the present book will find the topic developed much more fully in the writings of Kohler, and also in those of Abir-Am and Kay. On the subject of molecular biology at Cambridge the reader should consult the promising work of de Chadarevian.[6]

A number of the topics dealt with in this book have since been examined at some length by the scientists involved. Thus the impact of the identification of the transforming principle as DNA has been discussed by Maclyn McCarty[7] and Rollin Hotchkiss.[8] Although they come to somewhat different conclusions as to the extent of interest in bacterial transformation that was generated by the Avery-MacLeod-McCarty paper, both authors have added greatly to our understanding of the subject. Hotchkiss has provided the fullest collation of citations yet. Returning to this topic twenty years later I am impressed by the strength of the tendency to interpret DNA's role as mutagenic. Thus as late as 1951 Dobzhansky wrote:

> It is possible that a pneumococcus cell contains a population of as many kinds of plasmagenes as there are potential types, and that one or the other of them may be stimulated to reproduce by the transforming principle. On the other hand, a qualitative change in some self-reproducing bodies may be involved, specifically directed by the kind of nucleic acids with which these bodies come in contact. That chemically relatively simple substances may act as "transforming principles" has been demonstrated in several cases.[9]

My analysis of the earlier history, and, in particular, the conception of transformation by Griffith, has not, as far as I know, been challenged, but I claim that the neo-Lamarckian account I gave is important for our understanding of the whole transformation story. Griffith's plan to produce a change of type experimentally was quite deliberate, and any suggestion of serendipity seems to me misleading.[10] The work of the phage group has also come in for further study. Ton van Helvoort has re-examined the work of the biochemists Northrop and Kunitz and has concluded that the historiography of the establishment of the one-step growth curve has been adversely influenced by Max Delbrück's view of the subject. Van Helvoort judges the biochemists' papers more favourably than do traditional accounts and he denies that a decision between the two interpretations of bacteriophage multiplication was clear-cut.[11] My response to this is to remind the reader of the many difficulties that the biochemists had to face with their autocatalytic theory of phage multiplication. These difficulties were separate from concern over the interpretation of the quantitative data, and they affected the credibility of the concept of viral autocatalysis, as has been explained in Chapter 10 of this book.

Although I have made some minor changes in the part of that chapter dealing with the plant viruses, its underlying message still stands. This is that evidence from a variety of sources accumulated in the 1930s and 'forties could not be squared with the protein conception of the determination of viral and chromosomal specificity. The manner in which accessory hypotheses were invented to contain this situation illustrates beautifully T. S. Kuhn's conception of working within a paradigm. If changing the chemical constitution of parts of the viral protein did not seriously affect infectivity then such parts of the molecule were "not important to the basic reactions of virus reproduction".[12] Further, when progeny virus particles did not reflect these alterations in the parent viral proteins it was suggested that the host had changed the infecting particles back to their normal constitution.[13] The former of these results could be squared with the results of testing the effect of changing the chemical constitution of enzymes. Here Herriott and Northrop's study of acetylated pepsin showed that many chemical groups in the protein have no effect on the biological activity of the enzyme.[14] Yet these researchers knew that altering the viral RNA, as when treated with nitrous acid, inactivated the virus.[15] Just how well awarc were Miller and Stanley of the biological implications of these researches is evident in the following passage:

> If the infecting molecules served as exact models for reproduction, one would expect to reisolate the virus derivatives from plants so infected. Furthermore, the disease produced by such an altered virus might be different from the normal disease, since it is known that the nature of a virus disease varies with the strain of the virus and that strains of a virus differ from one another in their chemical properties. It is possible, therefore, that one might cause structural changes in vitro which would, in effect, correspond to the mutation of a virus. If, on the other hand, the inoculation of the virus derivative resulted in the production of normal virus, it might be concluded that the structural changes were reversed within the cells of the host or that that portion of the molecules involved in the structural change was unimportant and played a subordinate rôle in virus reproduction.[16]

In this analysis of the situation we note no suggestion that the protein might not be responsible for the determination of specificity at all.

Over the last three decades Kuhn's conception of the nature of science has received both criticism and praise.[17] In this book I did not simply apply his ideas in an uncritical manner. I have never accepted many of the claims that have been made for his conception of *incommensurability*. It certainly does not mean for me unintelligibility or incommunicability between scientists working within different paradigms. At the same time I have a clear sense of discontinuity in the history I have described, the underlying basis of which is deep-seated.

Horace Judson, whose account of the history of molecular biology overlaps chronologically with mine, takes a different view. For him Kuhn's "celebrated, fashionable discourses on scientific revolutions" are not well informed on the practices of scientists. Judson prefers to treat scientific development in terms of the "correspondence principle",[18] in which the inherent conservatism of science is stressed. A new theory has to explain all that was explained by the old one. In this way the cumulative nature of scientific knowledge is preserved. The problem with this view is that it fails to accommodate rejected knowledge. Such knowledge has to be treated as anomalous, ambiguous, mistaken. What is not explained is treated as if it does not belong to science; yet it *did once* belong. Nor need we exclude tentatively held and vague theories simply because they were advanced with less conviction than were other more successful theories. If they influenced the direction of research and its interpretation, then they are important to the historian. Thus, for Kuhn, what is shared by a scientific community—the paradigm—does not have to be a fully articulated theory, it can be a set of assumptions—for example that the molecule cannot be larger than the unit cell of its crystals, or that repeats in helical molecules must be integral. Then again, the paradigm could be associated with a technique that constrained or promoted a particular interpretation. Thus Svedberg's classification of the proteins on the basis of their sedimentation data yielded suggestive support for his conception of their construction on modular principles. Many of these features of scientific activity may be unacknowledged. They are covert, yet they function in directing research and in selecting and rejecting data.

In the 1950s, molecular biology rejected repeating sequences in proteins and nucleic acids. Yet such a conception of sequence was widely assumed among biochemists at one time. This assumption reflected underlying conceptions of the nature of protein synthesis and of the structural patterns of biologically important macromolecules. The result of such a process of rejection means that there are not conceptual and empirical parallels between all features of an old paradigm and a new one. Thus there is no *precise* equivalent in the protein era for the 1950s conception of the determination of genetic specificity. This does not mean that those earlier conceptions were merely vague and unscientific. They were advocated sufficiently strongly to generate both vigorous debate and the development of adequate techniques for their solution—chief amongst them the paper chromatography of the proteins and nucleic acids. Elsewhere I have described how interested was the protein chemist, A. C. Chibnall, in the question of amino-acid sequence, and how, through the scientific committee of the International Wool Secretariat, funds were made available for R. L. M. Synge to develop and apply his separation technique to the problem of the amino-acid composition of wool, whereby he established chromatography for proteins.[19]

Although the major break in the establishment of 1950s molecular biology concerned the tradition of protein supremacy, the implications that flowed from the baptism of the nucleic acids as the fount of specificity were many and diverse. Thus the notion of a chemical code, though already present in the literature of protein chemistry, did not contain the process of translation. Since proteins were thought to be made by proteins, no change in the chemical language was required. Once nucleic acids were given the role of determining protein specificities a translation process became necessary. In 1974 I stressed, however, that the move from the protein to the nucleic-acid theory of genetic specificity was not accomplished in one decisive step, but through an intermediary position in which specificity was seen as requiring both nucleic acids and proteins in a two-way process. It was the continued advocacy of such theories after 1953 that stimulated Francis Crick to enunciate the Central Dogma.

The importance that the Central Dogma has assumed serves to underline how different were the views of even those who recognised that the nucleic acids played an important role in genetic replication. Thus Darlington expressed the scepticism that others also felt as to the all-sufficiency of DNA as the material basis of heredity. Writing of the Watson–Crick theory of heredity he remarked:

> On this basis DNA would appear as a self-sufficient structure with the protein at least mechanically subordinate. We do not need to take such an extreme view: equality and reciprocity are also conceivable.[20]

Darlington's view is of particular interest because of the hopes he had entertained before the 1950s for discovering the material basis of heredity and gene replication. He was a key figure, too, in the network that embraced men like Caspersson, Sir John Randall and Wilkins. It was the results of his study of nucleic acid starvation that predisposed him to underestimate the importance of the nucleic acids.[21] On rereading Chapter 7 of the first edition of the present book, on the nucleoprotein theory of the gene, I feel confirmed in the view I took there, namely that the cytologists did have a sketch of a theory of the material basis of heredity epitomised by Darlington in the term "Midwife Molecule", which he coined to describe the role of DNA in the replication of the gene.[22] But their experimental methods were all indirect, and where direct methods were used, as in the chromosome studies of Mirsky and the Steadmans, the results were not clear-cut. Even when direct methods were used on bacteriophage there was room for doubt, but by that time the work of Avery, MacLeod and McCarty was well known, and the manner in which the DNA content of cells changed in time with the chromosome cycle was well established. Thus Hewson Swift concluded his paper on the DNA content of plant nuclei with the following chief points:

> (1) The amount [of DNA] per nucleus shows a marked steplike occurrence. (2) There is a duplication with mitosis and a reduction with meiosis. (3) Since species and strains have characteristic amounts of DNA it is apparent that the quantities involved are directly associated with the genotype. At least for the present these factors seem best interpreted by considering DNA as a component of the gene.[23]

"Component of the gene" and "at least for the present" are suitably cautious phrases, but the point I sought to make in this book in 1974 has been borne out by subsequent discussions, namely that the change from the protein gene to the DNA gene was brought about by a number of events and not by one or two "crucial" experiments alone. These several experimental results served to create a climate in which the outcome of the so-called crucial experiments could be interpreted in support of this radical change of view.

It is clear from the first five chapters of this book that the protein era of the 1930s and 'forties was strongly coloured by the kind of organic compounds upon which most research was done. These were chiefly substances of economic importance. Following World War I, government and industry put money into scientific research on textiles, rubber, collagen and synthetic polymers. The results of this work were to establish structures consisting of long-chain ribbons rather than helices, the sequences of which repeated a regular pattern throughout the chain. For silk and collagen these sequences were not far from those established today, but they were too easily taken to represent the type of sequences to be found in most proteins. Fortunately the medical interest in blood had assured support for a strong tradition in research into haemoglobin, a protein that did not conform to the repeating-sequence model. As for molecular structure, later chapters document how the Astbury model for alpha keratin, which was modelled on that for cellulose, continued to cast its shadow in Britain over subsequent discussions of the nucleic acids as well as the proteins.

The text of Section V, on the structure of DNA, has been corrected for minor typographical errors. One such error was particularly unfortunate because of its importance—the omission of the "greater than" sign from the number of chains per lattice point in Rosalind Franklin's notes (see p. 349). I have also added some footnote comments based on a detailed letter which I received from Professor Wilkins in 1986. There is now an entry for this letter in the main bibliography.

Finally, no author who has the pleasure of seeing his twenty-year-old book reprinted can refrain from commenting on how the science that he described has grown in its authority, sophistication, power and practical importance. Gone are the days when it was necessary to hand out apologetics to medical audiences for the lack of applications of the new knowledge. Thus Crick is on record as having frankly admitted in 1965 that:

> The medical reader will naturally wonder how much of this [research in molecular biology] will directly affect medical practice. Here one must confess that so far there have been no dramatic cases where the new knowledge has led to a revolution in medicine. On the other hand our insight into medical problems has gained new depth, and the influence on those actively engaged in medical research is all pervading. Many medical problems are genuinely problems at the molecular level, whether they be genetic abnormalities such as sickle-cell haemoglobin, or infective entities such as poliovirus. We can always hope that the problem can be solved by a happy accident, actively pursued, as in the case of penicillin but, failing that, we must try to understand what has gone wrong as a preliminary to seeing how it can be put right. Many of the conveniences of modern life are possible because of the earlier advances in chemistry, especially organic chemistry. In the same way we may hope to see, in the next decades, the increasing application of molecular biology to outstanding medical problems such as the diseases of auto-immunity, the treatment of viral infections and the nature and cure of cancer.[24]

Joshua Lederberg gave early warning of the technical difficulties of attempting to control and manipulate the genetic material. We would have to start with "biological reagents", he said, but synthetic chemistry is "challenged to produce model polymers that can emulate the essential features of genetic systems".[25] By contrast Sir Ernst Chain was deeply sceptical, and he expressed himself with characteristic force in the warnings he gave to the Medical Research Council in 1973 when his style of research was being set aside. Working out the molecular structure of enzymes, he declared, would not enable us to "obtain tailor-made drugs", and he added:

> The same disappointment, to even a greater degree, awaits us if we believe the prognostications of some molecular biologists in the field of genetics that it will become possible to cure genetical diseases by "genetic engineering"...This is science fiction which does harm rather than good to the image of science, and impedes rather than advances its progress.[26]

It was the developments from the late 'seventies that gave rise to recombinant DNA technology that both transformed our views of the machinery of gene expression and made possible the applications about which we hear so much today.[27] Just as the computer has penetrated into the marketplace and on into the office and home, and is now a key part of everyday living for us all, so has the science of DNA spread its influence and become a household word. We may not yet need our DNA fingerprints in order to withdraw money from the bank—let us hope we never will—but we are well aware of the ways in which recombinant DNA technology is being used to study genetic diseases, establish parentage and identify murderers and rapists, and how it offers a powerful set of research tools in many branches of biological research.[28] Sidney Brenner has described how these tools offer the evolutionist reliable methods for studying the phylogenetic history of species. Like the modern radio telescopes that enable us to "see" into the past history of the universe, so recombinant DNA technology allows us to peer into the past history of a species, by revealing the base sequences it shares with other, often much more primitive species.[29] With these techniques, phylogenetic studies, long the graveyard of rejected speculations, can be pursued with a reasonable hope of achieving more reliable results.

The advent of recombinant DNA technology, as we all know, not only has given us a foretaste of the benefits that the biological sciences and society may hope to reap from molecular biology but also has raised concerns. Is the molecular approach being privileged over and above what is its due? Is it not already fostering the hope that replacing genes in the germ tract will be a panacea for many of society's social problems and above all for human diseases?[30] Certainly it has encouraged a tendency to claim that if we can identify a DNA sequence in the human genome responsible for a given susceptibility to behave in a socially troublesome manner or to catch certain diseases, we should be able to eradicate the social problem and banish the disease by gene therapy. But as James van Neel warned, "The potential miracle of gene therapy must not be allowed to unduly divert attention from the rest of a balanced genetic program".[31] As the human genome project progresses, debate on these matters is likely to continue.

This is not the place to offer an account of the impact of molecular genetics on other fields. It suffices to remind the reader that in addition to the general impact of the new techniques upon many fields of the life sciences, the success of the Watson–Crick model very rapidly inspired certain among the neurobiologists to attempt to bring together the disparate disciplines studying the brain and cognition in the hope of achieving personal interaction and an intellectual synthesis. Foremost among these was F. O. Schmitt, founder of the Neurosciences Research Program in 1962. In his first report on the program he wrote:

> The breakthrough to precise knowledge in molecular genetics and immunology—"breaking the molecular code"—resulted from the productive interaction of physical and chemical sciences with the life sciences. It now seems possible to achieve similar revolutionary advances in understanding the human mind.[32]

Those who lived through the scientific excitement of the 1960s will know how enthusiastically did many scientists work to produce evidence that RNA not only is involved metabolically in the function of memory storage, but also serves to store the memories themselves in a chemical code much as DNA was being shown to store the information of the genes. One of the enthusiasts in this movement, John Gaito, dedicated his book *Molecular Psychobiology* to Francis Crick, "whose *Scientific American* articles of 1954 and 1957 suggested to me the possibility of integrating Psychology and Molecular Biology as Molecular Psychobiology".[33] By the end of the 'sixties memory molecules were no more, and one wonders at the willingness of editors to publish, and granting bodies to finance, much of this work. Take into consideration the willingness of many scientists at this time to find memory molecules, however, and the behaviour of the scientific community becomes more understandable. This I offer as just one example of the enthusiastic impact of the scientific achievement the history of which has been described in this book. In the case of these memory molecules, hopes were indeed dashed, but elsewhere—in embryology, phylogeny and evolutionary theory—the application of molecular biology is having remarkable results. When we look back over the century there seems little doubt that the Watson–Crick model for DNA with its associated Central Dogma[34] will still stand alongside the principle of natural selection as one of only two unifying concepts in biology. The more remarkable is it that the Watson–Crick structure has not been definitively established by the direct methods of X-ray crystallography applied to DNA from the cell, because native DNA does not form single crystals. Rather, direct methods have been applied to synthetic oligo- and polynucleotides, and the structures thus produced, without having to make any assumptions, were found to agree in all major features with the Watson–Crick model. In this way confirmation of the Watson–Crick structure for the B form has been achieved and all other known naturally occurring forms of DNA have been eliminated as contenders for the role of chromosomal DNA in the cell.

Before this happy state of affairs was reached, however, supporters of the Watson–Crick model had to fight off an attempt in 1976 to supplant it with the "side-by-side" or "zipper" structure.[35] This had been introduced in order to lessen the topological problems associated with the unwinding of the helix. Supporters of orthodoxy replied that after the discovery of enzymes for "nicking" and rejoining stretches of DNA the unwinding problem seemed far less difficult than it had twenty-five years before. But they added:

> ...we consider it unwarranted to rely solely on the details of exact model building [for hard evidence either way], our knowledge of stereochemistry, though now fairly good, may not be adequate to provide firm answers, nor is it advisable to put one's faith completely on the fine details of X-ray diffraction patterns. That of the B form has always been rather poor and may not yield a clear, unambiguous decision between the two alternative types of structure. One must turn to evidence of quite a different type.[36]

The SBS structure has now long been left behind, but in the 1970s it served its purpose. Reviewing the state of knowledge in 1978 David Freifelder recalled Nils Barricelli's office door upon which were posted the words "To say I believe is good religion but bad science". Freifelder added: "Perhaps this motto is true, but it is the belief, the hunch, or the disbelief that is frequently, if not usually, the driving force behind good experiments".[37]

Nucleic acid structures not found in nature of course can be produced. Albert Eschenmoser and his colleagues have synthesized a number in Zurich. They have shown how the properties of these alternative forms highlight the biological superiority of desoxyribonucleic acid and the Watson–Crick pairing:[38] stability, but not too much, mutability, not too little or too much. And as the years roll by the wonder created in the minds of biologists by this remarkable structure has increased rather than diminished. Familiarity has not bred contempt.

April 1994

R. O.
Visiting Professor
Rockefeller University
New York

Notes

[1]Stent, G. S., 1968, 'That was the Molecular Biology that Was', *Science*, **160**, 390–395.

[2]Abir-Am, P., 1982, 'The Discourse of Physical Power and Biological Knowledge in the 1930s: A Reappraisal of the Rockefeller Foundation's "Policy" in Molecular Biology', *Social Studies of Science*, **12**, 341–382.

[3]Kay, L., 1993, *The Molecular Vision of Life. Caltech, the Rockefeller Foundation, and the Rise of the New Biology*, Oxford University Press, New York & Oxford.

[4]Sinsheimer, R. L., & Horowitz, N. H., 1973, 'Biology at Caltech', *Science*, **261**, 1505–1506. Olby, R., 'Response', *Ibid.*, p. 1506.

[5]Kohler, R. E., 1991, *Partners in Science. Foundations and Natural Scientists 1900–1945*, University of Chicago Press, Chicago & London.

[6]Chadarevian, S. de, 1994, 'Architektur der Proteine, Strukturforschung am "Laboratory of Molecular Biology" in Cambridge', in: *Experimentalsysteme in den biologischen Wissenschaften II,* Berlin; and 1994, 'Molekularbiologie. Disziplin oder transdisziplinäre Bewegung?', *Biologisches Zentralblatt* (in press).

[7]McCarty, M., 1985, *The Transforming Principle: Discovering that Genes are Made of DNA,* Norton, New York. Also his (1994) 'A Retrospective Look: How We Identified the Pneumococcal Transforming Substance as DNA', *J. exp. Med.,* **179**, 381–394.

[8]Hotchkiss, R., 1979, 'The Identification of Nucleic Acids as Genetic Determinants', *Ann. New York Acad. Sci.,* **325**, 321–342.

[9]Dobzhansky, T., 1951, *Genetics and the Origin of Species,* Columbia University Press, New York & London.

[10]Pollock, M., 1970, 'The Discovery of DNA: An Ironic Tale of Chance, Prejudice and Insight', *J. gen. Microbiol.,* **63**, 1120.

[11]Helvoort, T. van, 1992, 'The Controversy between John H. Northrop and Max Delbrück on the Formation of Bacteriophage', *Annals of Science,* **49**, 545–575.

[12]Schramm, G., & Müller, H., 1940, 'Zur Chemie des Tabakmosaik virus ...', *Hoppe-Seyler's Z. physiol. Chem.,* **266**, 51.

[13]Miller, G. L., & Stanley, W. M., 1942, 'Derivatives of Tobacco Mosaic Virus. l. Acetyl and Phenylureido Virus', *J. biol. Chem.,* **141**, 905–920.

[14]Herriott, R. M., & Northrop, J. H., 1935, 'Crystalline Acetyl Derivatives of Pepsin', *J. gen. Physiol.,* **18**, 35–67.

[15]Stanley, W. M., 1936, 'The Inactivation of Crystalline Tobacco-Mosaic Virus Protein', *Science,* **83**, 626–627.

[16]Miller, G. L., & Stanley, W. M., 1942, *op.cit.,* p. 905.

[17]See Kuhn, T. S., 1970, 'Postscript', *The Structure of Scientific Revolutions,* 2nd ed., University of Chicago Press, Chicago, pp. 174–210. Also the essays in Imre Lakatos & Alan Musgrave (eds.), *Criticism and the Growth of Knowledge,* Cambridge University Press, Cambridge, 1970.

[18]Judson, H. F., 1980, 'Reflections on the Historiography of Molecular Biology', *Minerva,* **18**, 421.

[19]Olby, R., 1986, 'The Recasting of the Sciences: The Case of Molecular Biology', in: *La Ristrutturazione delle Scienze tra le Due Guerre Mondiali,* edited by G. Battimelli, M. de Maria & A. Rossi, Roma, 1986, vol. ii, pp. 237–247. For a discussion of the biochemists' conception of protein structure see my (1982) 'The Origins of Modern Biochemistry. A Retrospect on Proteins', *Hist. Phil. Life Sci.,* **4**, 159–168; also Synge, R. M. L., 1952, 'Biological Aspects of Proteins in the light of Recent Chemical Studies', *Roy. Inst. chem. Lect.. Monog. Rep.,* **1**.

[20]Darlington, C. D., 1955, 'The Chromosome as a Physico-Chemical Entity', *Nature,* **176**, 1143.

[21]Darlington, C. D., 1940, 'Nucleic Acid Starvation of Chromosomes in *Trillium*', *J. Genet.,* **40**, 185–213.

[22]Darlington, C.D., 1955, 'Nucleic Acid: The Midwife Molecules', *Adv. Sci.*, **12**, 355.

[23]Swift, H., 1950, 'The Constancy of Desoxyribose Nucleic Acid in Plant Nuclei', *Proc. natn. Acad. Sci. U.S.A.*, **36**, 652.

[24]Crick, F. H. C., 1965, 'Recent Research in Molecular Biology: Introduction', *Brit. Med. Bull.*, **21**, 186.

[25]Lederberg, J., 1959, 'A View of Genetics' [Nobel Lecture], *Stanford Medical Bulletin*, **17**, 129.

[26]Chain, Sir E., 1973, 'Future Contributions from Biochemistry to the Advancement of Medical Research', attached to letter to the Secretary of the Medical Research Council, dated November 20. Chain Papers, E 93, Archives, Wellcome Institute for the History of Medicine.

[27]For accounts of these developments see Wills, C., 1991, *Exons, Introns, and Talking Genes*, Basic Books, New York.

[28]See Kevles, D. K., & Hood, L., 1992, *The Code of Codes: Scientific and Social Issues in the Human Genome Project*, Harvard University Press, Cambridge, Mass.; also (1 October 1993) 'Genome Issue', *Science*, 262.

[29]Brenner, S., 1993, 'Lecture given to the workshop of the Mellon Program in the History of Biology', 30 April, MIT, Boston.

[30]Strohman, R. C., 1993, 'Ancient Genomes, Wise Bodies, Unhealthy People: Limits of a Genetic Paradigm in Biology and Medicine', *Perspectives Biol. Med.*, **113**, 112–144.

[31]Neel, J. van, 1991, 'Priorities in the Application of Genetic Principles to the Human Condition: A Dissident View', *Perspectives Biol. Med.*, **35**, 66.

[32]Schmitt, F. O., 1963, 'Progress Report', as cited in Swazey, J. P., 'Forging a Neuroscience Community: A Brief History of the Neurosciences Research Program', in: *The Neurosciences: Paths of Discovery*, edited by F. G. Worden, J. P. Swazey & G. Adelman, MIT Press, Cambridge, Mass., 1975, p. 530. See also: Rose, Steven, 1993, *The Making of Memory: From Molecules to Mind*, Doubleday, New York (Chap. VIII).

[33]Gaito, J., 1966, *Molecular Psychobiology: A Chemical Approach to Learning and other Behavior*, Charles Thomas, Springfield, Ill.

[34]For an account of the post-1953 history of molecular biology see Judson, H., 1978, *The Eighth Day of Creation*, Simon & Schuster, New York; also his informative essay (1992) 'A history of the Science and Technology behind Gene Mapping and Sequencing', in: Kevles & Hood (eds.), *op. cit.*, pp. 37–80.

[35]Rodley, G. A., Scobie, R. S., Bates, R. H. T., & Lewitt, R. M., 'A Possible Conformation for Double-stranded Polynucleotides', 1976, *Proc. natn. Acad. Sci. U.S.A.*, **73**, 2959–2963. See also: Sasisekheren, V., & Pattabiraman, N., 1976, 'Double-stranded Polynucleotides—II. Typical Alternative Conformations for Nucleic Acids', *Curr. Sci.*, **45**, 779–783; Cyriax, B., & Gäth, R., 1978, 'The Conformation of Double-stranded DNA', *Naturwissenschaften*, **65**, 106–108.

[36]Crick, F. H. C., Wang, J. C., & Bauer, W. R., 1979, 'Is DNA Really a Double Helix?', *J. mol. Biol.*, **129**, 451.

[37]Freifelder, D., 1978, *The DNA Molecule: Structure and Properties. Original Papers, Analyses, and Problems*, Freeman, San Francisco, p. 483.

[38]Eschenmoser, A., 1993, 'From Homo-DNA to Pyranosyl-RNA: Towards a Chemical Etiology of Nature's Nucleic Acids Structure', Rockefeller University Lecture (29 October).

Bibliography

Manuscript, oral and published sources have been listed in one alphabetical sequence. Wherever possible serial publications have been referred to by their abbreviations in the *World List of Scientific Periodicals*. Where it helps the reader to explain that the year of presentation and of publication differ a note has been added in brackets. For all manuscripts the source of the original copy (sometimes the top copy, sometimes a carbon copy) is given, and the author has a complete xerox set of such documents. Where the dating of a manuscript or paper is uncertain a query has been placed after the year. For further information on the sources see p. xiii.

Abbé, E., 1873, 'Beiträge zur Theorie des Microscops und der microscopischen Wahrnehmung', *Arch. mik. Anat.*, **9,** 413–468.

Adair, G. S., 1925, 'The Osmotic Pressure of Haemoglobin in the Absence of Salts', *Proc. R. Soc.*, **109 A,** 292–300.

——, 1968, Recorded Interview with the Author, Cambridge, July.

Alexander, J., & Bridges, C. B., 1929, 'Physicochemical Aspects of Life, Mutation and Evolution', *Science,* **70,** 508–510.

Allison, V. W., 1969, Letter to Pollock, dated 30 September. [Original with Pollock.]

Alloway, J. L., 1932, 'The Transformation *in vitro* of R Pneumococci into S Forms of Different Specific Types by the Use of Filtered Pneumococcus Extracts', *J. exp. Med.*, **55,** 91–99.

——, 1933, 'Further Observations on the Use of Pneumococcus Extracts in Effecting Transformation of Type *in vitro*', *J. exp. Med.*, **57,** 265–278.

Altenburg, E., 1933, 'The Production of Mutations by Ultra-violet Light', *Science,* **78,** 587.

Ambrose, E. J., & Hanby, W. E., 1949, 'Evidence of Chain Folding in a Synthetic Polypeptide and in Keratin', *Nature,* **163,** 483–484.

Anderson, E. S., 1953, 'Up the Crick with Watson'. [This is the first of the five verse doggerel written by Anderson and performed by Roy Markham and T. F. Anderson at the 1953 Cold Spring Harbor Symposium. Original with E. S. Anderson.]

——, 1973, Letter to the Author, dated 28 September. [Describes the circumstances in which his doggerel was performed.]

Anderson, T. F., 1949, 'The Reactions of Bacterial Viruses with their Host Cells', *Bot. Rev.*, **15,** 464–505.

Andrewes, F. W., 1906, 'The Horace Dobell Lecture on the Evolution of the Streptococci', *Lancet,* **ii,** 1415–1420.

Anschütz, L., 1927, 'Ueber aromatische Abkommlinge der Phosphorsäure und der hypothetischen Orthophosphorsäure $P(OH)_5$, insbesondere Verbindungen mit phosphorhaltigen Heterocyclen', *Justus Liebigs Annln Chem.*, **454,** 71–120.

Araki, T., 1903, 'Ueber enzymatische Zersetzung der Nucleinsäure', *Hoppe-Seylers Z. physiol. Chem.*, **38,** 84–97.

Arkwright, J. A., 1921, 'Variation in Bacteria in Relation to Agglutination both by Salts and by Specific Serum', *J. Path. Bact.*, **24,** 36–60.

Astbury, W. T., 1929, See: Leeds University, Clothworkers' Depts.

——, 1930, See: Leeds University, Clothworkers' Depts.

——, 1931, See: Leeds University, Clothworkers' Depts.

——, 1933, *Fundamentals of Fibre Structure,* London.

——, 1936, 'Recent Advances in the X-ray Study of Protein Fibres', *J. Text. Inst.*, **27,** 282–290.

——, 1938, Contribution to the Discussion, *Cold Spring Harb. Symp. quant. Biol.*, **6,** 118–120.

——, 1939a, Letter to Stanley, dated 13 October. [Original in Biophysics Dept. Leeds University.]

——, 1939b, 'Protein and Virus Studies in Relation to the Problem of the Gene', *Int. Conf. Genet.*, **7,** 49–51. [These Proceedings were not published until 1940.]

——, 1940a, Letter to Neurath, dated 8 January. [Biophysics Dept. Archive.]

——, 1940b, 'The Molecular Structure of the Fibres of the Collagen Group', *J. int. Soc. Leath. Trades Chem.*, **24,** 69–92. [First Proctor Memorial Lecture.]

——, 1940c, *Textile Fibres under the X-rays,* Birmingham.

——, 1941, 'Proteins', *Chemy Ind.*, **60,** 491–497.

——, 1942a, Letter to Fankuchen, dated 24 April. [American Institute of Physics Archive.]

——, 1942b, Letter to Darlington, dated 14 November. [Original with Darlington.]

——, 1945, 'Structure of Alginic Acid', *Nature,* **155,** 667.

——, 1947, 'X-ray Studies of Nucleic Acids', *Symp. Soc. exp. Biol.*, **1,** 66–76.

——, 1948a, 'The Evolution and Physical Interpretation of Synthetic Fibres', in: *Symposium on Coal, Petroleum and their Newer Derivatives,* London, 99–108.

——, 1948b, 'Science in Relation to the Community', *Sch. Sci. Rev.*, No. 109, 268–280.

——, 1948?, 'Nucleic Acids and the Patterns of Life'. [Undated reprint labelled 'As 55' in the late Dame Kathleen Lonsdale's collection.]

——, 1949, 'Structure of Polyglycine', *Nature,* **163,** 722.

——, 1950, 'Adventures in Molecular Biology', *Harvey Lect.*, Series **46,** 3–44. [Publication date of whole volume was 1952.]

——, 1951, Contribution to the Discussion in: 'Symposium on Submicroscopical

Structure of Protoplasm', *Pubbl. Staz. zool. Napoli,* **23,** Suppl vol., 113–114.

——, 1952, 'The Great Adventure of Fibre Structure', *J. Text. Inst.,* **44,** 81–97.

——, 1955, 'In Praise of Wool', *Int. Wool Text. Res. Conf.,* 220–237.

Astbury, W. T. & Bell, F. O., 1938, 'Some Recent Developments in the X-ray Study of Proteins and Related Structures', *Cold-Spring Harb. Symp. quant. Biol.,* **6,** 109–118.

——, 1939, 'X-ray Data on the Structure of Natural Fibres and other Bodies of High Molecular Weight', *Tabul. Biol.,* **17,** 90–112.

——, 1941, 'Nature of the Intramolecular Fold in α-Keratin and α-Myosin', *Nature,* **147,** 696–699.

Astbury, W. T. & Street, A., 1931, 'X-ray Studies of the Structure of Hair, Wool, and Related Fibres', *Phil. Trans. R. Soc.,* **230A,** 75–101.

Astbury, W. T. & Woods, H. J., 1931, 'The Molecular Weights of Proteins', *Nature,* **127,** 663–665. [Also see Astbury's report on Textile Physics in: Leeds University: Clothworkers' Depts, 1931, p. 21.]

——, 1934, 'X-ray Studies of the Structure of Hair, Wool, and Related Fibres. II. The Molecular Structure and Elastic Properties of Hair Keratin', *Phil. Trans. R. Soc.,* **232A,** 333–394.

Austrian, R., & MacLeod, C. M., 1949, 'Acquisition of M Protein by Pneumococci through Transformation Reaction', *J. exp. Med.,* **89,** 451–460.

Avery, O. T., 1943, Letter to Roy Avery, dated 13 May. [As published by Hotchkiss in Cairns *et al.*, 1966, pp. 185–187. Original at The University of Vanderbilt Archive.]

Avery, O. T., & Dubos, R. J., 1930, 'The Specific Action of Bacterial Enzyme on Pneumococci of Type III', *Science,* **72,** 151–152.

Avery, O. T., & Heidelberger, M., 1923, 'Immunological Relationships of Cell Constituents of Pneumococcus', *J. exp. Med.,* **38,** 81–85.

Avery, O. T., MacLeod, C. M., & McCarty, M., 1944, 'Studies on the Chemical Transformation of Pneumococcal Types', *J. exp. Med.,* **79,** 137–158.

Avery, O. T., & Morgan, H. J., 1925, 'Immunological Reactions of the Isolated Carbohydrate and Protein of Pneumococcus', *J. exp. Med.,* **42,** 347–353.

Bacon, G. E., 1966, *X-ray and Neutron Diffraction,* Oxford etc.

Baillie, Sir J., 1928, Letter to Astbury dated 13 June. [University of Leeds Archive.]

Bamberger, E., 1903, 'Ueber das Verhalten para-alkylirter Phenole gegen das Caro'sche Reagens', *Ber. dt. chem. Ges.,* **36,** 2028–2041.

Bamford, C. H., Hanby, W. E., & Happey, F., 1949a, 'Evidence for α-Protein Structure in Polypeptides', *Nature,* **164,** 138–139.

——, 1949b, 'The α-β Transformation in a Polypeptide', *Nature,* **164,** 751–752.

——, 1951, 'The Structure of Synthetic Polypeptides. I. X-ray Investigation', *Proc. R. Soc.,* **205A,** 30–47.

Bang, I. C., 1898, 'Die Guanylsäure der Pankreasdrüse und deren Spaltungsprodukte', *Hoppe-Seyler's Z. physiol. Chem.*, **26,** 133–159.

Barker, A. E., 1927, 'The Development of Research in the Department of Textile Industries, Leeds', Memorandum to the Vice-Chancellor dated 22 June. [University of Leeds Archive.]

Barreswil, M., 1861, 'Sur le blanc d'ablette qui sert à la fabrication des perles fausses', *C.r. hebd. Séanc. Acad. Sci., Paris,* **53,** 246.

Barton-Wright, E., & M'Bain, A., 1933, 'Possible Chemical Nature of Tobacco Mosaic Virus', *Nature,* **132,** 1003–1004.

Bateson, W., 1906, 'The Progress of Genetic Research', *Third Conference on Hybridization and Plant Breeding,* London, pp. 90–97.

——, 1909, *Mendel's Principles of Heredity,* Cambridge.

Bateson, W., & Saunders, E. R., 1902, *Reports to the Evolution Committee of the Royal Society,* London, **l,** 1–160.

Bauer, J. H., and Pickles, E. G., 1936, 'A High Speed Vacuum Centrifuge Suitable for the Study of Filterable Viruses', *J. exp. Med.,* **64,** 503–528.

Baumann, E., & Frankel, S., 1895, 'Ueber die Synthese der Homogentisinsäure', *Hoppe-Seyler's Z. physiol. Chem.,* **20,** 219–224.

Baur, E., 1924, 'Untersuchungen über das Wesen, die Entstehung und die Vererbung von Rassenunterschieden bei *Antirrhinum majus', Biblthca genet.,* **4,** 1–170. [Baur wrote: 'It seems very likely that in this series (of colour varieties) one has to do with a *Polymerization* in the sense of organic chemistry. . . .Thus one can picture each chromomere type as a kind of giant molecule, indeed a molecule capable of growth and division.']

Baurhenn, W., 1932, 'Experimente Untersuchungen zur Variabilität und zur Analyse der R-S Umwandlung von Pneumokokken', *Zentbl. Bakt. Parasitkde,* Abt. 1, **126,** 68–92.

Bawden, Sir F. C., 1939, *Plant Viruses and Virus Diseases,* Waltham, Mass. [Pagination from the 2nd ed., 1943.]

Bawden, Sir F. C., & Pirie, N. W., 1937, 'The Isolation and some Properties of Liquid Crystalline Substances from Solanaceous Plants Infected with Three Strains of Tobacco Mosaic Virus', *Proc. R. Soc.,* **123B,** 274–320.

Bawden, Sir F. C., Pirie, N. W., Bernal, J. D., & Fankuchen, I., 1936, 'Liquid Crystalline Substances from Virus-infected Plants', *Nature,* **138,** 1051–1052.

Beadle, G. W., 1939, 'Genetic Control of the Production and Utilization of Hormones', *Int. Conf. Genet.,* **7,** 58–61.

——, 1945, 'Genetics and Metabolism in Neurospora', *Physiol. Rev.,* **25,** 643–663.

——, 1946, 'Genes and the Chemistry of the Organism', *Am. Scient.,* **34,** 31–53. [Reprinted in Taylor, J. H., 1965, 17–49.]

——, 1948, 'Genes and Biological Enigmas', *Am. Scient.,* **36,** 71–74.

——, 1951, 'Chemical Genetics', in: Dunn, L. C. (ed.), 1951, 221–239.

——, 1958, 'Genes and Chemical Reactions in *Neurospora*', *Nobel Lectures... Physiology or Medicine,* 1942–1962, Amsterdam etc., 1964, 587–599.

Beadle, G. W., Pauling, L. C. & Sturtevant, A. H., 1946, 'Memorandum to the Rockefeller Foundation'. [Caltech Archive.]

Beadle, G. W., & Tatum, E., 1941, 'Genetic Control of Biochemical Reactions in Neurospora', *Proc. natn. Acad. Sci. U.S.A.,* **27,** 499–506.

Bechhold, H., 1907, 'Kolloidstudien mit der Filtrations-methode', *Z. phys. Chem.,* **60,** 257–318.

Bechhold, H., & Schlesinger, M., 1931, 'Die Grossenbestimmung von subvisiblem Virus durch Zentrifugieren. Die Grösse des Pockenvaccine- und Hühnerpesterregers', *Biochem. Z.,* **236,** 387–414.

Bell, F. O., 1939, *X-ray and Related Studies of the Structure of the Proteins and Nucleic Acids,* Leeds. [Ph.D. Thesis.]

Bergmann, M., 1926, 'Allgemeine Strukturchemie der komplexen Kohlenhydrate und der Proteine', *Ber. dt. chem. Ges.,* **59,** 2973–2981.

——, 1936, 'Proteins and Proteolytic Enzymes', *Harvey Lect.,* 37–56.

——, 1939, 'Some Biological Aspects of Protein Chemistry', *J. Mt. Sinai Hosp.,* **6,** 171–186.

Bergmann, M., & Niemann, C., 1937, 'Newer Biological Aspects of Protein Chemistry', *Science,* **86,** 187–190.

Bernal, J. D., 1930a, Contribution to the Discussion on the Report of the Syndicate upon the Position of Mineralogy in the Studies of the University, *Cambridge University Reporter,* 11 February 1930, 653–654.

——, 1930b, 'The Place of X-ray Crystallography in the Development of Modern Science', *Radiology,* **15,** 1–12.

——, 1931, 'The Crystal Structure of the Natural Amino Acids and Related Compounds', Z. *Krystallogr. Miner.,* **78,** 363–369.

——, 1935a, Undated Memorandum. [Needham Archive, Gonville & Caius College, Cambridge.]

——, 1935b, Letter to Fankuchen, dated 31 October. [American Institute of Physics Archive.]

——, 1939a, *The Social Function of Science, London.*

——, 1939b, Letter to Edwin Cohn, dated 12 September. [American Institute of Physics Archive.]

——, 1939c, Letter to Haurowitz, dated 12 September. [Am. Inst. Phys. Archive.]

——, 1939d, Letter to Fankuchen, dated 1 December. [Am. Inst. Phys. Archive.]

——, 1939e, 'Structure of Proteins', *Proc. R. Instn Gt Br.,* **30,** 541–557.

——, 1940a, Letter to Fankuchen, undated but probably early 1940. [Am. Inst. Phys. Archive.]

——, 1940b, Letter to Fankuchen, dated 13 April. [Am. Inst. Phys. Archive.]

——, 1940c, Letter to Fankuchen, dated 25 July. [Am. Inst. Phys. Archive.]

——, 1940d, *The Cell and Protoplasm,* Washington.

——, 1945, Letter to Fankuchen, dated 27 November. [Am. Inst. Phys. Archive.]

——, 1951, *The Physical Basis of Life,* London.

——, 1962, 'My Time at the Royal Institution, 1923-27', in: Ewald, P. (ed), 1962, 522–525.

——, 1963, 'William Thomas Astbury 1898-1961', *Biogr. Mem. Fellows R. Soc.,* **9,** 1–35.

——, 1964, 'Professor Isidore Fankuchen', *Nature,* **203,** 916–917.

——, 1966, Opening Remarks to: 'Principles of Biomolecular Organization', *Ciba Foundation Symposium,* London, 1–6.

——, 1968, 'The Material Theory of Life', *Labour Monthly,* July 1968, 324–326. [Review of *The Double Helix.*]

Bernal, J. D., & Fankuchen, I., 1937, 'Structure Types of Protein "Crystals" from Virus-infected Plants', *Nature,* **139,** 923–924.

——, 1941, 'X-ray and Crystallographic Studies of Plant Virus Preparations Pts. I, II, & III', *J. gen. Physiol.,* **25,** 111–165.

Bernal, J. D., Fankuchen, I., & Perutz, M., 1938, 'An X-ray Study of Chymotrypsin and Haemoglobin', *Nature,* **141,** 523–524.

Bernal, J. D., & Fowler, R. H., 1933, 'A Theory of Water and Ionic Solution, with Particular Reference to Hydrogen and Hydroxyl Ions', *J. chem. Phys.,* **1,** 515–548.

Bohr, C., Hasselbalch, K. A., & Krogh, A., 1904, 'Ueber einen in biologischer Beziehung wichtigen Einfluss, den die Kohlensäurespannung des Blutes auf dessen Säurestoffbindung übt', *Skand. Arch. Physiol.,* **16,** 401–412.

Bohr, N., 1933, 'Light and Life', *Nature,* **131,** 421–423.

——, 1958, *Atomic Physics and Human Knowledge,* London, vol. i, 94–101.

——, 1963, 'Light and Life Revisited', *I.C.S.U. Rev.,* **5,** 194–199.

Boivin, A., 1941, 'Sur la réversion de la forme rugueuse (rough) en forme lisse (smooth) et sur la stabilité des types sérologiques (antigène 0) chez les *Salmonella*', *C.r. Séanc. Soc. Biol.,* **135,** 796–799.

——, 1947, 'Directed Mutation in Colon Bacilli, by an Inducing Principle of Desoxyribonucleic Nature: Its Meaning for the General Biochemistry of Heredity', *Cold Spring Harb. Symp. quant. Biol.,* **12,** 7–17.

——, 1948, 'Les acides nucléiques dans la constitution cytologique et dans la vie de la cellule bactérienne', *C.r. Séanc. Soc. Biol.,* **142,** 1258–1273.

Boivin, A., Delaunay, A., Vendrely, R., & Lehoult, Y., 1945, 'L'acide thymonucléique polymérisé, principe susceptible de determiner la specificité sérologique et l'équipement enzymatique des bactéries. Signification pour la biochemie de l'hérédité', *Experientia,* **1,** 334–335.

Boivin, A., Vendrely, R., & Tulasne, R., 1949, 'La spécificité des acides nucléiques chez les êtres vivants, specialement chez les bactéries', *Colloques int. C.N.R.S.,* No. 8, 67–78. [Held June/July 1948.]

Boivin, A., Vendrely, R., & Vendrely, C., 1948, 'L'acide désoxyribonucléique du noyau cellulaire, dépositaire des caractères héréditaires; arguments d'ordre analytique', *C.r. hebd. Séanc. Acad. Sci., Paris,* **226,** 1061–1063.

Bonner, D., 1946, 'Biochemical Mutations in *Neurospora', Cold Spring Harb. Symp. quant. Biol.,* **11,** 14–24.

Boveri, T., 1904, *Ergebnisse über die Konstitution der chromatischen Substanz des Zellkerns,* Jena.

Bragg, Sir W. H., 1915, 'X-rays and Crystal Structure', *Phil. Trans. R. Soc.,* **215A,** 253–274.

——, 1921, 'The Structure of Organic Crystals', *Proc. phys. Soc. Lond.,* **34,** 33–50. [Presidential Address, November 1921.]

——, 1922, 'The Significance of Crystal Structure', *J. chem. Soc.,* **121,** 2766–2787.

——, 1925, *Concerning the Nature of Things. Six Lectures Delivered at the Royal Institution,* London.

——, 1926, *Old Trades and New Knowledge. Six Lectures Delivered... Christmas* 1925, London.

——, 1928, Letter to Barker, dated 19 May. [Leeds University Archive.]

——, 1930, 'Cellulose in the Light of the X-rays', *Nature,* **125,** 315–323.

——, 1933, 'Crystals of the Living Body', *Nature,* **132,** 11–13; 50–53.

Bragg, Sir W. L., 1929, 'The Determination of Parameters in Crystal Structures by Means of Fourier Series', *Proc. R. Soc.,* **123A,** 537–559.

——, 1949a, Letter to Lipson, dated 14 January. [Royal Institution Archive.]

——, 1949b, Letter to Harington, dated 7 February. [R.I. Archive.]

——, 1949c, Letter to Harington, dated 4 May. [R.I. Archive.]

——, 1949d, Letter to Heilbron, dated 4 May. [R.I. Archive.]

——, 1965, 'First Stages in the X-ray Analysis of Proteins', *Rep. Prog. Phys.,* **28,** 1–14.

——, 1967a, Letter to Weaver, dated 17 May. [Original with Weaver.]

——, 1967b, Recorded interview with the author, Royal Institution, December.

Bragg, Sir W. L., Kendrew, J. C., & Perutz, M. F., 1950, 'Polypeptide Chain Configuration in Crystalline Proteins', *Proc. R. Soc.,* **203A,** 321–357.

Bridges, C. B., 1925, 'Sex in Relation to Chromosomes and Genes', *Am. Nat.,* **59,** 127–137.

Brill, R., 1923, 'Ueber Seidenfibroin. I', *Justus Liebigs Annln. Chem.,* **434,** 204–217.

Brill, R., & Halle, F., 1938, 'Ueber das kautschukähnliche Verhalten eines Kunststoffes (Oppanol) im Röntgenlicht', *Naturwissenschaften,* **26,** 12–13.

Brink, R. A., & Styles, E. D. (eds.), 1967, *Heritage from Mendel,* Madison etc. [Proceedings of the Mendel Centennial Symposium, Genetics Society of America.]

Broomhead, J. M., 1948, 'The Structures of Pyrimidines and Purines. II. A Determination of the Structure of Adenine Hydrochloride by X-ray Methods', *Acta Crystallogr.,* **1,** 324–329.

——, 1950, *An X-ray Investigation of Certain Sulphonates and Purines,* Cambridge. [Ph.D. Thesis.]
——, 1951, 'The Structure of Pyrimidines and Purines. IV. The Crystal Structure of Guanine Hydrochloride and its Relation to that of Adenine Hydrochloride', *Acta Crystallogr.,* **4,** 92–100.
Brown, D. M., & Todd, Sir A. R., 1952, 'Nucleotides. Pt.X. Some Observations on the Structure and Chemical Behaviour of the Nucleic Acids', *J. chem. Soc.,* 52–58.
Brown, G. L., 1969, Minuted Interview with the Author, London, January.
Buck, J. B., & Melland, A. M., 1942, 'Methods for Isolating, Collecting and Orienting Salivary Gland Chromosomes for Diffraction Analysis', *J. Hered.,* **33,** 173–183.
Bunn, C. W., 1939, 'The Crystal Structure of Long-chain Normal Paraffin Hydrocarbons. The 'Shape' of the CH_2 Group', *Trans. Faraday Soc.,* **35,** 482–491.
——, 1942, 'Molecular Structure and Rubber-like Elasticity. I, II, & III', *Proc. R. Soc.,* **180A,** 40–99.
——, 1945, *Chemical Crystallography. An Introduction to Optical and X-ray Methods,* Oxford.
Burian, R., 1904, 'Chemie der Spermatozoen. I', *Ergebn. Physiol.,* physiol. Abth., **3,** 48–106.
——, 1906, 'Chemie der Spermatozoen. II', *Ergebn. Physiol.,* physiol. Abth., **5,** 768–846.
Burnet, Sir M., 1944, 'Some Borderlines of Microbiology, Genetics and Biochemistry', *Aust. J. Sci.,* **7,** 1–6.
——, 1959, *The Clonal Selection Theory of Acquired Immunity,* Nashville.
——, 1968, *Changing Patterns; An Atypical Biography,* Melbourne & London.

Cairns, J., Stent, G. S., & Watson, J. D. (eds.), 1966, *Phage and the Origins of Molecular Biology,* Cold Spring Harbor.
Caldwell, J., 1934, 'Possible Chemical Nature of Tobacco Mosaic Virus', *Nature,* **133,** 177.
Caldwell, P. C., & Hinshelwood, Sir C., 1950a, 'The Nucleic Acid Content of *Bact. lactis aerogenes*', *J. chem. Soc.,* 1415–1418.
——, 1950b, 'Some Considerations on Autosynthesis in Bacteria', *J. chem. Soc.,* 3156–3159.
Carlisle, C. H., 1968, Minuted Interview with the Author, London, January.
Carlson, E. A., 1966, *The Gene: A Critical History,* Philadelphia & London.
——, 1971, 'An Unacknowledged Founding of Molecular Biology: H. J. Muller's Contribution to Gene Theory, 1910–1936, *J. Hist. Biol.,* **4,** 149–170.
——, 1972, Letter to the Author, dated 28 February.
Carpenter, G. B., & Donohue, J., 1950, 'The Crystal Structure of N-Acetylglycine', *J. Am. chem. Soc.,* **72,** 2315–2328.

Carr, J. G., 1947?, Letter to Darlington, undated. [Original with Darlington.]

Caspersson, T., 1934, 'Druckfiltierung von Thymonucleinsäure', *Biochem. Z.*, **270,** 161–163.

——, 1936, 'Ueber den chemischen Aufbau der Strukturen des Zellkernes', *Acta Med. Skand.*, **73,** Suppl. 8, 1–151.

——, 1940, 'Nukleinsäureketten und Genvermehrung', *Chromosoma,* **1,** 605–619.

——, 1941, 'Studien uber den Eiweissumsatz der Zelle', *Naturwissenschaften,* **29,** 33–43.

——, 1967, Recorded Interview with the Author, Stockholm, September.

Caspersson, T., Hammarsten, E., & Hammarsten, H., 1935, 'Interactions of Proteins and Nucleic Acids', *Trans. Faraday Soc.*, **31,** 367–389.

Caspersson, T., & Schultz, J., 1938, 'Nucleic Acid Metabolism of the Chromosomes in Relation to Gene Reproduction', *Nature,* **142,** 294–295.

Cavalieri, L. F., Bendich, A., Tinker, J. F., & Brown, G. B., 1948, 'Ultraviolet Absorption Spectra of Purines, Pyrimidines and Triazolopyrimidines', *J. Am. chem. Soc.*, **70,** 3875–3880.

Chargaff, E., 1947, 'On the Nucleoproteins and Nucleic Acids of Microorganisms', *Cold Spring Harb. Symp. quant. Biol.*, **12,** 28–34.

——, 1950a, 'Chemical Specificity of Nucleic Acids and the Mechanism of their Enzymatic Degradation', *Experientia,* **6,** 201–209.

——, 1950b, 'Structure and Function of Nucleic Acids as Cell Constituents', *Fed. Proc.*, **10,** 654–659.

——, 1951, 'Some Recent Studies on the Composition and Structure of Nucleic Acids', *J. cell. comp. Physiol.*, **38,** Suppl. 1, 41–58.

——, 1957, 'The Base Composition of Desoxyribonucleic Acid and Pentose Nucleic Acid in Various Species', in: McElroy, W. D., & Glass, B. (eds.), 1957, 521–527.

——, 1963, *Essays on Nucleic Acids,* Amsterdam.

——, 1968a, 'A Quick Climb up Mount Olympus', *Science,* **159,** 1448–1449. [Review of The Double Helix.]

——, 1968b, Recorded Interview with the Author, New York, September.

——, 1970, Letter to Pollock, dated 2 February. [Original with Pollock.]

——, 1973, Minuted Interview with the Author, New York, January.

Chargaff, E., & Davidson, J. N. (eds.), 1955, *The Nucleic Acids: Chemistry and Biology,* vol. ii, New York.

Chargaff, E., Magasanik, B., Doniger, R., & Vischer, E. M., 1949, 'The Nucleotide Composition of Ribonucleic Acids', *J. Am. chem. Soc.*, **71,** 1513–1514.

Chargaff, E., Vischer, E. M., Doniger, R., Green, C., & Misani, F., 1949, 'The Composition of the Desoxypentose Nucleic Acids of Thymus and Spleen', *J. biol. Chem.*, **177,** 405–416. [Referred to as Chargaff *et al.*, 1949.]

Chargaff, E., Zamenhof, S., & Green, C., 1950, 'Composition of Human Desoxypentose Nucleic Acid', *Nature,* **165,** 756-757.

Child, C. M., 1915, *Senescence and Rejuvenescence,* Chicago.

Childs, B., 1970, 'Sir Archibald Garrod's Conception of Chemical Individuality—A Modern Appreciation', *New Engl. J. Med.*, **282,** 71–77.

Churchill, F. B., 1973, 'Cytology and Mendelism: The Preparation for and the Convergence of Two Fields. [Paper read at the one-day meeting on the history of genetics, Washington, D.C., December 1972.]

Clark, R., 1968, *The Life of J. B. S. Haldane,* London.

Clark, V. M., Todd, Sir A. R., & Zussman, J., 1951, 'Nucleotides. Pt. VIII. *cyclo* Nucleoside Salts. A Novel Rearrangement of some Toluene-p-sulphonylnucleosides', *J. chem. Soc.*, 2952–2958.

Clews, C. J. B., & Cochran, W., 1948, 'The Structure of Pyrimidines and Purines. I. A Determination of the Structures of 2-amino-4-chloropyrimidine and 2-amino-4,6-dichloropyrimidine by X-ray Methods', *Acta Crystallogr.*, **1,** 4–11.

——, 1949, 'The Structure of Pyrimidines and Purines. III. An X-ray Investigation of Hydrogen Bonding in Aminopyrimidines', *Acta Crystallogr.*, **2,** 46–57.

Coburn, A. F., 1969, 'Ostwald Theodore Avery and DNA', *Perspect. Biol. Med.*, **12,** 623–630.

Cochran, W., 1951, 'The Structure of Pyrimidines and Purines. V. The Electron Distribution in Adenine Hydrochloride', *Acta Crystallogr.*, **4,** 81–92.

——, 1968, Letter to the Author, dated 9 July.

Cochran, W., & Crick, F. H. C., 1952, 'Evidence for the Pauling-Corey α-Helix in Synthetic Polypeptides', *Nature,* **169,** 234–235.

Cohen, S. S., 1948, 'Synthesis of Bacterial Viruses; Synthesis of Nucleic Acid and Protein in *Escherichia coli* Infected with $T2r^+$ Bacteriophage', *J. biol. Chem.*, **174,** 295–303.

——, 1968, *Virus Induced Enzymes,* New York.

——, 1973a, Recorded Interview with the Author, Denver, January.

——, 1973b, Letter to the Author, dated 6 March.

Cohen, S. S., & Stanley, W. M., 1942, 'The Molecular Size and Shape of the Nucleic Acid of Tobacco Mosaic Virus', *J. biol. Chem.*, **144,** 589–598.

Cohn, E. J., & Edsall, J. T., 1943, *Proteins, Amino Acids and Peptides as Ions and Dipolar Ions. . . ,* New York.

Cohn, E. J., & Hendry, J. L., 1922/23, 'Studies in the Physical Chemistry of the Proteins. II. . . ', *J. gen. Physiol.*, **5,** 521–554.

Cohn, F., 1875, *Beiträge zur Biologie der Pflanzen,* Breslau.

Committee of the Privy Council for Scientific & Industrial Research (C.S.I.R.), 1928, *Report of the Year* 1926–27, London.

Commoner, B., 1964, 'Deoxyribonucleic Acid and the Molecular Basis of Self-duplication', *Nature,* **203,** 486–491.

——, 1968, 'Failure of the Watson–Crick Theory as a Chemical Explanation of Inheritance', *Nature,* **220,** 334–340.

Cooper, K., 1955, Discussion following Hotchkiss' paper on 'Bacterial Transformation', in: 'Symposium on Genetic Recombination', *J. cell. comp. Physiol.*, **45,** Suppl. 2, 18–20. [Held in 1954.]

Corey, R. B., 1938, 'The Crystal Structure of Diketopiperazine', *J. Am. chem. Soc.*, **60,** 1598–1604.

——, 1940, 'Interatomic Distances in Proteins and Related Substances', *Chem. Rev.*, **26,** 227–236. [Lecture delivered September 1939.]

——, 1948, 'X-ray Studies of Amino Acids and Peptides', *Adv. Protein Chem.*, **4,** 385–406.

——, 1953, Letter to Franklin, dated 13 April. [Original with Klug.]

Corey, R. B., & Donohue, J., 1950, 'Interatomic Distances and Bond Angles in the Polypeptide Chain of Proteins', *J. Am. chem. Soc.*, **72,** 2899–2900.

Cork, J. M., 1927, 'The Crystal Structure of Some of the Alums', *Phil. Mag.*, **4,** 688–698.

Corner, G. W., 1964, A *History of the Rockefeller Institute* 1901–1953: *Origins and Growth,* New York.

Correns, C. F. J. E., 1903, 'Ueber die dominierenden Merkmale der Bastarde', *Ber. dt. bot. Ges.*, **21,** 133–147.

——, 1904, 'Ein typisch spaltender Bastard zwischen einer einjährigen und einer zweijährigen Sippe der *Hyoscyamus niger'*, *Ber. dt. bot. Ges.*, **22,** 517–524.

——, 1919, 'Vererbungsversuche mit buntblättrigen Sippen I. *Capsella bursa pastoris albovariabilis* und *chlorina'*, *Sber. preuss. Akad. Wiss.*, math. nat. Kl., **2,** 585–610.

Coulson, C., 1951, Letter to Franklin, dated 27 February. [Original with Coulson.]

Crew, F. A. E., 1939, 'Seventh International Genetical Congress', *Nature,* **144,** 496–498.

Crick, F. H. C., 1947, Application to the M.R.C. for a Fellowship, as reported in a Letter to the Author from Mrs. Anne Sanderson. [Medical Research Council, Archive, London.]

——, 1951, Comments on the Wines Listed in the Catalogue of Hocks Moselles for the Wine Tasting on 31 October. [Matthew & Son Ltd.]

——, 1952, 'Experiments to Determine if [there is] any attraction between Pairs of Nucleic Acid Bases. [Crick's Laboratory Notebook. Also contains Experiments with Nucleosides.]

——, 1953a, Letter to Michael Crick, dated 19 March. [Original with Crick.]

——, 1953b, Letter to Franklin, enclosing a 5-page critique of her first two *Acta Crystallographica* papers, dated 5 June. [Original with Klug.]

——, 1953c, *Polypeptides and Proteins: X-ray Studies,* Cambridge. [Ph.D. Thesis, submitted July.]

——, 1954, 'The Structure of the Synthetic-Polypeptides', *Sci. Prog., London,* **42,** 205–219.

——, 1955, 'On Degenerate Templates and the Adaptor Hypothesis'. [17-page Cyclostyled Note for the RNA Tie Club, undated. Original with Brenner.]

——, 1958, 'On Protein Synthesis', *Symp. Soc. exp. Biol.*, **12**, 138–163. [Symposium held in 1957.]

——, 1962, '*The Prizewinners*'. [Transcript of BBC Television Programme.]

——, 1965, 'Recent Research in Molecular Biology: Introduction', *Br. med. Bull.*, **21,** 183–186.

——, 1966, *Of Molecules and Men,* Seattle and London.

——, 1967a, Letter to Watson, dated 13 April. [Critique of 'Honest Jim', subsequently *The Double Helix.*]

——, 1967b, 'The Discovery of the Structure of DNA', Public Lecture, Oxford, May. [Crick's MS Lecture Notes, Original with Crick.]

——, 1968, Recorded Interview with the Author, Cambridge, 8 March.

——, 1969, Letter to the Author, dated 5 June.

——, 1970a, Letter to the Author, dated 15 January. [Comments on the Author's paper: 'Schrödinger's Problem: What is Life?'—Olby, 1971.]

——, 1970b, 'Central Dogma of Molecular Biology', *Nature,* **227,** 561–563.

——, 1970c, Letter to the Author, dated 5 August. [Contains comments on the Author's Paper: 'Francis Crick, DNA and the Central Dogma—Olby 1970. On page 2 he says: 'I realized some years ago that my definition of the sequence hypothesis... in the SEB paper is back-to-front. It should be stated as saying that all (almost all) amino acid sequences in protein are coded for by a nucleic acid sequence'.]

——, 1972, Recorded Interviews with the Author, Cambridge, 7 August.

——, 1973, Comments on the Typescript of *The Path to the Double Helix* enclosed with his letter to the Author dated 13 April.

Crick, F. H. C., & Hughes, A. A. W., 1950, 'The Physical Properties of Cytoplasm, a Study by Means of the Magnetic Particle Method, Pt. I: Experimental, Pt. II: Theoretical Treatment', *Expl. Cell Res.,* **1,** 37–64, 505–542.

Crick, F. H. C., & Watson, J. D., 1951, 'The Structure of Sodium Thymonucleate. A Possible Approach: Summary'. [8-page MS in Crick's hand, Original with Crick.]

——, 1954, 'The Complementary Structure of Deoxyribonucleic Acid', *Proc. R. Soc.,* **223A,** 80–96.

Crompton, H., 1912, 'The Possible Limitation of Molecular Magnitude', *Proc. chem. Soc.,* **28,** 193–194.

Crowther, J. A., 1924, 'Some Considerations Relative to the Action of X-rays on Tissue Cells', *Proc. R. Soc.,* **96B,** 207–211.

——, 1926, 'The Action of X-rays on *Colpidium colpoda*', *Proc. R. Soc.,* **100B,** 390–404.

Crowther, J. G., 1968, *Scientific Types,* London.

Cuénot, L., 1903, 'L'hérédité de la pigmentation chez les souris', *Archs Zool. exp. gén.,* **1,** 33–41. [4th sér.]

Cunningham, J. T., & MacMunn, C. A., 1893, 'On the Coloration of the Skins of Fishes, Especially of Pleuronectidae', *Phil. Trans. R. Soc.*, **184,** 765–812.

Cyon, E. von, 1910, 'Eduard Pflüger: Ein Nachruf', *Pflügers Arch. ges. Physiol.*, **132,** 1–19.

Dakin, 1920, 'Amino Acids of Gelatin', *J. biol. Chem.*, **44,** 499–529.

Dale, Sir H., 1946, 'Address of the President', *Proc. R. Soc.*, **185A,** 127–143. [Read November, 1945.]

Danielli, J. F., 1969, Minuted Interview with the Author, London, February.

Daniels, M., 1972, 'Tautomerism of Uracil in Aqueous Solutions: Spectroscopic Evidence', *Proc. natn. Acad. Sci. U.S.A.*, **69,** 2488–2491.

Darlington, C. D., 1931, 'Meiosis', *Biol. Rev.*, **6,** 221–264.

——, 1932, *Recent Advances in Cytology*, 1st ed., London.

——, 1935, 'The Internal Mechanics of the Chromosomes. III. Relational Coiling and Crossing-over in *Fritillaria*', *Proc. R. Soc.*, **118B,** 74–96. [First statement on the Molecular Spiral.]

——, 1937, *Recent Advanccs in Cytology*, 2nd ed., London.

——, 1939a, 'Cytology', *Nature*, **144,** 816–817.

——, 1939b, Letter to Astbury, dated 3 December. [Original with Darlington.]

——, 1939c, *The Evolution of Genetic Systems*, Edinburgh & London.

——, 1940a, Letter to Caspersson, dated 17 January. [Original with Darlington.]

——, 1940b, 'Pure Experiment', *Nature*, **145,** 477.

——, 1942, 'Chromosomes Chemistry and Gene Action', *Nature*, **149,** 66–69.

——, 1945, 'The Chemical Basis of Heredity and Development', *Discovery*, **6,** 79–86.

——, 1947, 'Nucleic Acid and the Chromosomes', *Symp. Soc. exp. Biol.*, **1,** 252–269.

——, 1955, 'Nucleic Acid: the Midwife Molecule', *Adv. Sci.*, **12,** 355. [For Darlington's attitude to the Watson–Crick model see: Darlington, 'The Chromosome as a Physico-Chemical Entity', *Nature*, **176** (1955), 1139–1144.]

——, 1964, *Genetics and Man*, London.

——, 1968, 'Determined but Lonely', *Nature*, **220,** 933–934. [Review of *The Life of J. B. S. Haldane*, by Ronald Clark.]

Darlington, C. D., & La Cour, L., 1940, 'Nucleic Acid Starvation of Chromosomes in *Trillium*', *J. Genet.*, **40,** 185–213.

——, 1973, Letter to the Author, dated 19 February.

Darwin, G., 1875, 'Marriages between First Cousins in England and their Effects', *J. stat Soc .*, **38,** 153–184.

Davidson, J. N., 1947a, 'Some Factors Influencing the Nucleic Acid Content of Cells and Tissues', *Cold Spring Harb. Symp. quant.. Biol.*, **12,** 50–59.

——, 1947b, 'The Distribution of Nucleic Acids in Tissues', *Symp. Soc. exp. Biol.*, **1,** 77–85.

——, 1953, *The Biochemistry of the Nucleic Acids*, 2nd ed., London. [First ed. 1950.]

Davidson, J. N., & Chargaff, E., 1955, Introduction to: Chargaff & Davidson (eds.), 1955, vol. i, pp. 1–8.

Davies, M., 1967, Letter to the Author, dated 22 May.

Dawson, M. H., 1928, 'The Interconvertibility of 'R' and 'S' Forms of *Pneumococcus*', *J. exp. Med.*, **47,** 577–591.

——, 1930, 'The Transformation of Pneumococcal Types. I', *J. exp. Med.*, **51,** 99–122; Part II, pp. 123–147. [Part I received 17 July 1929, Part II on 15 October 1929.]

Dawson, M. H., & Sia, R. H. P., 1931, 'In Vitro Transformation of Pneumococcal Types. Parts I & II', *J. exp. Med.*, **54,** 681–699, 701–710.

Day, H. B., 1930, 'The Preparation of Antigenic Specific Substance from *Streptococci* and *Pneumococci* (Type I)', *Br. J. exp. Med.*, **11,** 164–173.

Debye, P., & Scherrer, P., 1916, 'Interferenzen an regellos orientierten Teilchen im Röntgenlicht', *Nachr. k. Ges. Wiss. Göttingen,* 1–26. [Submitted 4 December 1915. Date of Volume, 1916.] Also same title (1916) in: *Phys. Z.*, **17,** 277–283.

Delbrück, M., 1940, 'Radiation and the Hereditary Mechanism', *Am. Nat.*, **74,** 350–362.

——, 1941, 'A Theory of Autocatalytic Synthesis of Polypeptides and its Application to the Problem of Chromosome Reproduction', *Cold Spring Harb. Symp. quant. Biol.*, **9,** 122–124.

——, 1942, 'Bacterial Viruses (Bacteriophages)', *Adv. Enzymol.*, **2,** 1–32.

——, 1945, 'Interference between Bacterial Viruses. III. The Mutual Exclusion and the Depressor Effect', *J. Bact.*, **50,** 151–170.

——, 1946, 'Experiments with Bacterial Viruses (Bacteriophages)', *Harvey Lect.*, **41,** 161–187.

——, 1948, Letter to Luria, dated 30 November. [Caltech Archive.]

——, 1949, 'A Physicist Looks at Biology', *Trans. Connecticut Acad. Sci.*, **38,** 173–190 [Pagination taken from the Reprint in Cairns *et al.*, 1966, pp. 9–22.]

——, 1950, Letter to Muller, dated 1 June. [MS Dept., Lilly Library, Bloomington. In this letter Delbrück reports on a Seminar at Caltech on vibratory long-range forces and concludes: 'I think it is quite unfortunate that the paper (Jehle's) was published in its present form'.]

——, 1952a, Letter to Hershey, dated 2 April. [Caltech Archive.]

——, 1952b, Letter to Workington [N.F.I.P.], dated 4 June. [National Foundation Archive.]

——, 1953a, Letter to Watson, dated 5 March. [Caltech Archive.]

——, 1953b, Letter to Watson, dated 19 March. [Caltech Archive.]

——, 1953c, 'Introductory Remarks about the Programme', *Cold Spring Harb. Symp. quant. Biol.*, **18,** 1–2.

——, 1953d, Letter to Watson, dated April. [Caltech Archive.]

——, 1953e, Letter to Bohr, dated 14 April. [Caltech Archive.]

——, 1955, Letter to Rich, dated 9 November. [This followed a hoax letter from Gamow to Watson, then in Europe. As Watson was not in Cambridge at the time, the Delbrück letter was opened first. Original with Rich.]
——, 1967a, *Personal Records Questionnaire of the Royal Society.* [Caltech Archive.]
——, 1967b, Letter to Weaver, dated 11 April. [Original with Weaver.]
——, 1970, 'A Physicist's Renewed Look at Biology: Twenty Years Later', *Science,* **168,** 1312–1315. [Nobel Lecture.]
——, 1971a, 'Homo scientificus According to Beckett', *Chemistry and Society Lecture Series at the California Institute of Technology,* Pasadena.
——, 1971b, Letter to the Author, dated 24 June.
——, 1972, Comments on the Author's chapter xv.
Dellaporte, B., 1939, 'Sur les acides nucléiques des levures et leur localisation', *Rev. gén. Biol.,* **51,** 449–482.
Dessauer, F., 1922, 'Ueber einige Wirkungen von Strahlen', Z. *Physik.,* **12,** 38–47.
Dhéré, C., 1906, 'Sur l'absorption des rayons ultra-violets par l'acide nucléique extrait de la levure de bière', *C.r. Séanc. Soc. Biol.,* **60,** 34.
D'Hérelle, F., 1917, 'Sur un microbe invisible antagoniste des bacilles dysentériques', *C.r. hebd. Séanc. Acad. Sci. Paris,* **165,** 373.
Dobzhansky, T., 1941, *Genetics and the Origin of Species,* New York.
——, 1951, *Genetics and the Origin of Species,* 3rd edition, revised, New York.
——, 1968, Recorded Interview with the Author, New York, September.
Dochez, A. R., & Avery, O. T., 1917, 'The Elaboration of Specific Soluble Substance by *Pneumococcus* during Growth', *J. exp. Med.,* **26,** 477–493.
Donohue, J., 1952, 'The Hydrogen Bond in Organic Crystals', *J. phys. Chem.,* **56,** 502–510. [Presented at the Cleveland Meeting of the American Chemical Society, April 1951.]
——, 1953, 'Hydrogen Bonded Helical Configurations of the Polypeptide Chain', *Proc. natn. Acad. Sci. U.S.A.,* **39,** 470–478.
——, 1956, 'Hydrogen-bonded Helical Configurations of Polynucleotides', *Proc. natn. Acad. Sci. U.S.A.,* **42,** 60–65. [It is ironic that the third hydrogen bond between guanine and cytosine, intentionally omitted by Watson and Crick, may be considered to confer extra stability on the Watson–Crick pairing scheme as compared with all others—see p. 65.]
——, 1969a, 'Fourier Analysis and the Structure of DNA', *Science,* **165,** 1091–1096.
——, 1969b, Letter to the Author, dated 28 October.
——, 1970, 'Fourier Series and Difference Maps as Lack of Structure Proof: DNA is an Example', *Science,* **167,** 1700–1702.
——, 1973, Comments on the Author's chapter xvii, dated 23 January.
Drake, J. W., 1969, 'Mutagenic Mechanisms', *Ann. Rev. Genet.,* **3,** 247–268.
Driesch, H. A. E., 1894, *Analytische Theorie der organischen Entwicklung,* Leipzig.

Duane, W., 1925, 'The Calculation of the X-ray Diffracting Power at Points in a Crystal., *Proc. natn. Acad. Sci. U.S.A.,* **11,** 489–493.

Dubos, R. J., 1937, 'The Decomposition of Yeast Nucleic Acid by a Heat-resistant Enzyme', *Science,* **85,** 549–550.

——, 1956, 'Ostwald Theodore Avery', *Biogr. Mem. Fellows R. Soc.,* **2,** 35–48.

——, 1957, Recorded Interview with Saul Benison, Transcript vol. i, p. 39. [Columbia University Oral History Collection.]

——, 1969, Personal Communication, Serbelloni, September.

——, 1972, Letter to the Author, dated 14 February.

Dubos, R. J., & MacLeod, C. M., 1937, 'Effects of a Heat-resistant Enzyme Upon the Antigenicity of Pneumococci', *Proc. Soc. exp. Biol. Med.,* **36,** 696–697.

Dubos, R. J., & Thompson, R. H. S., 1938, 'The Decomposition of Yeast Nucleic Acid by a Heat-resistant Enzyme', *J. biol. Chem.,* **124,** 501–510.

Dulbecco, R., 1949, 'Reactivation of UV-inactivated Bacteriophage by Visible Light', *Nature,* **163,** 949–950.

Dunn, L. C., 1951, *Genetics in the Twentieth Century: Essays on the Progress of Genetics During its First Fifty Years,* New York.

——, 1962, 'Cross-currents in the History of Human Genetics', *Am. J. hum. Genet.,* **14,** 1–13.

Dvorak, M., 1927, 'The Effect of Mosaic on the Globulin of Potato', *J. infect. Dis.,* **41,** 215–221.

Dyer, H. B., 1951, 'The Crystal Structure of Cysteylglycine-Sodium Iodide', *Acta Crystallogr.,* **4,** 42–50. [I am told there is a correction note on this paper, but I have failed to find it.]

Eastwood, A., 1922, 'A Brief Review of Recent Bacteriological Work on Pneumococci', *Rep. public Health and med. Subjects,* **13,** 2–19.

——, 1923, 'Bacterial Variation and Transmissible Autolysis; The Relation of Bacterial Enzymes to Bacterial Structure', *Rep. public Health and med. Subj.,* **18,** 14–34.

Edlbacker, S., 1928, 'Albrecht Kossel zum Gedächtnis', *Hoppe-Seyler's Z. physiol. Chem.,* **177,** 1–14.

Edsall, J. T., 1962, 'Proteins as Macromolecules: An Essay on the Development of the Macromolecule Concept and some of its Vicissitudes', *Arch. Biochem. Biophys.,* Suppl. 1, 12–20.

Eigen, M., & de Maeyer, L., 1966, 'Chemical Means of Information Storage and Readout in Biological Systems', *Naturwissenschaften,* **53,** 50–57.

Elford, W. J., & Andrewes, C. H., 1932, 'The Sizes of Different Bacteriophages', *Brit. J. exp. Path.,* **13,** 446–456.

Elliott, S. D., 1970, Letter to Pollock, dated 3 March. [Original with Pollock.]

Ellis, E., & Delbrück, M., 1939, 'The Growth of Bacteriophage', *J. gen. Physiol.,* **22,** 365–384.

Ephrussi, E., 1938, Letter to Muller, dated 13 October. [MS dept. Lilly Library, Bloomington.]

——, 1942, 'Chemistry of "Eye Color Hormones" of *Drosophila*', *Quart. Rev. Biol.*, **17,** 327–338.

Eriksson-Quensel, I.-B., & Svedberg, T., 1936, 'Sedimentation and Electrophoresis of the Tobacco-Mosaic-Virus Protein', *J. Am. chem. Soc.*, **58,** 1863–1867.

Ewald, P. P., 1923, *Kristalle und Röntgenstrahlen,* Berlin.

——, (Ed.), 1962, *Fifty Years of X-ray Crystallography,* Utrecht.

Ewles, J., & Speakman, J. B., 1930, 'Examination of the Fine Structure of Wool by X-ray Analysis', *Proc. R. Soc.*, **105B,** 600–607.

Fankuchen, Mrs D., 1970, Recorded Interview with the Author, Brooklyn, June.

Felix, K., 1955, 'Albrecht Kossel: Leben und Werk', *Naturwissenschaften,* **42,** 473–477.

Feulgen, R., 1914, 'Ueber b-Nukleinsäure', *Hoppe-Seyler's Z. physiol. Chem.*, **91,** 165–173.

——, 1923, *Chemie und Physiologie der Nukleinstoffe nebst Einfuhrung in die Chemie der Purinkörper,* Berlin.

——, 1935, 'Ueber a- und b-Thymonucleinsäure und das die a-Form in die b-Form überführende Ferment (Nucleogelase)', *Hoppe-Seyler's Z. physiol. Chem.*, **237,** 261–267.

——, 1936, 'Die Darstellung der b-Thymonucleinsäure mittels Nucleogelase', *Hoppe-Seyler's Z. physiol. Chem.*, **238,** 105–110.

Feulgen, R., & Rossenbeck, H., 1924, 'Mikroskopisch-chemischer Nachweis einer Nucleinsäure vom Typus der Thymonucleinsäure und die darauf beruhende elektive Färburg von Zellkernen in mikroskopischen Präparaten', *Hoppe-Seyler's Z. physiol. Chem.*, **135,** 203–248.

Fischer, A., 1899, *Fixirung, Färbung und Bau des Protoplasmas, Kritische Untersuchungen über Technik und Theorie in der neueren Zellforschung.* Jena.

Fischer, E., 1906, *Untersuchungen über Aminosäuren, Polypeptide und Proteine,* Berlin.

——, 1913, 'Synthese von Depsiden, Flechtenstoffen und Gerbstoffen', *Ber. dt. chem. Ges.*, **46,** 3253–3289.

——, 1916, 'Isomerie der Polypeptide', *Sber. preuss. Akad. Wiss.*, Halbbd. **2,** 990–1008.

Fisher, E. A., 1923, 'Some Moisture Relations of Colloids—II. Further Observations on the Evaporation of Water from Clay and Wool', *Proc. R. Soc.*, **103A,** 664–675.

Fleming, D., 1968, 'Emigré Physicists and the Biological Revolution', *Perspect. Am. Hist.*, **2,** 176–213. [Reprinted in Fleming, D., & Bailyn, B. (eds.), 1969, *The Intellectual Migration: Europe and America,* 1930–1960, Cambridge, Mass., pp. 152–189.]

Flexner, S., 1929, Letter to P. A. Levene, dated 14 June. [Original in American Philosophical Society Library.]

Florkin, M., 1972, *A History of Biochemistry,* Amsterdam etc. [Comprehensive Biochemistry, vol. 30.]

Fosdick, R. B., 1952, *The Story of the Rockefeller Foundation, New York.*

——, 1956, *John D. Rockefeller Jr.: A Portrait,* New York.

Fraenkel-Conrat, H., 1956, 'The Role of the Nucleic Acid in the Reconstitution of Active Tobacco Mosaic Virus', *J. Am. chem. Soc.,* **78,** 882–883.

Frank, P., 1950, *Modern Science and its Philosophy,* Cambridge.

Franklin, R. E., 1950, Letter to Coulson, dated 12 March. [Original with Coulson.]

——, 1951a, 'Crystalline Growth in Graphitizing and Nongraphitizing Carbons', *Proc. R. Soc.,* **209A,** 196–218.

——, 1951b, Notes entitled 'Colloquium, November 1951'. [Original with Klug.]

——, 1952a, Laboratory Notebook. [Original with Klug.]

——, 1952b, 'Interim Annual Report: January 1 1951 to January 1 1952', King's College London. [5-page MS dated 7 February. Original with Klug.]

——, 1953, 'Interim Annual Report: January 1952 to January 1953. [11-page MS. Original with Klug.]

Franklin, R. E., & Gosling, R. G., 1953a, 'A Note on Molecular Configuration of Sodium Thymonucleate'. [5-page MS dated 17 March. Original with Klug.]

——, 1953b, 'The Structure of Sodium Thymonucleate Fibres. I. The Influence of Water Content', *Acta Crystallogr.,* **6,** 673–677.

——, 1953c, 'Molecular Configuration in Sodium Thymonucleate', *Nature,* **171,**740–741.

——, 1953d, 'Evidence for Two-chain Helix in Crystalline Structure of Sodium Deoxyribonucleate', *Nature,* **172,** 156–157.

——, 1953e, 'The Structure of Sodium Thymonucleate. II. The Cylindrically Symmetrical Patterson Function', *Acta Crystallogr.,* **6,** 678–685.

Fraser, R. D. B., 1950, 'Infra-red Microspectrometry with a 0.8 N.A. Reflecting Microscope', *Faraday Soc. Disc.,* **9,** 378–383.

Frazer, F., 1936, 'Obituary of Garrod', *Lancet,* **i,** 807–809.

Fredga, A., 1953, 'Presentation Speech to Hermann Staudinger', in *Nobel Lectures. . . Chemistry,* 1942–1962, Amsterdam etc., pp. 393–396.

Freundlich, H., 1916, 'Die Doppelbrechung des Vanadinpentoxydsols', *Z. Electrochem. angew. phys. Chem.,* **22,** 27–31.

Frey-Wyssling, A., 1938, *Submikroskopische Morphologie des Protoplasmas und seiner Derivate, Berlin.* [*Protoplasmatologia,* **ii,** A2.]

——, 1964, 'Frühgeschichte und Ergebnisse der submikroskopischen Morphologie', *Mikroskopie,* **19,** 2–12.

Friedrich,W., 1913, 'Eine neue Interferenzerscheinung bei Röntgenstrahlen', *Phys. Z.,* **14,** 317–319.

——, 1922, 'Die Geschichte der Auffindung der Röntgenstrahlinterferenzen' *Naturwissenschaften,* **10,** 363–366.

Friedrich, W., Knipping, P., & Laue, M., 1912, 'Interferenz-Erscheinungen bei Röntgenstrahlen', *Sber. bayer. Akad. Wiss.*, math.-phys. Kl., 303–322. [English trans. in Bacon, 1966, pp. 89–108.]

Friedrich-Freksa, H., 1940, 'Bei der Chromosomenkonjugation wirksame Kräfte und ihre Bedeutung für die identische Verdopplung von Nucleoproteinen', *Naturwissenschaften,* **28,** 376–379.

Fruton, J. S., 1972, *Molecules and Life: Historical Essays on the Interplay of Chemistry and Biology,* New York.

Fuller, W., Wilkins, M. H. F., Wilson, H. R., and Hamilton, L. D., 1957, 'The Molecular Configuration of Desoxyribonucleic Acid. IV. X-ray Diffraction. Study of the A Form', *J. mol. Biol.,* **12,** 60–80.

Funke, O., 1858, *Atlas der physiologischen Chemie,* Leipzig.

Furberg, S., 1949a, 'Crystal Structure of Cytidine', *Nature,* **164,** 22.

——, 1949b, *An X-ray Study of Some Nucleosides and Nucleotides,* London. [Ph.D. Thesis, submitted August.]

——, 1950a, An X-ray Study of the Stereochemistry of the Nucleosides', *Acta chem. Scand.,* **4,** 751–761.

——, 1950b, 'X-ray Studies on the Decomposition Products of the Nucleic Acids', *Trans. Faraday Soc.,* **46,** 791.

——, 1952, 'On the Structure of Nucleic Acids', *Acta. chem. Scand.,* **6,** 634–640.

——, 1967, Recorded Interview with the Author, Oslo, October.

——, 1968, Letter to the Author, dated 29 January.

Gamow, G., 1954a, 'Possible Relation between Deoxyribonucleic Acid and Protein Structure', *Nature,* **173,** 318.

——, 1954b, Letter to Crick dated 8 March. [Original with Crick.]

——, 1966, *Thirty Years that Shook Physics. The Story of Quantum Theory,* London.

Garnier, L., and Voirin, G., 1892, 'De l'alcaptonurie. Charactéres distinctifs de la matière alcaptonique et de la glucose dans les urines', *Arch Physiol.,* Ser. 5, **4,** 225–232.

Garrod, Sir A. E., 1899, 'A Contribution to the Study of Alkaptonuria', *Medico-Chirurgical Trans.,* **82,** 367–394.

——, 1902a, 'About Alkaptonuria', *Medico-Chirurgical Trans.,* **85,** 69–77.

——, 1902b, 'The Incidence of Alkaptonuria: A Study in Chemical Individuality', *Lancet,* **ii,** 1616–1620. [Pagination from the reprint in Harris, H., 1963, pp. 110–119.]

——, 1909, *Inborn Errors of Metabolism,* London. [Pagination from Harris, 1963, pp. 1–50.]

Gates, F. L., 1928, 'On Nuclear Derivatives and the Lethal Action of Ultraviolet Light', *Science,* **68,** 479–480.

Gierer, A., & Schramm, G., 1956, 'Die Infektiosität der Nucleinsäure aus Tabakmosaikvirus', *Z. Naturf.,* **llB,** 138–142.

Glass, B., 1965, 'A Century of Biochemical Genetics', *Proc. Am. phil. Soc.,* **109,** 227–236.

Glum, F., 1921, 'Zehn Jahre Kaiser-Wilhelm Gesellschaft zur Förderung der Wissenschaft', *Naturwissenschaften,* **9,** 293–300.

Goldschmidt, R. B., 1928, 'The Gene', *Quart. Rev. Biol.,* **3,** 307–324.

——, 1932, 'Bemerkungen zur Kritik der quantitativen Natur multipler Allele', *Bull. Lab. Genet. Leningrad,* **9,** 129–134.

——, 1938, *Physiological Genetics,* New York & London. [There is no suggestion of the chain-molecule nature of the gene in the original German edition: *Physiologische Theorie der Vererbung. . .,* Berlin, 1927.]

——, 1954, 'Different Philosophies of Genetics', *Int. Conf. Genet.,* **9,** i, 83–99.

——, 1955, *Theoretical Genetics,* Berkeley & Los Angeles.

——, 1956, *Portraits from Memory; Recollections of a Zoologist,* Seattle.

Goldsmith, M., & Mackay, A., 1966, 'Introduction' to *The Science of Science,* Harmondsworth, pp. 7–17.

Gortner, R. A., 1938, *Outlines of Biochemistry,* 2nd ed., New York.

Gosling, R. G., 1954, *X-ray Diffraction Studies of Desoxyribose Nucleic Acid,* London. [Ph.D. Thesis, February 1954.]

Gowen, J. W., 1939, 'Behaviour of Viruses and Genes under Similar Stimuli', *Int. Conf. Genet.,* **7,** 133–134.

Gray, Sir J., 1931, *A Textbook of Experimental Cytology,* Cambridge.

Gregory, R. P., 1903, 'Seed Characters of *Pisum sativum*', *New Phytologist,* **2,** 226.

Griffith, F., 1922, 'Types of Pneumococci Obtained from Cases of Lobar Pneumonia', *Rep. Public Health med. Subj.,* **13,** 20–45.

——, 1923, 'The Influence of Immune Serum on the Biological Properties of Pneumococci', *Rep. Public Health med. Subj.,* **18,** 1–13.

——, 1928, 'The Significance of Pneumococcal Types', *J. Hygiene,* **27,** 135–159.

Groth, P. H. von, 1919, 'Ueber den krystallisierten und amorphen Zustand organischer Verbindungen und über die sogenannten flüssigen Kristalle', *Naturwissenschaften,* **7,** 648–652.

Gulick, A., 1938, 'What are Genes? I. The Genetic and Evolutionary Picture; II. The Physio-Chemical Picture; Conclusions', *Quart. Rev. Biol.,* **13,** 1–18; 140–168.

Gulland, J. M., 1938, 'Nucleic Acids', *J. chem. Soc.,* 1722–1734.

——, 1939, Letters to Astbury, dated 1 May and 25 October. [Leeds University Dept. of Biophysics Archive.]

——, 1944, 'Some Aspects of the Chemistry of Nucleotides', *J. chem. Soc.,* 208–217.

——, 1947, 'The Structure of Nucleic Acids', *Symp. Soc. exp. Biol.,* **1,** 1–14.

Gulland, J. M., Barker, G. R., & Jordan, D. O., 1945, 'The Chemistry of the Nucleic Acids and Nucleoproteins', *A. Rev. Biochem.,* **14,** 175–206.

Gulland, J. M., Holliday, E. R., & Macrae, T. F., 1934, 'Constitution of the Purine Nucleosides. Pt. II', *J. chem. Soc.*, 1639–1644.

Gulland, J. M., & Jordan, D. O., 1947, 'The Macromolecular Behaviour of Nucleic Acids', *Symp. Soc. exp. Biol.*, **1,** 56–65.

Gulland, J. M., Jordan, D. O., & Taylor, H. F. W., 1947, 'Deoxypentose Nucleic Acids. Pt. II. Electrometric Titration of the Acidic and the Basic Groups of the Deoxypentose Nucleic Acid of Calf Thymus', *J. chem. Soc.*, 1131–1141.

Hadley, P., 1927, 'Microbic Dissociation', *J. infect. Dis.*, **40,** 1–312.

Hagedoorn, A., 1911, 'Autocatalytic Substances: The Determinants for the Inheritable Characters', *Vorträge und Aufsätze über Entwickelungsmechanik der Organismen* (Leipzig), Hft. 12, 1–35.

Hahn, F. E., 1973, 'Reverse Transcription and the Central Dogma', *Prog. mol. subcell. Biol.*, **3,** 1–14.

Haldane, J. B. S., 1929, 'Mr. J. B. S. Haldane's Report', *Ann. Rep. John Innes hort. Inst.*, p. 6. [For the year 1928.]

——, 1937a, Foreword to Darlington's *Recent Advances in Cytology*, 2nd ed.

——, 1937b, 'The Biochemistry of the Individual', in: Needham, J., and Green, D. E. (eds.), 1937, pp. 1–10.

——, 1937c, 'Professor Haldane's Report', *An. Rep. John Innes hort. Inst.*, p. 8. [For the year 1936.]

——, 1941, *New Paths in Genetics*, London.

Hammarsten, E., 1920, 'Eine "gekoppelte" Nucleinsäure aus Pankreas', *Hoppe-Seyler's Z. physiol. Chem.*, **109,** 141–165.

——, 1924, 'Zur Kenntnis der biologischen Bedeutung der Nucleinsäureverbindungen', *Biochem. Z.*, **144,** 383–466.

——, 1939, 'The Function of Thymonucleic Acid in Living Cells', *J. Mt. Sinai Hosp.*, **6,** 115–125.

Hammarsten, E., & Hevesy, G., 1946, 'Rate of Renewal of Ribo- and Desoxyribonucleic Acids', *Acta Physiolog. Scand.*, **11,** 225–343.

Hammarsten, E., & Jorpes, E., 1922, 'Eine "gekoppelte" Nucleinsäure aus Pankreas', *Hoppe-Seyler's Z. physiol. Chem.*, **118,** 224–232.

Hardy, Sir W. B., 1903, 'Colloidal Solutions, the Globulin System', *J. Physiol.*, **29,** xxvi–xxix.

——, 1928, 'Living Matter', *Colloid Symp. Monog.*, **6,** 7–16.

Harker, D., 1936, 'The Application of the Three-dimensional Patterson Method and the Crystal Structure of Proustite, Ag_3AsS_3, and Pyorargyrite', *J. chem. Phys.*, **4,** 381–390.

Harnack, A. von, 1963, 'Erinnerungen an die Gründungszeit der "Kaiser-Wilhelm-Gesellschaft zur Förderung der Wissenschaften"', *Naturw. Rdsch., Stuttg.*, **16,** 435–438.

Harries, C. R., 1904, 'Abbau und Konstitution des Parakautschuks vermittelst Ozon', *Ber. dt. chem. Ges.*, **37**, 2708–2711.

Harris, H., 1963, *Garrod's Inborn Errors of Metabolism. Reprinted with a Supplement by H. Harris,* London.

Harris, F. I., & Hoyt, H. S., 1917, 'The Possible Origin of the Toxicity of Ultraviolet Light', *Science,* **46,** 318–320.

Harris, J. I., and Knight, C. A., 1952, 'Action of Carboxypeptidase on Tobacco Mosaic Virus', *Nature,* **170,** 613.

Hastings, A. B., 1970, 'Contribution to the Discussion', *Proc. Conference on the History of Biochemistry and Molecular Biology,* American Academy of Arts and Sciences, Brookline, Mass.

Haurowitz, F., 1950, *Chemistry and Biology of Proteins,* New York.

Haurowitz, F., & Crampton, C. F., 1950, 'The Role of the Nucleus in Protein Synthesis', *Exp. Cell Res.*, Suppl. **2,** 45–54.

Haworth, W. N., & Machemer, H., 1932, 'Polysaccharides. Pt. X. Molecular Structure of Cellulose', *J. chem. Soc.,* 2270–2273.

Heidelberger, M., & Avery, O. T., 1923, 'The Soluble Specific Substance of Pneumococcus', *J. exp. Med.,* **38,** 73–85.

Heidenhain, M., 1907, *Plasma und Zelle,* Jena.

Heimann, P., 1970, 'Molecular Forces, Statistical Representation, and Maxwell's Demon', *Studies in History and Philosophy of Science,* **1,** 189–211.

Heisenberg, W., 1927, 'Ueber den anschaulichen Inhalt der quantentheoretischen Kinematik und Mechanik', *Z. Physik.,* **43,** 172–198.

Heitz, E., 1928, 'Das Heterochromatin der Moose', *Jahrb. wiss. Bot.,* 69.

——, 1933, 'Die somatische Heteropyknose bei *Drosophila melanogaster* und ihre genetische Bedeutung', *Z. Zellf. mikr. Anat.,* **20,** 237–287.

Hengstenberg, J., 1927, 'Röntgenuntersuchungen über die Struktur der Polymerisatsprodukte des Formaldehyds', *Ann. Physik.,* **84,** 245–278.

——, 1928, 'Röntgenuntersuchungen über den Bau der C-Ketten in Kohlenwasserstoffen (C_n H_{2n+2})', *Z. Krystallogr. Miner.,* **67,** 583–594.

Hengstenberg, J., & Mark, H., 1928, 'Ueber Form und Grosse der Mizelle von Zellulose und Kautschuk', *Z. Krystallogr. Miner.,* **69,** 271–284.

Herriott, R., 1951a, 'Nucleic-acid-free T2 Virus 'Ghosts' with Specific Biological Action', *J. Bacteriol.,* **61,** 752–754.

——, 1951b, Letter to Hershey, dated 16 November. [Original with Hershey; a part of this letter has appeared in Cairns *et al.,* 1966, p. 102.]

Herrmann, K., Gerngross, O., & Abitz, W., 1930, 'Zur röntgenographischen Strukturerforschung des Gelatinemicells', *Z. physikalische Chemie*, **10,** 371–394.

Hershey, A. D., 1951a, Letter to Delbrück, dated 3 March. [Caltech Archive.]

——, 1951b, Letter to Herriott, dated 20 November. [Original with Hershey.]

——, 1953a, 'Intracellular Phases in the Reproductive Cycle of Bacteriophage T2', *Ann. Inst. Pasteur.,* **85,** 99–112. [I.U.B.S., 'Le Bactériophage: Premier Colloque international Royaumont, 1952.]

——, 1953b, 'Functional Differentiation within Particles of Bacteriophage T2', *Cold Spring Harb. Symp. quant. Biol.,* **18,** 135–139.

——, 1966, The Injection of DNA into Cells by Phage', in: Cairns *et al.* (eds.), 1966, pp. 100–108.

Hershey, A. D., & Chase, M., 1952, 'Independent Functions of Viral Proteins and Nucleic Acid in Growth of Bacteriophage', *J. gen. Physiol.,* **36,** 39–56.

Herzog, R. O., 1910, 'Zur Kenntnis der Losungen', *Z. Elektrochem. angew. phys. Chem.,* **16,** 1003–1004.

——, 1924, 'Einige Arbeiten über den Feinbau der Faserstoffe', *Naturwissenschaften,* **12,** 955–960.

——, 1925, 'Depolymerisation oder Dispergierung der Cellulose', *Naturwissenschaften,* **13,** 1040–1042.

——, 1928, 'Zur Erkenntnis der Skleroproteine', *Helv. chim. Acta.,* **11,** 529–533.

Herzog, R. O., & Jancke, W., 1920, 'Ueber den physikalischen Aufbau einiger hochmolekularer organischer Verbindungen', *Ber. dt. chem. Ges.,* **53b,** 2162–2164.

Herzog, R. O., & Kasernovsky, H., 1908, 'Ueber die Diffusion von Kolloiden II', *Biochem Z.,* **11,** 173–176.

Hess, K., 1920, 'Ueber die Konstitution der Zellulose. I. Die Acetolyse der Äthylzellulose', *Z. Elektrochem. angew. phys. Chem.,* **26,** 232–251.

Hess, L., 1970, 'Origins of Molecular Biology', *Science,* **168,** 664–669.

Hevesy, G. von, & Otteson, J., 1943, 'Rate of Formation of Nucleic Acid in the Organs of the Rat', *Acta physiol. Scand.,* **5,** 237–247.

Heyroth, F. F., & Loofbourow, J. R., 1931, 'Changes in the Ultra-violet Absorption Spectrum of Uracil and Related Compounds under the Influence of Radiations', *J. Am. chem. Soc.,* **53,** 3441–3453.

Hill, D. W., 1947, *Co-operative Research in Industry,* London.

Himsworth, Sir Harold, 1968, Letter to the Author, dated 17 May.

Hodgkin, D. C., & Riley, D. P., 1968, 'Some Ancient History of Protein X-ray Analysis', in : Rich and Davidson (eds.), 1968, pp. 15–28.

Hörmann, G., 1899, *Die Kontinuität der Atomverkettung; ein Strukturprinzip der lebendigen Substanz,* Jena.

Hoeseh, K., 1920, 'Emil Fischer. Sein Leben und sein Werk', *Ber. dt. chem. Ges.,* Sonderhft. 54 Jahrgang, 1–478.

Hofmeister, F., 1893, 'Ueber die Zusammensetzung des krystallinischen Eieralbumins, *Hoppe-Seyler's Z. physiol. Chem.,* **16,** 187–191.

——, 1902, 'Ueber Bau und Gruppierung der Eiweisskörper', *Ergebn. Physiol.* (Biochemie), **1,** 759–802.

Holiday, E. R., 1930, 'The Characteristic Absorption of Ultra-violet Radiation by Certain Purines', *Biochem. J.,* **24,** 619–625.

Hollaender, A., & Emmons, C. W., 1939, 'The Action of Ultraviolet Radiation on Dermatophytes', *J. cell. comp. Physiol.,* **13,** 391–402.

——, 1941, 'Wavelength Dependence of Mutation Production in the Ultraviolet with Special Emphasis on Fungi', *Cold Spring Harb. Symp. quant. Biol.,* **9,** 179–185.

Holton, G. (ed.), 1972, *The Twentieth-Century Sciences. Studies in the Biography of Ideas,* New York.

Hoogsteen, K., 1959, 'The Structure of Crystals Containing a Hydrogen-bonded Complex of 1-methylthymine and 9-methyladenine', *Acta Crystallogr.,* **12,** 822–823.

Hopkins, F. G., 1896, 'The Pigments of the Piridae: A Contribution to the Study of Excretory Substances which Function in Ornament', *Proc. R. Soc.,* **186B,** 661–682.

——, 1913, 'The Dynamic Side of Biochemistry', *Rep. Br. Ass. Advmt. Sci.,* 83rd. Meeting, 652–668.

——, 1933, 'Some Chemical Aspects of Life', *Rep. Br. Ass. Advmt Sci.,* 103rd. Meeting, 1–24. [Reprinted in Needham and Baldwin (eds.), 1949, pp. 242–263.]

Hotchkiss, R. D., 1949, 'Etudes chimiques sur le facteur transformant du pneumocoque', *Colloques int. C.N.R.S.,* No. 8, 57–65. [Held June/July 1948.]

——, 1951, 'Transfer of Penicillin Resistance in Pneumococci by the Desoxyribonucleate Derived from Resistant Cultures', *Cold Spring Harb. Symp. quant Biol.,* **16,** 457–460.

——, 1952, 'The Role of Desoxyribonucleates in Bacterial Transformation', in: McElroy and Glass (eds.), 1952, pp. 426–436. [Discussion pp. 437–439.]

——, 1955a, 'The Biological Role of Desoxypentose Nucleic Acids', in Chargaff and Davidson (eds.), 1955, vol. ii, pp. 435–473.

——, 1955b, 'Bacterial Transformation', in: 'Symposium on Genetic Recombination', *J. cell. comp. Physiol.,* Suppl. 2, **45,** 1–14. [Discussion, pp. 20–21.]

——, 1965, 'Oswald T. Avery', *Genetics,* **51,** 1–10.

——, 1966, 'Gene, Transforming Principle, and DNA', in: Cairns *et al.* (eds.), 1966, pp. 180–200.

——, 1972, Letter to the Author, dated 10 March.

Hotchkiss, R. D., & Marmur, J., 1954, 'Double Marker Transformations as Evidence of Linked Factors in Desoxyribonucleate Transforming Agents', *Proc. natn. Acad. Sci. U.S.A.,* **40,** 55–60.

Huggins, M., 1931, 'Principles Determining the Arrangement of Atoms and Ions in Crystals', *J. Phys. Chem.,* **35,** 1270–1280.

——, 1937, 'Hydrogen Bridges in Organic Compounds', *J. organ. Chem.,* **1,** 407–456.

——, 1943, 'The Structure of Fibrous Proteins', *Chem. Rev.,* **32,** 195–218.

——, 1969a, Letter to the Author, dated 18 June.

——, 1969b, Recorded Interview with the Author, London Airport, August .

Hughes, A., 1959, *A History of Cytology,* London and New York.

Hughes, E. W., & Moore, W. J., 1942, 'The Structure of β-Glycylglycine', *J. Am. chem. Soc.,* **64,** 2236–2237. For the full account see:

——, 1949, 'The Crystal Structure of β-Glycylglycine', *J. Am. chem. Soc.*, **71,** 2618–2623.

Hunger, F. W. T., 1905, 'Neue Theorie zur Ätiologie der Mosaikkrankheit des Tabaks', *Ber. dt. bot. Ges.*, **23,** 415–418.

Hunter, L., 1945, 'Mesohydric Tautomerism', *J. chem. Soc.*, 806–809.

Huppert, K. H. H., 1895, *'Ueber die Erhalten der Arteigenschaften,* Prague. [Rectoral Address. Xerox copy supplied by Dr. V. Orel, Brno.]

Hutchinson, A., 1929, 'Memorandum on the History, Development, and Present Condition of the Department of Mineralogy', *Cambridge University Reporter,* Dec. 10, pp. 390–395.

——, 1930, 'Contribution to the Discussion of the Report of the Syndicate upon the Position of Mineralogy in the Studies of the University', *Cambridge University Reporter,* Feb. 11, pp. 645–648.

Inoko, Y., 1894, 'Ueber die Verbreitung der Nucleinbasen in den thierischen Organen', *Hoppe-Seyler's Z. physiol. Chem.*, **18,** 540–544.

Jacquet, A., 1890, 'Beiträge zur Kenntnis des Blutfarbstoffes', *Hoppe-Seyler's Z. physiol. Chem.*, **14,** 289–296.

Jahnke, E., & Emde, F., 1933, *Tables of Functions with Formulae and Curves,* 2nd ed., Leipzig and Berlin.

Jehle, H., 1950, 'Specificity of Interaction between Identical Molecules', *Proc. natn. Acad. Sci. U.S.A.*, **36,** 238–246.

Jones, M. E., 1953, 'Albrecht Kossel, a Biographical Sketch'. *Yale J. Biol. Med.*, **26,** 80–97.

Jones, W., 1920, *Nucleic Acids; Their Chemical Properties and Physiological Conduct,* London.

Jones, W., & Read, B. E., 1917, 'Adenine-Uracil Dinucleotide and the Structure of Yeast Nucleic Acid', *J. biol. Chem.*, **29,** 111–126.

Jordan, P., 1938, 'Zur Frage einer spezifischen Anziehung zwischen Genmolekulen', *Phys. Z.*, **39,** 711–714.

Jordan, D. O., 1952, 'Physicochemical Properties of the Nucleic Acids', *Prog. Biophys.*, **2,** 51–89.

——, 1955, 'The Physical Properties of Nucleic Acids', In: Chargaff and Davidson (eds.), 1955, vol. i, pp. 447–492.

Jorpes, E., 1924, 'Zur Frage nach den Pankreasnucleinsäuren', *Biochem. Z.*, **151,** 227–245.

——, 1928, 'Zur Analyse der Pankreasnucleinsäuren', *Acta med. Skand.*, **68,** 503–573.

Kalckar, H. M., 1966, 'High Energy Bonds: Optional or Obligatory', in Cairns *et al.*, 1966, pp. 43–49.

Karrer, P., 1920, 'Zur Kenntnis der Polysaccharide. I. Methylierung der Starke', *Helv. chim. Acta.,* **3,** 620–625.

Katz, J. R., 1925, 'Röntgenspektrographische Untersuchungen am gedehnten Kautschuks'. . . , *Naturwissenschaften,* **13,** 410–416.

Kausche, G. A., Pfankuch, E., & Ruska, H., 1939, 'Die Sichtbarmachung von pflanzlichem Virus im Übermikroskop', *Naturwissenschaften,* **18,** 292–299.

Kavenoft, R., & Zimm, B., 1973, 'Chromosom-sized DNA Molecules from *Drosophila', Chromosoma,* **48,** 1–27. [They give the value 20 to 80 × 10^9. This is represented in the graph on p. 95 by $\log_{10}$ of 5 × 10^{10}.]

Kekulé, A., 1858, 'Ueber die Constitution und Metamorphosen der chemischen Verbindungen und über die chemische Natur des Kohlenstoffs', *Justus Liebigs Annln. Chem.,* **106,** 129–159.

——, 1878, 'The Scientific Aims and Achievements of Chemistry', *Nature,* **18,** 210–213.

Kelner, A., 1949, 'Effects of Visible Light on the Recovery of *Streptomyces griseus* conidia from ultraviolet Irradiation Injury', *Proc. natn. Acad. Sci.,* **35,** 73–79.

Kendrew, J. C., 1970, 'Some Remarks on the History of Molecular Biology', *Biochem. Soc. Symp.,* **30,** 5–10.

——, 1967, 'How Molecular Biology Started', *Scient. Am.,* **216,** No. 3, 141–143.

Kennaway, E., 1952, 'Some Recollections of Albrecht Kossel, Professor of Physiology in Heidelberg, 1901–1924', *Ann. Sci.,* **8,** 393–397.

Keynes, R. D., 1972, Letter to the Author, dated 13 July. [Keynes wrote: 'The Hardy Club was really a University Biophysics Club....The quality of discussion at most of our early meetings was far superior to anything I have ever experienced anywhere else. Later it became increasingly difficult to get the physiologists and molecular biologists to take a really critical interest in each other's work.']

Kiliani, H., 1895, 'Ueber Digitoxin', *Arch. Pharm. Berlin,* **233,** 311–320.

——, 1896, 'Ueber den Nachweis der Digitalis-Glycoside und ihrer Spaltungsprodukte durch eisenhaltige Schwefelsäure', *Arch. Pharm. Berlin.,* **234,** 273–277.

Kloss, G., 1968, 'University Reform in West Germany: The Burden of Tradition', *Minerva,* **6,** 323–353.

Klug, A., 1968, 'Rosalind Franklin and the Discovery of the Structure of DNA', *Nature,* **219,** 808–810, 843–844, *Corrigenda,* **879,** 1192, *Correspondence,* 880.

Klug, A., & Caspar, D. L. D., 1960, 'The Structure of Small Viruses', *Adv. Virus Res.,* **7,** 225–325.

Knapp, E., Reuss, A., Risse, O., & Schreiber, H., 1939, 'Quantitative Analyse der mutationsauslösenden Wirkung monochromatischen UV-Lichtes', *Naturwissenschaften,* **27,** 304.

Knight, C. A., & Lauffer, M. A., 1942, 'A Comparison of the Alkaline Cleavage Products of two Strains of Tobacco Virus', *J. biol. Chem.,* **144,** 411–417.

Knox, E., 1958, 'Sir Archibald Garrod's Inborn Errors of Metabolism', *Am. J. hum. Genet.,* **10,** 95–124.

Köhler, A., 1904, 'Mikrophotographische Untersuchungen mit ultraviolettem Licht', *Z. wiss. Mikroskopie,* **21,** 129–165, 273–304.

Koltzoff, N. K., 1928, 'Physikalisch-chemische Grundlage der Morphologie', *Biol. Zbl.,* **48,** 345–369.

Kossel, K. M. L. A., 1885, 'Ueber das Adenin', *Ber. dt. chem. Ges.,* **18,** 1928–1930.

——, 1887, 'Ueber das Adenin III', *Ber dt. chem. Ges.,* **20,** 3356–3358.

——, 1893, 'Ueber die Nucleinsäure', *Archiv Anat. Physiol.,* 157–164.

——, 1908, 'Die Probleme der Biologie', *Akademische Rede,* Heidelberg.

——, 1910, 'Ueber die Beschaffenheit des Zellkerns', translated in: *Nobel Lectures including Presentation Speeches and Laureates' Biographies: Physiology or Medicine* 1901–1921, Amsterdam, 1967, pp. 394–405.

——, 1911, 'The Chemical Composition of the Cell', *Harvey Lect.,* 33–51.

——, 1912, 'Lectures on the Herter Foundation', *Johns Hopkins Hosp. Bull.,* **23,** 65–76.

——, 1913, 'Beziehungen der Chemie zur Physiologie', in: Meyer, E. von (ed.), *Die Kultur der Gegenwart, ihre Entwicklung und ihre Ziele: Chemie,* Leipzig and Berlin, pp. 376–412.

——, 1928, *The Protamines and Histones,* London. English translation of the German text later published as *Protamine und Histone,* Leipzig & Wien, 1929.

Kossel, K. M. L. A., & Neumann, A., 1893, 'Ueber das Thymin, ein Spaltungsproduct der Nucleinsäure', *Ber. dt. chem. Ges.,* **26,** 2753–2756.

Kozloff, L. M., 1966, 'Transfer of Parental Material to Progeny', in: Cairns, *et al.,* 1966, pp. 109–115.

Kozloff, L. M., & Putnam, F. W., 1950, 'Biochemical Studies of Viral Reproduction. III. The Origin of Virus Phosphorus in the *Escherichia coli* T6 Bacteriophage System', *J. biol. Chem.,* **182,** 229–242.

Krebs, Sir H. A., 1970, 'Intermediary Metabolism of Animal Tissue between 1911 and 1969', *Biochem. Soc. Symp.,* No. 30, 123–136.

Krueger, A. P., 1937, 'The Mechanism of Bacteriophage Production', *Science,* **86,** 379–380.

Kunitz, M., 1940, 'Crystalline Ribonuclease', *J. gen. Physiol.,* **24,** 15–32.

——, 1948, 'Isolation of Crystalline Desoxyribonuclease', *Science,* **108,** 19–20.

La Du, B. N., Zannoni, V. G., Laster, L., & Seegmiller, J. E., 1958, 'The Nature of the Defect in Tyrosine Metabolism in Alkaptonuria', *J. biol. Chem.,* **230,** 251–260.

Laland, S. G., & Zimmer, T.-L., 1973, 'The Protein Thiotemplate Mechanism of Synthesis for the Peptide Antibiotics Produced by *Bacillus brevis', Essays in Biochem.,* **9,** 31–57.

Landsteiner, K., 1919, 'Ueber die Bedeutung der Proteinkomponente bei den Präcipitinreaktionen der Azoproteine, XIII. Mitteilung über Antigene', *Biochem. Z.,* **93,** 106–118.

Langmuir, I., & Wrinch, D., 1938, 'Vector Maps and Crystal Analysis', *Nature,* **142,** 581–583.

Latarjet, R., 1949, 'Mutation induite chez un Virus par irradiation ultraviolette de cellules infectées', *C.r. hebd. Séanc. Acad. Sci., Paris,* **228,** 1354–1357.

Laue, M. von, 1913, 'Les phénomènes des rayons de Röntgen produits par le réseau tridimensional des cristaux', in: *La Structure de la Matière,* Institut international de Physique Solvay, October 1913, pp. 75–102. [Published in Paris, 1921.]

——, 1944, *Mein physikalischer Werdegang. Eine Selbstdarstellung,* Berlin [1952].

Lauffer, M. A., 1938, 'The Molecular Weight and Shape of Tobacco-Mosaic Protein', *Science,* **87,** 469–470.

Lauffer, M. A., & Stanley, W. M., 1938, 'Stream Double Refraction of Virus Proteins', *J. Biol. Chem.,* **123,** 507–525.

Leathes, J. B., 1926, 'Function and Design', *Science,* **64,** 387–394.

Lederberg, J., 1947, Letter to Boivin, dated 17 October. [Original with Lederberg. This contains the statement: 'I have not been able to pick out a suitable C2-R strain with which to duplicate your experiments.']

——, 1949, 'Problems in Microbial Genetics', *Heredity,* **2,** 145–198.

——, 1951, *Papers in Microbial Genetics: Bacteria and Bacterial Viruses,* Madison. (Introduction, pp. ix-xx.)

——, 1959, 'Genes and Antibodies', *Science,* **129,** 1649–1653.

——, 1972a, Letter to the Author, dated 11 October.

——, 1972b, 'Reply to H. V. Wyatt', *Nature,* **239,** 234.

——, 1973, Recorded Interview with the Author, Stanford, January.

Leeds University: Advisory Committee on Cloth-Workers' Grants, 1927, Memorandum entitled 'Textile Research' and dated 8th July, 1927.

Leeds University: Clothworkers' Departments, 1929, 1930, 1931. 'Report of the work done under the research scheme established in 1928 with the aid of a Special Grant from the Worshipful Company of Clothworkers', Session 1928–29; 1929–30; 1930–31.

Leeds University: Department of Textile Industries, 1937, *Annual Report.*

Leeds University: Textiles Subcommittee on Clothworkers' Grants, 1929. Minutes of the meeting held on the 15th February, 1929. [University of Leeds Archive.]

Lepage, G. A., & Heidelberger, C., 1951, 'Incorporation of Glycine-2-C^{14} into Proteins and Nucleic Acids of the Rat', *J. biol. Chem.,* **188,** 593–602.

Levene, P. A., 1919, 'The structure of yeast nucleic acid', *J. biol. Chem.,* **40,** 415–424.

——, 1921, 'On the structure of thymus nucleic acid and on its possible bearing on the structure of plant nucleic acid', *J. biol. Chem.,* **48,** 119–125.

——, 1925, Letter to Flexner dated 30th March.

Levene, P. A., & Bass, L. W., 1931, *Nucleic Acids,* New York.

Levene, P. A., & Dillon, R. T., 1930, 'Intestinal Nucleotidase', *J. biol. Chem.,* **88,** 753–769.

Levene, P. A., & Jacobs, W. A., 1909a, 'Uber die Inosinsäure, *Ber. dt. chem. Ges.*, **42,** 335–338.

——, 1909b, 'Ueber die Hefe-Nucleinsäure', *Ber. dt. chem. Ges.*, **42,** 2474–2478.

——, 1909c, 'Ueber Hefenucleinsäure', *Ber. dt. chem. Ges.*, **42,** 2703–2706.

——, 1911, 'Ueber die Inosinsäure', *Ber. dt. chem. Ges.*, **44,** 746–753.

——, 1912, 'On the structure of Thymus Nucleic Acid', *J. biol. Chem.*, **12,** 411–420.

Levene, P. A., & Jorpes, E., 1930, 'A method of Separation of ribonucleotides from thymonucleic acid and on the conditions for a quantitative separation of the Purine Bases from the Ribopolynucleotides', *J. biol. Chem.*, **86,** 389–415.

Levene, P. A., & London, E. S., 1929, 'The Structure of Thymonucleic Acid', *J. biol. Chem.*, **83,** 793–802.

Levene, P. A., & Mori, T., 1929. 'Ribodesose and Xylodesose and their bearing on the Structure of Thyminose', *J. biol. Chem.*, **83,** 803–816.

Levene, P. A., & Schmidt, G., 1938, 'Ribonucleodepolymerase (the Jones-Dubos Enzyme)', *J. biol. Chem.*, **126,** 423–434.

Levene, P. A., Schmidt, G., & Pickles, E. G., 1939, 'Enzymatic Dephosphorylation of Desoxyribonucleic Acids of Various Degrees of Polymerization', *J. biol. Chem.*, **127,** 251–259.

Levene, P. A., & Simms, H. S., 1925, 'The dissociation Constants of Plant Nucleotides and Nucleosides and their Relation to Nucleic Acid Structure', *J. biol. Chem.*, **65,** 519–534.

——, 1926, 'Nucleic Acid Structure as determined by Electrometric Titration Data', *J. biol. Chem.*, **70,** 327–341.

Levene, P. A., & Tipson, R. S., 1932, 'The Ring Structure of Adensonine', *J. biol. Chem.*, **94,** 809–819. For Guanosine see: *Ibid.*, **97,** 491–495.

Levy, H. A., & Corey, R. B., 1941, 'The Crystal Structure of dl Alanine', *J. Am. Chem. Soc.*, **63,** 2095–2108.

Lewis, G. N., 1923, *Valence and the Structure of Atoms and Molecules,* New York (Pagination from Dover paperback, 1966).

——, 1939, Letter to Pauling, dated 25 August. [College of Chemistry Papers, Bancroft Library, Berkeley.]

Leydig, F., 1889, 'Die Pigmente der Hautdecke und der Iris', *Verh. phys. med. Ges. Würzburg.*, **22,** (241)–(265).

Linderstrøm-Lang, K., 1935, 'Some Electrochemical Properties of a Simple Protein', *Trans. Faraday Soc.*, **31,** 324–335.

Lipson, H. S., 1970, *Crystals and X-rays,* London & Winchester.

Lipson, H. S., & Beevers, C. A., 1936, 'An improved Numerical Method of Two-dimensional Fourier Synthesis for Crystals', *Proc. phys. Soc.*, **48,** 772–780. Reprinted in Bacon, G. E., 1966, pp. 172–185.

Loeb, J., 1906, *The Dynamics of Living Matter,* New York.

——, 1922, *Proteins and the Theory of Colloidal Behaviour,* New York and London.

Lonsdale, Dame K., 1928, 'The Structure of the Benzene Ring', *Nature,* **122,** 810.

——, 1948, *Crystals and X-rays,* London.

——, 1968, Recorded Interview, London, May.

Loring, H. S., 1942, The Reversible Inactivation of Tobacco Mosaic Virus by Crystalline Ribonuclease', *J. gen. Physiol.,* **25,** 497–505.

Luria, S. E., 1947, 'Reactivation of Irradiated Bacteriophage by Transfer of Self-reproducing Units', *Proc. natn. Acad. Sci.* U.S.A., **33,** 253–264.

——, 1950, 'Bacteriophage: An Essay on Virus Reproduction', *Science,* **111,** 507–511.

——, 1951a, Letter to Weiss, dated 20 October. [Caltech Archive.]

——, 1951b, 'The Frequency Distribution of Spontaneous Bacteriophage Mutants as Evidence for the Exponential Rate of Phage Reproduction', *Cold Spring Harb. Symp. quant. Biol.,* **16,** 463–470.

——, 1952a, Letter to Hershey, dated 3 March. [Original with Hershey. This letter contains the statement: 'I still think there is a pretty good chance that a good deal of the protein enters and acts genetically.']

——, 1952b, 'An Analysis of Bacteriophage Multiplication', *Symp. Soc. gen. Microbiol.,* **2,** 99–113.

——, 1968, Recorded Interview, Cambridge, Mass., September.

Luria, S. E., & Anderson, T. F., 1942, 'The Identification and Characterization of Bacteriophages with the Electron Microscope', *Proc. natn. Sci. U.S.A.,* **28,** 127–130.

Luria, S. E., & Dulbecco, R., 1949, 'Genetic Recombinations Leading to Production of Active Bacteriophage from Ultraviolet Inactivated Bacteriophage Particles', *Genetics,* **34,** 93–125.

Lwoff, A. (ed.), 1949, 'Unités biologiques douées de continuité génétique', *Colloques int. C.N.R.S.,* No. 8. [June-July, 1948.]

——, 1953, The Nature of Phage Reproduction, *Sym. Soc. gen. Microbiol.,* **2,** 149–169.

Lythgoe, B., 1949, 'Some Aspects of Pyrimidine and Purine Chemistry', *Quart. Rev. chem. Soc.,* **3,** 181–207. On p. 202 Lythgoe referred to the lack of evidence that the pyrimidines were 'truly tautomeric'. [In a conversation with the author, Lythgoe explained that tautomerism was a sort of 'magic word' used for a wide range of compounds, some like the pyrimidines little understood. The purines and pyrimidines should not therefore have been lumped with acetoacetic ester as if they were known to give measurable quantities of both tautomers in an equilibrium mixture in solution.]

Maaløe, O., 1952, Letter to Hershey dated 2 April. [Caltech Archive.]

——, 1974, Letter to the Author, dated 22nd January. [Original with the Author.]

Maaløe, O., & Watson, J. D., 1951, 'The Transfer of Radioactive Phosphorus from Parental to Progeny Phage', *Proc. natn. Acad. Sci. U.S.A.,* **37,** 507–513. [The text before editing by Delbrück is a 10-p. MS preserved in the Delbrück papers, Caltech Archive.]

MacArthur, I., 1943, 'Structure of α-Keratin', *Nature,* **152,** 38–41.

——, 1961, 'Prof. W. T. Astbury, F.R.S.', *Nature,* **191,** 331–332.

McCarty, M., 1946a, 'Purification and Properties of Desoxyribonuclease Isolated from Beef Pancreas', *J. gen. Physiol.,* **29,** 123–139.

——, 1946b, 'Biochemical and Biophysical Studies on Viruses'. [Unpublished Lecture cited in a letter to the author, dated 1 March, 1972.]

——, 1946c, 'Chemical Nature and Biological Specificity of the Substance Inducing Transformation of Pneumococcal Types', *Bact. Rev.,* **10,** 63–71.

——, 1968, Recorded Interview with the Author, New York, September.

——, 1970, Letter to the Author, dated 10 March.

——, 1972, Letter to the Author, dated 1 March.

McCarty, M., & Avery, O. T., 1946a, 'Studies on the Chemical Nature of the Substance Inducing Transformation of Pneumococcal Types', *J. exp. Med.,* **83,** 89–96.

——, 1946b, 'Studies... Part III', *Ibid.,* 97–104.

McElroy, W. D., & Glass, B. (eds.), 1952, *Phosphorus Metabolism. A Symposium on the Role of Phosphorus in the Metabolism of Plants and Animals,* Baltimore, 2 vols.

——, 1957, *A Symposium on the Chemical Basis of Heredity,* Baltimore. [Held June 1956.]

McKinney, H. H., 1939, 'Virus Genes', *Int. Conf. Genet.,* **7,** 200–203.

Macleod, C M., 1967, Letter to the Author, dated 14 April.

——, 1968, Recorded Interview with the Author, New York, September.

Mandel, P., Mandel, L., & Jacob, M., 1948, 'Sur le comportement comparé, au cours du jeûne protéique prolongé, des deux acides nucléiques des tissus animaux et sur sa signification', *C.r. hebd. Séanc. Acad. Sci., Paris,* **226,** 2019–2021.

Mark, H., 1965, 'Polymers—Past, Present and Future', in an unpublished symposium of the Welch Foundation on Polymer Science.

Markham, R., Matthews, R. E. F., & Smith, K. M., 1948, 'Specific crystalline Protein and Nucleoprotein from a Plant Virus having Insect Vectors', *Nature,* **162,** 88–90.

Markham, R., & Smith, J. D., 1954, 'Nucleoproteins and Viruses' in: Neurath, H., & Bailey, K. (eds.), *The Proteins, Chemistry, Biological Activity, and Methods,* New York, 1–122. [This essay is not found in the second and subsequent editions.]

Marshall, J. R., and Walker, J., 1951, 'An Experimental Study of some Potentially Tautomeric 2- and 4(6)-Substituted Pyrimidines', *J. chem. Soc.,* 1004-1017.

Mason, M., & Weaver, W., 1929, *The Electromagnetic Field,* Chicago.

Massini, R., 1907, 'Ueber einem in biologischer Beziehung interessanten Kolistamm *(B. coli mutabile)*', *Arch. Hyg.,* **61,** 250–292.

Mathews, A. P., 1897, 'Zur Chemie der Spermatozoen', *Hoppe-Seyler's Z. physiol. Chem.*, **23**, 399–411.

——, 1924, 'Some general Aspects of the Chemistry of Cells', in: Cowdry, E. V. (ed.), *General Cytology: A Textbook of Cellular Structure and Function for Students of Biology and Medicine,* Chicago, pp. 15–95.

——, 1927, 'Professor Albrecht Kossel', *Science,* **66,** 293.

——, 1936, *Principles of Biochemistry,* London.

Maxwell, J. C., 1875, 'Atom' in: *The Encyclopaedia Britannica,* 9th ed., Edinburgh, vol. 3, pp. 36–49.

——, 1890, *The Scientific Papers of James Clerk Maxwell,* ed. W. D. Niven, Cambridge.

Mazia, D., 1952, 'Physiology of the Cell Nucleus', in: Barron, E. S. G. (ed.), *Modern Trends in Physiology and Biochemistry,* New York, pp. 77–122. [Woods Hole Lectures dedicated to the Memory of Leonor Michaelis, held Summer 1950.]

Mazia, D., & Jaeger, L., 1939, 'Nuclease Action, Protease Action and Histo-chemical Tests on Salivary Chromosomes of *Drosophila'*, *Proc. natn. Acad. Sci. U.S.A.*, **25,** 456–461.

Medical Research Council, 1972, Letter to the Author dated 4 August, enclosing a Memorandum, 'Proposals for Research on Biophysics', August, 1946. [M.R.C. Archive.]

Melchers, G., 1939, 'Neuere Untersuchungen über die Physiologie der Genwirkung an Pflanzen', *Int. Conf. Genet.*, **7,** 213–214.

——, 1948, 'Phytopathogene Viren', *Fiat Review of German Science* 1939–1946. *Biologie,* Pt. I, pp. 111–129.

——, 1953, 'Fritz von Wettstein (1895–1945)', *Mitt. Max-Planck-Ges. Förd Wiss.*, **6,** 11–15.

——, 1962, 'Geschichte der Max-Planck Institut für Biologie in Tübingen'. [Photocopy sent to the author by G. Melchers.]

——, 1965, 'Contributions of Plant Virus Research to Molecular Genetics', *Proc. Gregor Mendel Memorial Symposium* 1865–1965, Prague, pp. 119–136.

——, 1971, Letter to the Author dated 14 July.

Mellanby, Sir E., 1948, Letter to Perutz, dated 18 June. [M.R.C. Archive.]

Melland, A. M. [Mrs. Young], 1975, Letter to the Author dated 5 October. [Original with the author.]

Meyer, H. K., 1936, 'Inorganic Substances with Rubber-like Properties', *Trans. Faraday Soc.*, **32,** 148–152.

——, 1942, *Natural and Synthetic High Polymers,* New York.

Meyer, H. K., & Mark, H., 1928a, 'Ueber den Bau des krystallisierten Anteils der Cellulose', *Ber. dt. chem. Ges.*, **61,** 593–614.

——, 1928b, 'Ueber den Aufbau des Seiden-Fibroins', *Ber. dt. chem. Ges.*, **61B,** 1932–1936.

——, 1930, *Der Aufbau der hochpolymeren organischen Naturstoffe,* Leipzig.

Miescher, J. F. (Jr.), 1897, *Die histochemischen und physiologischen Arbeiten von Friedrich Miescher. Gesammelt und herausgegeben von seinen Freunden,* Leipzig.

Miller, G. L., & Stanley, W. M., 1942, 'Derivatives of Tobacco Mosaic Virus. 1. Acetyl and Phenylureido Virus', *J. Biol. Chem.,* **141,** 905–920.

Millikan, R. A., 1910, 'Das Isolieren eines Ions, eine genaue Messung der daran gebundenen Elektrizitätsmenge und die Korrektion des Stokesschen Gesetzes', *Phys. Z.,* **11,** 1097–1109.

Mirsky, A. E., 1947a, Contribution to the discussion of Boivin's paper, *Cold Spring Harb. Symp. quant. Biol.,* **12,** 15–16.

——, 1947b, 'Chemical Properties of Isolated Chromosomes', *Cold Spring Harb. Symp. quant. Biol.,* **12,** 143–146.

——, 1966, 'Chairman's Introduction', to *Histones; their Role in the Transfer of Genetic Information* (Ciba Foundation Study Group No. 24). London, pp. 1–3.

——, 1968, 'The Discovery of DNA', *Scient. Am.,* **218,** 78–88.

Mirsky, A. E., and Pollister, A. W., 1942, 'Nucleoproteins of Cell Nuclei', *Proc. natn. Acad. Sci. U.S.A.,* **28,** 344–352.

——, 1946, 'Chromosin, a Desoxyribose Nucleoprotein Complex of the Cell Nucleus', *J. gen. Physiol.,* **30,** 117–148.

Mirsky, A. E., & Ris, H., 1949, 'Variable and Constant Components of Chromosomes', *Nature,* **163,** 666–667.

Mitchell, H. K., & Nye, J. F., 1948, 'Hydroxyanthranylic Acid as a Precursor of Nicotinic Acid in *Neurospora', Proc. natn. Acad. Sci. U.S.A.,* **34,** 1–5.

Montgomery, E., 1885, 'Ueber das Protoplasma einiger 'Elementarorganismen', *Jenaische Zeitschriften für Naturwissenschaften,* **17,** 677–712.

Morgan, T. H., 1934, 'The Relation of Genetics to Physiology and Medicine', in: *Nobel Lectures...Physiology and Medicine* 1922–1941, Amsterdam etc., 1965, pp. 313–328.

Mosse, G. L. (ed.), 1966, *Nazi Culture: Intellectual, Cultural and Social Life in the Third Reich,* London.

Mudd, S., 1938, Contribution to the discussion, *Cold Spring Harb. Symp. quant. biol.,* **6,** 118–119.

Müller, A., 1928, 'A Further X-Ray Investigation of Long Chain Compounds (n-Hydrocarbons)', *Proc. R. Soc. London.,* **120A,** 437–459.

Muller, H. J., 1916, 'The Mechanism of Crossing-over II', *Am. Nat.,* **50,** 284–305.

——, 1922, 'Variation due to Change in the Individual Gene', *Am. Nat.,* **56,** 32–50.

——, 1926, 'The Gene as a Basis of Life', *Proc. int. Congr. Plant. Sci.,* **1,** 897–931. [Page reference is from the reprint in: Muller, 1962, pp. 188–204. Publication of the Congress was 1929.]

——, 1936, 'Physics in the Attack on the Fundamental Problems of Genetics', *Sci. Monthly,* **44,** 210–214.

——, 1939, 'Gene and Chromosome Theory', *Heredity,* **144,** 814–816.
——, 1941, 'Résumé and Perspectives of the Symposium on Genes and Chromosomes', *Cold Spring Harb. Symp. quant. Biol.,* **9,** 290–308.
——, 1946, Letter to Darlington, dated 2 March. [Original with Darlington. The contents of this report on the Conference will be found in a footnote to Muller's Pilgrim Trust Lecture—Muller, 1947, p. 23.]
——, 1947a, 'Genetic Fundamentals: The Work of the Genes', in: Muller, H. J., Little, C. C., & Snyder, L. H., *Genetics, Medicine, and Man,* New York, pp. 1-34. [Messenger Lectures delivered in the Fall of 1945.]
——, 1947b, 'The Gene. Pilgrim Trust Lecture', *Proc. R. Soc.* B, **134,** 1–37. Read November 1945. Footnote added later—see Muller, 1946.
——, 1951, 'The Development of the Gene Theory', in: Dunn, L. C. (ed.), 1951, pp. 77–99.
——, 1962, *Studies in Genetics. The Selected Papers of Hermann Joseph Muller,* Bloomington.
——, 1965, 'The Gene Material as the Initiator and the Organizing Basis of Life', in: Brink, R. A., & Styles, E. D. (eds.), 1967, pp. 419–447.
Mullins, N. C., 1972, 'The Development of a Scientific Speciality', *Minerva,* **10,** 51–82.
Mulvania, M., 1926, 'Studies on the Nature of the Virus of Tobacco Mosaic Virus', *Phytopath.,* **16,** 853–871. ['The evidence... seems to point to the virus as a non-living thing, possibly a very simple colloid, possibly also of protein nature and having enzymatic characteristics' (p. 870).]
Muthmann, W., 1893, 'Beiträge zur Volumtheorie der krystallisierten Körper', *Z. Krystall. ogr. Miner.,* **22,** 497–551.
Myrbäck, K. D. R., & Jorpes, E., 1935, 'Die freie Diffusion von Nucleinsäuren und Mononucleotiden als Mittel zur Bestimmung ihrer Molekülgrösse', *Hoppe-Seyler's Z. physiol. Chem.,* **237,** 159–164.

Nägeli, C. von, 1877, *Die niederen Pilze in ihren Beziehungen zu den Infektionskrankheiten und der Gesundheitspflege,* Munich.
Nature, 1970, 'News and Views: Central Dogma Reversed', *Nature,* **226,** 1198–1199 [Editorial].
Needham, J., 1935, Letter to council, dated 12 January. [Needham Archive, Cambridge.]
Needham, J., & Baldwin, E. (eds.), 1949, *Hopkins and Biochemistry* 1861–1947, Cambridge.
Needham, J., & Green, D. E., 1937, *Perspectives in Biochemistry: Thirty-one Essays Presented to Sir Frederick Gowland Hopkins,* Cambridge.
Neisser, M., 1906, 'Ein Fall von Mutation nach de Vries bei Bakterien', *Centralb. Bakteriol.,* 1 Ref., **38,** 98–102.
Nernst, W., 1895, *Theoretical Chemistry from the Standpoint of Avogadro's Rule and Thermodynamics,* London and New York.

Neubauer, O., & Flatow, L., 1907, 'Synthesen von Alkaptonsäuren', *Hoppe-Seyler's Z. physiol. Chem.*, **52,** 375–398.

Neufeld, F., & Levinthal, W., 1928, 'Beiträge zur Variabilität der Pneumokokken', *Z. Immunitätsforschung und exp. Therapie,* Jena, **55,** 324–340.

Neumann, A., 1899, 'Verfahren zur Darstellung der Nucleinsäure *a* und *b* und der Nucleothyminsäure', *Arch. Anat. Physiol.* (Physiol. Abth.), Suppl. vol., 552–555.

Neurath, H., 1939, Letter to Astbury, dated 8 December. [Leeds University Dept. of Biophysics Archive.]

Newman, G., 1923, 'Prefatory Note by the Chief Medical Officer', *Rep. Public Health med. Subj.*, **18,** iii–iv.

Nishikawa, S., & Ono, S., 1913, 'Transmission of X-rays through Fibrous, Lamellar and Granular Substances', *Proc. Math. Phys. Soc., Tokyo.*, **7,** 131–138.

Noethling, W., & Stubbe, H., 1934, 'Untersuchungen über experimentelle Auslösung von Mutationen bei *Antirrhinum majus. V. . .* ', *Z. indukt. Abstamm.–u. VererbLehre,* **67,** 152–172.

Northrop, J., 1937, 'Chemical Nature and Mode of Formation of Pepsin, Trypsin and Bacteriophage', *Science,* **86,** 479–483.

——, 1938, Concentration and Purification of Bacteriophage, *J. gen. Physiol.*, **21,** 335–366.

Oishi, M., & Cosloy, S. D., 1972, 'The Genetic and Biochemical Basis of the Transformability of *Escherichia coli* K 12', *Biochem. biophys. Res. Comm.*, **49,** 1568–1572.

Olby, R. C., 1966, *The Origins of Mendelism,* London and New York. (2nd ed., Chicago, 1988.)

——, 1970a, 'The Macromolecular Concept and the Origins of Molecular Biology', *J. chem. Ed.*, **47,** 168–174.

——, 1970b, 'Francis Crick, DNA and the Central Dogma', *Daedelus,* Reprinted with a Postscript in: Holton, G., 1972, pp. 227–280.

——, 1971a, 'The Influence of Physiology on Hereditary Theories in the Nineteenth Century', *Folia Mendeliana,* **6,** 99–103. [Gregor Mendel Colloquium.]

——, 1971b, 'Schrödinger's Problem: What is Life?', *J. Hist. Biol.*, **4,** 119–148.

——, 1971c, 'Carl F. J. E. Correns', *Dictionary of Scientific Biography,* New York, vol. iii, pp. 421–423.

——, 1972, 'Avery in Retrospect', *Nature,* **239,** 295–296.

Olby, R. C., & Posner, E., 1967, 'An Early Reference to Genetic Coding', *Nature,* **215,** 556.

Osborne, T. B., & Harris, I. F., 1902, 'Die Nucleinsäure des Weizenembryos', *Hoppe-Seyler's Z. physiol. Chem.*, **36,** 85–133.

Osborne, W. A., 1903, 'A New Synthesis of Homogentisic Acid', *J. Physiol.*, **29,** xiii–xiv.

Oster, G., & Pollister, A. W. (eds.), 1955, *Physical Techniques in Biological Research,* vol. i, Optical Techniques, New York.

Ostwald, W., 1908, 'Ueber die zeitlichen Eigenschaften der Entwicklungsvorgänge', *Vorträge und Aufsätze über Entwicklungsmechanik des Organismus,* ed. W. Roux, Leipzig, Hft 5.

Painter, T. S., 1933, 'A New Method for the Study of Chromosome Rearrangements and the Plotting of Chromosome Maps', *Science,* **78,** 585–586.

Patterson, A. L., 1934, 'A Fourier Method for the Determination of the Components of Interatomic Distances in Crystals', *Phys. Rev.,* **46,** 372–376.

——, 1962, 'Experiences in Crystallography—1924 to Date', in: Ewald, 1962, pp. 612–622.

Pauling, L., 1927, 'The Sizes of Ions and the Structure of Ionic Crystals', *J. Am. Chem. Soc.,* **49,** 765–790.

——, 1928a, 'The Shared-Electron Chemical Bond', *Proc. natn. Acad. Sci.,* **14,** 359–362.

——, 1928b, 'The Coordination Theory of the Structure of Crystals', *Festschrift zum 60. Geburtstage Arnold Sommerfelds,* Leipzig, pp. 11–17.

——, 1932, 'Interatomic Distances in Covalent Molecules and Resonance between Two or More Lewis Electronic Structures', *Proc. natn. Acad. Sci. U.S.A.,* **18,** 293–297.

——, 1948, 'Molecular Architecture and the Processes of Life', *21st Sir Jesse Boot Foundation Lecture.*

——, 1949, 'On the Stability of the S_8 Molecule and the Structure of Fibrous Sulfur', *Proc. natn. Acad. Sci. U.S.A.,* **35,** 495–499.

——, 1953a, Letter to Peter Pauling, dated 4 February. [Original with Peter Pauling.]

——, 1953b, Letter to Peter Pauling, dated 18 February. [Original with Peter Pauling.]

——, 1953c, Letter to Watson, dated 5 March. [Caltech Archive.]

——, 1953d, Letter to Peter Pauling, dated 10 March. [Original with L. Pauling.]

——, 1953e, 'Discussion des rapports de MM. L. Pauling et L. Bragg', *Rep. Institut international de Chimie Solvay,* pp. 111–112.

——, 1965, 'Fifty Years of Physical Chemistry in the California Institute of Technology, *A. Rev. Phys. Chem.,* **16,** 1–14.

——, 1967, Letter to the Author, dated 13 March.

——, 1968, Recorded Interview, San Diego, November.

——, 1970, 'Fifty Years of Progress in Structural Chemistry and Molecular Biology', *Daedalus,* **99,** 988–1014.

——, 1973a, Letter to the Author, dated 5 February.

——, 1973b, Letter to the Author, dated 15 March.

Pauling, L., Brockway, L. O., & Beach, J. Y., 1935, 'The Dependence of Interatomic Distance on Single Bond–Double Bond Resonance', *J. Am. chem. Soc.,* **57,** 2705–2709.

Pauling, L., & Corey, R. B., 1950, 'Two Hydrogen-bonded Spiral Configurations of the Polypeptide Chain', *J. Am. chem. Soc.,* **71,** 5349.

——, 1951a, 'Atomic Coordinates and Structure Factors for Two Helical Configurations of Polypeptide Chains', *Proc. natn. Acad. Sci. U.S.A.,* **37,** 235–240.

——, 1951b, 'The Structure of Synthetic Polypeptides', *Proc. natn. Acad. Sci. U.S.A.,* **37,** 241–250.

——, 1951 c, 'The Pleated Sheet, a New Layer Configuration of Polypeptide Chains', *Proc. natn. Acad. Sci. U.S.A.,* **37,** 251–256.

——, 1951d, 'The Structure of Feather Rachis Keratin', *Ibid.,* 256–261.

——, 1951e, 'The Structure of Hair, Muscle, and Related Proteins', *Ibid.,* 261–271.

——, 1951f, 'The Structure of Fibrous Proteins of the Collagen-Gelatin Group', *Ibid.,* 272–281.

——, 1951g, 'The Polypeptide-Chain Configuration in Hemoglobin and other Globular Proteins', *Ibid.,* 282–285.

——, 1953, 'A Proposed Structure for the Nucleic Acids', *Proc. natn. Acad Sci. U.S.A.,* **39,** 84–97.

Pauling, L., Corey, R. B., & Branson, H. R., 1951, 'The Structure of Proteins: Two Hydrogen-Bonded Helical Configurations of the Polypeptide Chain', *Proc. natn. Acad. Sci. U.S.A.,* **37,** 205–211.

Pauling, L., & Delbrück, M., 1940, 'The Nature of the Intermolecular Forces Operative in Biological Systems', *Science,* **92,** 77–79.

Pauling, L., & Huggins, M., 1934, 'Covalent Radii of Atoms and Interatomic Distances in Crystals containing Electron-Pair Bonds', *Z. Krystallogr. Miner.,* **87,** 205–238.

Pauling, L., Itano, H. A., Singer, S. J., & Wells, I. C., 1949, 'Sickle Cell Anemia, A Molecular Disease', *Science,* **110,** 543–548.

Pauling, L., & Niemann, C., 1939, 'The Structure of Proteins', *Science,* **61,** 1860–1867.

Pauling, L., & Schomaker, V., 1952a, 'On a Phospho-tri-anhydride Formula for the Nucleic Acids', *J. Am. chem. Soc.,* **74,** 1111.

——, 1952b, 'On a Phospho-tri-anhydride Formula for the Nucleic Acids', *J. Am. chem. Soc.,* **74,** 3712–3713.

Pauling, L., & Sherman, J., 1933, 'The Nature of the Chemical Bond. VI. The Calculation from Thermochemical Data of the Energy of Resonance of Molecules among Several Electronic Structures', *J. chem. Phys.,* **1,** 606–617.

Pauling, P., 1967, Recorded Interview with the Author, dated 6 May.

Payne, F., 1973, Letter to the Author, dated 14 January.

Pederson, K. O., 1940, 'The Protein Molecule', in: Svedberg, T., & Pederson, K. O. (eds.), 1940, pp. 406–415.

Perrin, J., 1908, 'L'agitation moléculaire et le mouvement brownien', *C.r. hebd. Séanc. Acad. Sci., Paris,* **146,** 967–970.

Person, C., & Suzuki, T., 1968, 'Chromosome Structure—A Model Based on DNA Replication', *Canad. J. Genet. Cytol.*, **10,** 627–647.

Perutz, M. F., 1937, 'The Iron-Rhodonite from Slag', *Min. Mag.*, **24,** 573–576.

——, 1948, 'An X-ray Study of Horse Methaemoglobin. II', *Proc. R. Soc.*, **195A,** 474–499.

——, 1949a, 'Recent Developments in the X-ray Study of Haemoglobin', in: Roughton, F. J. W., & Kendrew, J. C. (eds.), *Haemoglobin. A Symposium based on a Conference held at Cambridge in June* 1948 *in Memory of Sir Joseph Barcroft*, London pp. 135–147.

——, 1949b, 'An X-ray Study of Horse Methaemoglobin. II', *Proc. R. Soc.* **195A,** 474–499. [But see MacArthur, 1943, for earlier record of the 1.49Å reflexion.]

——, 1949c, Letter to Mellanby dated 9 June. [Medical Research Council Archive.]

——, 1951, 'New X-ray Evidence on the Configuration of Polypeptide Chains', *Nature,* **167,** 1053–1054.

——, 1970, 'Sir Lawrence Bragg: An Appreciation by Max Perutz, FRS', *Chemy Brit.*, **6,** 152–153. [See also *Acta Crystallogr.*, 1970, **26A,** 183–185.]

Peterlin, A., 1953, 'Modèle statistique des grosses molécules à chaines courtes. V. Diffusion de la lumière', *J. Poly. Sci.*, **10,** 425–436.

Pfankuch, E., 1940, 'Ueber die Spaltung von Virusproteinen der Tabakmosaik-Gruppe', *Biochem. Z.,* **306,** 125–129.

Pfankuch, E., and Kausche, G. A., 1938, 'Ueber Darstellung, Eigenschaften und quantitative Bestimmung von Tabakmosaik-Virus und Kartoffel-X-Virus und ihre physikochemische Differenzierung', *Biochem. Z.,* **299,** 334–345.

Pfankuch, E., Kausche, G. A., & Stubbe, H., 1940, 'Ueber die Entstehung, die biologische und physikalisch-chemische Charakterisierung von Röntgen- und γ-Strahlen induzierten Mutationen des Tabakmosaik-virus-proteins', *Biochem. Z.,* **304,** 238–258.

Pfankuch, E., & Piekenbrock, F., 1943, 'Zur Spaltung von Virusproteinen der Tabakmosaik-Gruppe', *Naturwissenschaften,* **31,** 94.

Pflüger, E., 1875, 'Ueber die physiologische Verbrennung in den lebendigen Organismen', *Pflüger's Arch. ges. Physiol.,* **10,** 251–367.

Pitt, G. J., 1948, 'The Crystal Structure of 4,6-dimethyl-2-hydroxypyrimidine. I', *Acta Crystallogr.,* **1,** 168–174.

Planck, M., 1958, *Physikalische Abhandlungen und Vorträge,* 3 vols., Braunschweig.

Pohl, J., & Spiro, K., 1923, 'Franz Hofmeister, sein Leben und Wirken', *Ergebn. Physiol.,* **22,** 1–31; bibliography, 39–50.

Polanyi, M., 1921, 'Faserstruktur im Röntgenlichte', *Naturwissenschaften,* **9,** 337–340.

——, 1962, 'My Time with X-rays and Crystals' in: Ewald, P. (ed.), 1962, pp. 629–636.

——, 1968, 'Life's Irreducible Structure', *Science,* **160,** 1308–1312.

Polanyi, M., & Weissenberg, K., 1922, 'Das Röntgen-Faserdiagramm', *Z. Phys.*, **9,** 123–130.

Pollister, A. W., and Mirsky, A. E., 1946, 'The Nucleoprotamine of Trout Sperm', *J. gen. Physiol.*, **30,** 101–115.

Pollister, A. W., Swift, H., & Alfert, M., 1951, 'Studies on the Desoxypentose Nucleic Acid Content of Animal Nuclei', *J. cell. comp. Physiol.*, **38,** Suppl. 1, 101–119.

Pollock, M. R., 1970, 'The Discovery of DNA: An Ironic Tale of Chance, Prejudice and Insight', *J. gen. Microbiol.*, **63,** 1–20.

Pontecorvo, G., 1968, 'Hermann Joseph Muller', *Ann. Rev. Genet.*, **2,** 1–10.

Popjak, G., 1970, 'Lessons Learnt from Small Molecules', *Biochem. Soc. Symp.*, **30,** 137–153.

Porter, A. B., 1906, 'On the Diffraction Theory of Microscopic Vision', *Phil. Mag.*, 6th ser., **11,** 154–166.

Preston, C., 1968, Letter to the Author, dated 22 May.

Preston, R. D., 1968, Recorded Interview with the Author, Leeds, February.

Preyer, T. W., 1871, *Die Blutkrystalle*, Jena.

Price, J. R., 1938, Biochemistry Department's Contribution to the 28th *Annual Report of the John Innes Horticultural Institution* for the year 1937, Cambridge.

Przibram, H., 1929, 'Quanta in Biology', *Proc. R. Soc. Edinb.*, **49,** 224–231.

Purdy, H. A., 1929, 'Immunological Reactions with Tobacco Mosaic Virus', *J. exp. Med.*, **49,** 919–935.

Randall, Sir J. T., 1950, Letter to Franklin, dated 4 December. [Original with Klug.]

——, 1951, 'An Experiment in Biophysics', *Proc. R. Soc.*, **208A,** 1–24.

——, 1952, Report to the Biophysics Committee of the M.R.C. for their visit on 15 December 1952. [Original in M.R.C. Archive.]

——, 1974, Letter to the Author. [Original with the Author.]

Ravin, A. W., 1969, 'The Genetics of Transformation', *Adv. Genetics*, **10,** 62–163.

——, 1974, Letter to the Author.

Redei, G. P., 1971, 'A Portrait of Lewis John Stadler 1896–1964', *Stadler Symposia*, University of Missouri, Columbia, pp. 5–20.

Reichmann, M. E., Bunce, B. H., and Doty, P., 1953, 'The Changes Induced in Sodium Desoxyribonucleate by Dilute Acid', *J. Polymer Sci.*, **10,** 109–119.

Reiman, H. A., 1929, 'The Reversion of R. to S Pneumococcus', *J. exp. Med.*, **49,** 237–249.

Rich, A., & Davidson, N., 1968, *Structural Chemistry and Molecular Biology*, San Francisco and London.

Riley, D. P., & Oster, G., 1951a, 'Some Theoretical and Experimental Studies of X-ray and Light Scattering by Colloidal and Macromolecular Systems', *Faraday Soc. Disc.*, **11,** 107–116. [Published 1952.]

——, 1951b, 'An X-ray Diffraction Investigation of Aqueous Systems of Desoxyribonucleic Acid (Na Salt)', *Biochim. biophys. Acta.*, **7,** 526–546.

Ris, H., & Mirsky, A. E., 1949, 'Quantitative Cytochemical Determination of Desoxyribonucleic Acid with the Feulgen Nucleal Reaction', *J. gen. Physiol.*, **33,** 125–146.

Robertson, J. M., 1935, 'An X-ray Study of the Phthalocyanines. Part I. The Metal-free, Nickel, Copper, and Platinum Compounds', *J. chem. Soc.*, 615–621.

——, 1936, 'An X-ray Study of the Phthalocyanines. Part II. Quantitative Structure Determination of the Metal-free Compound', *J. chem. Soc.*, 1195–1209.

——, 1962, 'Personal Reminiscences', in Ewald, 1962, pp. 637–641.

Robertson, J. M., & Woodward, I., 1937, 'An X-ray Study of the Phthalocyanines. Part III. Quantitative Structure Determination of Nickel Phthalocyanine', *J. chem. Soc.*, 219–230.

Robinow, C. F., 1942, 'A Study of the Nuclear Apparatus of Bacteria', *Proc. R. Soc.*, **130B,** 299–324.

Rolleston, J. D., 1949, 'Garrod, Sir Archibald Edward (1857–1936)', *Dictionary of National Biography,* vol. for 1931–40, London, pp. 308–309.

Ronwin, E., 1951, 'A Phospho-tri-anhydride Formula for the Nucleic Acids', *J. Am. chem. Soc.*, **73,** 5141–5144.

Rosenberg, J. M., Seeman, N. C., Kim, J. J. P., Suddath, F. L., Nicholas, H. B., & Rich, A., 1973, 'Double Helix at Atomic Resolution', *Nature,* **243,** 150–154.

Rosenfeld, L., 1963, 'Niels Bohr's Contribution to Epistomology', *Physics Today,* **16,** 47–52.

——, 1967, 'Niels Bohr in the Thirties: Consolidation and Extension of the Conception of Complementarity', in: Rozental, S. (ed.), *Niels Bohr: his life and work as seen by his Friends & Colleagues,* Amsterdam, 1967, pp. 114–136.

Royal Society, 1940–1945, 'Minutes of the Meeting held 4 November 1943', *Minutes of Council,* **16,** 190–193; Appendix A, 'The Needs of Research after the War', 284–286.

Ruska, H., 1941, 'Ueber ein neues bei der Bakteriophagen Lyse im Uebermikroskop', *Naturwissenschaften,* **29,** 367–368.

Sabanejèv, A. P., & Alexandrov, N. A., 1891, 'Cryoscopic Investigations of Colloids...III. On the Molecular Weight of Egg Albumin', *Russ. phys. Chem. Soc. J.*, **23,** 7–19.

Sauter, E., 1933, 'Ein Modell der Hauptvalenzkette im Makromolekülgitter der Polyoxymethylene', *Z. phys. Chem.*, **21,** 186–197.

Sawyer, F. O. [née Bell], 1967, Letter to the Author, dated November.

Schaffner, K., 1969, 'The Watson–Crick Model and Reductionism', *British J. Phil. Sci.*, **20,** 325–348.

Schepartz, B., and Gurin, S., 1949, 'The Intermediary Metabolism of Phenylalanine Labelled with Radioactive Carbon', *J. biol. Chem.*, **180,** 663–673.

Scherrer, P., 1918, 'Bestimmung der Grösse und der inneren Struktur von Kolloidteilchen mittels Röntgenstrahlen', *Nachr. Ges. Wiss. Göttingen, Math.-phys. Kl.*, 98–100.

——, 1920, 'Bestimmung der inneren Struktur und der Grösse von Kolloidteilchen mittels Röntgenstrahlen', in Zsigmondy, 1920, pp.386–409.

Schlesinger, M., 1932, 'Die Bestimmung von Teilchengrösse und spezifischem Gewicht des Bakteriophages durch Zentrifugierversuche', *Z. Hyg. Infekt. Krankh.*, **114,** 161–176.

Schmidt, G., & Levene, P. A., 1938, 'The Effect of Nucleophosphatase on "Native" and Depolymerized Thymonucleic Acid', *Science,* **88,** 172–173.

Schmidt, W. J., 1928, 'Der submikroskopische Bau des Chromatins. I. Mitteilung: Ueber die Doppelbrechung des Spermienkopfes', *Zool. Jb., Abt. allg. Zool. Physiol.*, **45,** 177–216.

——, 1932, 'Die Doppelbrechung der α-Thymonukleinsäure im Hinblick auf die Doppelbrechung des Chromosoms', *Naturwissenschaften,* **20,** 658.

Schmidt-Ott, F., 1921, 'Hochverehrte Exzellenz!' *Naturwissenschaften,* **9,** 293–294. [Tribute to Harnack.]

Schmiederberg, O., 1896, 'Physiologisch-chemische Untersuchungen über die Lachsmilch', *Arch. exp. Path. Pharmak.*, **37,** 100–155.

——, 1900, 'Ueber die Nukleinsäure aus der Lachsmilch', *Archiv exp. Path. Pharm.*, **43,** 57–83.

Schomaker, V., 1968, Recorded Interview with the Author, Seattle, November.

Schönefeldt, M., 1935, 'Entwicklungsgeschichtliche Untersuchungen bei *Neurospora tetrasperma* and *N. sitophila*', *Z. indukt. Abstamm.-u. VererbLehre,* **69,** 193–209.

Schramm, G., 1941, 'Enzymatische Abspaltung der Nucleinsäure aus den Tabakmosaikvirus', *Ber. dt. chem. Ges.*, **74,** 532–536.

——, 1943, 'Ueber die Spaltung des Tabakmosaikvirus in niedermolekulare Proteine und die Rückbildung hochmolekularen Proteins aus den Spaltstücken', *Naturwissenschaften,* **31,** 94–96.

——, 1948, 'Zur Chemie des Mutationsvorganges beim Tabakmosaikvirus', *Z. Naturf.*, **3B,** 320–327.

Schramm, G., & Müller, H., 1940, 'Zur Chemie des Tabakmosaikvirus. Ueber die Einwirkung von Keten and Phenylisocyanat auf das Virusprotein', *Hoppe-Seyler's Z. physiol. Chem.*, **266,** 43–56.

Schramm, G., & Rebensburg, L., 1942, 'Zur vergleichenden Charakterisierung einiger Mutanten des Tabakmosaikvirus', *Naturwissenschaften,* **30,** 48–50.

Schrödinger, E., 1933, 'The Fundamental Idea of Wave Mechanics', *Nobel Lectures, Physics* 1922–1941, Amsterdam, pp. 305–316.

——, 1944, *What is Life? The Physical Aspect of the Living Cell,* Cambridge [Referred to as 'Life'].

——, 1951, *Science and Humanism: Physics in our Time,* Cambridge.

Schrötter, H., 1906, 'Beitrag zur Mikrophotographie mit ultraviolettem Lichte nach Köhler', *Virchow's Arch. path. Anat.,* **183,** 343–376.

Schultz, J., 1941, 'The Evidence of the Nucleoprotein Nature of the Gene', *Cold Spring Harb. Symp. quant. Biol.,* **9,** 55–65.

Schultze, W., 1939, *Erste Reichstagung der Wissenschaftlichen Akademien des NSD Dozentenbundes,* Munich.

Scott-Moncrieff, R., 1936, 'A Biochemical Survey of some Mendelian Factors for Flower Colour', *J. Genet.,* **32,** 117–170.

——, 1937, 'The Biochemistry of Flower Colour Variation', in: Needham, J., & Green, D. E. (eds.), 1937, pp. 230–243.

Seeds, W. E., & Wilkins, M. H. F., 1949, 'A Simple Reflecting Microscope', *Nature,* **164,** 228–229.

——, 1950, 'Ultraviolet Microspectrographic Studies of nucleoproteins and crystals of Biological Interest', *Faraday Soc. Disc.,* **9,** 417–423, Paper given at Cambridge, Sept. 25–28, 1950.

Sevag, M. G., 1952, Discussion of Paper by K. G. Stern, *Exp. Cell Res.* Supp. **2,** 14–15.

Shearer, G., 1925, 'On the Distribution of the Intensity in the X-ray Spectra of Certain Long-Chain Organic Compounds', *Proc. R. Soc.,* **108A,** 655–666.

Signer, R., 1968, Recorded Interview with the Author, Bern, September.

Signer, R., Caspersson, T., & Hammarsten, E., 1938, 'Molecular Shape and Size of Thymonucleic Acid', *Nature,* **141,** 122.

Smith, K. M., 1935, *Plant Viruses,* London.

Snow, C. P., 1934, *The Search,* London.

——, 1966, 'J. D. Bernal, a Personal Portrait', in: Goldsmith, M., and Mackay, A. (eds.), 1966, pp. 19–31.

Sørensen, S. P. L., 1917, 'Proteinstudien', *C.r. Lab. Carlsberg.,* **12,** 1–364.

——, 1925, *Proteins,* New York.

Soret, J. L., 1883, 'Recherches sur l'absorption des rayons ultra-violets par diverses substances', *Arch. Sci. Phys. Nat.,* 3rd Periode, **10,** 429–494.

Speakman, J. B., 1927, 'The Development of Research in the Textile Industries', Memorandum to the Vice-Chancellor, dated 15 June 1927.

——, 1968, Recorded Interview with the Author, Leeds, July.

Sponsler, O. L., & Dore, W. H., 1926, 'The Structure of Ramie Cellulose as Derived from X-ray Data', *Colloid Symp. Monogr.,* **4,** 174–202.

Stacey, M., 1947, 'Bacterial Nucleic Acids and Nucleoproteins', *Symp. Soc. exp. Biol.,* **1,** 86–100.

Stadler, L. J., 1939, 'Genetic Studies with Ultraviolet Radiation', *Int. Conf. Genet.,* **7,** 262–276.

Stadler, L. J., & Sprague, G. F., 1936, 'Genetics Effects of Ultra-violet Radiation in Maize. III...', *Proc. natn. Acad. Sci. U.S.A.*, **22,** 584–591.

Stanley, W. M., 1935, 'Isolation of a Crystalline Protein Possessing the Properties of Tobacco-Mosaic Virus', *Science,* **81,** 644–645.

——, 1936, 'The Inactivation of Crystalline Tobacco-Mosaic Virus Protein', *Science,* **83,** 626–627.

——, 1937, 'Chemical Studies on the Virus of Tobacco Mosaic', *J. biol. Chem.,* **117,** 325–340. [Received for publication October 1936.]

——, 1938, 'The Reproduction of Virus Proteins', *Am. Nat.,* **72,** 110–123.

——, 1939, Letter to Astbury dated 13 October. [Leeds University, Biophysics Archive.]

——, 1946, 'The Isolation and Properties of Crystalline Tobacco Mosaic Virus', in: *Nobel Lectures... Chemistry* 1942–1962, Amsterdam etc., 1964, pp. 137–157.

——, 1955, Personal communication to Melchers, cited in Melchers, 1965, p. 120.

Staudinger, H., 1922, 'Ueber die Hydrierung des Kautschuks und über seine Konstitution', *Helv. Chim. Acta.,* **5,** 785–806.

——, 1924, 'Ueber die Konstitution des Kautschuks', *Ber. dt. chem. Ges.,* **57,** 1203–1208.

——, 1926, 'Die Chemie der hochmolekularen organischen Stoffe im Sinne der Kekuléschen Strukturlehre', *Ber. dt. chem. Ges.,* **59,** 3019–3043.

——, 1943, *Makromolekulare Chemie und Biologie,* Basel.

——, 1961, *Arbeitserinnerungen,* Heidelberg.

——, 1970, *From Organic Chemistry to Macromolecules,* New York etc. [English translation of the *Arbeitserinnerungen.*]

Staudinger, H., & Schulz, G. V., 1935, 'Vergleich der osmotischen und viscosmitreschen Molekulargewichtsbestimmungen an polymerhomologen Reihen', *Ber. dt. chem. Ges.,* **68,** 2320–2335.

Staudinger, M., 1969, Recorded Interview with the Author, Freiburg-im-Breisgau, September.

Stedman, E., and Stedman, E., 1943, 'Chromosomin, a Constituent of Chromosomes', *Nature,* **152,** 267–269.

——, 1947, 'The Function of Desoxy-ribose-Nucleic Acid in the Cell Nucleus', *Symp. Soc. exp. Biol.,* **1,** 232–251.

Stent, G. S., 1966, 'Waiting for the Paradox', in Cairns *et al.* (eds.), 1966, pp. 3–8.

——, 1968, 'That was the Molecular Biology that Was', *Science,* **160,** 390–395.

——, 1972, 'Prematurity and Uniqueness in Scientific Discovery', *Scient. Am.,* **227,** 84–93.

Stern, C., 1967, 'Richard Benedict Goldschmidt', *Biogr. Mem. natn. Acad. Sci.,* **39,** 141–192.

Stern, K. G., 1947, 'Nucleoproteins and Gene Structure', *Yale J. Biol. Med.,* **19,** 937–949.

——, 1952, 'Problems in Nuclear Chemistry and Biology', *Exp. Cell Res.*, Suppl. 2, 1–12. [Discussion pp. 12-15.]

Steudel, H., 1912, 'Ueber den Bau der Nucleinsäure aus der Thymusdrüse', *Hoppe-Seyler's Z. physiol. Chem.*, **77**, 497–507.

Stimson, M. M., & Reuter, M. A., 1943, 'Ultraviolet Absorption Spectra of Nitrogenous Heterocycles (VI). Effect of *pH* on Spectrum of Uracil-5-Carboxylic Acid', *J. Am. Chem. Soc.*, **65**, 151–152. See also companion paper on pp. 153–155.

Stokes, A R., 1951, Letter to Crick, dated 19 December. [Original with Stokes.]

——, 1967a, Letter to the Author, dated 23 March.

——, 1967b, Letter to the Author, dated 28 January.

Strasburger, E., 1909, 'The Minute Structure of Cells in Relation to Heredity', in: Seward, A. C. (ed.), *Darwin and Modern Science*, Cambridge, pp. 102–111.

Strauss, B. S., 1960, *An Outline of Chemical Genetics*, Philadelphia and London.

Sturtevant, A. H., 1920, 'The Vermilion Gene and Gynandromorphism', *Proc. Soc. exp. Biol. Med.*, **17**, 70–71.

——, 1965, *A History of Genetics*, New York.

Svedberg, T., 1924, *Colloid Chemistry: Wisconsin Lectures*, New York. [The page number is from the 2nd ed. of 1928.]

——, 1939, 'Opening Address', a discussion on the Protein Molecule, 17 November 1938, *Proc. R. Soc.*, **170A**, 40–56.

Svedberg, T., & Chirnoaga, E., 1928, 'The Molecular Weight of Hemocyanin', *J. Am. chem. Soc.*, **50**, 1399–1411.

Svedberg, T., & Estrup, K., 1911, 'Bestimmung der Häufigkeitsverteilung der Teilchengrössen in eventuell dispersen System', *Kolloid Z.*, **9**, 259–261.

Svedberg, T., & Fåhraeus, R., 1926, 'A New Method for the Determination of the Molecular Weight of the Proteins', *J. Am. chem. Soc.*, **48**, 430–438.

Svedberg, T., & Nichols, J. B., 1926, 'The Molecular Weight of Egg Albumen, I. In Electrolyte-free Condition', *J. Am. chem. Soc.*, **48**, 3081–3092.

Svedberg, T., & Pederson, K. O., 1940, *The Ultracentrifuge*, Oxford.

Svedberg, T., & Sjögren, B., 1928, 'The Molecular Weight of Serum Albumin and of Serum Globulin', *J. Am. chem. Soc.*, **50**, 3318–3332.

Swift, H. H., 1950a, 'The Desoxyribose Nucleic Acid Content of Animal Nuclei', *Physiol. Zool.*, **23**, 169–198.

——, 1950b, 'The Constancy of Desoxyribose Nucleic Acid in Plant Nuclei', *Proc. natn. Acad. Sci. U.S.A.* , **36**, 643–654.

Takahashi, H., 1932, 'Ueber fermentative Dephosphorierung der Nukleinsäure', *J. Biochem. Japan.*, **16**, 463–481. [The Cyclic structure is shown on p. 471.]

Takahashi, W. N., & Rawlins, T. E., 1932, 'Methods of Determining Shape of Colloidal Particles; Application in Study of Tobacco Mosaic Virus', *Proc. Soc. exp. Biol. Med.*, **30**, 155–157.

——, 1933, 'Rod-shaped Particles in Tobacco Mosaic Virus Demonstrated by Stream Double Refraction', *Science,* **77,** 26–27.

——, 1937, 'Stream Double Refraction of Preparations of Crystalline Tobacco-Mosaic Protein', *Science,* **85,** 103–104.

Taylor, H. S., 1941, 'Large Molecules through Atomic Spectacles', *Proc. Am. Phil. Soc.,* **85,** No. 1.

Taylor, J. H., 1965, *Selected Papers on Molecular Genetics,* New York and London.

Temin, H. M., 1964, 'Nature of the Provirus of Rous Sarcoma', *Natnl. Cancer Monogr.* 17 [International Conference of Avian Tumour Viruses], Bethesda, pp. 557–570.

Temin, H., & Mizutani, S., 1970, 'RNA-dependent DNA Polymerase in Virions of Rous Sarcoma Virus', *Nature,* **226,** 1211–1213. [Temin implied that RNA can make DNA six years before. See: Temin, 1964.]

Thompson, R. H. S., & Dubos, R. J., 1938, 'The Isolation of Nucleic Acid and Nucleoprotein Fractions from Pneumococci', *J. biol. Chem.,* **125,** 65–74.

Thomson, W. (Lord Kelvin), 1893, 'On the Elasticity of a Crystal according to Boscovich', *Proc. R. Soc.,* **54,** 59–75.

Tilley, C. E., 1930, 'Contribution to the Discussion of the Report of the Syndicate upon the Position of Mineralogy in the Studies of the University', *Cambridge University Reporter,* Feb. 11, 648–650.

Times, The, 1910, Report published on October 12, p. 6.

Timoféeff-Ressovsky, N. W., 1934, 'The Experimental Production of Mutations', *Biol. Rev.,* **9,** 411–457.

Timoféeff-Ressovsky, N. W., Zimmer, K. G., & Delbrück, M., 1935, 'Ueber die Natur der Genmutation und der Genstruktur', *Nachr. Ges. Wiss. Göttingen. math.-phys. Kl.,* Fachgr. **6,** 1, 189–245.

Tipson, R. S., 1957, 'Phoebus Aaron Theodor Levene, 1869–1940', *Adv. Carbohydrate Chem.,* **12,** 1–12.

Tiselius, A., & Claesson, S., 1967, 'The Svedberg and Fifty Years of Physical Chemistry in Sweden', *A. Rev. phys. Chem.,* **18,** 1–8.

Troland, L. T., 1917, 'Biological Enigmas and the Theory of Enzyme Action', *Am. Nat.,* **51,** 321–350.

Tulasne, R., 1947, 'Sur la mise en évidence du noyau des cellules bactériennes', *C.r. Séanc. Soc. Biol.,* **141,** 411–413.

Tulasne, R., & Vendrely, R., 1947, 'Demonstration of Bacterial Nuclei with Ribonuclease', *Nature,* **160,** 225–226. [Also see *C.r. Séanc. Soc. Biol.,* **141,** 674–676.]

Twort, F., 1915, 'An Investigation on the Nature of the Ultramicroscopic Viruses', *Lancet,* **ii,** 1241.

Van Slyke, D. D., & Jacobs, W. A., 1945, 'Phoebus Aaron Theodor Levene (1869–1940)', *Biogr. Mem. natn. Acad. Sci.,* **23,** 75–126.

Vendrely, R., 1947, 'La libération des deux acides nucléiques au cours de l'autolyse des bactéries et sa signification', *Experientia,* **3,** 196–198.
——, 1972, Letter to the Author, dated 10 April.
——, & Lipardy, J., 1946, 'Acides nucléiques et noyaux bactériens', *C.r. hebd. Séanc. Acad. Sci., Paris,* **223,** 342–344.
——, & Vendrely, C., 1949, 'La teneur de noyau cellulaire en acide désoxyribonucléique à travers les organes, les individus et les espèces animales', *Experientia,* **5,** 327–329.
Vischer, E., 1970, Recorded Interview with the Author, dated September.
——, & Chargaff, E., 1948, 'The Separation and Quantitative Estimation of Purines and Pyrimidines in Minute Amounts', *J. biol. Chem.,* **176,** 703–734.
——, Zamenhof, S., & Chargaff, E., 1949, 'Microbial Nucleic Acids: The Desoxypentose Nucleic Acids of Avian Tubercle Bacilli and Yeast', *J. biol. Chem.,* **177,** 429–438.

Wackernagel, W., 1973, 'Genetic Transformation in *E. coli:* The Inhibiting Role of the recBC DNase', *Biochem. biophys. Res. Comm.,* **51,** 306–311.
Waddington, C. H., 1939, *Introduction to Modern Genetics,* London.
——, 1969, 'Some European Contributions to the Prehistory of Molecular Biology', *Nature,* **221,** 318–321.
Walker, P. M. B., & Yates, H. B., 1952, 'Nuclear Components of Dividing Cells', *Proc. R. Soc.,* **140B,** 274–299.
Watson, J. D., 1949a, 'Lysozyme (with some comments about Avidin)', Bloomington, Indiana. [22-p. MS, original with Haurowitz.]
——, 1949?b, 'The Genetics of *Chlamydomonas* with Special Regard to Sexuality', Bloomington, Indiana. [25-p. MS, original with Professor Sonneborn.]
——, 1950a, *The Biological Properties of X-ray Inactivated Bacteriophage,* Bloomington, Indiana. [Ph.D. Thesis.]
——, 1950b, 'The Properties of X-ray-inactivated Bacteriophage', *J. Bact.,* **60,** 697–718.
—— , 1951a, Application to the National Foundation for Infantile Paralysis, dated 25 August. [National Foundation Archive.]
——, 1951b, Letter to Delbrück, dated 9 December. [Caltech Archive.]
——, 1952a, Letter to Delbrück, dated 20 May. [Original in Caltech Archive.]
——, 1952b, Contributions to the Discussions in *Symp. Soc. gen. Microbiol.,* **2,** 113–116, 171–172.
——, 1953a, Letter to Delbrück, dated 20 February. [Caltech Archive.]
——, 1953b, Letter to Delbrück, dated 12 March. [Caltech Archive.]
——, 1953c, Letter to Delbrück, dated 22 March. [Caltech Archive.]
——, 1953d, Letter to Delbrück, dated 21 May. [Caltech Archive.]
——, 1953e, 'Some Remarks on Desoxyribonucleic Acid'. [Talk to Hardy Club, 17 April. No surviving MS. No list of those attending is given in the Club's Minute Book.]

——, 1953f, 'Method of Devising a Possible Structure for DNA', Protein Conference, Pasadena, 24 September. [No surviving MS traced.]

——, 1953g, Letter to Crick, dated 9 October. [Original with Crick.]

——, 1954, Letter to Crick, dated 13 February. [Original with Crick.]

——, 1965, *The Molecular Biology of the Gene.*

——, 1966, 'Growing up in the Phage Group', in Cairns *et al.* (eds.), pp. 239–245.

——, 1968, *The Double Helix,* London. [Referred to as *Helix.*]

Watson, J. D., & Crick, F. H. C., 1953a, Letter to Pauling, dated 21 March. [Caltech Archive.]

——, 1953b, 'Molecular Structure of Nucleic Acids. A Structure for Deoxyribose Nucleic Acid', *Nature,* **171,** 737–738.

——, 1953c, 'Genetical Implications of the Structure of Desoxyribonucleic Acid', *Nature,* **171,** 964–967.

——, 1953d, 'The Structure of DNA', *Cold Spring Harb. Symp. quant. Biol.,* **18,** 123–131

Watson, J. D., & Maaløe, O., 1953, 'Nucleic Acid Transfer from Parental to Progeny Bacteriophage', *Biochim. Biophys. Acta.,* **10,** 432–442.

Weaver, W., 1933, 'Natural Science—Program and Policy', Extract from Agenda for Special Meeting of the Trustees of the Rockefeller Foundation, 11 April 1933, p. 76. [Quoted in: Fosdick, 1952, p. 157.]

——, 1934, 'The Natural Sciences', *Rep. Rockefeller Found.,* 125–141.

——, 1938, 'The Natural Sciences', *Rep. Rockefeller Found.,* 203–225.

——, 1949, Letter to H. M. H. Carson, dated 17 June 1949. [As published in: Fosdick, 1952, p. 166.]

——, 1970a, 'Molecular Biology: Origin of the Term', *Science,* **170,** 581–582.

——, 1970b, *Scene of Change,* New York.

——, 1971, Interview with Henahan at the Salk Institute. [The passage comes from Tape 3.]

——, 1972, Letter to the Author, dated 16 September.

Weiss, P., 1973, Letter to the Author, dated 25 April.

Weissenberg, K., 1922, '"Spiralfaser" und "Ringfaser" im Röntgendiagramm' *Z. Physik.,* **8,** 20–31.

——, 1926, 'Die geometrischen Grundlagen der Stereochemie', *Ber. dt. chem. Ges.,* **59,** 1526–1542.

Werner, A., 1902, 'Ueber Haupt- und Nebenvalenzen und die Constitution der Ammoniumverbindungen', *Justus Liebigs Annln.,* **322,** 261–296.

Westgren, A., 1913, 'Ueber die kinetische Energie der Teilchen in kolloiden Lösungen', *Ark. Mat. Astr. Fys.,* **9,** No. 5, pp. 1–36.

——, 1915, *Untersuchungen über die brownsche Bewegung, besonders als Mittel zur Bestimmung der Avogadroschen Konstant,* Uppsala. [Dissertation.]

Wettstein, D. von, 1974, Letter to the Author, dated 7 January.

Wettstein, F. von., 1938, 'Carl Erich Correns', *Ber. dt. bot. Ges.,* **56,** 2, 140–160.

Wheeler, H. L., & Johnson, T. B., 1903, 'On Cytosine or 2-Oxy-6-Amino-pyrimidine from Tritico-Nucleic Acid', *Am. chem. J.*, **29,** 505–511.

Wheldale, M., 1907, 'Inheritance of Flower Colour in *Antirrhinum majus', Proc. R. Soc.,* **79B,** 288–305.

——, 1910, 'Plant Oxydases and the Chemical Interrelationships of colour varieties', *Prog. rei Bot.,* **3,** 457–473.

——, 1916, *The Anthocyanin Pigments of Plants,* 1916, Cambridge.

Wiener, O., 1912, 'Die Theorie des Mischkörpers für das Feld der stationären Strömung', *Abh. sächs. Akad. Wiss., math.-phys. Kl.,* **32,** 507–609.

Wigand, A., 1862, 'Einige Sätze über die physiologische Bedeutung des Gerbstoffes und der Pflanzenfarbe', *Bot. Ztg.,* **20,** 121–125.

Wilkins, M. H. F., 1948?, Letter to Crick, undated. [Original with Crick.]

——, 1950a, Letter to Markham, dated 15th June. [Original with Markham.]

——, 1950b, Letter to Markham, dated 18th July. [Original with Markham.]

——, 1950c, Letter to Markham, dated 16 August. [Original with Markham.]

——, 1950d, Letter to Markham, dated 20th October. [Original with Markham.]

——, 1951a, 'Ultraviolet Dichroism and Molecular Structure in Living Cells. . . .', *Pubbl. Staz. Zool. Napoli,* **23,** Suppl., pp. 104–114 . [Symposium on Submicroscopical Structure of Protoplasm, May 22–25, 1951, Naples.]

——, 1951b, Letter to Markham, dated 6th February. [Original with Markham.]

——, 1951c, 'Contribution to the discussion', *Faraday Soc. Disc.,* **11,** 214–215.

——, 1951d, Letter to Franklin, undated. [Original with Klug.]

——, 1952, Letter to Crick, undated. [Original with Crick.]

——, 1953a, Letter to Crick, simply dated 'Thursday'. [Original with Crick.]

——, 1953b, Letter to Crick, simply dated 'Saturday'. [Original with Crick.]

——, 1953c, Letter to Crick, dated 18th March. [Original with Crick.]

——, 1953d, Letter to Crick, dated 'Thursday'. [Original with Crick.]

——, 1963, 'The Molecular Configuration of Nucleic Acids', *Les Prix Nobel en* 1962, Stockholm, pp. 126–154. [Also in *Nobel Lectures. . . Physiology or Medicine* 1942–1962, Amsterdam etc., 1964, pp. 754–782.]

——, 1968, Letter to the Author, dated 28th May.

——, 1969, Letter to Klug, dated 19 June. [Original with Klug.]

——, 1970, Letter to the Author, dated 12th February.

——, 1972, Letter to the Author, dated 18th December. [The typescript of the Author's chapters XIX, XX, and XXI were returned with Wilkins' annotations.]

——, 1986, Letter to the Author, dated 27th May. [Original with the Author.]

——, Gosling, R. G., & Seeds, W. E., 1951, 'Physical Studies of Nucleic Acids', *Nature,* **167,** 759–760.

——, & Randall, Sir J. T., 1953, 'Crystallinity in Sperm Heads. Molecular Structure of Nucleoprotein *in vivo', Biochim. Biophys. Acta,* **10,** 192–193.

——, Seeds, W. E., Stokes, A. R., & Wilson, H. R., 1953, 'Helical Structure of Crystalline Deoxypentose Nucleic Acid', *Nature,* **172,** 759–762.

Williams, R. C., 1952, 'Electron Microscopy of Sodium Desoxyribonucleate by use of a New Freeze-Drying Method', *Biochim. Biophys. Acta,* **9,** 237–239.

Williamson, A. R., 1972, 'Extent and Control of Anti-body Diversity', *Biochem. J.,* **130,** 325–333. [See especially the section entitled 'Multiple germ-line V-genes' on pp. 331–332.]

Wilson, E. B., 1925, *The Cell in Development and Heredity,* 3rd ed., New York.

Willstätter, R., 1965, *From my Life: the Memoirs of Richard Willstätter,* New York & Amsterdam. [English trans. of German edition of 1949.]

Wolkow, M., & Baumann, E., 1891, 'Ueber das Wesen der Alkaptonurie', *Hoppe-Seyler's Z. physiol. Chem.,* **15,** 228–285.

Woods, A. F., 1900, 'Inhibiting Action of Oxidase upon Diastase', *Science,* **11,** 17–19.

Woods, H. J., 1955, *Physics of Fibres, An Introductory Survey,* London.

——, 1968, Recorded interview, Leeds, July.

Wright, S., 1941, 'The Physiology of the Gene', *Physiol. Rev.,* **21,** 487–527.

——, 1973, Letter to Lederberg, dated 19th September. [Original with Lederberg. Wright wrote that he discussed Avery's work in his course on 'Physiological Genetics' in Chicago from 1944 onwards, and 'Watson audited it a few years later as an undergraduate'.]

Wrinch, D. M., 1934, 'Chromosome Behaviour in Terms of Protein Pattern', *Nature,* **134,** 978–979.

——, 1935, 'Contribution to the Discussion', *Trans. Faraday Soc.,* **31,** 389. [Meeting held in 1934.]

——, 1936, 'On the Molecular Structure of Chromosomes', *Protoplasma,* **25,** 550–569.

——, 1938, 'The Structure of the Insulin Molecule', *Science,* **88,** 148–149.

——, 1941, 'The Native Protein Theory of the Structure of Cytoplasm', *Cold Spring Harb. Symp. quant. Biol.,* **9,** 218–235.

Wülker, H., 1935, 'Untersuchungen über Tetradenaufspaltung bei *Neurospora sitophila* Shear et Dodge', *Z. indukt. Abstamm.-u. VererbLehre,* **69,** 210–248.

Wyatt, G. R., 1950, *Studies on Insect Viruses and Nucleic Acids,* Cambridge. [Ph.D. Thesis.]

——, 1952a, 'Specificity in the Composition of Nucleic Acids', *Exp. Cell Res.,* Suppl. 2, 201–215. [Held 1951. Discussion on pp. 215–217.]

——, 1952b, 'The Nucleic Acids of some Insect Viruses', *J. gen. Physiol.,* **36,** 201–205.

——, 1968, Recorded Interview with the Author, Yale, September.

Wyatt, G. R., & Cohen, S. S., 1952, 'Nucleic Acids of Rickettsiae', *Nature,* **170,** 846.

——, 1953, 'The Bases of the Nucleic Acids of Some Bacterial and Animal Viruses: The Occurrence of 5-Hydroxymethyl-cytosine', *Biochem. J.,* **55,** 774–782.

Wyatt, V., 1972, 'When does Information become Knowledge?', *Nature,* **235,** 86–89.

Wyckoff, R. W. G., 1962, 'The Development of X-ray Diffraction in the U.S.A.', in: Ewald, P. (ed.), 1962, pp. 430-433.

——, Biscoe, J., & Stanley, W. M., 1937, 'An Ultracentrifugal Analysis of the Crystalline Virus Proteins Isolated from Plants Diseased with different strains of Tobacco Mosaic Virus', *J. Biol. Chem.*, **177,** 57–71.

Yoshida, H., 1883, 'Chemistry of Lacquer (urushi)', *J. chem. Soc.*, **43,** 472–486.

Zahn-Harnack, A., 1936, *Adolf von Harnack*, Berlin.

Zamenhof, S., 1952, 'Newer Aspects of the Chemistry of Nucleic Acids', in: McElroy and Glass (eds.), 1952, vol. ii, pp. 301–328.

——, & Chargaff, E., 1949, 'Evidence of the Existence of a Core in Desoxyribonucleic Acids', *J. biol. Chem.*, **178,** 531–532.

——, & Chargaff, E., 1950, 'Dyssymetry in Nucleotide Sequence of Desoxypentose Nucleic Acids', *J. biol. Chem.*, **187,** 1–14.

——, Shettles, L. B., & Chargaff, E., 1950, 'Human Deoxypentose Nucleic Acid', *Nature*, **165,** 756.

Zimmer, K. G., 1941, 'Zur Berücksichtigung der biologischen Variabilität bei der Treffertheorie der biologischen Strahlenwirkung', *Biol. Zbl.*, **61,** 208.

——, 1950, *Otchet, Fondy Ural. Fil. A. N. SSSR*, Sverdlovsk.

——, 1961, *Studies on Quantitative Radiation Biology*, Edinburgh and London.

——, 1966, 'The Target Theory', in: Cairns *et al.* (eds.), 1966, pp. 33–42.

Zinoffsky, O., 1886, 'Ueber die Grösse des Hämoglobin-moleküls', *Hoppe-Seyler's Z. physiol. Chem.*, **10,** 16–34.

Zinsser, H., & Parker, J. T., 1923, 'Further Studies on Bacterial Hypersusceptibility. II', *J. exp. Med.*, **37,** 275–302.

Zsigmondy, R., 1920, *Kolloidchemie: ein Lehrbuch*, 3rd ed., Leipzig.

Index

NOTE (1994): For the Dover Edition the Postscript has been indexed, and that index has been integrated with the original one. In addition, death dates have been added (except where, after a reasonable effort, the information could not be found) to the entries for those people who have died since the original edition was published in 1974. A number of corrections have also been made in the original index.